MICROTECHNOLOGY AND MEMS

MICROTECHNOLOGY AND MEMS

Series Editor: H. Fujita D. Liepmann

The series Microtechnology and MEMS comprises text books, monographs, and state-of-the-art reports in the very active field of microsystems and microtechnology. Written by leading physicists and engineers, the books describe the basic science, device design, and applications. They will appeal to researchers, engineers, and advanced students.

Please view available titles in *Microtechnology and Mems*
on series homepage http://www.springer.com/series/4526

Arno Lenk
Rüdiger G. Ballas
Roland Werthschützky
Günther Pfeifer

Electromechanical Systems in Microtechnology and Mechatronics

Electrical, Mechanical and Acoustic Networks, their Interactions and Applications

With 313 Figures

Springer

Professor Dr.-Ing habil Arno Lenk
Dresden University of Technology
Faculty of Electrical and Computer Engineering
Helmholtzstraße 10
01069 Dresden
Germany

Dr.-Ing Rüdiger G. Ballas
Karl Mayer
Textile Machinery
Bruehlstr. 25
63179 Obertshausen
Germany
rballas@karlmayer.de

Professor Dr.-Ing habil Roland Werthschützky
Darmstadt University of Technology
Institute for Electromechanical Design
Merckstr. 25
64283 Darmstadt
Germany
werthschuetzky@emk.tu-darmstadt.de

Professor Dr.-Ing habil Günther Pfeifer
Dresden University of Technology
Faculty of Electrical and
Computer Engineering
Helmholtzstraße 10
01069 Dresden
Germany

ISSN 1615-8326
ISBN 978-3-642-26526-6 ISBN 978-3-642-10806-8 (eBook)
DOI 10.1007/978-3-642-10806-8
Springer Heidelberg Dordrecht London New York

Cover design: WMXDesign GmbH, Heidelberg

Printed on acid-free paper

Springer is part of Springer Science+Business Media (www.springer.com)

Preface

Within the wide field of technical information processing, *electromechanical systems* consisting of coupled electrical and mechanical functional elements have a significant importance. Both the design of interfaces between human and information processing mechanisms and the design of interfaces with the material process during metrological data acquisition and actuatory influence of process variables is made possible by these electromechanical systems. Examples for realization of electromechanical systems in the form of devices, assemblies or components are:

- peripheral devices of data processing systems like printers, scanners, disk drives and data memories,
- electroacoustic devices like loudspeakers, microphones and ultrasonic transducers,
- sensors for medicine, automotive and process measurement engineering,
- actuators in the form of small drives and precision positioning systems.

The list mentioned above is increasingly extended by direct coupled sensor-actuator-systems with integrated data processing. Thus, the result is a smooth transition to more complex electromechanical systems of *mechatronics*.
The production of electromechanical systems results from enhanced precision engineering methods and modern technologies of microtechnology and microsystems technology. In addition, used materials like high-grade steels, ceramics, glasses, silicon and quartz, are subjected to continuous further development.

In the phase of industrial development of electromechanical systems, the *design process* based on a solution concept provides a fundamental stage. Here, geometrical, electrical and technological system parameters are defined being based on a physical model considering special design criteria and technological limitations. The closed dynamic design of the overall system is made more difficult by different subsystems consisting of electronic, mechanical, acoustic and fluid elements.

The main objective of this book is the obtaining of a physically clear design method for complex electromechanical systems. This design method is based on the *network theory,* electrical engineers and engineers of information technology are familiar with. The total electromechanical system is described in the form of a common technical circuit representation of different subsystems including their interactions by means of network theory. Clear physical functions are assigned to either lumped or distributed elements of the network. The advantages of this design method are the application of clear analytical methods of electrical networks, the possibility of the closed design of physically different subsystems and the use of existing circuit simulation software. The structuring of electromechanical systems according to electrical, mechanical and acoustic elementary networks and the introduction of passive transducers as two-port networks which describe the loss-free linear interactions between the subsystems, are the fundamental conditions for the application of network theory.

The main features of this book are based on the structure-oriented theory of electromechanical systems developed by Arno Lenk in the 60s to 90s. The results were summarized in the books „Elektromechanische Systeme – Systeme mit konzentrierten Parametern" [1], „Elektromechanische Systeme – Systeme mit verteilten Parametern" [2] and „Elektromechanische Systeme – Systeme mit Hilfsenergie" [3] which were published in the 70s in the Verlag Technik.

The book is suitable for students of information technology, measurement and automation engineering, mechatronics, technical acoustics as well as microsystems technology and precision engineering. The book enables the electrical engineer being familiar with network theory to get started quickly with the solution of many dynamic problems concerning the design of coupled electrical, mechanical, acoustic and fluid systems. In addition, this book is also suitable for mechanical engineers in order to get started with the efficient and practice-oriented design method for *mechatronic systems.* The necessary basic knowledge of network theory is summarized in an extra chapter.

We gratefully acknowledge Stephan Sindlinger, Stefan Leschka, Eric Starke and Uwe Marschner, whose current research results are presented in the sections concerning the finite network elements (Leschka, Sect. 6.3.1 and Sindlinger, Sect. 6.3.2), the combination of FEA and network theory (Starke, Sect. 6.4) and the application of electrodynamic and piezomagnetic actuators (Marschner, Sect. 8.1.2, Sect. 8.3.3 and Sect. 8.3.4).

Finally, our thanks go to Eva Hestermann-Beyerle and Birgit Kollmar-Thoni of Springer-Verlag, who offered an excellent cooperation and continuous support while we were writing this book.

Dresden and Darmstadt, July 2010

Arno Lenk, Rüdiger G. Ballas, Roland Werthschützky, Günther Pfeifer

Contents

List of Symbols xv

Part I Focus of the Book

1 Introduction 3
1.1 Focus of the Book 4
1.2 Fields of Application and Examples for Electromechanical Systems 6
1.3 Design of Electromechanical Systems 9
1.4 Simulation Methods for Electromechanical Systems 10
1.4.1 Historical Overview 10
1.4.2 Design Methods 13

2 Electromechanical Networks and Interactions 15
2.1 Signal Description and Signal Transmission in Linear Networks 16
2.1.1 The Circular Function as Basic Module for Time Functions of Linear Networks 16
2.1.2 Fourier Expansion of Time Functions 20
2.1.3 The Fourier Transform 25
2.1.4 The Laplace Transform 33
2.2 Electrical Networks 36
2.3 Mechanical Networks 40
2.4 Interactions 44
2.4.1 Mechanical Interactions 44
2.4.2 Electromechanical Interactions 46
2.5 Structured Network Representation of Linear Dynamic Systems 56
2.6 Basic Equations of Linear Networks 58

Part II Network Representation of Systems with Lumped and Distributed Parameters

3 Mechanical and Acoustic Networks with Lumped Parameters ... 61
3.1 Mechanical Networks for Translational Motions ... 62
3.1.1 Arrangements ... 62
3.1.2 Coordinates ... 64
3.1.3 Components ... 66
3.1.4 Rules of Interconnection ... 74
3.1.5 Isomorphism between Mechanical and Electrical Circuits ... 77
3.1.6 Representation of Transient Characteristics of Mass Point Systems in the Frequency Domain (BODE diagram) ... 79
3.1.7 Network Representation of Mass Point Systems ... 85
3.1.8 Sample Applications ... 88
3.2 Mechanical Networks for Rotational Motions ... 99
3.2.1 Coordinates ... 100
3.2.2 Components and System Equations ... 101
3.2.3 Isomorphism between Mechanical and Electrical Circuits ... 102
3.2.4 Sample Application for a Rotational Network ... 106
3.3 Acoustic Networks ... 107
3.3.1 Coordinates ... 108
3.3.2 Acoustic Components ... 109
3.3.3 Network Representation of Acoustic Systems ... 110
3.3.4 Real Acoustic Components ... 114
3.3.5 Isomorphism between Acoustic and Electrical Circuits ... 120
3.3.6 Sample Applications ... 120

4 Abstract Linear Network ... 129
4.1 Coordinates ... 129
4.2 Components ... 130
4.3 Nodal and Loop Rules ... 132
4.4 Characteristics of the Abstract Linear Network ... 132

5 Mechanical Transducers ... 137
5.1 Translational-Rotational Transducer ... 137
5.1.1 Rigid Rod ... 137
5.1.2 Bending Rod ... 141
5.2 Mechanical-Acoustic Transducer ... 146
5.2.1 Ideal and Real Mechanical-Acoustic Piston Transducers 147
5.2.2 General Elastomechanical-Acoustic Plate Transducer ... 149

5.3 Characteristics of Selected Mechanical-Acoustic Transducers . . 155

6 Mechanical and Acoustic Networks with Distributed Parameters 165
6.1 Representation of Mechanical Systems as one-dimensional Waveguides 165
6.1.1 Extensional Waves within a Rod 166
6.1.2 Approximate Calculation of the Input Impedance 172
6.1.3 Approximate Representation of an Impedance at Resonance 177
6.1.4 Approximated two-port Network Representation at Resonance 178
6.1.5 Flexural Vibrations within a Rod 183
6.2 Network Representation of Acoustic Systems as Linear Waveguides 192
6.3 Modeling of Transducer Structures with Finite Network Elements 195
6.3.1 Ultrasonic Microactuator with Capacitive Diaphragm Transducer 195
6.3.2 Fluid-filled Pressure Transmission System of a Differential Pressure Sensor 198
6.4 Combined Simulation with Network and Finite Element Methods 202
6.4.1 Applied Combination of Network Methods and Finite Element Methods 204
6.4.2 Combined Simulation using the Example of a Dipole Bass Loudspeaker 209
6.4.3 Combined Simulation using the Example of a Microphone with Thin Acoustic Damping Fabric 216

Part III Electromechanical Transducers

7 Electromechanical Interactions 229
7.1 Classification of Electromechanical Interactions 229
7.2 Network Representation of Electromechanical Interactions . . . 233

8 Magnetic Transducers 247
8.1 Electrodynamic Transducer 247
8.1.1 Derivation of the Two-Port Transducer Network 247
8.1.2 Sample Applications 251
8.2 Electromagnetic Transducer 267
8.2.1 Derivation of the Two-Port Transducer Network 268
8.2.2 Sample Applications 275
8.3 Piezomagnetic Transducer 285

8.3.1 Derivation of the Two-Port Transducer Network 286
8.3.2 Sample Applications 296
8.3.3 Piezomagnetic Unimorph Bending Elements 302
8.3.4 Example of a Parametric Magnetoelastic Bending Sensor 308

9 Electrical Transducers 313
9.1 Electrostatic Transducer 313
9.1.1 Electrostatic Plate Transducer 313
9.1.2 Sample Applications 323
9.1.3 Electrostatic Diaphragm Transducer 331
9.1.4 Sample Applications 334
9.1.5 Electrostatic Solid Body Transducers 339
9.1.6 Sample Application 341
9.2 Piezoelectric Transducers with Lumped Parameters 345
9.2.1 Model Representation of the Piezoelectric Effect 345
9.2.2 Piezoelectric Equations of State and Circuit Diagram for Longitudinal Coupling 348
9.2.3 General Piezoelectric Equations of State 350
9.2.4 Piezoelectric Transducers and Corresponding Equivalent Parameters 353
9.2.5 Piezoelectric Bending Bimorph Elements 358
9.2.6 Piezoelectric Materials 360
9.2.7 Sample Applications 365
9.3 Piezoelectric Transducer as one-dimensional Waveguide 370
9.3.1 Transition from Lumped Parameters to the Waveguide using the Example of an Accelerometer 371
9.3.2 Piezoelectric Longitudinal Oscillator as Waveguide 375
9.3.3 Piezoelectric Thickness Oscillator as Waveguide 375
9.3.4 Sample Applications of Piezoelectric Longitudinal and Thickness Oscillators 381
9.3.5 Piezoelectric Beam Bending Element as Waveguide 392
9.3.6 Sample Applications of Piezoelectric Beam Bending Elements 393

10 Reciprocity in Linear Networks 413
10.1 Reciprocity Relations in Networks with only One Physical Structure 413
10.2 Reciprocity Relations in General Linear Two-Port Networks 415
10.3 Electromechanical Transducers 417
10.4 Mechanical-Acoustic Transducers 420

Part IV Appendix

A Characteristics of Selected Materials 425
A.1 Material Characteristics of Crystalline Quartz 425
A.2 Piezoelectric Constants of Sensor Materials 426
A.3 Characteristics of Metallic Structural Materials 427
A.4 Material Characteristics of Silicon and Passivation Layers 428
A.4.1 Comparison of Main Characteristics of Silicon, Silicon Dioxide and Silicon Nitride Layers 428
A.4.2 Characteristics of Silicon Dioxide Layers 429
A.4.3 Characteristics of Silicon Nitride Layers 430
A.5 Characteristics of Ceramic Structural Materials 431
A.6 Material Characteristics of Selected Polymers 432
A.7 Characteristics of Plastics as Structural Materials 433
A.8 Composition and Material Characteristics of Selected Glasses . 434
A.9 Material Characteristics of Metallic Solders and Glass Solders . 435
A.10 Sound Velocity and Characteristic Impedance 436

B Signal Description and Transfer within Linear Networks .. 437
B.1 Fourier Expansion of Time Functions 437
B.1.1 Estimate of Approximation Error with Numerical Analyses of Fourier Series 437
B.1.2 Sample Application for the Periodic Iteration of Singular Processes 440
B.2 Ideal Impulse and Step Functions 442
B.2.1 Problem Definition 442
B.2.2 Ideal Impulses and their System Response 443
B.2.3 The Ideal Step Function and its System Response 448
B.3 The Convolution Integral 449

References 453

Index 459

List of Symbols

Symbol	Quantity	Unit
A	cross-sectional area, pole face	m^2
$\underline{\mathbf{A}}$	two-port matrix	
A_{el}	electrically active area	m^2
A_{F}	cross-sectional area (femur)	m^2
A_i	area, rod segment area	m^2
$\underline{A}_{ij}$	two-port matrix elements	
A_{K}	piston area	m^2
A_{P}	cross-sectional area (prosthesis)	
A_{mech}	mechanically active area	m^2
$a(t)$	acceleration	$\mathrm{m\,s}^{-2}$
$\underline{a}$	complex acceleration (excitation)	$\mathrm{m\,s}^{-2}$
$a(\omega), b(\omega)$	coefficients (Fourier integral transform)	s
a_i, b_i, c_i	Fourier coefficients	s
$\mathbf{B}$	vector of magnetic induction	T
$\underline{B}(p)$	complex transfer function (Laplace transform)	
$\underline{B}(\omega)$	complex transfer function (Fourier transform)	
$\underline{B}$	complex amplitude of transfer function	
B_0	magnetic induction	T
B_k	component of magnetic induction	T
B_{max}	maximum magnetic induction	T
B_S	saturation induction	T
b	width	m
C	capacitance	F
C_0	reference capacitance	F
C_{b}	capacitance (mechanically locked state)	F
C_{K}	cable capacitance	F
C_{nm}	capacitance matrix elements	F

Symbol	Quantity	Unit
$c(\omega)$	spectral densities	s
$\underline{c}(\omega)$	coefficient (Fourier integral transform)	s
c	wave velocity	$\mathrm{m\,s^{-1}}$
$\underline{c}_i$	complex Fourier coefficient	s
$\underline{c}_i^*$	complex conjugate Fourier coefficient	s
$c_{ij}^{(Q)}$	elastic coefficient for $Q = \text{const.}$	$\mathrm{N\,m^{-1}}$
c_{ij}^{E}	elastic coefficient for $E = 0$	$\mathrm{N\,m^{-2}}$
c_{ij}^{H}	elastic coefficient für $H = 0$	$\mathrm{N\,m^{-2}}$
c_l	extensional wave velocity	$\mathrm{m\,s^{-1}}$
c_{L}	sound propagation velocity (air)	$\mathrm{m\,s^{-1}}$
c_p	specific heat capacity for $p = \text{const.}$	$\mathrm{J\,kg^{-1}\,K^{-1}}$
c_V	specific heat capacity for $V = \text{const.}$	$\mathrm{J\,kg^{-1}\,K^{-1}}$
c_{W}	sound propagation velocity (water)	$\mathrm{m\,s^{-1}}$
D	electric displacement	$\mathrm{C\,m^{-2}}$
$\mathbf{D}$	vector of electric displacement	$\mathrm{C\,m^{-2}}$
D_0	electric reference displacement	$\mathrm{C\,m^{-2}}$
D_{el}	electrically generated electric displacement	$\mathrm{C\,m^{-2}}$
D_{mech}	mechanically generated electric displacement	$\mathrm{C\,m^{-2}}$
D_n	component of electric displacement	$\mathrm{C\,m^{-2}}$
d_{ij}	piezoelectric coefficient	$\mathrm{m\,V^{-1}}$
d_{ji}	piezomagnetic coefficient	$\mathrm{m\,A^{-1}}$
d_{eff}	effective distance	m
E	YOUNG's modulus	$\mathrm{N\,m^{-2}}$
	electric field strength	$\mathrm{V\,m^{-1}}$
$\underline{E}$	complex YOUNG's modulus	$\mathrm{N\,m^{-2}}$
$\mathbf{E}$	vector of electric field	$\mathrm{V\,m^{-1}}$
E_0	electric reference field strength	$\mathrm{V\,m^{-1}}$
E_{33}^{B}	YOUNG's modulus (magnetostrict. material)	$\mathrm{N\,m^{-2}}$
E_{c}	coercive field strength	$\mathrm{V\,m^{-1}}$
E_{F}	YOUNG's modulus (femur)	$\mathrm{N\,m^{-2}}$
E_{P}	YOUNG's modulus (prosthesis)	$\mathrm{N\,m^{-2}}$
E_{Poly}	YOUNG's modulus (polymer)	$\mathrm{N\,m^{-2}}$
E_m	component of electric field vector	$\mathrm{V\,m^{-1}}$
$E_{\max}$	maximum electric field strength	$\mathrm{V\,m^{-1}}$
$\overline{EI}$	average bending stiffness	$\mathrm{N\,m^{2}}$
e_{nj}	piezoelectric modulus	$\mathrm{N\,V^{-1}}$
e_{jn}	piezomagnetic modulus	$\mathrm{N\,A^{-1}\,m^{-1}}$
$\mathbf{e}_x, \mathbf{e}_y, \mathbf{e}_z$	unit vectors (Cartesian coordinates)	
$F(t)$	force	N
$F(p)$	Laplace transform	
$F(\omega)$	Fourier transform	
$\underline{F}, \underline{F}_i$	complex force	N
$\hat{F}$	force amplitude	N
$\underline{F}_0$	excitational force, source force	N

Symbol	Quantity	Unit
$\underline{F}_{\mathrm{B}}$	bottom force	N
$\underline{F}_{\mathrm{d}}$	distal excitation force	N
F_{el}	Coulomb force	N
$\mathbf{F}_{\mathrm{el}}$	vector of electric force	N
$\mathbf{F}_{\mathrm{el},n}$	vector of electric force at reference point n	N
$\mathbf{F}_i$	vector of affecting force	N
F_{K}	short circuit force	N
F_m	inertia force	N
F_{mag}	magnetic force	N
$\mathbf{F}_{\mathrm{mag}}$	vector of magnetic force	N
$\mathbf{F}_{\mathrm{mag},n}$	vector of magnetic force at reference point n	N
F_{max}	maximum force	N
F_{mech}	mechanical force	N
$\mathbf{F}_{\mathrm{mech},n}$	vector of mechanical force at reference point n	N
F_n	spring force	N
$\underline{F}_n$	complex spring force	N
$\mathbf{F}_n$	vector of force at reference point n	N
$\hat{F}_n$	spring force amplitude	N
F_r	frictional force	N
$\underline{F}_r$	complex frictional force	N
$\hat{F}_r$	frictional force amplitude	N
$\underline{F}_{\mathrm{W}}$	transducer force	N
F_x, F_y, F_z	components of force vector	N
f	frequency	Hz
$\underline{f}\,(x)$	complex load per unit length	$\mathrm{N\,m^{-1}}$
f_0	resonant frequency, assigned frequency	Hz
f_{g}	cut-off frequency	Hz
$f_{\mathrm{P}}, f_{\mathrm{p}}$	parallel-resonant frequency	Hz
f_{r}	resonant frequency	Hz
$f_{\mathrm{S}}, f_{\mathrm{s}}$	series-resonant frequency	Hz
G	conductance	S
	shear modulus	$\mathrm{N\,m^{-2}}$
G_i	proportionality factor	
$\underline{G}_{mn}$	leg conductance	S
$g\,(t)$	normalized impulse response (weighting function)	1
H	enthalpy	J
	magnetic field strength	$\mathrm{A\,m^{-1}}$
$\underline{\mathbf{H}}$	admittance matrix	
H_{app}	applied magnetic field	$\mathrm{A\,m^{-1}}$
H_{in}	magnetic field in material	$\mathrm{A\,m^{-1}}$
H_m	component of magnetic field	$\mathrm{A\,m^{-1}}$
H_{max}	maximum magnetic field strength	$\mathrm{A\,m^{-1}}$
h	height	m

Symbol	Quantity	Unit
h	translational admittance	$\mathrm{m\,s^{-1}\,N^{-1}}$
$\underline{h}(\omega)$	complex admittance	
h_D	wave admittance	$\mathrm{s\,kg^{-1}}$
h_i	layer thickness (layer i)	m
$\underline{h}_{ij}$	admittance matrix elements	
h_R	rotational admittance	$\mathrm{J^{-1}\,s^{-1}}$
I	geometrical moment of inertia	$\mathrm{m^4}$
	integral of DIRAC delta function	s
	direct current	A
I_0	supply current	A
I_F	moment of inertia (femur)	$\mathrm{m^4}$
I_m	magnetic flux rate	$\mathrm{Wb\,s^{-1}}$
I_P	moment of inertia (prosthesis)	$\mathrm{m^4}$
$i(t)$	current	A
$\hat{i}$	current amplitude	A
$\underline{i}$	complex current	A
i	summation index	
$\underline{i}_\mathrm{K}$	short-circuit current	A
$\underline{i}_\mathrm{W}$	transducer current	A
J	magnetic polarization	T
j	imaginary unit	
$\mathrm{K, M, N}$	summation limits	
$K_\mathrm{el}, K'_\mathrm{el}$	coefficients	
$K_\mathrm{mag}, K'_\mathrm{mag}$	coefficients	
$K_\mathrm{mag,r}$	rotational transduction coefficient	$\mathrm{N\,m\,Wb^{-1}}$
K_i	constant	1
$K_{km}^{(\xi)}$	reciprocal inductance oefficient for $\xi = \mathrm{const.}$	$\mathrm{H^{-1}}$
k	coupling factor, transformation ratio	1
k_el	electric coupling factor	1
k_mech	mechanical coupling factor	1
$k_{mn,i}$	coupling factor (layer i)	1
L	inductance	H
L_0	reference inductance, air coil inductance	H
L_Air	inductance in air	H
L_b	inductance (mechanically locked state)	H
L_m	inductance in magnetic layer	H
L_∞	inductance of planar coil	H
l	length, lever length, beam bender length	m
l_0	reference length, neutral position	m
l_el	electrically active length	m
l_i	rod segment length	m
l_mech	mechanically active length	m
M	torsional moment	N m
M_0	magnetostrictive generated moment	N m

Symbol	Quantity	Unit
$\underline{M}_0$	excitational moment	N m
$\mathbf{M}_i$	vector of torsional moment	N m
$\underline{M}_i$	complex torsional moment	N m
$\underline{M}_\mathrm{W}$	transducer moment	N m
$M_\mathrm{a}, M_{\mathrm{a,S}i}, M_{\mathrm{a}i}$	acoustic masses	$\mathrm{kg\,m^{-4}}$
$M_\mathrm{a,L}$	acoustic mass of moved air	$\mathrm{kg\,m^{-4}}$
$M_\mathrm{a,M}$	acoustic mass of a strip diaphragm	$\mathrm{kg\,m^{-4}}$
M_a0	acoustic reference mass	$\mathrm{kg\,m^{-4}}$
m^*	effective mass	kg
m	mass	kg
m, n	reference points	
m', μ	mass per unit length	$\mathrm{kg\,m^{-1}}$
m_ers	equivalent mass	kg
N	transducer factor	1
$N_\mathrm{a}, N_{\mathrm{a}i}$	acoustic compliance (adiab. change of cond.)	$\mathrm{m^5\,N^{-1}}$
$N_\mathrm{a,iso}$	acoustic compliance (isoth. change of cond.)	$\mathrm{m^5\,N^{-1}}$
$N_\mathrm{a,K}$	acoustic short-circuit compliance	$\mathrm{m^5\,N^{-1}}$
$N_\mathrm{a,L}$	acoustic open-circuit compliance	$\mathrm{m^5\,N^{-1}}$
$N_\mathrm{a,M}$	acoustic compliance (diaphragm)	$\mathrm{m^5\,N^{-1}}$
$N_\mathrm{a,P}$	acoustic compliance (plate)	$\mathrm{m^5\,N^{-1}}$
N_a0	acoustic reference compliance	$\mathrm{m^5\,N^{-1}}$
N_d	demagnetization factor	1
n	mechanical compliance	$\mathrm{m\,N^{-1}}$
n'	compliance per unit length	$\mathrm{N^{-1}}$
n_0	translational compliance	$\mathrm{m\,N^{-1}}$
n_C	field compliance	$\mathrm{m\,N^{-1}}$
n_ers	equivalent compliance	$\mathrm{m\,N^{-1}}$
n_K	short-circuit compliance,	$\mathrm{m\,N^{-1}}$
	compliance of piezoceramics	$\mathrm{m\,N^{-1}}$
n_L	compliance (electric open-circuit)	$\mathrm{m\,N^{-1}}$
n_mech	mechanical compliance	$\mathrm{m\,N^{-1}}$
n'_R	rotational compliance per unit length	$\mathrm{N^{-1}\,m^{-2}}$
n_R	rotational compliance	$\mathrm{N^{-1}\,m^{-1}}$
n_RK	rotational short-circuit compliance	$\mathrm{N^{-1}\,m^{-1}}$
n_stat	static compliance	$\mathrm{m\,N^{-1}}$
P	power	W
$P_\mathrm{a}, P_{\mathrm{a}k}$	radiated acoustic power	W
P_el	electric power	W
P_mech	mechanical power	W
$\mathbf{P}$	vector of polarization	$\mathrm{C\,m^{-2}}$
P_i	internal polarization	$\mathrm{C\,m^{-2}}$
P_r	remanent polarization	$\mathrm{C\,m^{-2}}$
p	complex frequency	$\mathrm{s^{-1}}$
$p\,(t)$	pressure	Pa

Symbol	Quantity	Unit
$\hat{p}$	pressure amplitude	Pa
p_0	reference pressure	Pa
$\underline{p}_i$	complex pressure	Pa
$\underline{p}_\mathrm{W}$	transducer pressure	Pa
Q	charge	C
Q_0	reference charge	C
Q	Q-factor	1
Q_el	electrically generated charge	C
Q_{m0}	reference point charge	C
Q_mech	mechanically generated charge	C
Q_n	charge at reference point n	C
$q\,(t)$	volumetric flow	$\mathrm{m^3\,s^{-1}}$
$\underline{q}_0$	excitational volumetric flow	$\mathrm{m^3\,s^{-1}}$
$\underline{q}_\mathrm{W}$	volumetric flow of transducer	$\mathrm{m^3\,s^{-1}}$
R	general gas constant	$\mathrm{J\,mol^{-1}\,K^{-1}}$
R, r	radius	m
R	resistance	Ω
R_i	internal resistance	Ω
R_mag	magnetic resistance	$\mathrm{H^{-1}}$
R_{mn}	leg resistance	Ω
r	friction impedance	$\mathrm{N\,s\,m^{-1}}$
r_a	coefficient of friction per unit length	$\mathrm{N\,s\,m^{-2}}$
$\mathbf{r}_i$	position vector (reference point i)	m
r_R	torsional friction impedance	$\mathrm{N\,m\,s}$
S	mechancial strain	1
S_A	area strain	1
$\underline{S}_i$	complex mechanical strain (layer i)	1
S_i	component of mechanical strain	1
S_max	maximum mechanical strain	1
S_r	remanent mechanical strain	1
S_S	saturation magnetostriction	1
$s\,(t)$	normalized step function	1
s	elastic constant	$\mathrm{m^2\,N^{-1}}$
s_{ij}^E	elastic compliance for $E = 0$	$\mathrm{m^2\,N^{-1}}$
s_{ij}^H	elastic compliance for $H = 0$	$\mathrm{m^2\,N^{-1}}$
T	mechanical stress	$\mathrm{N\,m^{-2}}$
	oscillation period	s
T_0	fundamental oscillation period	s
	mechanical prestressing	$\mathrm{N\,m^{-2}}$
	reference temperature	K
T_E	additional mechanical stress	$\mathrm{N\,m^{-2}}$
T_j	component of mechanical stress	$\mathrm{N\,m^{-2}}$
$\underline{T}$	complex mechanical stress	$\mathrm{N\,m^{-2}}$
T_M	MAXWELL stress	$\mathrm{N\,m^{-2}}$

Symbol	Quantity	Unit
$T_{\max}$	maximum mechanical stress	$\mathrm{N\,m^{-2}}$
t	time	s
U	direct current voltage	V
U_0	supply voltage	V
$u(t)$	electrical voltage	V
$\underline{u}_0$	source voltage	V
$\underline{u}_{\mathrm{L}}$	complex open-circuit voltage	V
$\underline{u}_{\mathrm{W}}$	transducer voltage	V
V	volume	$\mathrm{m^3}$
V_0	reference volume	$\mathrm{m^3}$
V_{A}	armature volume	$\mathrm{m^3}$
V_m	magnetic voltage	A
$V_{m,\mathrm{Air}}$	magnetic voltage (air)	A
$V_{m,\mathrm{m}}$	magnetic voltage (ferromag. layer)	A
$v(t)$	velocity	$\mathrm{m\,s^{-1}}$
$\underline{v}$	complex velocity	$\mathrm{m\,s^{-1}}$
$\underline{v}_{\mathrm{d}}$	distal velocity	$\mathrm{m\,s^{-1}}$
$\underline{v}_{\mathrm{L}}$	open-circuit velocity	$\mathrm{m\,s^{-1}}$
$\underline{v}_{\mathrm{S}}$	velocity (generator)	$\mathrm{m\,s^{-1}}$
$\underline{v}_{\mathrm{W}}$	transducer velocity	$\mathrm{m\,s^{-1}}$
W	internal energy	J
W_{el}	electrical field energy	J
W_{kin}	kinetic energy	J
W_{mag}	magnetic field energy	J
W_{mech}	mechanical energy	J
w	number of turns	1
	energy density	$\mathrm{J\,m^{-3}}$
w_{el}	electical field energy density	$\mathrm{J\,m^{-3}}$
w_{mag}	magnetic field energy density	$\mathrm{J\,m^{-3}}$
X	transformer-like transducer constant	
X_m	eigenmode m	1
$\underline{x}$	complex input variable	1
$\underline{x}(t), \underline{\tilde{x}}(t)$	complex time functions	1
$\tilde{x}(t)$	periodic approximate function	1
$\underline{\tilde{x}}^*(t)$	complex conjugate time function	1
x, y, z	coordinate axes	
$\hat{x}$	amplitude of input variable	1
$\bar{x}$	arithmetic mean	1
x_0	reference position	m
x_1, x_2, x_3	coordinate axes	
Y	gyrator-like transducer constant	
Y^*	imaginary transducer constant	
$\underline{y}$	complex output variable	1
$\hat{y}$	amplitude of output variable	1

Symbol	Quantity	Unit
$\underline{Z}$	electrical impedance	Ω
$\underline{Z}_{\mathrm{a}}$	acoustic impedance	$\mathrm{N\,s\,m^{-5}}$
$Z_{\mathrm{a,L}}$	acoustic friction of moved air	$\mathrm{N\,s\,m^{-5}}$
$Z_{\mathrm{a},r}$	acoustic friction	$\mathrm{N\,s\,m^{-5}}$
Z_{a0}	acoustic reference friction	$\mathrm{N\,s\,m^{-5}}$
$\underline{z}$	mechanical impedance	$\mathrm{kg\,s^{-1}}$
z_0	reference wave impedance	$\mathrm{kg\,s^{-1}}$
z_{W}	wave impedance	$\mathrm{kg\,s^{-1}}$
α	coefficient of thermal expansion	$\mathrm{K^{-1}}$
$\alpha(\omega), \beta(\omega)$	coefficients (Fourier integral transform)	s
α, β, γ	abstract components	
α_f	temperature coefficient of resonant frequency	$\mathrm{K^{-1}}$
α_i, β_i	Fourier coefficients	1
β	wave number	$\mathrm{m^{-1}}$
$\underline{\gamma}$	propagation constant	$\mathrm{m^{-1}}$
	abstract admittance	
$\underline{\gamma}_{mn}, \underline{\rho}_{mn}$	matrix elements	
ΔA	surface segment	$\mathrm{m^2}$
ΔC	capacitance change	F
ΔC_{b}	capacitance change (mechanically locked state)	F
$\Delta F, \Delta F_i$	force change	N
ΔF_{mag}	change of magnetic field force	N
Δf_{r}	resonant frequency change	Hz
Δi	current change	A
Δl	change in length	m
ΔL	inductance change	H
Δm	mass element	kg
Δn	spring element	$\mathrm{N^{-1}\,m^{-1}}$
Δn_{I}	interface compliance	$\mathrm{N^{-1}\,m^{-1}}$
Δn_{R}	rotational compliance (beam segment)	$\mathrm{N^{-1}\,m^{-1}}$
Δn_{RK}	rotational short-circuit compliance (beam segment)	$\mathrm{N^{-1}\,m^{-1}}$
Δp	pressure change	Pa
ΔP	power change	W
ΔQ	charge change	C
Δr	equivalent viscous damping (differential beam element)	$\mathrm{N\,s\,m^{-1}}$
ΔR	resistance change	Ω
Δt	time difference	s
ΔT	temperature change	K
Δu	voltage change	V
ΔV	volume change	$\mathrm{m^3}$
ΔW	energy change	J

Symbol	Quantity	Unit
ΔW_{el}	electric field energy change	J
Δx	rod element length, position change	m
$\Delta\mu_{\mathrm{r}}$	permeability change	1
$\Delta\varphi$	angle change	rad
$\Delta\Phi$	magnetic flux change	Wb
$\Delta\omega$	angular frequency change	s^{-1}
δ	error of Fourier spectral density	1
$\delta(t)$	Dirac delta function	s^{-1}
ε	permittivity	
ε_0	electric constant	$\mathrm{F\,m^{-1}}$
ε_{mn}^{S}	permittivity for $S = 0$	$\mathrm{F\,m^{-1}}$
ε_{mn}^{T}	permittivity for $T = 0$	$\mathrm{F\,m^{-1}}$
ε_{r}	dielectric constant	1
η	loss factor, efficiency factor	1
θ	moment of inertia	$\mathrm{kg\,m^2}$
ϑ	temperature	°C
$\vartheta_{\mathrm{Curie}}$	Curie temperature	°C
κ	adiabatic exponent	1
λ	heat conductivity	$\mathrm{W\,m^{-1}\,K^{-1}}$
$\lambda, \lambda_{\mathrm{B}}, \lambda_{\mathrm{D}}$	wavelength	m
$\underline{\lambda}_m$	abstract flow coordinate	1
$\hat{\lambda}$	amplitude of abstract flow coordinate	1
μ	differential coordinate	1
	permeability	$\mathrm{H\,m^{-1}}$
	voltage integral	$\mathrm{V\,s}$
	viscosity	$\mathrm{Pa\,s}$
$\hat{\mu}$	amplitude of abstract differential coordinate	1
μ_0	magnetic constant	$\mathrm{H\,m^{-1}}$
μ_{ij}^{H}	permeability for $H = 0$	$\mathrm{H\,m^{-1}}$
μ_{air}	dynamic viscosity of air	$\mathrm{Pa\,s}$
$\underline{\mu}_n$	differential coordinate	1
μ_{nm}^{T}	permeability for $T = 0$	$\mathrm{H\,m^{-1}}$
μ_{r}	relative permeability	1
ν	Poisson ratio	1
$\xi(t)$	deflection, displacement	m
$\underline{\xi}$	complex displacement	m
ξ_{g}	limit deflection	m
$\xi_{\max}$	maximum deflection	m
ρ	density	$\mathrm{kg\,m^{-3}}$
$\underline{\rho}$	abstract impedance	
$\rho(x)$	charge density function	$\mathrm{C\,m^{-3}}$
ϱ	unbalance amplitude	m
ρ_0	reference density	$\mathrm{kg\,m^{-3}}$
ρ_{F}	density (femur)	$\mathrm{kg\,m^{-3}}$

Symbol	Quantity	Unit
ρ_K	density (piezoelectric ceramics)	$kg\,m^{-3}$
ρ_L	density (air)	$kg\,m^{-3}$
ρ_P	density (prosthesis)	$kg\,m^{-3}$
ρ_W	density (water)	$kg\,m^{-3}$
σ	conductance	S
τ	impulse time	s
Φ	magnetic flux	Wb
Φ_e	flux in magnetic circuit	Wb
Φ_0	constant flux component	Wb
φ	phase angle	rad
φ_0	reference angle	rad
$\varphi_B(\omega)$	phase angle of transfer function	rad
φ_i	phase angle of harmonic oscillating component	rad
φ_n	phase angle of spring force	rad
φ_r	phase angle of frictional force	rad
φ_v	phase angle of velocity	rad
φ_x	phase angle of input quantity	rad
φ_y	phase angle of output quantity	rad
χ	porosity, magnetic susceptibility	1
Ω	angular velocity	s^{-1}
$\underline{\Omega}$	complex angular velocity	s^{-1}
$\underline{\Omega}_0$	excitational angular velocity	s^{-1}
$\underline{\Omega}_W$	transducer angular velocity	s^{-1}
ω	angular frequency	s^{-1}
ω'	reference frequency	s^{-1}
ω_0	resonant frequency	s^{-1}
ω_g	cut-off angular frequency	s^{-1}
ω_i	characteristic frequency, angular frequency of harmonic oscillating component	s^{-1}
ω_k	discrete angular frequency	s^{-1}

Part I

Focus of the Book

1 Introduction

The presented book mainly addresses engineers and students of electrical engineering and information technology, who want to deduce the dynamic design of mechatronic devices within the scope of product development or research work. It concerns technical tasks which are closely interconnected with electronic, mechanical and acoustic functional elements. These electromechanical systems comprise many issues which can be solved effectively and well structured with the methods presented in this book. Thereby, tools of handling problems with network methods and methods of handling dynamic processes are primarily used, the electrical engineer is familiar with. The transducer elements working between the electrical and mechanical parts of a system and the general presentation of the system beyond transducer elements form the focus of the book. Particularly, the functional understanding of feedback mechanisms, e.g. from mechanical or acoustic subsystems into electrical subsystems, is enabled without any difficulty. Just these considerations are generally not obvious for the student of electrical engineering.
On the one hand, the structured representation enables a helpful notional decomposition of the system in coupled individual assembly units, on the other hand, it enables a fast calculability of the dynamic behavior. In addition, the modeling forces the user to focus on the system's core. Thereby, the necessary restrictions by assumptions and approximations are reduced only so far as it is necessary for the solution of the special problem. Thus, the model keeps a problem-oriented optimal size.
Linear or approximately linearizable relations between physical quantities are assumed. The presented structuring and calculation methods are less suitable for processes which primarily use a distinct nonlinear effect. However, they can be used as starting basis for iterative solutions.

A. Lenk et al., *Electromechanical Systems in Microtechnology and Mechatronics*,
Microtechnology and MEMS, DOI 10.1007/978-3-642-10806-8_1,

1.1 Focus of the Book

The topic of this book are systems of precision engineering and microtechnology like devices, assemblies and components consisting of interconnected electrical, mechanical and acoustic functional components. The interactions between these different domains – electrical, mechanical and acoustic subsystems – are caused by electrical, magnetic or mechanical transformation mechanisms. The reversibility of signal processing directions, thus from the mechanical to the electrical side and vice versa, is remarkable. The essential structure of electromechanical systems considered in this book is shown in Fig. 1.1.

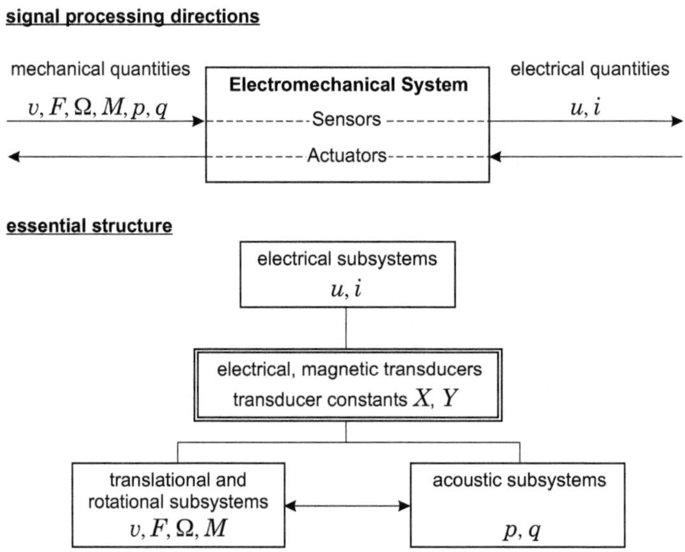

Fig. 1.1. Signal processing directions and essential structure of electromechanical systems

Concerning the design and realization, electromechanical systems as branch of the mechatronics are based on synergies of different theoretically and technologically characterized technical disciplines. Thus, the design methods are characterized by the fundamentals of the signal and system theory, the theory of electric and magnetic fields, the acoustics and mechanics as well as the network theory. In order to realize these systems, technologies of precision engineering and microtechnology, of microsystems technology, optomechanics and optoelectronics as well as of semiconductor electronics and circuit technology are employed. The fields of application and examples for products are specified in Sect. 1.2.

In the presented book the analysis and synthesis of electromechanical systems focus on the determination of the time response of physical quantities (coordinates), thus on the *dynamic system modeling*, for different excitations. The *network theory* well-known within the field of electrical engineering is used as description method. The reason for that are the possibility of a well structured and clear network representation of different subsystems, the existing sophisticated and comfortable solution and representation methods as well as the easy access for electrical engineers. The fundamentals of network representation and interactions with electric and magnetic fields are summarized in Chap. 2.

First of all, Sect. 2.1 deals with the basic knowledge of signal description and signal transmission within *linear networks*. Afterwards, this general description is applied to electrical (Sect. 2.2) and mechanical networks (Sect. 2.3). Subsequently, Sect. 2.4 deals with mechanical and electromechanical interactions. These representations are consciously formulated in general terms in order to show the reader already having knowledge in handling of coupled electrical and mechanical systems the separation of methodology used in this book compared to alternative representations. Finally, in Sect. 2.5 the structuring of the network representation of electromechanical systems takes place.

The network representation of mechanical-translational, mechanical-rotational and acoustic subsystems is introduced and demonstrated by several examples in Chap. 3. In the following, Chap. 4 deals with the *general linear network* on a higher abstraction level. The models are also transferable to linear magnetic and thermal networks.

The mechanical transducers discussed in Chap. 5 desribe the coupling relations between mechanical-translational and mechanical-rotational subsystems as well as between mechanical and acoustic systems.

So far, the network analysis was solely effected for systems with lumped parameters. In Chap. 6, the network analysis is extended on systems with distributed, thus position-dependent parameters. This extension is necessary for higher frequencies of extensional waves their wavelengths reach the scale of component dimensions. The systems are described as linear waveguides or finite network elements. Finally, the extension to a combined simulation of FEA computations of components and of dynamic analysis by means of network representation is presented.

The description of electromechanical transducers is effected in Chap. 7 up to Chap. 9. On the basis of the fundamentals of electromechanical interactions presented in Chap. 7, in Chap. 8 and Chap. 9 the two-port network representations of magnetic and electrical transducers are derived. These loss-free, linear transducer two-port networks are applied to selected examples afterwards.

Finally, the reciprocity principle belonging to linear passive transducers is discussed in Chap. 10.

Parameters of materials being especially important for electromechanical systems are assorted in Appendix A. In addition to Sect. 2.1, Appendix B contains more fundamentals concerning the field of signal description and signal transmission within linear networks.

1.2 Fields of Application and Examples for Electromechanical Systems

An overview of important fields of application of electromechanical systems with assigned examples is shown in Table 1.1.

Table 1.1. Fields of application and typical examples of electromechanical systems

fields of application	examples
process engineering	flow rate and pressure sensors (process measurement engineering), electromagnetic actuating element
automotive engineering	silicon angular rate sensor, small-power motors, piezoelectric fuel injection valves
automotive engineering (utility vehicles, rail vehicles, ships, airplanes)	active sound absorbers, hydraulic coupling systems, thin film pressure sensors, optical and ultrasonic distance sensors
mechanical engineering	piezoelectric pneumatic valves, electrodynamic positioning systems, vibration dampers, electro-dynamic vibration generators
communication technology	microphones, headphones, loudspeakers, laser printers, hard disk drives, ink-jet printers, LCD projectors, camera object lenses
household appliances	level sensors, small-power motors, heat controllers
medical engineering	ultrasonic transducers, miniaturized pressure and force sensors, micro valves, prosthetics

Table 1.1 shows that the main applications of electromechanical systems are realized in the form of devices, assemblies and integrated components in actuating elements, e.g. small-power motors and positioning systems, and in sensor

systems, e.g. single sensors and sensor arrays, as well as in directly coupled sensor-actuator-systems with integrated signal processing.
The dimensions of electromechanical systems with simultaneous improvement of scale integration has been made possible by microtechnology and microsystems technology supplementing the traditional precision engineering. Typical structural dimensions associated with these technologies are presented in Table 1.2.

Table 1.2. Component sizes and structural dimensions of important manufacturing technologies for electromechanical systems

manufacturing technology	machining process	typical component size	smallest structural dimensions
precision engineering	milling and turning, cutting, injection molding, electrical discharge machining, laser welding	a few cm^2	>50 µm
microtechnology	lithographic structuring, thin film technology by means of vapor deposition and sputtering, microelectroplating, microwelding, three-dimensional isotropic and anisotropic etching	>10 mm^2	<50 µm
microsystems technology	silicon bulk or surface micromechanics with integrated microelectronics or integrated optics	<10 mm^2	<5 µm

Simoultaneously, with the introduction of the microtechnology and microsystems technology new materials have been applied to sensors and actuators. Especially silicon, borosilicate glasses and special ceramics like high purity aluminum ceramics and mechanically machinable Low Temperature Cofired Ceramics (LTCC) are among these new materials. These materials are particularly characterized both by their extremely low viscoelastic and viscoplastic behavior and by easy integration possibilities of electrical and optical components. The appropriate realizations are denoted as *Micro Electromechanical Systems* (MEMS). In order to get an introduction to the fundamentals of microtechnology and microsystems technology within the field of electromechanical systems manufacturing, it is referred to [4–7].

In combination with the introduction of these new technologies and materials, in the last few years a technological product change of electromechanical systems toward MEMS realizations has taken place. In addition to the reduction of dimensions, the product change is characterized by an increasing range of functions as it can be seen in Table 1.3.

Table 1.3. Product change in electromechanical systems by using microtechnology and microsystems technology as well as new materials

application	precision engineering device	micro engineering device
printer	dot matrix printer	ink-jet print head (bubble-jet principle)
data memory	magnetic read/write head, hard disk drive	thin film read/write head (GMR read sensor), micro hard disk drive
information rendering	long-playing record, magnetic tape record	DVD position sampling system MP3 player
light projection	projector with rotation mirror system	projector with micromirror array
automotive sensors	piezoelectric airbag acceleration sensors, resistance temperature detector, potentiometric level sensors	large-scale integrated silicon angular rate sensor, microwave and ultrasonic distance sensors, energy self-sufficient silicon tire pressure sensors
process sensors	force compensated differential pressure transducer, potentiometric level transducer	BAW silicon resonance differential pressure sensors, self-monitoring ultrasonic level sensor
electrical circuit switching	elektromagnetic relay	electrostatic silicon-microrelay, silicon-microrelay array
final controlling equipment	electromagnetic control valve	automotive fuel injection valve with piezoelectric stack actuator, electrostatic polymer actuators

For the explanation of the design method „network representation" both traditional and new MEMS realizations are used as examples for electromechanical systems in the following Chaps. 3, 6, 8 and 9.

1.3 Design of Electromechanical Systems

The aim of modeling is the calculation of the parameters – design parameters – of the designed solution for a technical product. Thus, the design phase is of central importance within the product development process of mechatronic, electromechanical and microelectromechanical systems. The design phase is especially distinct, both with „the V-Model" [8] of mechatronic systems and the development models of microsystems technology [4, 9]. In Table 1.4, the phases of the product development process of an electromechanical system are specified as rough stage model according to [10].

Table 1.4. Classification of design in the technical product development process

technical task		
stage of development		result
clarification of technical task	→	list of requirements
conception	→	• determination of subproblems • specification of solution principles • selection of partial solutions • rough concept of the total system
design	→	determination of design parameters of components and of the total solution
construction design	→	design of the total solution
prototyping and test	→	detailed definition of the rough concept, design and construction
documentation	→	drawings and manufacturing technologies
technical release for production		

The design phase follows the conception of the total solution and forms the basis of the subsequent preparation of constructional documents. A physical model of the designed solution forms the basis of the conception (quantitative technical parameter definition for the designed solution). The calculation of the model parameters – design parameters – is based on different description methods (simulation methods) (Sect. 1.4). The aims of these simulations are either the covering of predetermined characteristics – design characteristics – of the electromechanical system or the fulfillment of special optimization criteria like minimum energy consumption, minimum available space and mass or maximum operating frequency range, respectively. The appraisal – verification – of simulation results with the parameters of the requirement specification completes the design phase in Fig. 1.2. If the result differs strongly from the target values with respect to the defined limits, the simulation will be repeated with changed parameter sets. If no convergence can be obtained, a validation of the model approach will be necessary.

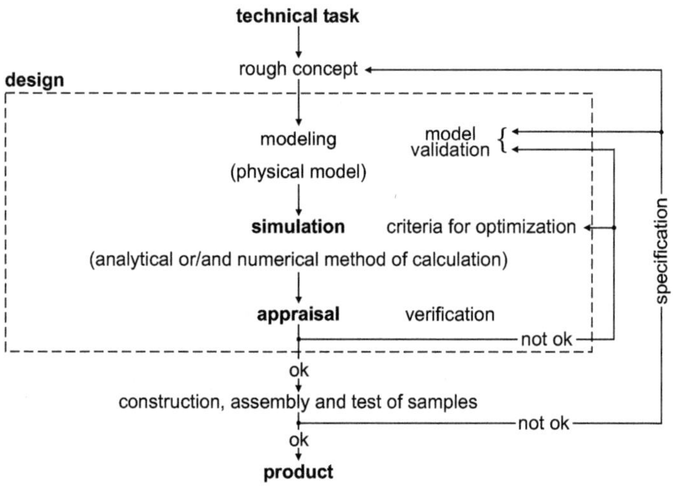

Fig. 1.2. Design workflow for electromechanical systems

The improvement of the design process results in a decrease of experimental testings in order to evaluate development parameters. Simultaneously, the number of samples is reduced resulting in shorter periods of development and lower development costs. The practical relation, the applicability for engineers and the detailed description of simulation methods are decisive for the design improvement that is to be always aspired.

1.4 Simulation Methods for Electromechanical Systems

1.4.1 Historical Overview

The theoretical basics for the description of mechanical and electrical dynamic systems, the LAGRANGE-HAMILTON methods for calculation of dynamic systems including the fluiddynamics, gasdynamics and thermodynamics as well as

the FARADAY-MAXWELL electrodynamics are part of, were nearly completed in the mid-19th century. However, in order to apply these fundamentals practically, a solid mathematical and physical knowledge was essential. Due to the craft and commercial experience at that time, mathematical and physical knowledge was only available to a limited extent.
In the second half of the 19th century, a kind of production developed which caused methods of thought and calculation enabling a broad educated class of technicians and engineers to apply independently the results of the theories mentioned above to their real technical environment.
The completion of electrical network theory based on the papers of KIRCHHOFF and HELMHOLTZ was an important step. Practial experiences in the application of this theory have shown that the relevant technical requirements and processes were describable at least in the first approximation by means of linear relations between physical quantities. Therefore, the electrical network theory was preferably developed with this condition for linearity. The associated possible analysis and synthesis methods are described with a foresighted thinking in the principles of CAUER [11] published 1940. The book proved to be the central source of this field of knowledge for several decades. With the completion of a signal theory adapted to linear systems, a system theory of processes in general electrical configurations developed at the same time. It was finished by KÜPFMÜLLER regarding the aspect of application of electrical communications engineering and was summarized 1952 for the first time under this designation [12].
In the period of time considered above, the electrical engineering increasingly developed into an interdisciplinary special field. Interactions with both the mechanical and precision engineering and the measurement and process engineering caused a mutual influence of subject-specific methods of thought and calculation. At first, in a heuristic way it was obvious that isomorphic system equations of corresponding linear electrical networks could be assigned to linear mechanical system equations. In this way, the so-called electromechanical analogies were generated. In the books by BARKHAUSEN [13], WAGNER [14], HECHT [15], REICHARDT [16] and CREMER [17] their development can be traced from the first beginnings. On the one hand, they enabled the electrical engineers to handle mechanical problems with the network methods they are familiar with. On the other hand, these analogies represented a possible access to mechanics to understand the processes within electrical networks. The latter aspect has a long prehistory. Already FARADAY and MAXWELL [18] tried to assign mechanical models to electrical and magnetic fields which were hard to understand in a literal sense. These attempts had to fail, because the electrodynamic processes just have not been a copy of mechanical processes. Only many years later, one realized which accesses exist in order to compare electrodynamic with mechanical systems.
A possible way is the attempt to find LAGRANGE functions which result in electrical network equations and MAXWELL equations, respectively. The solution of this problem is presented in summary in the concluding chapter of the

book „Theoretische Elektrotechnik" by SIMONYI [19]. It is found that such a LAGRANGE function will exist, if the potential and kinetic energy of mechanics are replaced by the electric and magnetic field energy, respectively. Thereby, the charges have the meaning of position coordinates, the currents have the meaning of impulse coordinates and the potentials have the meaning of generalized forces.

A second way consists in a structural-oriented interpretation of the LAGRANGE equations of first kind which is outlined in the Sects. 2.2 up to 2.6. This description method is based on the work of LENK which he presented in detail in his text books [1–3] in the 70s. Essential parts of these presentations form the basis for the description of linear electromechanical systems in the presented edition.

Although the method of differential equation of state is not used in this book, it has to be mentioned in connection with the description of general dynamic systems. It was generated as problem-oriented description method in connection with the development of control engineering in the mid-19th century. Beyond the linear network-theoretical approaches mentioned before, it enabled additionally the consideration of nonlinear system characteristics. Due to its origin from the energy-oriented HAMILTON-JACOBI differential equations, it is not restricted to mechanical systems.

In the last fifty years the computer science development enabled the application of structural- and network-oriented analytical methods which are summarized under the generic term „Finite Element Analysis (FEA)". At first, the FEA was a result of analytical problems in the field of solid body mechanics. A solid body that has to be analyzed is divided into partial volumes – finite elements – with triangular or rectangular shaped boundary surfaces that are only interconnected at the corners (nodes). The corners of a partial volume are considered to be reference points of a LAGRANGE point system. If quasi-static processes are assumed at first, the equilibrium forces can be determined at the nodes by means of nodal displacements or vice versa. A condition for that it is the knowledge of the stress-strain relations and the assumption of a displacement function within the partial volume interpolating the nodal displacements. By means of balances of forces at the nodes and by equating the displacements of the interconnected nodal points, the individual partial volumes are joined to an overall system. Thus, all nodal displacements can be determined from the equilibrium forces at the nodes or vice versa by means of these coupling conditions. Shortly afterwards, this method was extended to other field problems, e.g. to acoustic, electrostatic and magnetic fields as well as to their interconnections. Nowadays, software programs are available enabling the numerical handling of all technically relevant continuum problems. These software programs result from the development mentioned above.

In this context the question is, where the limits and advantages of applicability of the two mentioned structural-oriented descriptions of dynamic systems are. Here, the following aspects can be identified:

- Finite element methods (e.g. ANSYS, ABAQUS, NASTRAN, COMSOL) will be unrevaled, if structure and parameter field of configuration being analyzed and optimized are known in sufficiently narrow limits.
- Supposing an existing less a priori knowledge and supposing a wide field of possible structures and active principles, in the first step a network description seems to be suitable. In combination with that, calculable analytical solutions are usually attainable easily and often enable to formulate closed solutions for optimization problems or to find constructional or physical invariants. For more complex networks, comfortable numerical simulation systems, e.g. OrCAD CAPTURE, MICRO CAP, can be advantageously applied.

In contrast to the advantages of better physical clarity and closed analytical solubility, the network description distinguishes itself by the disadvantage of limited model accuracy. Reason for that are approximations for used elements and their arrangement within the network. The Finite Element Analysis is able to fill this gap. Thus, an iterative improvement of a network model is possible resulting in a deeper insight into existing active mechanisms. With respect to the individual application, it can help to identify the dominant system elements or transmission paths.

1.4.2 Design Methods

In the following section, an overview of currently available design processes for electromechanical systems will be given. Thereby, it is distinguished between systems with *lumped* and *distributed* (position-dependent) parameters. Thus, it is also to be referred to alternatives of network representations being applicable to both groups.

- **Systems with lumped parameters**

  Static analysis:
  Calculation of electrodynamic interacting forces and deflections on the basis of energy and power balances by means of *Differential Algebraic Equations* (DAE). Differential equations are solved with respect to specific initial and boundary conditions. The solutions are usually nonlinear equations that must be solved by means of iterative methods [9, 20].

  Dynamic analysis:
  Network analysis on the basis of harmonic excitations and calculation of transfer functions (amplitude and phase response; see Sect. 2.1.1). In order to consider nonlinear electrostatic and electromagnetic interacting forces,

the operating point calculation and following small-signal consideration around the operating point are realized by means of the static analysis. In this book, it is focused on this method which is described in detail in the Chaps. 2 up to 10 [21].
Method of *Differential Equations of State* on the basis of ordinary differential equations considering also nonlinear system characteristics in contrast to the linear network analysis. This method also results in system transfer functions which can also be defined for non-decaying excitation functions (LAPLACE transform; see Sect. 2.1.4). Particularly, this method is used within the field of automation engineering in order to describe general technical systems [22].

- **Systems with distributed parameters**
In order to describe electromechanical systems as continuum – position-dependent stress-strain relations, extensional wave propagation – the network method with concentrated elements is extended to *wave processes* or discretization in the form of *finite network elements.* Also here, the interacting forces have to be linearized. These two extended network methods are described in detail in Chap. 6, sample applications are discussed in the Chaps. 6, 8 and 9.
Numerical methods on the basis of the *Finite Element Method* (FEM) or the *Boundary Element Method* (BEM) on the basis of commercial programs such as ANSYS, ABAQUS or NASTRAN. With these programs also nonlinear interactions between mechanical, electrical, magnetic, acoustic, fluid and thermal subsystems can be described. As already mentioned before in Sect. 1.3, best simulation accuracies can be achieved with that. However, the limited physical clarity and more difficult engineering approach to interpretation as well as the additional optimization work are disadvantageous.
In order to use the advantages of network methodes and numerical methods, the combination of both methods is used as *combined simulation* more and more. Thus, the static component characteristics are calculated by means of FE-analyses and the dynamic system characteristics are represented by means of network modeling. But also the reverse way is successful. Here, „substituting structures" being suitable for the FEA can be derived from network structures. These substituting structures are implemented into the FE-model and solved together. In this way, the performance of existing FE-programs is extended. Examples for the combined simulation are given in Sect. 6.4. In [4] an overview of design processes especially for microsystems is shown in detail.

2

Electromechanical Networks and Interactions

The network description of a physical structure requires the definition of

- network coordinates
- fundamental one-port components
- coupling two-port networks
- ideal sources and
- balance equations of coordinates.

The signal and system theory of electrical engineering provide a basis for network theory. Thus, the functional transformations between stationary sinusoidal time functions and complex amplitudes of a linear system are described at first. By means of *Fourier series*, the transfer of circular functions by linear systems can be extended on general periodic functions. The transition to the *Fourier transform* is carried out for singular bandlimited signals.
This description method is subsequently applied to the well-known linear electrical networks. The transition to the application of this method to mechanical and acoustic networks is based on the corresponding basic structure of *ordinary differential equations*. Being founded on the theoretical mechanics, the derivation of fundamentals of the network description for mechanical and acoustic systems is specified in Sect. 2.3. A detailed network description of mechanical and acoustic systems is carried out in Chap. 3.
Between the different subsystems interactions exist. Their characteristics are discussed afterwards. Furthermore, their derivation from electrostatic and magnetostatic field equations is represented in Sect. 2.4. In the Chaps. 7 up to 9, both the classification of electromechanical interactions and their network description are carried out in detail.
Finally, the network description of the considered subsystems and their interactions provide a basis for a structured network description of complex electromechanical systems. In the presented book, it is discussed starting from Chap. 3.

A. Lenk et al., *Electromechanical Systems in Microtechnology and Mechatronics*,
Microtechnology and MEMS, DOI 10.1007/978-3-642-10806-8_2,

2.1 Signal Description and Signal Transmission in Linear Networks

2.1.1 The Circular Function as Basic Module for Time Functions of Linear Networks

The dynamic characterization of electrical, mechanical and acoustic subsystems including their interactions results in system equations between *flow coordinates* and *difference coordinates*. The details are discussed in the Sects. 2.2 up to 2.6. These system equations enable the derivation of differential equations which allow for response determination of individual system coordinates to the excitation of other system coordinates by means of well-known *excitational time functions*. In order to solve such tasks of analysis, the restriction of excitational time functions to *model time functions* has proved to be reasonable.
On the one hand, these models should enable for a solution of the above mentioned task as simple as possible, on the other hand, it should be possible to generate real existent time functions as sums of such model building blocks. The *circular function*

$$x(t) = \hat{x} \cos(\omega t + \varphi_x) \tag{2.1}$$

is particularly suitable as model building block for the analysis of linear networks for the following reasons:

- By means of a *Fourier series*, general periodic functions can be generated from harmonic circular functions.
- The *circular function* is invariant with respect to operations such as *differentiation, addition* and *multiplication* with a constant. These operations are found in the system equations of linear networks discussed in the Sects. 2.2 up to 2.6. Also *linear combinations* of circular functions feature the same characteristics.

From this follows that in case of an excitation of a linear network with a circular function according to (2.1), all remaining network coordinates can also be characterized by circular functions of same frequency ω. Using this particular excitation, the analysis task can be reduced to the determination of amplitudes and phase angles of the remaining coordinates.

The approach

$$\mu(t) = \hat{\mu} \cos\left(\omega t + \varphi_\mu\right) \qquad \text{and} \qquad \lambda(t) = \hat{\lambda} \cos(\omega t + \varphi_\lambda) \tag{2.2}$$

results in a set of equations for the determination of the unknown quantities λ and μ in case of excitation by λ_0 and μ_0. An example is shown in Fig. 2.1.

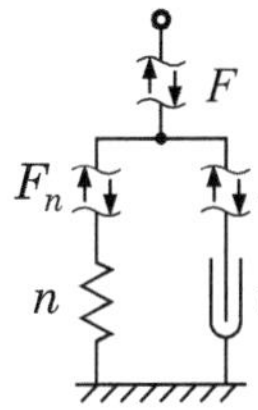

$$\mu_1 = v(t) = \hat{v}\cos(\omega t + \varphi_v) \qquad v(t) = n\frac{\mathrm{d}F_n}{\mathrm{d}t}$$
$$\lambda_1 = F_n(t) = \hat{F}_n\cos(\omega t + \varphi_n) \qquad v(t) = \frac{1}{r}F_r(t)$$
$$\lambda_2 = F_r(t) = \hat{F}_r\cos(\omega t + \varphi_r) \qquad F(t) = F_n(t) + F_r(t)$$
$$\lambda_3 = F(t) = \hat{F}\cos(\omega t)$$

Fig. 2.1. Analysis of a shock absorber by means of a cosine approach

On the basis of the differential equation of the network shown in Fig. 2.1 according to

$$n\frac{\mathrm{d}F_n(t)}{\mathrm{d}t} = \frac{1}{r}F_r(t) = \frac{1}{r}\left(F(t) - F_n(t)\right) \tag{2.3}$$

and thus

$$F(t) = rn\frac{\mathrm{d}F_n(t)}{\mathrm{d}t} + F_n(t), \tag{2.4}$$

by using circular functions for $F(t)$, $F_n(t)$, $F_r(t)$

$$\hat{F}\cos(\omega t) = rn(-\omega)\hat{F}_n\sin(\omega t + \varphi_n) + \hat{F}_n\cos(\omega t + \varphi_n)$$

$$\Leftrightarrow \hat{F}\cos(\omega t) = \hat{F}_n(-\omega nr\cos\varphi_n - \sin\varphi_n)\sin(\omega t) + \hat{F}_n(-\omega nr\sin\varphi_n + \cos\varphi_n)\cos(\omega t) \tag{2.5}$$

with

$$a = -\omega nr\cos\varphi_n - \sin\varphi_n \quad \text{and} \quad b = -\omega nr\sin\varphi_n + \cos\varphi_n$$

the amplitude $\hat{F}_n$ and the phase angle φ_n of the searched force $F_n(t)$ with

$$a = 0 \Rightarrow \omega nr = -\tan\varphi_n \tag{2.6}$$

and

$$b = \sqrt{1+\tan^2\varphi_n} \Rightarrow \hat{F}_n = \frac{1}{\sqrt{1+\tan^2\varphi_n}}\hat{F} \tag{2.7}$$

can be determined.

Due to the more complex addition and differentiation rules for circular functions compared to exponential functions of imaginary arguments, it is common practice and appropriate to use the *complex* time function $\underline{x}(t)$ and the accordingly derived *complex amplitude* $\underline{x}$ instead of the *real* time function in (2.1). The *real* time function in (2.1) can be described as a *complex* time function in the form of

$$\underline{x}(t) = \hat{x}\left[\cos(\omega t + \varphi_x) + \mathrm{j}\sin(\omega t + \varphi_x)\right] = \hat{x}\,\mathrm{e}^{\mathrm{j}(\omega t + \varphi_x)}. \tag{2.8}$$

With respect to (2.8), the *complex amplitude* $\underline{x}$ results in

$$\underline{x} = \hat{x}\,\mathrm{e}^{\mathrm{j}\varphi_x}\,. \tag{2.9}$$

Resulting relations between time- and frequency-domain are represented in Table 2.1.

Table 2.1. Functional transformation between stationary sinusoidal time functions and complex amplitudes

time-domain				frequency domain
$w(t) = \mathrm{d}x(t)/\mathrm{d}t$				$\underline{w} = \mathrm{j}\omega\underline{x}$
$w(t) = \int\limits_a^t x(t)\,\mathrm{d}t$	$x(t) = \Re\{\underline{x}\,\mathrm{e}^{\mathrm{j}\omega t}\}$	$\Longleftrightarrow$	$\underline{x} = \hat{x}\,\mathrm{e}^{\mathrm{j}\varphi}$	$\underline{w} = \frac{1}{\mathrm{j}\omega}\underline{x}$
$w(t) = x_1(t) + x_2(t)$				$\underline{w} = \underline{x}_1 + \underline{x}_2$

The corresponding equations between the complex amplitudes result from the introduction of the approach according to (2.9) into the basic operations of the set of equations of the linear networks in the Sects. 2.2 up to 2.6.

$$\underline{w} = \mathrm{j}\omega\underline{x} \tag{2.10}$$

$$\underline{w} = \frac{1}{\mathrm{j}\omega}\underline{x} \tag{2.11}$$

$$\underline{w} = \underline{x}_1 + \underline{x}_2 \tag{2.12}$$

In this way, a *functional transformation* between variables and equations of the time- and frequency-domain is defined which is described by

$$x(t) = \Re\{\underline{x}\,\mathrm{e}^{\mathrm{j}\omega t}\} \tag{2.13}$$

and

$$\underline{x} = \hat{x}\,\mathrm{e}^{\mathrm{j}\varphi}\,. \tag{2.14}$$

The application of this functional transformation to the system equations in the Sects. 2.2 up to 2.6 transfers the systems of differential equations into algebraic sets of equations between complex amplitudes.

Figure 2.2 shows the simplification of the analysis of Fig. 2.1 using complex amplitudes. Details and examples as well as graphical interpretations to this

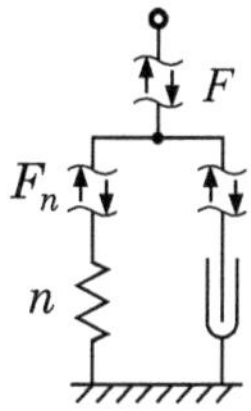

$$\left.\begin{aligned} \underline{v} &= \mathrm{j}\omega n \underline{F}_n \\ \underline{v} &= \frac{1}{r}\underline{F}_r \\ \underline{F} &= \underline{F}_n + \underline{F}_r \end{aligned}\right\} \Rightarrow \left\{\begin{aligned} \mathrm{j}\omega n \underline{F}_n &= \frac{1}{r}\underline{F}_r \\ \mathrm{j}\omega n \underline{F}_n &= \frac{1}{r}\left(\underline{F} - \underline{F}_n\right) \\ \underline{F}_n &= \frac{1}{1+\mathrm{j}\omega n r}\underline{F} \end{aligned}\right.$$

Fig. 2.2. Analysis of a shock absorber with complex amplitudes

calculation method are contained in standard text books of electrical engineering [23–26].

In combination with both relations (2.13) and (2.14), the time-dependent force progression $F_n(t)$ of the shock absorber shown in Fig. 2.2 can be calculated according to

$$F_n(t) = \Re\left(\frac{\hat{F}}{1+\mathrm{j}\omega n r}\,\mathrm{e}^{\mathrm{j}\omega t}\right)$$

$$\Leftrightarrow F_n(t) = \Re\left(\frac{\hat{F}}{\sqrt{1+(\omega n r)^2}}\,\mathrm{e}^{\mathrm{j}\varphi}\,\mathrm{e}^{\mathrm{j}\omega t}\right)$$

$$\Leftrightarrow F_n(t) = \frac{\hat{F}}{\sqrt{1+(\omega n r)^2}}\cos\left(\omega t - \arctan \omega n r\right) \qquad (2.15)$$

with

$$\tan\varphi = -\omega n r. \qquad (2.16)$$

The example shown in Fig. 2.2 suggests the generalization represented in Fig. 2.3 with the introduction of the *transfer function* $\underline{B}(\omega)$ between input and output variables $\underline{x}$ and $\underline{y}$ of a linear system.

$$\left.\begin{aligned} x(t) &= \hat{x}\cos(\omega t + \varphi_x) \\ \underline{x} &= \hat{x}\,\mathrm{e}^{\mathrm{j}\varphi_x} \end{aligned}\right\}$$

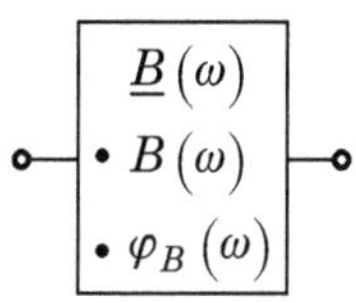

$$\left\{\begin{aligned} y(t) &= \hat{y}\cos(\omega t + \varphi_y) \\ \underline{y} &= \hat{y}\,\mathrm{e}^{\mathrm{j}\varphi_y} \end{aligned}\right.$$

$$\hat{y} = B(\omega)\hat{x} \qquad \underline{B}(\omega) = B(\omega)\mathrm{e}^{\mathrm{j}\varphi_B} \qquad \underline{y} = \underline{B}(\omega)\underline{x}$$

$$\varphi_y = \varphi_x + \varphi_B$$

Fig. 2.3. Transfer function of a linear system

It is remarkable that the conclusions mentioned above are also valid for approaches of the form

$$x(t) = \hat{x}\,\mathrm{e}^{\sigma t}\cos(\omega t + \varphi) \tag{2.17}$$

and

$$\underline{x}(t) = \hat{x}\,\mathrm{e}^{(\sigma + \mathrm{j}\omega)t + \mathrm{j}\varphi}\,. \tag{2.18}$$

Using (2.18), (2.10) up to (2.13) are modified in such a manner that instead of $\mathrm{j}\omega$ the complex frequency $p = \sigma + \mathrm{j}\omega$ must be introduced. These facts are important for stability analyses and enable far-reaching conclusions with respect to complex analytical characteristics of network functions of complex frequency p. Details about that are discussed in Chap. 4 „Abstract Linear Network".

2.1.2 Fourier Expansion of Time Functions

Explanation and Classification of the Fourier Expansion

The model for transferring circular functions by linear systems shown in Fig. 2.3 can be extended to periodic and time-limited („unique") functions by means of the *Fourier transform.*

At first, in (2.19) up to (2.22) the periodic function $\tilde{x}(t)$ of period T is described by *Fourier series* of the form

$$\tilde{x}(t) = \bar{x} + \sum_{i=1}^{\mathrm{M}} c_i \cos(\omega_i t + \varphi_i) \tag{2.19}$$

and

$$\tilde{x}(t) = \bar{x} + \sum_{i=1}^{\mathrm{M}} \left[a_i \cos(\omega_i t) + b_i \sin(\omega_i t)\right], \tag{2.20}$$

respectively with

$$\bar{x} = \int_0^T x(t)\,\mathrm{d}t, \qquad a_i = c_i \cos\varphi_i, \qquad b_i = c_i \sin\varphi_i \tag{2.21}$$

$$\omega_i = 2\pi\frac{1}{T}i, \qquad c_i = \sqrt{a_i^2 + b_i^2}, \qquad \varphi_i = -\arctan\frac{b_i}{a_i}. \tag{2.22}$$

If $x(t)$ is an input function of a linear system, with the methods presented in Sect. 2.1.1, Fig. 2.3 it can be supposed that every harmonic partial oscillation $x_i(t)$ runs separately through the system being not disturbed by the other partial oscillations due to the assumed *system linearity.* In accordance with

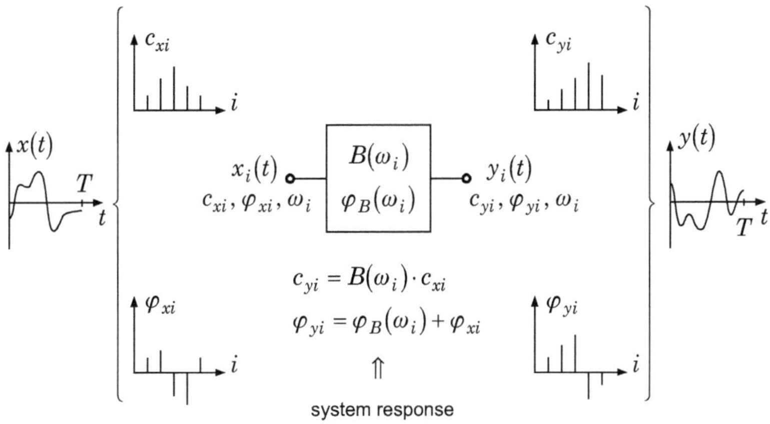

Fig. 2.4. General transfer model for periodic time functions

Fig. 2.3, the resulting partial responses $y_i\,(t)$ add up to the output function $y\,(t)$. The resulting transfer model is illustrated in Fig. 2.4.
KÜPFMÜLLER was the first who used this fundamental system model in order to analyze response characteristics of electrical circuits [12]. It forms the basis of spectral analysis methods of the signal and system theory. The facts being involved justify the central importance of the Fourier series expansions for this field of knowledge.

Fourier Series

The task of finding an *approximation function* $\tilde{x}\,(t)$ or the coefficients c_i, φ_i and a_i, b_i which is given by (2.19) up to (2.22) requires the definition of a suitable criterion. According to the tasks that are to be solved in combination with the approximation equation, two criterions are usual:

- The deviations between the base function $x\,(t)$ and the approximation functions $\tilde{x}\,(t)$ are to become minimal in the *root mean square*. This criterion results in the *approximative* form of Fourier series.
- At a given number N of equidistant sample values t_n, the approximation function $\tilde{x}\,(t)$ is to match the function values of the base function $x\,(t_n)$. This criterion results in the *interpolative* form of Fourier series.

Under restrictive conditions for the function $x\,(t)$ and for the number of sample values, the approximation functions $\tilde{x}\,(t)$ correspond with respect to both criterions.
The *interpolative Fourier series* and the derived Fourier transformations form the basis of *sample systems theory*. They are not used in this book, thus they are not further considered.

The *approximative Fourier series* require the knowledge of the piecewise continuous base function $x(t)$ at all possible, i.e. infinitely many points of the independent variable t. Therefore, they are especially useful for the expansion of analytically given functions.

Real Form of the Approximative Fourier Series

The *mean square deviation* is particularly suitable for the error measure of deviations of two periodic functions $x_1(t)$ and $x_2(t)$ with same periodic time T.

If $x_1(t)$ and $x_2(t)$ are functions that can be accurately mapped with same periodic time T and same mean value $\bar{x}$ by an approach in the form of (2.19) and (2.20), the PARSEVAL equations (2.23) and (2.24) are valid due to the orthogonality of circular functions:

$$\frac{1}{T}\int_0^T x_1^2(t)\,\mathrm{d}t = \frac{1}{2}\sum_{i=1}^{\mathrm{M}} a_i^2 + b_i^2 + \bar{x}_1^2 \tag{2.23}$$

$$\frac{1}{T}\int_0^T x_2^2(t)\,\mathrm{d}t = \frac{1}{2}\sum_{i=1}^{\mathrm{M}} \alpha_i^2 + \beta_i^2 + \bar{x}_2^2 \tag{2.24}$$

From this follows that the mean square deviation of two periodic functions the above mentioned conditions apply to is given by (2.25)

$$\frac{1}{T}\int_0^T (x_1(t) - x_2(t))^2\,\mathrm{d}t = \frac{1}{2}\sum_{i=1}^{\mathrm{M}} (a_i - \alpha_i)^2 + (b_i - \beta_i)^2\,. \tag{2.25}$$

If one of both functions is a sectionally continuous function $x(t)$ and the second function is defined by the approach of (2.19) and (2.20) respectively, the mean square deviation ε yields

$$\begin{aligned}
\varepsilon^2 &= \overline{(\tilde{x}(t) - x(t))^2} \\
&= \frac{1}{T}\int_0^T \left(\bar{x} + \sum_{i=1}^{\mathrm{M}} a_i \cos\left(2\pi i\frac{t}{T}\right) + b_i \sin\left(2\pi i\frac{t}{T}\right) - x(t)\right)^2 \mathrm{d}t \\
&= \overline{x(t)^2} - \bar{x}^2 - \frac{2}{T}\int_0^T x(t)\sum_{i=1}^{\mathrm{M}}\left[a_i \cos\left(2\pi i\frac{t}{T}\right) + b_i \sin\left(2\pi i\frac{t}{T}\right)\right] \mathrm{d}t \\
&\quad + \frac{1}{2}\sum_{i=1}^{\mathrm{M}} \left(a_i^2 + b_i^2\right).
\end{aligned} \tag{2.26}$$

The condition $\varepsilon \to \min$ results in the constitutive equations (2.27) and (2.28) for the coefficients a_i and b_i:

$$\frac{\partial \varepsilon}{\partial a_i} = -\frac{2}{T}\int_0^T x(t)\cos\left(2\pi i\frac{t}{T}\right)\mathrm{d}t + a_i \to 0 \tag{2.27}$$

$$\frac{\partial \varepsilon}{\partial b_i} = \frac{2}{T}\int_0^T x(t)\sin\left(2\pi i\frac{t}{T}\right)\mathrm{d}t + b_i \to 0 \tag{2.28}$$

In summary, (2.29) up to (2.32) result in the *real form* of the approximative Fourier series:

$$\tilde{x}(t) = \bar{x}_i + \sum_{i=1}^{\mathrm{M}}\left[a_i\cos\left(2\pi i\frac{t}{T}\right) + b_i\sin\left(2\pi i\frac{t}{T}\right)\right]$$

$$\tilde{x}(t) = \bar{x} + \sum_{i=1}^{\mathrm{M}} c_i\cos\left(2\pi i\frac{t}{T} + \varphi_i\right) \tag{2.29}$$

$$a_i = \frac{2}{T}\int_0^T x(t)\cos\left(2\pi i\frac{t}{T}\right)\mathrm{d}t, \qquad \forall i = 1\ldots\mathrm{M} \tag{2.30}$$

$$b_i = \frac{2}{T}\int_0^T x(t)\sin\left(2\pi i\frac{t}{T}\right)\mathrm{d}t, \qquad \forall i = 1\ldots\mathrm{M} \tag{2.31}$$

$$\tilde{x}(t) \to x(t) \qquad \text{for} \qquad \mathrm{M} \to \infty \tag{2.32}$$

First of all, with (2.29) up to (2.31) there is the question how the error ε depends on the number M of addends. An exact analysis of this fact results in the conclusion that ε converges uniformly against zero with increasing M.

From this the conclusion of (2.32) follows that both functions match for M $\to \infty$. The detailed analysis of the error characteristics of $x(t)$ is not simple. It is described in detail and with mathematical precision in technical literature [27, 28].

Complex Form of the Approximative Fourier Series

By introducing complex time functions according to (2.8), the transformation equations (2.29) up to (2.32) can be transformed in such a way that the application of the complex transfer function $\underline{B}(\omega_i)$ (see Fig. 2.3) according to the system model shown in Table 2.1 is possible:

$$\underline{\tilde{x}}(t) = \bar{x} + \sum_{i=1}^{\mathrm{M}} c_i\,\mathrm{e}^{\mathrm{j}\varphi_i}\,\mathrm{e}^{\mathrm{j}2\pi i\frac{t}{T}} \tag{2.33}$$

$$\Rightarrow \tilde{x}(t) = \Re\{\underline{\tilde{x}}(t)\} = \frac{1}{2}\left(\underline{\tilde{x}}(t) + \underline{\tilde{x}}^*(t)\right) \tag{2.34}$$

$$\Leftrightarrow \tilde{x}(t) = \bar{x} + \sum_{i=1}^{\mathrm{M}} \frac{1}{2} c_i \,\mathrm{e}^{\mathrm{j}\varphi_i}\, \mathrm{e}^{\mathrm{j}2\pi i \frac{t}{T}} + \sum_{i=1}^{\mathrm{M}} \frac{1}{2} c_i \,\mathrm{e}^{-\mathrm{j}\varphi_i}\, \mathrm{e}^{-\mathrm{j}2\pi i \frac{t}{T}} \tag{2.35}$$

$$\Leftrightarrow \tilde{x}(t) = \bar{x} + \sum_{\substack{i=-\mathrm{M} \\ i\neq 0}}^{\mathrm{M}} \underbrace{\frac{1}{2} c_i \,\mathrm{e}^{\mathrm{j}\varphi_i}}_{\underline{c}_i} \mathrm{e}^{\mathrm{j}2\pi i \frac{t}{T}} \tag{2.36}$$

with

$$\underline{c}_i = \underbrace{\frac{1}{2} c_i \cos\varphi_i}_{a_i} + \mathrm{j}\underbrace{\frac{1}{2} c_i \sin\varphi_i}_{-b_i} = \frac{1}{T}\int_0^T x(t)\,\mathrm{e}^{-\mathrm{j}2\pi i \frac{t}{T}}\,\mathrm{d}t \tag{2.37}$$

Taking $\underline{c}(i=0) = \bar{x}$ into consideration, finally the *complex form* of the approximated Fourier series according to (2.38) up to (2.42) is achieved:

$$\tilde{x}(t) = \sum_{i=-\mathrm{M}}^{\mathrm{M}} \underline{c}_i \,\mathrm{e}^{\mathrm{j}2\pi i \frac{t}{T}} \tag{2.38}$$

$$\underline{c}_i = \frac{1}{T}\int_0^T x(t)\,\mathrm{e}^{-\mathrm{j}\,2\pi i \frac{t}{T}}\,\mathrm{d}t \tag{2.39}$$

$$\underline{c}_i = \frac{1}{2}(a_i - \mathrm{j}b_i), \qquad \bar{x} = c_0 \tag{2.40}$$

$$\underline{c}_i^* = \underline{c}(-i) \tag{2.41}$$

$$\tilde{x}(t) \rightarrow x(t) \qquad \text{for} \qquad \mathrm{M} \rightarrow \infty \tag{2.42}$$

If such a described time function is transfered by a linear system defined by the complex transfer function $\underline{B}(\omega_i)$, according to Table 2.1 and Fig. 2.4 following relations will be achieved:

$$\tilde{y}(t) = \sum_{i=-\mathrm{M}}^{\mathrm{M}} \underline{c}_{ix} \underline{B}(\omega_i)\,\mathrm{e}^{\mathrm{j}\omega_i t} \tag{2.43}$$

$$\underline{c}_{ix} = \frac{1}{T}\int_0^T x(t)\,\mathrm{e}^{-\mathrm{j}\omega_i t}\,\mathrm{d}t, \qquad i = -\mathrm{M}\ldots\mathrm{M} \tag{2.44}$$

$$\tilde{y}(t) \rightarrow y(t) \qquad \text{for} \qquad \mathrm{M} \rightarrow \infty \tag{2.45}$$

When comparing (2.29) up to (2.32) with (2.43) up to (2.45), it has to be considered that $|c_i|$ according to (2.40) is identical with $c_i/2$ formulated in (2.29).
In textbooks and data reference books of electrical engineering as well as in books of the signal and system theory tables for the coefficients a_i and b_i of periodic model functions can be found which are useful for theoretical and experimental problems [29–32].

2.1.3 The Fourier Transform

Aspects of Application with the Spectral Representation of Single Processes

In the measurement technology and systems analysis the determination of system reactions on single time-limited processes is of same importance like that on periodic processes. The reaction on a single process comprises the condition that the considered system is in static state at the beginning of the single process. All energy storages must be empty (see. Fig. 2.5).

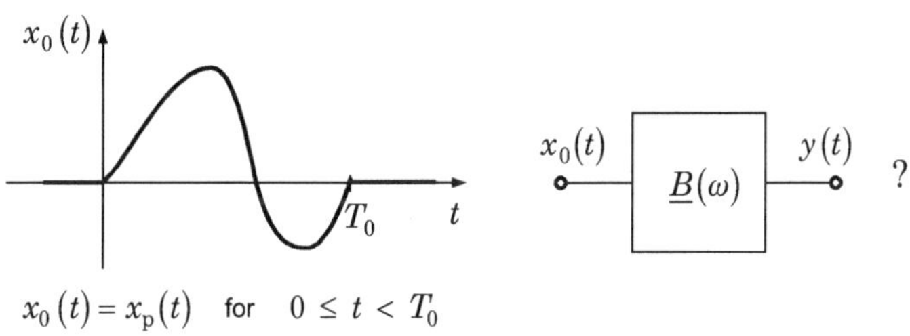

Fig. 2.5. Transfer of a single time-limited process by a linear system

In principle, the solution of this task is possible by the solution of the differential equations which can be derived from the system equations in the Sects. 2.2 up to 2.6. A suitable form of these equations are the *state differential equations*, whose application is described in detail in [33]. These problem-solving methods are not subject to restriction on linear systems which had to be made with the spectral approach.
The question is whether with restriction on linearity also for single processes in the above mentioned sense a spectral description can be found that enables the large performance of the calculation methods represented in the Sects. 2.1.1 and 2.1.2, in particular the use of complex transfer functions for system description.
The answer to this question results in the *Fourier integral transform.*

The practise in experimental system analysis and measurement technology is the starting point for such a concept. Here, the excitation of the system with the single process is periodically repeated after a sufficiently long time duration, i.e. for the suppression of disturbances. The resulting process is periodic and provides the application of all methods which are drawn up in the Sects. 2.1.1 and 2.1.2. Now it has to be analyzed whether the models of quasi-single processes being constructible in this way result in descriptions in the limit which correspond to the above mentioned demands.

Realization of Function Sequences for Transition to the Fourier Integral Transform

The periodic repetition of a time-limited process according to the considerations of the preceding section results in the model shown in Fig. 2.6.

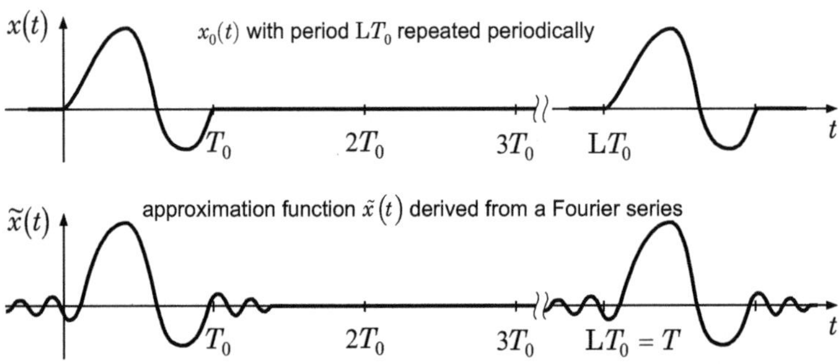

Fig. 2.6. Model of a function sequence for spectral representation of single processes

The time-limited function $x_0(t)$ with period T_0 is periodically repeated with period $T = \mathrm{L}T_0$. With respect to a finite upper limit of summation K, the Fourier series of the resulting function $\tilde{x}(t)$ will comprise the errors at the jump discontinuities and kink positions of $x(t)$ shown in Appendix B. Considerations to these figures result in the facts that the deviations $\Delta x(t)$ between $x(t)$ and $\tilde{x}(t)$ converge against zero with increasing K outside the neighborhood of these positions. This also applies to the period between T_0 and $\mathrm{L}T_0$. It can be expected that $\tilde{x}(t)$ approximates the function $x_0(t)$ shown in Fig. 2.5 with increasing K and L. The determination of a_k and b_k in the spectral representation of $\tilde{x}(t)$ according to

$$\tilde{x}(t) = \bar{x} + \sum_{i=1}^{\mathrm{K}} a_i \cos\left(2\pi i \frac{t}{T}\right) + b_i \sin\left(2\pi i \frac{t}{T}\right) \tag{2.46}$$

is possible with (2.30) and (2.31).

For further considerations it is suitable to use the spectral description by means of a Fourier series as analytical description of $x_0(t)$ within the range $0 \le t \le T_0$. It is achieved by periodic continuation from $x_0(t)$ out of this range as shown in Fig. 2.7.

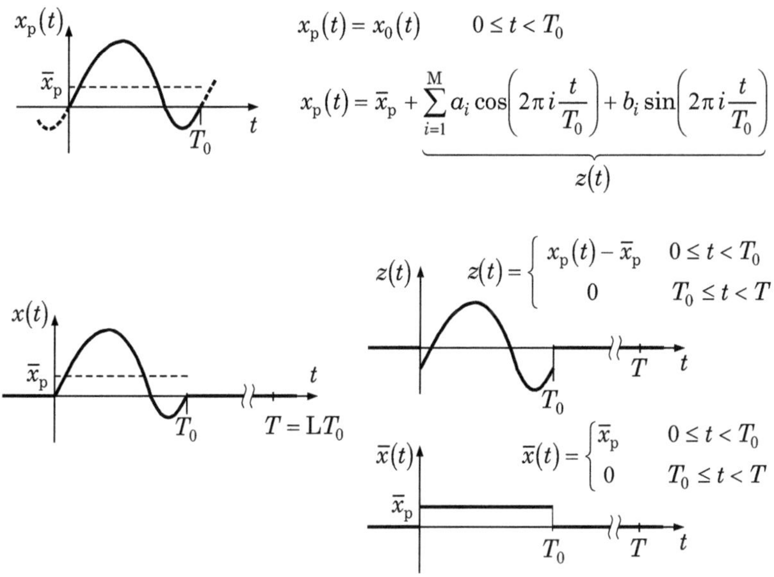

Fig. 2.7. Consideration of constant component of function $x_p(t)$ with expansion of function $x(t)$ according to (2.46)

At first, for further considerations it is to be assumed that $x_p(t)$ is bandlimited. Thus, the function $x_p(t)$ is to be correctly representable with finite M by (2.36).

If the mean value $\bar{x}_p$ of $x_p(t)$ is nonzero, special characteristics will result from the expansion of $x(t)$ according to (2.46) which are specified in Fig. 2.7. With the expansion of periodic functions, whose mean value $\bar{x}_p$ is different from zero over one period T_0, the determination of the coefficients a_i and b_i according to (2.30) and (2.31) always yields the values of the zero-mean function $z(t) = x_p(t) - \bar{x}_p$, regardless of whether $x(t)$ or $z(t)$ are inserted into the equations mentioned above.

Here, the function $x(t)$ must be divided up into the two components $z(t)$ and $x(t)$ over the whole interval T. The expansion of both components must be done separately.

The result is represented in (2.47) and (2.48). In addition to the coefficients of the function $z(t)$, the coefficients $\Delta\alpha_k$ and $\Delta\beta_k$ are generated:

$$\begin{Bmatrix} \alpha_k \\ \beta_k \end{Bmatrix} = \frac{2}{T}\int\limits_0^{T_0} x(t) \begin{Bmatrix} \cos \\ \sin \end{Bmatrix} \left(2\pi k \frac{t}{T}\right) \mathrm{d}t$$
$$= \frac{2}{T}\int\limits_0^{T_0} \left(\bar{x}_p + \sum_{i=1}^{\mathrm{M}} \left[a_k \cos\left(2\pi i \frac{t}{T_0}\right) + b_k \sin\left(2\pi i \frac{t}{T_0}\right)\right]\right)$$
$$\cdot \begin{Bmatrix} \cos \\ \sin \end{Bmatrix} \left(2\pi k \frac{t}{T}\right) \mathrm{d}t \tag{2.47}$$

$$\begin{Bmatrix} \alpha_k \\ \beta_k \end{Bmatrix} = \underbrace{\bar{x}_p \frac{2}{T}\int\limits_0^{T_0} \begin{Bmatrix} \cos \\ \sin \end{Bmatrix} \left(2\pi k \frac{t}{T}\right) \mathrm{d}t}_{\Delta\alpha_k, \Delta\beta_k}$$
$$+ \sum_{i=1}^{\mathrm{M}} \left[\underbrace{a_k \frac{2}{T}\int\limits_0^{T_0} \cos\left(2\pi i \frac{t}{T_0}\right) \begin{Bmatrix} \cos \\ \sin \end{Bmatrix} \left(2\pi k \frac{t}{T}\right) \mathrm{d}t}_{A_{ik}, C_{ik}} \right.$$
$$\left. + \underbrace{b_k \frac{2}{T}\int\limits_0^{T_0} \sin\left(2\pi i \frac{t}{T_0}\right) \begin{Bmatrix} \cos \\ \sin \end{Bmatrix} \left(2\pi k \frac{t}{T}\right) \mathrm{d}t}_{D_{ik}, B_{ik}} \right] \tag{2.48}$$

The evaluation of both the integrals A_{ik}, B_{ik}, C_{ik}, D_{ik} and the $\Delta\alpha_k$ and $\Delta\beta_k$ being found in (2.48) is shown in (2.49) up to (2.52):

$$z_k = \frac{k}{\mathrm{L}} = \frac{\omega_k}{2\pi/T}\frac{1}{\mathrm{L}} = \frac{\omega_k}{2\pi/T_0} = \frac{\omega_k}{\omega_0} \tag{2.49}$$

$$\Delta\alpha_k = 2\bar{x}_p \frac{1}{\mathrm{L}} \frac{\sin(2\pi z_k)}{2\pi z_k}, \qquad \Delta\beta_k = 2\bar{x}_p \frac{1}{\mathrm{L}} \frac{1 - \cos(2\pi z_k)}{2\pi z_k} \tag{2.50}$$

$$\begin{Bmatrix} A_{ik} \\ B_{ik} \end{Bmatrix} = \frac{\begin{Bmatrix} z_k \\ i \end{Bmatrix} \sin(2\pi z_k)}{\pi\left(z_k^2 - i^2\right)} \tag{2.51}$$

$$\begin{Bmatrix} C_{ik} \\ D_{ik} \end{Bmatrix} = \frac{\begin{Bmatrix} i \\ -z_k \end{Bmatrix} \left[\cos(2\pi z_k) - 1\right]}{\pi\left(z_k^2 - i^2\right)} \tag{2.52}$$

Thus, (2.53) up to (2.55) represent the results for the searched coefficients in (2.46):

$$\alpha_k = \frac{1}{\mathrm{L}} \sum_{i=1}^{\mathrm{M}} (A_{ik} a_i - C_{ik} b_i) + \Delta\alpha_k \tag{2.53}$$

$$\beta_k = \frac{1}{\mathrm{L}} \sum_{i=1}^{\mathrm{M}} (D_{ik} a_i - B_{ik} b_i) + \Delta\beta_k \tag{2.54}$$

$$\bar{x} = \frac{\bar{x}_p}{\mathrm{L}} \tag{2.55}$$

A more detailed analysis of A_{ik}, B_{ik}, C_{ik} and D_{ik} for the values $k = R\mathrm{L}i$ results in the values represented in (2.56) and (2.57):

$$A_{ik} = B_{ik} = \begin{Bmatrix} 1, & R = 1 \\ 0, & R \neq 1 \end{Bmatrix} \tag{2.56}$$

$$C_{ik} = D_{ik} = 0 \tag{2.57}$$

From this follows that the α_k and β_k assume the values specified in (2.58) at the points $k = i\mathrm{L}$:

$$\begin{Bmatrix} \alpha_k \\ \beta_k \end{Bmatrix} = \frac{1}{\mathrm{L}} \begin{Bmatrix} a_i \\ b_i \end{Bmatrix} \tag{2.58}$$

All in all, the situation in Fig. 2.8 shows the result for the transition from the a_i, b_i to the α_k, β_k.

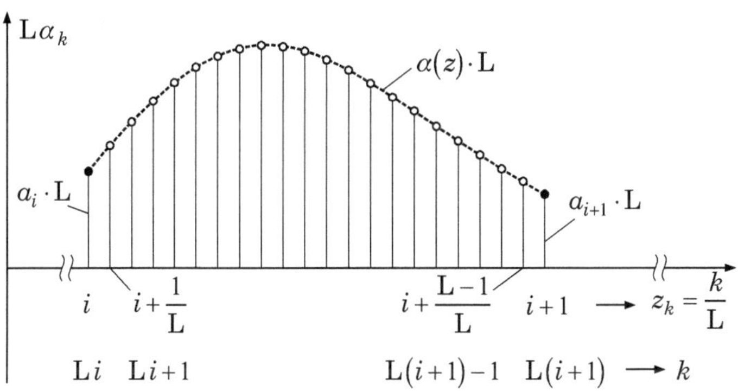

Fig. 2.8. Transition from the a_i, b_i to the α_k, β_k using the example of a section $k = \mathrm{L}i \ldots \mathrm{L}(i+1)$

Thereby, following equations are valid:

$$\alpha(z_k) = \frac{1}{\mathrm{L}} \sum_{i=1}^{\mathrm{M}} \left(\frac{z_k \sin(2\pi z_k)}{\pi(z_k^2 - i^2)} a_i + \frac{i[\cos(2\pi z_k) - 1]}{\pi(z_k^2 - i^2)} b_i \right) + \Delta\alpha_k \quad (2.59)$$

$$\beta(z_k) = \frac{1}{\mathrm{L}} \sum_{i=1}^{\mathrm{M}} \left(\frac{-z_k[\cos(2\pi z_k) - 1]}{\pi(z_k^2 - i^2)} a_i + \frac{i \sin(2\pi z_k)}{\pi(z_k^2 - i^2)} b_i \right) + \Delta\beta_k \quad (2.60)$$

By means of the operations with the function $x_p(t)$ shown in Fig. 2.7, additional *spectral lines* are generated inbetween the a_i, b_i matching with the initially present a_i, b_i at the positions $k = i\mathrm{L}$ except for the factor L.
The interpolating envelope curve for the α_k, β_k will be obtained, if instead of the discrete variable z_k the continuous variable z is inserted in (2.59) and (2.60). For further considerations it is remarkable that the envelope functions $\alpha(z)\mathrm{L}$, $\beta(z)\mathrm{L}$ are independent of L. Further on the relative differences between neighboring spectral lines decrease with increasing L due to the continuity of these functions.
In addition, it is remarkable that the α_k, β_k are completely defined by the a_i, b_i. That's why in the *frequency domain* (2.59) and (2.60) repesent the *analogon to the sample relation in the time domain* concerning bandlimited time functions.
At this point it can already be asked the question how the derived relations will change, if the function $x_p(t)$ is not bandlimited.
According to the considerations in Sect. 2.1.2 the upper summation limit M in (2.53) and (2.54) must converge to infinity. The α_k, β_k already become a limit value. By the way, from (2.59) and (2.60) it can be realized that the α_k, β_k are nonzero beyond the value $k = \mathrm{ML}$ even if $x_p(t)$ is bandlimited. That's why the summation over k in (2.46) must converge to infinity anyway. In case of abandonment of the condition of band limitation for $x_p(t)$, there will be no changes of the limit value characteristics in (2.46).

Transition to the Fourier Integral Transform

Taking Fig. 2.8 into account, the considerations in the preceding section suggest the substitution of the sum in (2.46) by an integral for large L. This results in combination with

$$\mathrm{L} \gg 1 \Rightarrow \omega_k \Rightarrow \omega, \quad z_k = \frac{k}{\mathrm{L}} \Rightarrow \frac{\omega}{\omega_0}, \quad \omega_0 = \frac{2\pi}{T_0}, \quad \Delta\omega = \frac{2\pi}{T_0}$$

in the following equations:

$$x(t) = \bar{x} + \frac{1}{\Delta\omega} \sum_{k=1}^{\mathrm{K}} [\alpha_k \cos(\omega_k t) + \beta_k \sin(\omega_k t) \Delta\omega] \quad (2.61)$$

$$\alpha_k \Rightarrow \alpha(\omega), \quad \beta_k \Rightarrow \beta(\omega) \quad (2.62)$$

$$x(t) \approx \bar{x} + \frac{1}{\pi} \int_0^{\omega_k} \left(\frac{T}{2} \alpha(\omega) \cos(\omega t) + \frac{T}{2} \beta(\omega) \sin(\omega t) \right) \mathrm{d}\omega \tag{2.63}$$

$$a(\omega) = \frac{T}{2} \alpha(\omega), \qquad b(\omega) = \frac{T}{2} \beta(\omega) \tag{2.64}$$

The Equations (2.30) and (2.31) turn into (2.65) and (2.67), respectively. (2.67) results from (2.51) up to (2.55):

$$\begin{Bmatrix} a(\omega) \\ b(\omega) \end{Bmatrix} = \int_0^{T_0} x(t) \begin{Bmatrix} \cos \\ \sin \end{Bmatrix} (\omega t) \, \mathrm{d}t \tag{2.65}$$

$$\bar{x} = \bar{x}_p \frac{T_0}{T} \tag{2.66}$$

$$\begin{aligned} \begin{Bmatrix} a(\omega) \\ b(\omega) \end{Bmatrix} = {} & \frac{1}{2\pi} \sum_{i=1}^{\mathrm{M}} \frac{1}{(\omega/\omega_0)^2 - i^2} \begin{Bmatrix} (\omega/\omega_0) \sin(2\pi\omega/\omega_0) \\ -(\omega/\omega_0) [\cos(2\pi\omega/\omega_0) - 1] \end{Bmatrix} a_i \\ & + \frac{1}{2\pi} \sum_{i=1}^{\mathrm{M}} \frac{1}{(\omega/\omega_0)^2 - i^2} \begin{Bmatrix} -i [\cos(2\pi\omega/\omega_0) - 1] \\ i \sin(2\pi\omega/\omega_0) \end{Bmatrix} b_i \\ & + \bar{x}_p T_0 \begin{Bmatrix} \sin(2\pi\omega/\omega_0) / (2\pi\omega/\omega_0) \\ [1 - \cos(2\pi\omega/\omega_0)] / (2\pi\omega/\omega_0) \end{Bmatrix} \end{aligned} \tag{2.67}$$

With respect to the function sequence defined in the preceding section, the transition to the limits $\mathrm{L} \to \infty$, $\mathrm{K} \to \infty$ and $\mathrm{M} \to \infty$ will be possible without any problem, if the condition of time limitation for $x_p(t)$ is maintained further on. The value $\bar{x}$ converges to zero in the limit. The result is summarized in (2.68) and (2.69) representing the Fourier integral transform:

$$x(t) = \frac{1}{\pi} \int_0^{\infty} [a(\omega) \cos(\omega t) + b(\omega) \sin(\omega t)] \, \mathrm{d}\omega \tag{2.68}$$

$$\begin{Bmatrix} a(\omega) \\ b(\omega) \end{Bmatrix} = \int_0^{\infty} x(t) \begin{Bmatrix} \cos \\ \sin \end{Bmatrix} (\omega t) \, \mathrm{d}t \tag{2.69}$$

$$\int_0^{\infty} |x(t)| \, \mathrm{d}t \to \text{limited} \tag{2.70}$$

The strict mathematical derivation of this functional transformation can be found in selected textbooks [27, 28]. The result of this proof shows that the demand for time limitation of $x(t)$ in (2.69) is not necessary. However, $x(t)$ must fulfill the condition in (2.70).

According to (2.33) up to (2.35) the relations in (2.68) and (2.69) can also be formulated in complex form. The Fourier integral transform in complex form results from these equations:

$$x(t) = \frac{1}{2\pi} \int_{-\infty}^{\infty} \underline{c}(\omega)\, e^{j\omega t}\, d\omega \tag{2.71}$$

$$\underline{c}(\omega) = \int_{0}^{\infty} x(t)\, e^{-j\omega t} dt \tag{2.72}$$

$$\underline{c}(\omega) = \frac{1}{2}\left[a(\omega) - jb(\omega)\right] \tag{2.73}$$

$$\underline{c}(-\omega) = -\underline{c}^{*}(\omega) \tag{2.74}$$

The relations (2.68) and (2.69) and (2.71) up to (2.74) are not necessarily restricted to positive values of t. Taking (2.70) into consideration $x(t)$ can assume values within the range $-\infty \leq t \leq \infty$. Therefore, the lower limit in (2.69) and (2.72) can also be assumed to be $-\infty$ as it is usually done with the definition of the Fourier integral transform.
Comprehensive tables of relations between time functions and their Fourier transforms can be found in textbooks of the signal and system theory [30–32, 34].
Finally, also the initially asked question can be answered whether with restriction to linearity for single processes a spectral description can be specified. If the view of sequence elements defined in Fig. 2.6 is taken, the Fourier integral transform according to (2.68) and (2.69) can be applied.

The integrand in (2.71) bears the meaning of the complex time function of a differential partial oscillation adding up to $x(t)$. The quantity $\underline{c}(\omega)\, d\omega$ denotes the complex amplitude of this partial oscillation. The integrand in (2.68) represents the appropriate differential partial oscillations in real form. By means of (2.61) and (2.63), the coefficients $\alpha(\omega)$ and $\beta(\omega)$ can be reduced to discrete Fourier coefficients of a sequence element in (2.46). If $\underline{c}_x(\omega)\, d\omega$ represents the differential Fourier coefficient of an input function of a linear system, the appropriate coefficients of the output function $y(t)$ will be received:

$$\underline{c}_y(\omega) = \underline{B}(\omega)\, \underline{c}_x(\omega) \tag{2.75}$$

Thus, the function $y(t)$ accordingly yields

$$y(t) = \frac{1}{2\pi} \int_{-\infty}^{\infty} \underline{c}_x \underline{B}(\omega) \, e^{j\omega t} \, d\omega \tag{2.76}$$

with

$$\underline{c}_x(\omega) = \int_{0}^{\infty} x(t) \, e^{-j\omega t} \, dt. \tag{2.77}$$

These facts are illustrated in Fig. 2.9.

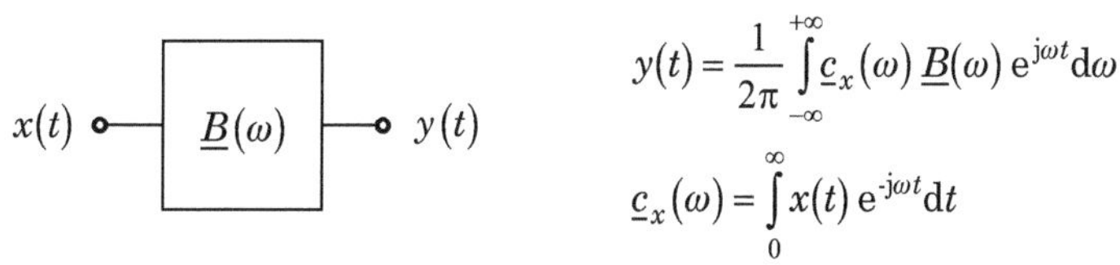

Fig. 2.9. Transfer of a single process by a linear system with transfer function $\underline{B}(\omega)$

2.1.4 The Laplace Transform

Concerning the Fourier transform described in Sect. 2.1.3, there are problems with convergence of functions which do not converge sufficiently strong against zero for $t \to \infty$. Using the characteristics of exponentially increasing and decreasing circular functions represented in (2.17) and (2.18), (2.71) and (2.72) can be rearranged in such a way that the above mentioned problems can be avoided. Thus, the Fourier transform is transfered to the *Laplace transform.*

If the function

$$u(t) = x(t) \, e^{-\sigma t} \tag{2.78}$$

instead of the function $x(t)$ is subjected to a Fourier transform according to (2.72), the result is

$$\underline{c}_u(\omega) = \int_{0}^{\infty} x(t) \, e^{-(\sigma + j\omega)t} \, dt. \tag{2.79}$$

In combination with the complex frequency

$$p = \sigma + j\omega \tag{2.80}$$

follows:

$$\underline{c}_u(\omega) = \int_0^{\infty} x(t)\,\mathrm{e}^{-pt}\,\mathrm{d}t = \mathcal{L}\{x(t)\} = F(p) \tag{2.81}$$

The inverse transform of $\underline{c}_u$ in combination with (2.71) yields $u(t)$:

$$u(t) = \frac{1}{2\pi}\int_{-\infty}^{+\infty} \underline{c}_u(p)\,\mathrm{e}^{-\mathrm{j}\omega t}\,\mathrm{d}\omega = \frac{1}{2\pi\mathrm{j}}\int_{\sigma-\mathrm{j}\infty}^{\sigma+\mathrm{j}\infty} F(p)\,\mathrm{e}^{-\mathrm{j}\omega t}\,\mathrm{d}p \tag{2.82}$$

The original function $x(t)$ is achieved in combination with (2.78):

$$x(t) = \frac{1}{2\pi\mathrm{j}}\int_{\sigma-\mathrm{j}\infty}^{\sigma+\mathrm{j}\infty} F(p)\,\mathrm{e}^{pt}\,\mathrm{d}p = \mathcal{L}^{-1}\{F(p)\} \tag{2.83}$$

The pair of equations (2.84) and (2.85) defines the Laplace transform:

$$\mathcal{L}\{x(t)\} = F(p) = \int_0^{\infty} x(t)\,\mathrm{e}^{-pt}\,\mathrm{d}t \tag{2.84}$$

$$x(t) = \mathcal{L}^{-1}\{F(p)\} = \frac{1}{2\pi\mathrm{j}}\int_{\sigma-\mathrm{j}\infty}^{\sigma+\mathrm{j}\infty} F(p)\,\mathrm{e}^{pt}\,\mathrm{d}p \tag{2.85}$$

The equations (2.79) up to (2.83) clearly describe its relation with the Fourier transform. The way back from this relatively high abstraction level to simple models and algebraically and numerically verifiable relations by means of the Fourier transform can be useful, if there is doubt with the application of such very efficient methods of analysis whether the made operations are allowed. The history of the application of the Laplace transform within the field of electrical engineering comprises examples of such invalid operations and conclusions.

In contrast to the usual definition of the Fourier transform, the lower limit of integration $t = 0$ in (2.79) is useful. For $-\infty \le t \le \infty$ the function $x(t)$ should meet unnecessary requirements due to the exponentially increasing factor $\mathrm{e}^{-\sigma t}$ in (2.86) with negative values for t.

$$\int_0^{\infty} \left|x(t)\,\mathrm{e}^{-\sigma t}\right|\mathrm{d}t \to \text{limited} \tag{2.86}$$

In addition, the Laplace transform has been developed as a tool for the transient analysis, for which $x(t) = 0$ is valid for $t \le 0$.

For mathematical and technically significant functions detailed tables of correspondences between $F(p) = \mathcal{L}\{x(t)\}$ and $x(t)$ can be found in textbooks as well as in mathematical formularies [23, 30–32, 35–37].

With the considerations concerning Fig. 2.9, the question can be answered how Laplace transforms are transfered by linear systems. The key to that are the facts that not only circular functions but also products of circular and exponential functions will not change their characteristics, if they are transfered by linear systems. These facts are also represented in (2.17) and (2.18).

In analogy with (2.72) equation (2.79) can be interpreted in such a way that the differential complex partial oscillations

$$\Delta \underline{x}(t) = \underline{c}_u(p)\, \mathrm{e}^{pt}\, \mathrm{d}t \tag{2.87}$$

add up to $x(t)$. When discussing about complex time functions and amplitudes in this context, the concept represented in Table 2.1 must also be extended to the functional type

$$\underline{x}(t) = \hat{x}\, \mathrm{e}^{\sigma t}\, \mathrm{e}^{\mathrm{j}(\omega t + \varphi)} = \hat{x}\, \mathrm{e}^{\mathrm{j}\varphi}\, \mathrm{e}^{pt}\,. \tag{2.88}$$

Here, the functional transformation shown in Table 2.1 will remain, if the complex frequency p is introduced instead of $\mathrm{j}\omega$. The transition of network differential equations (see Sects. 2.2 and 2.3) to equations with complex amplitudes causes the modified functional transformation shown in Table 2.1. Thus, e.g. the complex transfer function $\underline{F}_n/\underline{F}$ illustrated in Fig. 2.2 changes according to

$$\underline{B}(\omega) \rightarrow \underline{B}(p) = \frac{\underline{F}_n}{\underline{F}} = \frac{1}{1 + rnp} = \frac{1}{1 + rn\sigma + \mathrm{j}\omega nr}\,. \tag{2.89}$$

Under these circumstances, $\underline{c}_u(p)\,\Delta\omega$ can be considered as differential complex amplitude $\Delta\underline{x}$ of the complex time function $\Delta\underline{x}(t)$. If such a described time function affects a linear system characterized by the complex transfer function $\underline{B}(p)$, the differential complex amplitude of the output variable $\Delta\underline{y}$ will result in

$$\Delta\underline{y} = \underline{c}_u(p) \cdot \underline{B}(p) \tag{2.90}$$

which adds up to the output variable $y(t)$:

$$y(t) = \frac{1}{2\pi\mathrm{j}} \int\limits_{\sigma - \mathrm{j}\omega}^{\sigma + \mathrm{j}\omega} \underline{c}_u(p)\, \underline{B}(p)\, \mathrm{e}^{pt}\, \mathrm{d}p \qquad \text{with} \qquad \underline{c}_u(p) = \mathcal{L}\{x(t)\} \tag{2.91}$$

The application of the Laplace transform in the signal and system theory bears a special meaning due to following reasons:

- Concerning the behavior of $x(t)$ for $t \to \infty$, the restrictions existing with the Fourier transform are omitted.
- It is possible to use important functional theoretic characteristics of the transfer functions $\underline{B}(\omega)$ or $\underline{B}(p)$ of linear systems for system analysis.

These characteristics are drawn up in Appendix B.2 and B.3. They are based on the fact that transfer functions of spatially varying linear systems can be represented by rational functions dependent on $\mathrm{j}\omega$ and p, respectively. They are well-defined by their poles and zeros. Also the inverse transform formula of HEAVISIDE is based on this characteristic that will result in a closed-form solution of (2.88), if the poles and zeros of $\underline{B}(p)$ and $\underline{F}(p)$ are known. Many specialist books and text books of mathematics and electrical engineering contain detailed representations concerning the characteristics and applications of the Laplace transform [30, 32, 38–40].

2.2 Electrical Networks

In Table 2.2, the network coordinates, fundamental components, coupling two-port networks, sources and balance equations for electrical networks are shown. The following summarizing explanations of relations the electrical engineer is familiar with provide a better understanding of methodology concerning the network description of non-electrical subsystems.

Due to the requirement that the product of network coordinates must represent a *power quantity*, current and voltage are defined as network coordinates. This characteristic is the condition for the validity of fundamental relations of linear time-invariant networks. The time-invariant components C and L describe the ability of electromagnetic fields to store electrical and magnetic field energy. If the operational modes of components show nonlinear characteristics, both component equations and balance equations result in differential equations dependent on the network coordinates u and i.
In case of linear characteristics, these differential equations can be transformed into algebraic equations using complex amplitudes explained in Sect. 2.1. The component R describes the irreversible conversion of electrical energy into thermal energy.

The most general topological form of a network consisting of one-port components with a specified number N of nodes is represented in Fig. 2.10.

Table 2.2. Components and system equations of electrical networks

components	nonlinear	linear	network coordinates	complex amplitudes
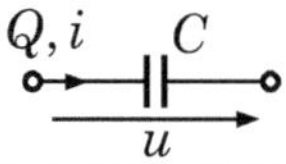	$Q = Q(u)$	$Q = C\,u$	$i = \frac{\mathrm{d}Q}{\mathrm{d}t} = C\frac{\mathrm{d}u}{\mathrm{d}t}$	$\underline{i} = \mathrm{j}\omega C\underline{u}$
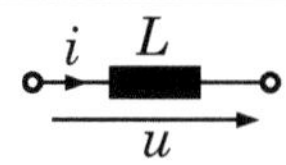	$\mu = \mu(i)$	$\mu = Li$	$u = \frac{\mathrm{d}\mu}{\mathrm{d}t} = L\frac{\mathrm{d}i}{\mathrm{d}t}$	$\underline{u} = \mathrm{j}\omega L\underline{i}$
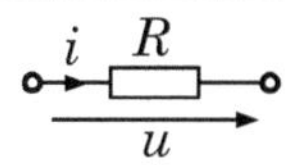	$u = u(i)$	$u = Ri$	$u = Ri$	$\underline{u} = R\underline{i}$

coupling element: ideal transformer

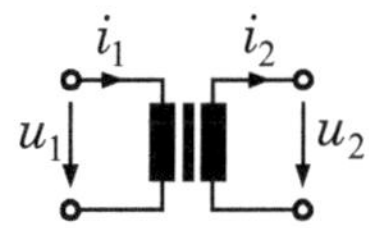

$$\begin{pmatrix} u_1 \\ i_1 \end{pmatrix} = \begin{pmatrix} k & 0 \\ 0 & 1/k \end{pmatrix} \begin{pmatrix} u_2 \\ i_2 \end{pmatrix}$$

ideal sources:

voltage source

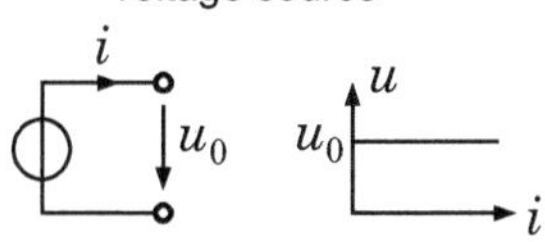

current source

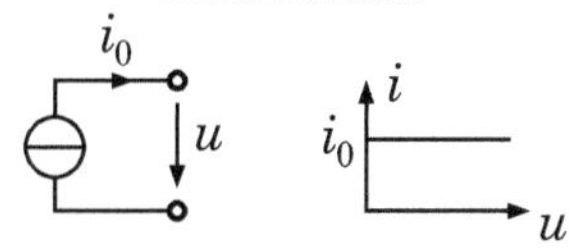

balance equations:

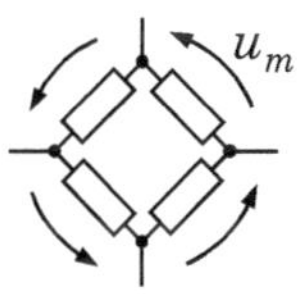

$$\sum_{\circlearrowright} u_m = 0$$

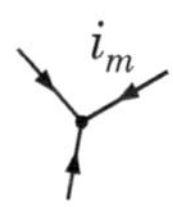

$$\sum_{*} i_m = 0$$

Following system equations are valid for the general network shown in Fig. 2.10:

$$\underline{u}_m = \sum_{n=1}^{\mathrm{N}} \underline{\rho}_{mn} \underline{i}_n \qquad \text{with} \qquad \underline{\rho}_{mn} = \underline{\rho}_{nm} \tag{2.92}$$

and

$$\underline{i}_n = \sum_{n=1}^{\mathrm{N}} \underline{\gamma}_{nm} \underline{u}_m \qquad \text{with} \qquad \underline{\gamma}_{nm} = \underline{\gamma}_{mn} \tag{2.93}$$

The network consists of a maximum of

$$Z = \frac{(\mathrm{N}-1)\,\mathrm{N}}{2} \tag{2.94}$$

components (legs). It is a fundamental task of network theory to derive how many of the overall existing currents and voltages (coordinates) in network

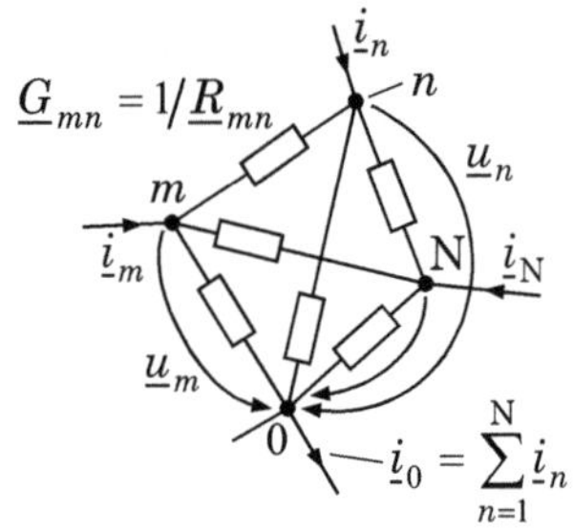

Fig. 2.10. General topology of an electrical network consisting of one-port components

can be independently selected in each case and how the remaining coordinates can be determined out of this set of independent coordinates. For the solution of this task, the component equations and balance equations represented in Table 2.2 are necessary.
If the networks consist of few components, this task can be solved by means of heuristic combination of the component equations with appropriate balance equations. However, in order to discover general network characteristics or to analyze larger networks with several sources, it necessitates a systematic approach. It is included in CAUER's fundamental book of electrical network theory [11] and in textbooks of electrical engineering e.g. [29] and [41].
Concerning linear electrical fundamental components, the assumption of node voltages and node currents represented as network variable in Fig. 2.10 results in a conceptually simple approach in order to solve the tasks mentioned above.

At first, it can be realized that the N − 1 node voltages u_n represent a set of independently selectable coordinates. The same applies for the N − 1 node currents i_m. Thus, for linear component equations the hypothesis of (2.93) is justified that the node currents are associated with the node voltages by N−1 linear equations. By means of the definition

$$\underline{\gamma}_{nm} = \left(\frac{\underline{i}_n}{\underline{u}_m}\right)_{\underline{u}_l=0} \quad \text{with} \quad l = 1\ldots m-1,\ m+1\ldots \mathrm{N}-1, \tag{2.95}$$

the coefficients $\underline{\gamma}_{nm}$ can be identified as negative leg conductances $-\underline{G}_{nm}$ in combination with the network's circuit:

$$\underline{\gamma}_{nm} = -\underline{G}_{nm} \tag{2.96}$$

This consideration also results in symmetry of the conductance matrix in (2.93). This reciprocity relation constitutes a fundamental characteristic of networks. In combination with the supposed component's linearity, it is a structural characteristic of the network and not the result of the existence of a state function of inner energy, as it is sometimes claimed. In fact, it results from the reciprocity condition of (2.93) with respect to the restriction that the network comprises only capacitances and inductances, respectively. According to their meaning the coefficients $\underline{\gamma}_{mm}$

$$\underline{\gamma}_{mm} = \left(\frac{\underline{i}_m}{\underline{u}_m}\right)_{\underline{u}_l=0} \quad \text{with} \quad l = 1 \ldots m-1,\ m+1 \ldots \mathrm{N}-1 \tag{2.97}$$

are defined as sum of the leg conductances starting from node m:

$$\underline{\gamma}_{mm} = \sum_{n=1}^{\mathrm{N}-1} \underline{G}_{mn} \quad \text{with} \quad n \neq m \tag{2.98}$$

Since the matrix $(\underline{\rho}_{mn})$ is the reciprocal matrix of $(\underline{\gamma}_{mn})$, the symmetry of the individual matrix elements $\underline{\gamma}_{mn}$ also results in the symmetry of the matrix elements $\underline{\rho}_{mn}$ in (2.92). For the $\underline{\rho}_{mn}$ not so simple relations to the leg resistances R_{mn} can be found, as it is possible in case of the conductance matrix. By means of the balance equations, the independent node voltages and node currents can be replaced by leg voltages and leg currents in both forms of (2.92) and (2.93).
A further modification of (2.93) can be achieved, if the node currents of selected nodes are set at zero. Thus, the associated node voltage is not an independent variable anymore and the modified network consists of only N − 1 ports. In this way, the equations (2.93) which can be easily derived from the structure of the complete network, can be reduced with respect to the available ports in such a way that only those ports will remain which are needed for the particular application.
For both mentioned modifications the reciprocity characteristics of the modified networks remain. Figure 2.11 shows the reduction of a circuit still containing sources for selected independent coordinates in addition to the network illustrated in Fig. 2.10. If two arbitrary nodes are led through this network, the complete circuit being located behind these two nodes can be replaced by one of the two neighboring active one-ports.

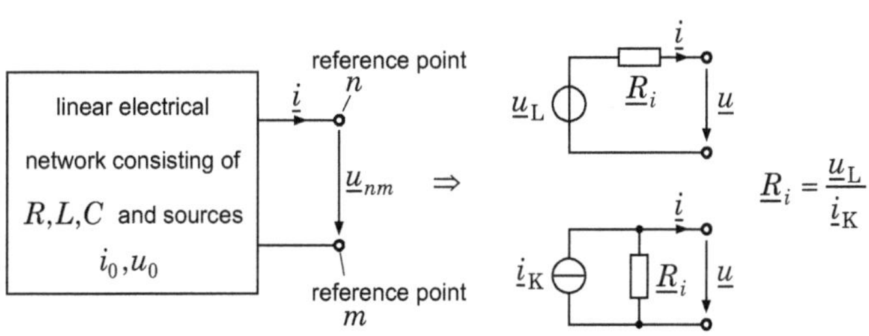

Fig. 2.11. Active electrical one-ports

The open-circuit voltage $\underline{u}_{\mathrm{L}}$ can be measured at the terminals n, m, if the current is set at $\underline{i} = 0$. The short-circuit current $\underline{i}_{\mathrm{K}}$ can be measured at the terminals, if the output voltage is set at $\underline{u} = 0$. The internal resistance R_i can be measured between the terminals m, n, if all internal voltage sources are replaced by a short-circuit and all internal current sources are removed.

2.3 Mechanical Networks

The explanations in this section help to show the origin of network descriptions of mechanical systems used in this book from the terminology of theoretical mechanics. If this aspect is not in the reader's interest, this section can be skipped. A closed problem-oriented description of mechanical networks is contained in Chap. 3.
A structural-oriented interpretation of the LAGRANGE *equations of first kind* provides a basis for the development of a mechanical network theory. These equations assume an amount of N + 1 reference points. The position coordinates of an amount N of these reference points ($m = 1 \ldots \mathrm{N}$) are defined by their position in a Cartesian coordinate system. Its point of origin coincides with the zeroth reference point. Generally mass elements are located at reference points, spring elements and and frictional elements are located between reference points. As with the electrical network, a mechanical network structure is defined which is characterized by N independent nodes (reference points) and N(N−1)/2 legs of components. N displacement components being independent from each other can be assigned to 3N reference points. For that purpose, 3N components of external equilibrium forces are necessary at the reference points which are clearly defined by the affecting displacements. The required system equations consist of balances of forces at the reference points, component equations and structural description of the network by means of the position coordinates of the reference points being not affected by external equilibrium forces.

The network decribed in such a way can be modified by inserting coupling systems. These coupling systems consist of K coupling functions interconnecting the position coordinates. They reduce the 3N primarily independent position coordinates to F coordinates. The same applies accordingly to the force coordinates.
If the displacements of the reference points among themselves are smaller than the distances of the reference points among themselves, in a first approximation the system equations are assumed to be *linear*.

By transition from the time functions of the components to complex amplitudes according to Sect. 2.1, these original differential equations can be transformed to algebraic equations. Based on these conditions, according to electrical networks a linear system of equations can be derived assigning the associated equilibrium forces to 3N independently selectable displacement components of the reference points:

$$\underline{F}_k = \sum_{i=1}^{3\mathrm{N}} \underline{\gamma}_{ki}\underline{\xi}_i, \qquad k = 1\ldots 3\mathrm{N} \tag{2.99}$$

$$\underline{\gamma}_{ki} = \underline{\gamma}_{ik} \tag{2.100}$$

Here, it is made use of the common agreement that the components of all force and displacement vectors are numbered independently of their assignment to the reference points 1...3N. Again, the matrices of this system are symmetric with respect to the main diagonals. Taking the topology of the spatial network into account, their coefficients can be determined similarly to the electrical network. However, the situation here is more complicated because angle information between coordinate axes and orientation of the component legs are additionally integrated into the relation matrix coefficient – component parameters.

Equation (2.99) can be modified in such a way that the displacement coordinates $\underline{\xi}_i$ are replaced by a new set of coordinates consisting of linear combinations of the $\underline{\xi}_i$:

$$\underline{\xi}'_i = \sum_{k=1}^{3\mathrm{N}} \underline{\alpha}_{ik}\underline{\xi}_k \tag{2.101}$$

It concerns the introduction of generalized coordinates with respect to the demand for linearity. Generalized coordinates have been introduced with the LAGRANGE *formalism*. Also the associated generalized force coordinates $\underline{F}'_m$ can be represented as linear combination of the original force coordinates:

$$\underline{F}'_m = \sum_{m=1}^{3\mathrm{N}} \underline{\alpha}'_{mk}\underline{F}_k \tag{2.102}$$

The resulting transformation matrix $(\underline{\alpha}'_{mk})$ will be well-defined, if it is postulated that the system matrix $(\underline{\gamma}'_{ki})$ of the transformed system equations

$$\underline{F}'_\nu = \sum_{m=1}^{3\mathrm{N}} \underline{\gamma}'_{\nu\mu}\underline{\xi}'_\mu \tag{2.103}$$

resulting from (2.99) is symmetric with respect to the main diagonal just like the output matrix $(\underline{\gamma}'_{ki})$ does:

$$\underline{\gamma}'_{\nu\gamma} = \underline{\gamma}'_{\gamma\nu} \tag{2.104}$$

The matrix $(\underline{\alpha}'_{mk})$ proves to be the inverse matrix of matrix $(\underline{\alpha}_{mk})$:

$$(\underline{\alpha}'_{mk}) = (\underline{\alpha}_{mk})^{-1} \tag{2.105}$$

Thus, the $\underline{\gamma}'_{\nu\mu}$ result in:

$$\underline{\gamma}'_{\nu\mu} = \sum_i \sum_k \underline{\alpha}'_{k\nu} \underline{\alpha}'_{i\nu} \underline{\gamma}_{ki} \tag{2.106}$$

Due to the agreed linearity restriction, the representation of general linear mechanical networks mentioned here in its essential results enables an assignment of oriented graphs to the characterizing systems of equations. Compared to the essentially more general LAGRANGE *theory* of dynamic systems, this characteristic offers special advantages concerning the treatment of technical and constructional tasks, respectively.
However, during an experience period lasting over many decades it proved to be useful to proceed in structuring the network description of mechanical systems presented in this book. The above mentioned introduction of generalized coordinates and coupling systems as well as a number of further restrictions represent the formal tools for that. Thereby, the first essential aspect is the definition of fundamental substructures which are represented in the Figs. 2.12 up to 2.14. They result from

- the introduction of one-dimensional networks characterized by only one direction of motion – *translational systems*;
- the restriction on systems which allow only for rotational motions around one spatial axis with simultaneous introduction of the generalized coordinates torsional moment and angle or angular velocity, respectively – *rotational systems*;
- the restriction on *linear fluid systems* with the generalized coordinates volumetric flow rate and pressure – *fluid-mechanical systems.*

The description of interactions between these subsystems represents the second aspect. It will be considered in more detail in Sect. 2.4.1.
Fig. 2.12 shows a *translational* network. The sketched guides are to enable the necessary one-dimensional motion. The lever in the lower figure part is a coupling element which establishes a linear relation between the three displacements of the points of force application. The task of analysis consists in the determination of both the displacements or displacement velocities of the reference points respectively and the forces within the connecting rods as

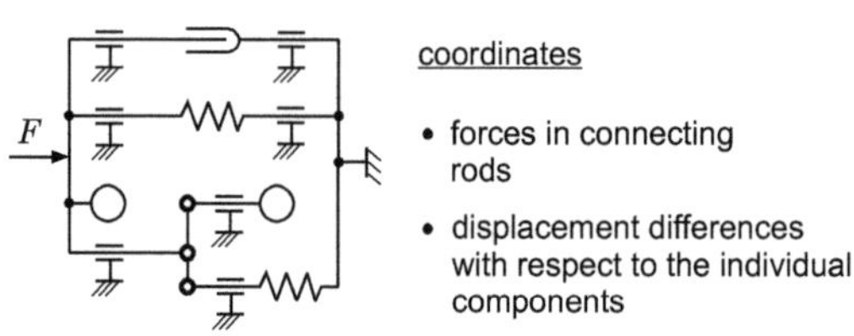

Fig. 2.12. Translational network

function of the excitational force F. Here, the connecting rods are considered to be massless and rigid. In Sect. 3.1 it will be shown how network coordinates can be defined for such arrangements, which balance equations exist for these coordinates and by which relations the network coordinates are connected with the components. According to the fundamental considerations at the beginning of this section, the derivation of these special system equations completely results from the rules being valid in mechanics. The result is a set of equations being *isomorphic* to those of electrical network shown in Table 2.2.
That means: the derived mechanical system equations and their assignment to the structures generating them differ from those of electrical networks only by the terms of their coordinates and parameters. These circumstances allow for extensive conclusions:
At first, the electrical restructuring and the associated calculation rules of electrical networks can be transfered to translational mechanical networks. For example, the derivation of active one-port parameters of a terminal pair of electrical networks shown in Fig. 2.11 is also valid for a pair of reference points of the mechanical network.
Furthermore, all results of the network theory, like e.g. methods of circuit synthesis for given transfer characteristics, function theoretical characteristics of general one-ports and transfer factors up to network analysis programs are also valid for the considered system structure „translational mechanical network".
Fig. 2.13 shows a *rotational* network. The system elements being arranged in the middle of the figure represent a massless gear transmission which couples the rotation angle of the lower axis with that of the upper axis in the sense of a coupling condition of generalized position coordinates.

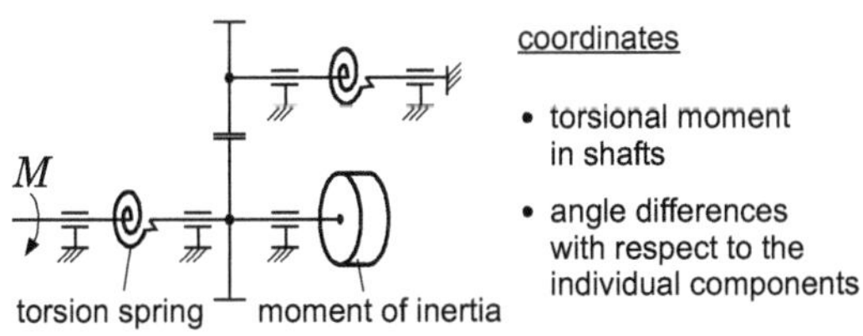

Fig. 2.13. Rotational network

The preceding considerations mentioned before will also be done for this type of network in Sect. 3.2. They also result in a set of system equations being *isomorphic* to those of the translational and electrical network.

Finally, the same situation is found with the *fluid-mechanical* network illustrated in Fig. 2.14. Due to the assumed linearity, the component descriptions

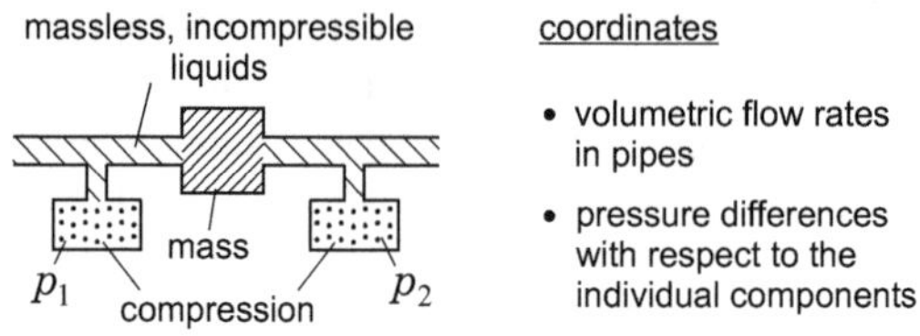

Fig. 2.14. Fluid-mechanical (acoustic) network

being valid here are only linear approximations of the in general strongly nonlinear equations of fluid dynamics around an operating point. The media located in the center (see Fig. 2.14) moves only with small displacements at local-independent velocity around the rest position. In both volumes, pressure fluctuations are generated by supplying a medium quantity that is small compared to the medium quantity being in the volume. Also the pressure fluctucations themselves are small compared to the static pressure in the volume with respect to a missing media inflow. The conversion of these system characteristics to a network description is presented in Sect. 3.3. It also results in equations and structures which are *isomorphic* to those of network types that have been considered before.

In summary, it can be noticed that it is possible to derive three substructures from the general linear network description of dynamic mechanical systems which enable the modeling of a large number of applications assigned to this book. The network description of these subsystems results in a corresponding set of equations which are mutually *isomorphic* and isomorphic to electrical networks. The equations definitely correspond to network structures which are equivalent to the geometrical and constructional system structure that is to be modeled.

2.4 Interactions

2.4.1 Mechanical Interactions

The networks considered in the Sects. 2.2 and 2.3 are interdependent. Interactions are able to exist between them. Between the translational network on the one hand shown in Fig. 2.12 and the rotational and acoustic network on the other hand shown in Fig. 2.13 and Fig. 2.14 respectively, this coupling is described, in the simplest case, by the assumption that a linear relation exists between the position coordinate x of a translational system point and the generalized position coordinates of the particular rotational and acoustic system point φ or V, respectively:

$$\Delta\varphi = K_{\mathrm{r}}\Delta x \tag{2.107}$$

$$\Delta V = K_a \Delta x \tag{2.108}$$

with

$$\Delta\varphi = \varphi - \varphi_0, \qquad \Delta V = V - V_0, \qquad \text{and} \qquad \Delta x = x - x_0$$

This coupling can be subjected to the condition that coupling elements can not store energy and can also not convert energy into heat. This condition is only fulfilled by systems which consist of massless and rigid compounds and frictionless guides and hinges:

$$F\Delta x + M\Delta\varphi = 0 \tag{2.109}$$

$$F\Delta x + p\Delta V = 0 \tag{2.110}$$

It is assumed that the generalized force coordinates at the considered system points have been defined in such a way that the two addends in (2.109) and (2.110) have the meanings of energies flowing into the coupling system.

In combination with the energy balances in (2.109) and (2.110), the kinematic conditions of (2.107) and (2.108) result in coupling equations (2.111) and (2.112) for generalized force coordinates:

$$F = -K_r M \tag{2.111}$$

$$F = -K_a p \tag{2.112}$$

Thus, a total of two fundamental structures results for the elementary coupling systems – transducers – between translational, rotational and acoustic systems. The elementary coupling systems are illustrated in Fig. 2.15.
The individual mechanical transducers in Fig. 2.15 can be described by following functional relations:

$$\begin{pmatrix} \Delta x \\ F \end{pmatrix} = \begin{pmatrix} 1/K_r & 0 \\ 0 & -K_r \end{pmatrix} \begin{pmatrix} \Delta\varphi \\ M \end{pmatrix} \tag{2.113}$$

$$\begin{pmatrix} \Delta x \\ F \end{pmatrix} = \begin{pmatrix} 0 & K_a \\ -1/K_a & 0 \end{pmatrix} \begin{pmatrix} \Delta V \\ p \end{pmatrix} \tag{2.114}$$

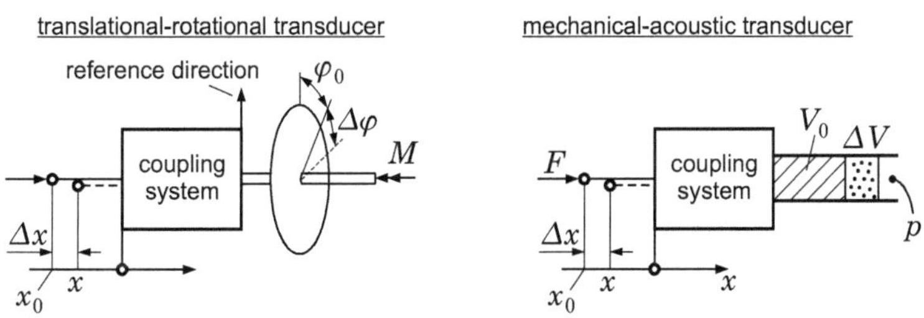

Fig. 2.15. Two-port network representation of mechanical transducers

Usually in case of the two-port network representation, the network coordinates are chosen in such a way that they describe an energy flow out of the system at one of the two ports. If this agreement is also applied to the coupling elements shown in Fig. 2.15, the minus sign in the coupling matrices in (2.113) and (2.114) will be omitted.
Figure 2.16 shows two elementary realizations of the formally introduced coupling systems comprising a rigid and massless rod and a rigid and massless plate inside of a cylinder as coupling elements.

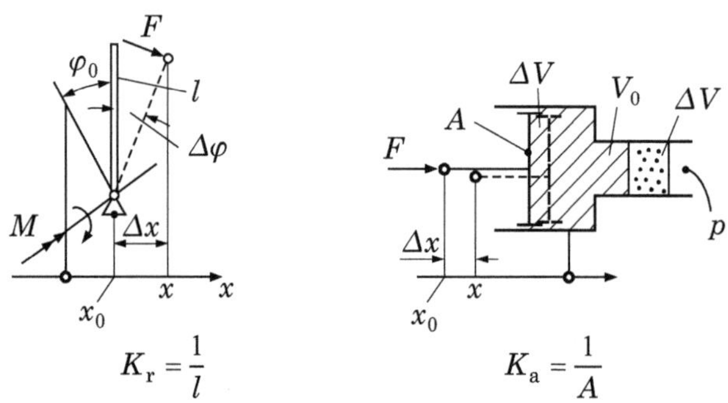

Fig. 2.16. Examples for translational-rotational and mechanical-acoustic transducers

The detailed definitions of these coupling elements and their application for the representation of concrete structures are described in Chap. 5. It is remarkable that the system equations in form of (2.113) and (2.114) can not be transformed into the form of the general network equations. In the sense of the coupling systems mentioned at the beginning of Sect. 2.3, they can be solely considered as system components reducing the number of independent position and force coordinates.

2.4.2 Electromechanical Interactions

The explanations in this section help to show the relation between the problem-oriented description of electromechanical interactions discussed in the Chaps. 7 up to 9 with the theory of electrodynamic fields. The reader, who is not superficially interested in this aspect, should read on in Sect. 2.5. The explanations in the Chaps. 7 up to 9 represent a self-contained mainly energy-based description of these facts.
The consideration of experimental basis experiences explaining the interconnection of mechanical and electrical network coordinates represents the starting point for the description of interconnection. There are two groups of relations. The first group describes the equilibrant forces which will additionally

be necessary in mechanical systems, if charges or elements of current loops are coupled with reference points. These forces are described by the COULOMB and BIOT-SAVART equations providing the starting points for the definition of the electrostatic and magnetostatic field equations at the same time [19].
The dependence of the component parameters L and C of electrical networks on their geometry is part of the second group of relations, i.e. the dependence on position coordinates of the mechanical system or, formulated more general, the dependence of electrostatic and magnetic field quantities as solution of a boundary value problem on the geometry of this problem.
Due to the characteristic of the COULOMB and BIOT-SAVART equations as logic source for field equations, it is not surprising that close relations exist between these two groups of relations. Furthermore, it may not be forgotten that the law of induction must be considered as third independent experience fact in order to define the magnetic field energy and to link the magnetic field quantities with the electric field strength and voltage, respectively.
The equivalence of galvanic currents and displacement currents concerning their coupling with magnetic field quantities as fourth experimental basis experience is not used here. Therefore, it is possible to treat the interactions separately for systems with electric and magnetic fields.
In addition, it is assumed that the electromechanical systems to be considered here only comprise spring elements and, alternatively, capacitances and inductances respectively or, if necessary, electric and, alternatively, magnetic fields. This system limitation is possible without loss of generality, since the separation of such subnetworks from respective electrical and mechanical networks out of the total network is possible.
Furthermore, the restriction to linear processes is expected to be valid. The fundamental nonlinearity of interaction forces mentioned above and, if necessary, of the geometry dependence of the component parameters is linearized by series expansion around a reference point.

Interactions with Electric Fields

Furthermore, the case will be considered that the electromechanical system which has to be modeled comprises only one spring network and components being based on electric fields. For that it is assumed that the spring network comprises electrical point charges Q_n at its reference points $n = 1 \dots \mathrm{N}$ which show potential differences with respect to reference point 0. Additionally, the COULOMB forces $\mathbf{F}_{\mathrm{el},n}$ caused by charges must be added to the mechanically generated equilibrant forces:

$$\mathbf{F}_n = \mathbf{F}_{\mathrm{el},n} + \mathbf{F}_{\mathrm{mech},n} \tag{2.115}$$

The electrical network equations are extended according to

$$u_n = u_n\left(Q_m, \boldsymbol{\xi}_j\right) = \sum_{m=1}^{\mathrm{N}} B_{nm}\left(\mathbf{r}_1, \ldots, \mathbf{r}_n\right) Q_m. \tag{2.116}$$

In order to enumerate the mechanical force, position and displacement components, in the following, the definition made in (2.99) is used, where all mechanical coordinates are counted from 1 up to 3N. The matrix (B_{nm}) represents the reciprocal matrix of the capacitance matrix (C_{nm}).
The series expansion of (2.115) and (2.116) with respect to the position coordinates and the charges Q_m around a reference point Q_{m0}, x_{j0} results in:

$$F_i - F_{i0} = \sum_{j=1}^{3\mathrm{N}} c_{ij}^{(Q)} \xi_j + \sum_{m=1}^{\mathrm{N}} \alpha_{im}\left(Q_m - Q_{m0}\right), \quad i = 1 \ldots 3\mathrm{N} \tag{2.117}$$

$$u_n - u_{n0} = \sum_{m=1}^{\mathrm{N}} B_{nm}^{(\xi)}\left(Q_m - Q_{m0}\right) + \sum_{j=1}^{3\mathrm{N}} \alpha'_{nj} \xi_j, \quad n = 1 \ldots \mathrm{N} \tag{2.118}$$

$$c_{ij}^{(Q)} = c_{ji}^{(Q)} \tag{2.119}$$

$$B_{nm}^{(\xi)} = B_{mn}^{(\xi)} \tag{2.120}$$

$$\alpha'_{nj} = \alpha_{jn} \tag{2.121}$$

Like before, the displacements ξ_j are defined by $x_j - x_{j0}$. The coefficients $c_{ij}^{(Q)}$ denote the stiffness coefficients which can be determined under constant charges. Due to the dependence of the electric field forces on the position coordinates according to (2.115), they differ from those which can be measured under constant node voltages. The same is valid for the difference of the reciprocal capacitance coefficients which can be measured under infinitesimal displacement ξ_i or under constant forces F_i. Now in combination with the abbreviations

$$F_i - F_{i0} = \Delta F_i, \quad Q_m - Q_{m0} = \Delta Q_m, \quad u_n - u_{n0} = \Delta u_n$$

the equations (2.115) and (2.116) can be combined to the matrix equation shown in Table 2.3.

By means of COULOMB's law and the resulting electrostatic field equations, it can be derived that the two partial matrices $(\underline{\alpha}_{im})$ and $(\underline{\alpha}'_{nj})$ are consistent with the symmetry condition (2.119). Such a corresponding pair of coefficients α'_{qr} and α_{rq} is listed in Table 2.3. Due to the symmetry of the partial matrices (c_{ij}) and (B_{nm}), the complete matrix is symmetric with respect to the main diagonal. Thereby, the variables ΔF_i and Δu_n and the variables ξ_i and ΔQ_m

Table 2.3. System matrix of an electromechanical system with interactions caused by electrostatic fields

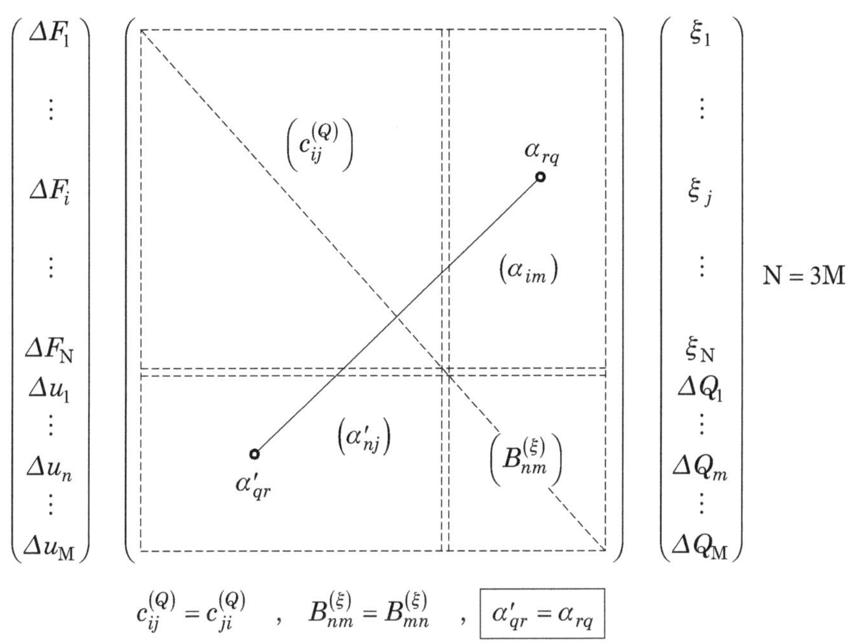

can be combined to a generalized vector $\mathbf{\Phi}$ and Ξ, respectively. Two experiments concerning a special two-dimensional point model with three reference points are represented in Fig. 2.17. The experiments allow for the determination of such a consistent pair of coefficients.
The equilibrant forces $\mathbf{F}_{i0}$ and charges Q_{m0} at the reference points represent the starting point. If a charge ΔQ_m is moved from point 0 to point 1, all equilibrant forces will change. The change of the vertical component of $\mathbf{F}_{i0}$ as function of a displacement ξ_4 results in a change of all node voltages u_n. The change of the node voltage u_1 as function of ξ_4 results in the coefficient α_{14}. From the point model illustrated in Fig. 2.17, the identity of both coefficients can directly be proved by means of COULOMB equations and field equations.

The general proof of these symmetry characteristics for technically real configurations calls for a comprehensive analysis of possible coordinate transformations as well as for the introduction of coupling conditions and constraints to elementary networks which form the basis of the system description represented in Table 2.3. Thus, generalized networks with problem-oriented, generalized and in numbers reduced coordinates could be generated by these elementary networks. Both the application of the transformations described

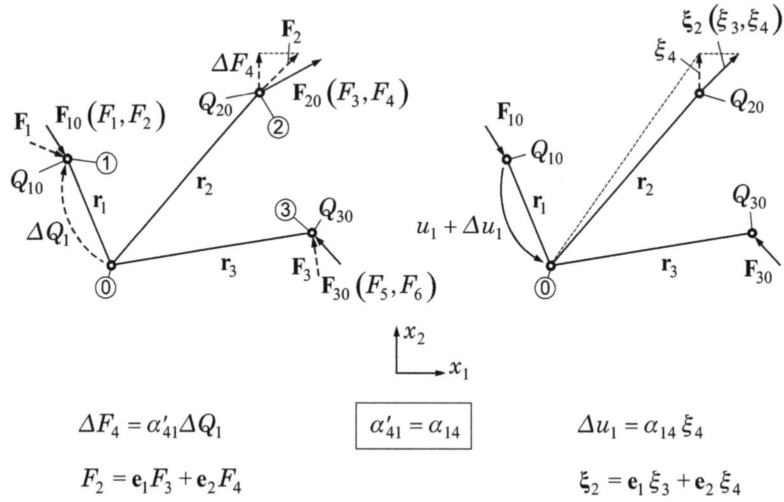

Fig. 2.17. Example of electromechanical interconnections in electrostatic fields

in (2.101) up to (2.106) and the introduction of generalized coupling systems of this kind described in the Figs. 2.15 and 2.16 provide the maintenance of the symmetry conditions in the generalized network.

The difficulties that may be expected with this direct way of derivation make it understandable that such considerations can be found in literature rudimentary only. In case no system model is available, also this approach would fail. It is about all kinds of dielectric materials. Their properties are described only by empirical material constants.

Due to the described facts, in literature another approach is chosen in order to describe the existing interaction. The approach is based on the assumption that electromechanical systems comprise N mechanical and M electrical reference points. The interaction of these electromechanical systems is assumed to be generated by means of electrostatic fields. Their state is described by N′ generalized position coordinates and M electric charge coordinates at these reference points. If the mechanical subsystem contains also coupling elements in addition to the assumed spring elements, then N′ can differ from 3N. Around a reference point Q_{m0}, x_{i0}, these coordinates are combined by linear relations according to (2.117) and (2.118).

Without knowing which structural system characteristics probably form the basis of these equations, it is axiomatically agreed that the total matrix is symmetric with respect to the main diagonal, thus (2.119), (2.120) and (2.121) are valid. Thus, all further conclusions are only valid for systems that fulfill this axiomatic condition. By the way, for a given concrete system the fulfillment of this condition can always be controlled by means of well-known structural specifications and field descriptions. If they are not available for

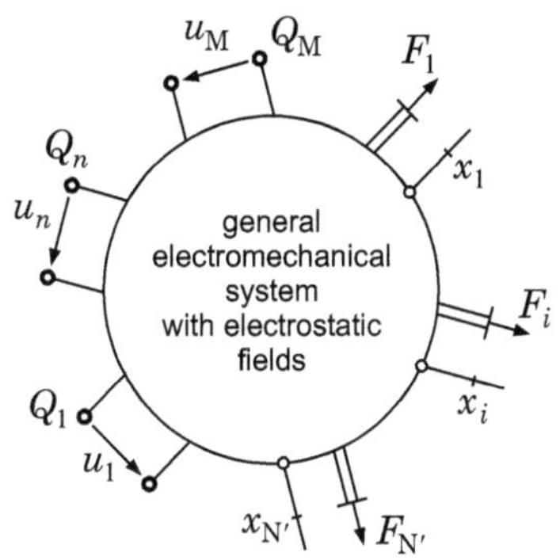

Fig. 2.18. Electromechanical system with electrostatic fields as 3N-port

the concrete system (e.g. in case of dielectrics), only the experimental testing will remain. Therefore, the system defined in such a way can be considered as $(\mathrm{N}' - \mathrm{M})$-port according to Fig. 2.18.
If the system equations are known, internal network structures and coupling structures can be assigned, if necessary backwards, to the system.

The existence of a state function *internal energy* W is the most important conclusion from this symmetry condition:

$$W = W\left(x_{j0}, Q_{m0}\right) + \Delta W\left(\xi_j, \Delta Q_m\right) \tag{2.122}$$

The change ΔW of total energy W results from changing the independent displacement coordinates and charge coordinates with the values ξ_j and ΔQ_m. Due to the dependencies of the F_i and u_n on the actual ξ_j and ΔQ_m each, an addition of partial energies $\mathrm{d}\left(\Delta W\right)$ in differential steps $\mathrm{d}\xi_j, \mathrm{d}\left(\Delta Q_m\right)$ along an initially specific path of the state vector $\xi_j, \Delta Q_m$ in the state space from initial point $\xi_j = 0, \Delta Q_m = 0$ to end point $\xi_j, \Delta Q_m$ is necessary in order to discover ΔW:

$$\mathrm{d}\left(\Delta W\right) = \sum_{i=1}^{\mathrm{N}'} F_i \,\mathrm{d}\xi_i + \sum_{n=1}^{\mathrm{M}} u_n \,\mathrm{d}\left(\Delta Q\right)$$

$$\Rightarrow \Delta W = \sum_{i=1}^{\mathrm{N}'} \int_0^{\xi_i} F_i \,\mathrm{d}\xi_i + \sum_{n=1}^{\mathrm{M}} \int_0^{\Delta Q_n} u_n \,\mathrm{d}\left(\Delta Q\right) \tag{2.123}$$

Taking the system equations (2.117) up to (2.121) with $3\mathrm{N} \to \mathrm{N}'$ into consideration the integration results in:

$$\Delta W = \sum_{i=1}^{N'} F_{i0}\xi_i + \sum_{n=1}^{M} u_{n0}\Delta Q_n + \sum_{i=1}^{N'}\sum_{j=1}^{N'} c_{ij}^{(Q)}\xi_i\xi_j$$
$$+ \sum_{m=1}^{M}\sum_{n=1}^{M} B_{nm}^{(\xi)}\Delta Q_n \Delta Q_m + \sum_{j=1}^{N'}\sum_{m=1}^{M} \alpha_{jm}\xi_j\Delta Q_m \qquad (2.124)$$

This expression is independent of the pathway leading from initial point $\xi_i = 0, \Delta Q_n = 0$ to end point $\xi_i, \Delta Q_n$ in the state space. These fundamental facts allow for referring to the function $\Delta W(\xi_i, \Delta Q_n)$ in (2.122) as state function of the variables ξ_i and ΔQ_n. They also constitute the generally existing representation axiomatically postulating the existence of a state function *internal energy* instead of the symmetry of the system matrix.
In case of nonlinear system equations

$$F_i = f(Q_m, x_j) \qquad (2.125)$$

$$u_n = \varphi(Q_m, x_j) \qquad (2.126)$$

resulting from (2.115) and (2.116), it can additionally be noticed that the path independence of energy $W(Q_n, x_i)$ defined by

$$\mathrm{d}W = \sum_{i=1}^{N'} F_i \,\mathrm{d}x_i + \sum_{n=1}^{M} u_n \,\mathrm{d}Q_n \qquad (2.127)$$

also exists between two system states $Q_n^{(1)}, x_i^{(1)}$ and a reference state $Q_n^{(0)}, x_i^{(0)}$. The reference state is usually described by

$$Q_n^{(0)} = 0, \quad F_i(x_{j0}) = F_i^{(0)} \quad \Rightarrow \quad u_n = 0, \quad W_0 = 0. \qquad (2.128)$$

In this context it is also appropriate to refer to the used differentiation between the so-called virtual, linear and general changes of state. By means of sufficiently small $\mathrm{d}x_i$ and $\mathrm{d}Q_n$, in (2.127) changes of state are possible, where the F_i and u_n generally depending on x_i, Q_m according to (2.125) and (2.126) can be replaced by their values in the reference point. Thus, $\mathrm{d}W$ becomes a linear function of the independent variable x_i and Q_n. However, these linear dependences exist only for those pairs of coordinates whose dependent coordinates are nonzero in the reference point. Pairs of coordinates for which these conditions are not fulfilled do not contribute to $\mathrm{d}W$. Such changes of state are called virtual changes of state and are described by $\delta(..)$ in this book.
If the changes of the Q_n and x_i are that significant that they do not meet the condition of virtual change of state, the energy W and $\mathrm{d}W$ respectively can only be determined by integration of (2.127), while path independence also exists for the nonlinear case. However, in case of linear system equations,

a closed expression for ΔW according to (2.124) results from the integration of (2.127). Their changes of coordinates relative to the point of reference are denoted by $\Delta(..)$.
The characteristic of W as state function and the characteristic of $\mathrm{d}W$ as complete differential in (2.127) respectively results in

$$F_i = \left.\frac{\partial W}{\partial x_i}\right|_{x_i, Q_n} \tag{2.129}$$

$$u_n = \left.\frac{\partial W}{\partial Q_n}\right|_{x_i, Q_n} \tag{2.130}$$

and thus in the identity (2.121) in case of linearity. In Chap. 9, the application of this concept is demonstrated by means of a multiplicity of technical real configurations. Particularly, the case is considered in more detail that only one electrical pair of coordinates is interconnected with one mechanical pair of coordinates. This consideration provides a basis for a special coupling element corresponding to the coupling elements illustrated in Fig. 2.15 and Fig. 2.16. It represents the possible basic module of electromechanical coupling within a structured network-oriented representation of linear dynamic systems according to Fig. 2.19.

Interactions with Magnetic Fields

In case of interconnection of mechanical spring networks with magnetic field elements, the preceding considerations can be repeated. However, the situation is essentially more complicated, because the BIOT-SAVART law represents an almost unclear combination of equilibrant forces at elements of current loops. The allocation to a mechanical reference point with three position coordinates is impossible due to the fact that an oriented current element, thus a rigid body, must be equilibrated. Rather the simplest rigid body must be used whose position is defined by five position coordinates.
Moreover, not only the motion of the complete current loop as rigid body but also the change of his geometrical form according to the mechanical coupling of the current ring elements must be allowed. This unclear multiplicity also excludes those simple models just like the one shown in Fig. 2.17. Therefore, only the corresponding relation to (2.115) of the addition of necessary equilibrant force consisting of mechanical and magnetic force at each mechanical reference point can be specified:

$$\mathbf{F}_n = \mathbf{F}_{\text{magn},n} + \mathbf{F}_{\text{mech},n} \tag{2.131}$$

$$\mathbf{F}_{\text{magn}} = f\left(i_k, \mathbf{r}_1 \ldots \mathbf{r}_{\mathrm{N}}\right) \tag{2.132}$$

The electrical network equations with the coordinates voltage integral μ_m and currents i_k are extended according to:

$$\mu_m = \mu_m\left(i_k, \mathbf{r}_1 \ldots \mathbf{r}_\mathrm{N}\right) = \sum_{k=1}^{\mathrm{M}} L_{mk}\left(\mathbf{r}_1 \ldots \mathbf{r}_\mathrm{N}\right) i_k \tag{2.133}$$

For further considerations all components of the independent position and force vectors are numbered again from 1...N′ in each case. (2.122) and (2.133) can be linearized again around the reference point $x_{j0}, F_{i0}, i_{k0}, \mu_{m0}$ by means of series expansion. For that the displacement $\xi_j = x_j - x_{j0}$ is introduced again. It is expected to be valid:

$$F_i - F_{i0} = \sum_{j=1}^{\mathrm{N}'} c_{ij}^{(i)} \xi_j + \sum_{m=1}^{\mathrm{M}} \beta_{im}\left(i_m - i_{m0}\right), \quad i = 1 \ldots \mathrm{N}' \tag{2.134}$$

$$\mu_m - \mu_{m0} = \sum_{k=1}^{\mathrm{M}} L_{mk}^{(\xi)}\left(i_k - i_{k0}\right) + \sum_{j=1}^{\mathrm{N}'} \beta'_{mj} \xi_j, \quad m = 1 \ldots \mathrm{M} \tag{2.135}$$

If clear combinations of spring networks and associated current loops are existent, the symmetry of the total matrix presented in (2.134) and (2.135) can be directly derived from the BIOT-SAVART forces and magnetic field forces or the network structure of the L_{mk} and c_{ij}, respectively. If these structural information is not existent, only the axiomatic assumption will remain that the (N′, M)-port defined by (2.134) and (2.135) and as shown in Fig. 2.18 possesses a state function *internal energy* W:

$$W\left(x_j, \mu_m\right) = W\left(x_{j0}, \mu_m\right) + \Delta W\left(\xi_j, \Delta\mu_m\right) \tag{2.136}$$

In order to infer the symmetry of a system matrix from this, two ways are possible:

The state function *enthalpy*

$$H = W - \sum_{m=1}^{\mathrm{M}} \mu_m i_m \tag{2.137}$$

is used instead of the internal energy W. By analogy with the process in the electrostatic case, the symmetry of the total matrix in (2.134) and (2.135) can be proven. In order to make use of the internal energy as state function, it is necessary to transform the system of equations (2.134) and (2.135) in such a way that the $\Delta\mu_m$ and ξ_j appear as independent variables:

$$\Delta F_i = \sum_{j=1}^{\mathrm{N}'} c_{ij}^{(\mu)} \xi_j + \sum_{m=1}^{\mathrm{M}} b_{im} \Delta\mu_m \tag{2.138}$$

$$\Delta i_k = \sum_{k=1}^{\mathrm{M}} K_{km}^{(\xi)} \Delta\mu_m + \sum_{j=1}^{\mathrm{N}'} b'_{mj} \xi_j \tag{2.139}$$

The $K_{km}^{(\xi)}$ denote the reciprocal inductance coefficients. The coefficients $c_{ij}^{(i)}$, β_{im}, β'_{mj} can be determined by means of the coefficients $c_{ij}^{(\mu)}$, b_{im}, b'_{mj}. With the meaning of ΔW according to (2.123)

$$\Delta W = \sum_{i=1}^{N'} \int_0^{\xi_i} F_i \, \mathrm{d}\xi_i + \sum_{m=1}^{M} \int_0^{\Delta\mu_m} i_m \, \mathrm{d}\mu_m \tag{2.140}$$

and by means of (2.138) and (2.139), (2.140) can be integrated. By means of the symmetry conditions

$$c_{ij}^{(\mu)} = c_{ji}^{(\mu)} \tag{2.141}$$

$$K_{km}^{(\xi)} = K_{mk}^{(\xi)} \tag{2.142}$$

$$b_{im} = b'_{mi} \tag{2.143}$$

ΔW proves to be independent of the path of integration:

$$\begin{aligned} \Delta W = {} & \sum_{i=1}^{N'} F_{i0}\xi_i + \sum_{m=1}^{M} i_{m0}\Delta\mu_m + \sum_{i=1}^{N'}\sum_{j=1}^{N'} c_{ij}^{(\mu)} \xi_i \xi_j \\ & + \sum_{m=1}^{M}\sum_{k=1}^{M} K_{km}^{(\xi)} \Delta\mu_k \Delta\mu_m + \sum_{j=1}^{N'}\sum_{m=1}^{M} b_{jm}\xi_j \Delta\mu_m \end{aligned} \tag{2.144}$$

On the other hand, the system equations (2.131) up to (2.133) result from:

$$F_i = \left. \frac{\partial W}{\partial x_i} \right|_{x_{i0},\mu_{m0}} \tag{2.145}$$

$$i_k = \left. \frac{\partial W}{\partial \mu_m} \right|_{x_{i0},\mu_{m0}} \tag{2.146}$$

The considerations in connection with Fig. 2.18 apply analogously. In combination with the relations between the coefficients mentioned in (2.134), (2.135), (2.138) and (2.139), the symmetry conditions (2.136) result in the conclusion that also the matrix in (2.134) and (2.135) is symmetric with respect to the main diagonal. Also here, the symmetry conditions are valid:

$$c_{ij}^{(i)} = c_{ji}^{(i)} \tag{2.147}$$

$$L_{mk}^{(\xi)} = L_{km}^{(\xi)} \tag{2.148}$$

$$\beta_{im} = \beta'_{mi} \tag{2.149}$$

Based on the described relations, a multiplicity of technical real configurations with magnetic field forces is described in Chap. 8.

2.5 Structured Network Representation of Linear Dynamic Systems

Based on the physical fundamentals quoted in this introduction, structured network descriptions of subdomains are developed in the Chaps. 3 up to 9. Their correlation becomes apparent in Fig. 2.19.
Three fundamental networks are generated which are *isomorphic* to each other and whose topological structures are consistent with geometrical and constructional structures of their technical originals. The resulting consequences are discussed in detail in Chap. 4. The fact, that two topologically determined forms of coordinates can be defined in all fundamental networks, represents an important conclusion.
It concerns, on the one hand, *flow coordinates* coinciding at both component ends and, on the other hand, *difference coordinates* defined between both component ends.
Between these fundamental networks coupling elements exist. They are also called *transducer two-port networks*. Concerning the transfer behavior, two different groups can be identified. One of these groups interconnects two flow coordinates and two difference coordinates each.
The other group interconnects one flow coordinate of one port with one difference coordinate of the other port. The introduced transducer two-port networks, in short *transducers*, show *reciprocity*. In Chap. 10, it is dealt with reciprocity in linear networks in detail.
The elements illustrated in Fig. 2.19 are incomplete from that point of view that they do not comprise linear thermodynamic systems with the components heat accumulator and heat conductor and the thermomechanical (CARNOT cycle) and thermoelectric transducer elements (PELTIER elements) according to the application spectrum of this book. In the same way their network-oriented description is possible with the coordinates temperature difference and entropy flow or relative temperature difference and heat flow, as with those of the remaining elements shown in Fig. 2.19.
By the way, it may not be forgotten that essential operations can not be covered by means of the system elements shown in Fig. 2.19 within the scope of the LAGRANGE-HAMILTON formalism. At first, the linearity restriction must be mentioned. However, for many real dynamic problems the linear approximate solution represents the appropriate starting point enabling the iterative derivation of nonlinear solutions.
The exclusion of such coordinate transformations comprising explicitly the time represents a further restriction. Thus, such effects like the occurrence of CORIOLIS forces and all gyroscope phenomena are impossible. However, the general physical fundamentals quoted in this introductive chapter hint at the methods being necessary for eliminating the specified restrictions.

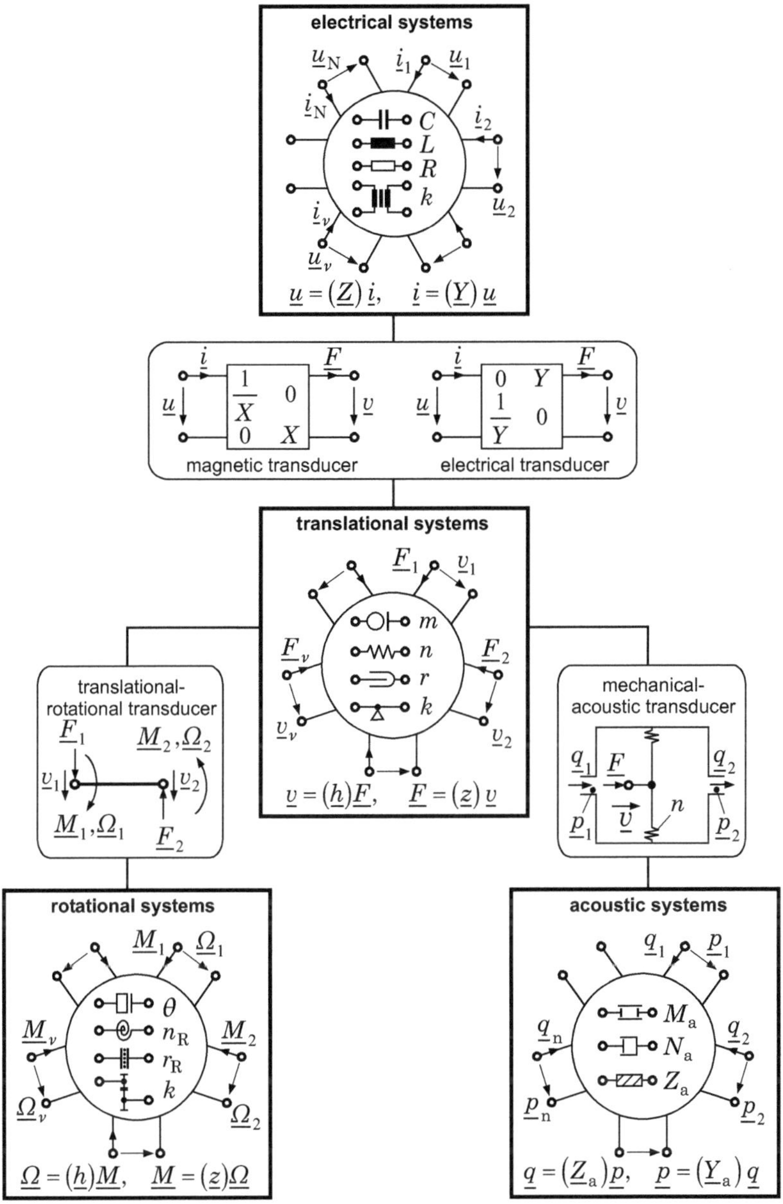

Fig. 2.19. Network structuring of linear dynamic systems

2.6 Basic Equations of Linear Networks

The spatially varying description of electrical and mechanical systems including their interactions allows for structuring with respect to fundamental networks and interacting components as it is shown in Fig. 2.19. In Chap. 3, it is shown how these equations can be found for each fundamental network by means of the components' characteristics and balance equations existing in each fundamental structure. In (2.150) up to (2.152), the *basic equations of linear networks* are presented in general form:

$$\mu_i = \alpha_i \frac{\mathrm{d}\lambda_i}{\mathrm{d}t}, \qquad \sum_{\text{loop}} \mu_i = 0 \tag{2.150}$$

$$\lambda_m = \beta_m \frac{\mathrm{d}\mu_m}{\mathrm{d}t}, \qquad \sum_{\text{node}} \lambda_j = 0 \tag{2.151}$$

$$\mu_n = \gamma_n \lambda_n \tag{2.152}$$

The quantity λ denotes a *flow coordinate*, the quantity μ denotes a *difference coordinate*, the quantities α, β and γ represent *components*.

By means of coupling elements in (2.153) and (2.154), interactions between fundamental networks of different physical structures can be described:

$$\begin{pmatrix} \mu_L \\ \lambda_L \end{pmatrix} = \begin{pmatrix} X & 0 \\ 0 & 1/X \end{pmatrix} \begin{pmatrix} \mu_K \\ \lambda_K \end{pmatrix} \tag{2.153}$$

$$\begin{pmatrix} \mu_L \\ \lambda_L \end{pmatrix} = \begin{pmatrix} 0 & Y \\ 1/Y & 0 \end{pmatrix} \begin{pmatrix} \mu_K \\ \lambda_K \end{pmatrix} \tag{2.154}$$

Coupling elements of same type can also be contained in fundamental networks of a physical structure. In the Chaps. 8 and 9 it is shown how the *transducer coefficients* X and Y can be determined from the available interactions.

Part II

Network Representation of Systems with Lumped and Distributed Parameters

3

Mechanical and Acoustic Networks with Lumped Parameters

The design and construction of electromechanical and electroacoustic systems necessitates the knowledge of the dynamic behavior during the design phase, since certain dynamic target parameters must be fulfilled among other requirements. Also the control of an existing system often requires the modeling of the dynamic behavior. This model then allows for simulations. The simulation results are important for further decisions. In order to think purposefully about a task or to discuss with partners, an available model is also useful. Models that focus on the important aspects of the dynamic behavior are especially suitable for this purpose.

Thus, in the following sections only those mechanical systems are to be considered in more detail which can be represented by the fundamental networks shown in Fig. 2.19 in Sect. 2.5. Compared with general mechanical systems, at first the restriction to linear relations between the coordinates is associated with that. The system can be described by linear differential equations.

As the most simple model stage it is often possible to consider the system parameters to be concentrated spatially. The partial space-time differential equations which are able to describe the physical system comprehensively, are thereby reducible to a system of ordinary time differential equations. The restriction on fundamental networks mentioned above allows for reducing of general solution methods used in theoretical mechanics to the definition of mechanical basic components and coordinates as well as to rules concerning the components' interconnection. These rules have the meaning of equilibrium of forces and kinematic compatibility conditions for the position coordinates. Thus, coordinates and components can be defined in such a way that the conditions of linear network theory are fulfilled.

As a result, the effective calculation methods of the linear system theory particularly common in electrical engineering and available network simulation programs can be used without further effort. That also represents a priority objective of the method. Generally this *network method* is significantly more effective than the direct calculation by means of differential equations. By means of abstraction of the real physical structure, a scheme is achieved

A. Lenk et al., *Electromechanical Systems in Microtechnology and Mechatronics*,
Microtechnology and MEMS, DOI 10.1007/978-3-642-10806-8_3,

which comprises only the information between the coordinates being necessary for calculation.
In the Chaps. 7, 8 and 9 it will be demonstrated that the electrical, mechanical and acoustic part of such a described system can be interconnected by means of transducers. Thus, a closed model description is generated which allows for convenient calculation of the dynamic relation between a source quantity and an observational quantity or between two observational quantities assuming the existence of a source quantity.
In the following sections, the network description of mechanical and acoustic systems is represented. The conditions specified there considerably limit the problem categories this instrument can be applied to. Alternatively or supplementary, it will be referred to calculation methods of *finite elements* and *boundary element methods*. However, they require a significantly higher effort for dynamic analyses. For suitable problems, a mixed method is able to provide dynamic solutions for complete models with electrical and mechanical-acoustic input and output high effectively. These dynamic solutions could not be generated by both methods separately (see Sect. 6.4).

3.1 Mechanical Networks for Translational Motions

3.1.1 Arrangements

In this section, it is dwelled on mechanical systems which consist of spatially concentrated mass, spring and frictional elements. By means of bearings, it is ensured that all regarded junction points can move only in parallel to a straight line (linear motion). In the following, the junction points are also called reference points. Figure 3.1 illustrates an example of such a system.
If external forces $\mathbf{F}_i$ affect the reference points of such a system, then constraining forces will be generated by the bearings resulting after summation in the fact that only force components have an effect which point to the possible direction of motion. Thus, the vector representation of the affecting forces $\mathbf{F}_i$ can be reduced to the components of the force vectors pointing to this direction. The problem can also be simplified by the fact that according to

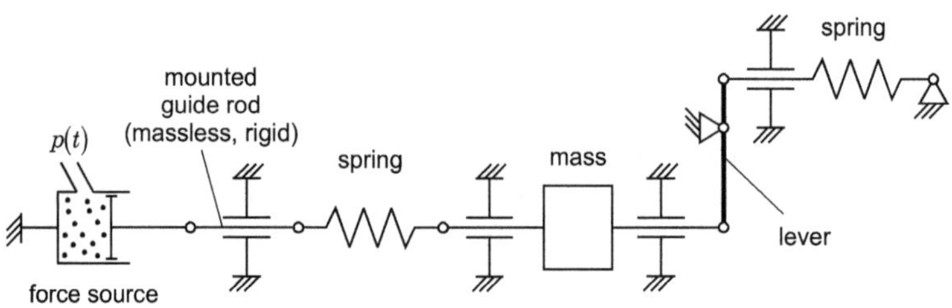

Fig. 3.1. Mechanical system with translational motions

the arrangement only external forces pointing to this direction are expected to be effective.

As an alternative to the vector representation of force, in mechanics it is common to express the force by a scalar quantity F (coordinate) along a drawn direction (represented by an arrow). If the coordinate F is positive, then the force will have the direction of the drawn arrow. If it is negative, then the force will act in opposite direction. In both cases, the force affects the reference point being located at the end of arrow or in front of arrowhead. The arrangements for representation of force are illustrated in Fig. 3.2.

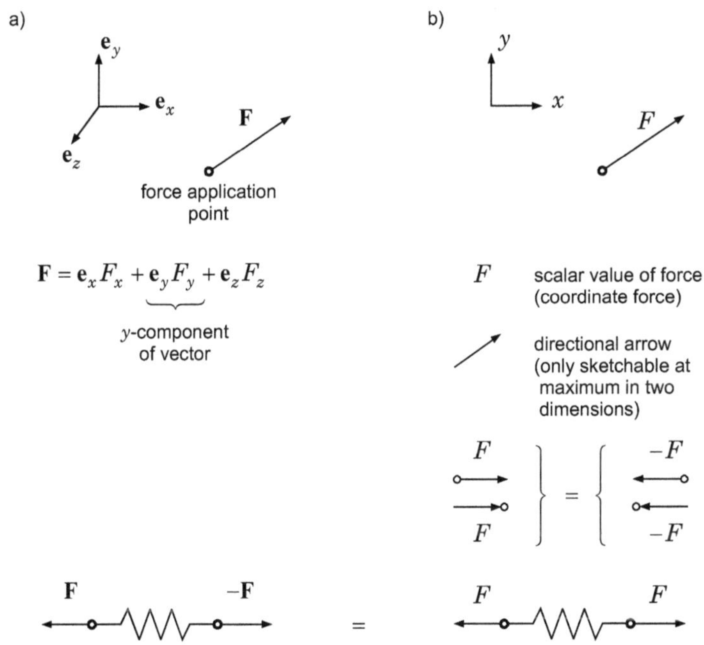

Fig. 3.2. Vector and coordinate representation of force
a) *representation of force by a vector*
b) *description of force by a coordinate F along a drawn direction*

For a system as shown in Fig. 3.1 e.g. the task to be solved can consist in calculation of the time function of the position coordinates of the junction points as function of affecting forces. For well-known time functions of position coordinates also the time functions of forces between the elements could be searched. Instead of position coordinates, velocities or accelerations of reference points can be of interest.
Among the already mentioned components spring and mass, the system illustrated in Fig. 3.1 comprises also the component lever with fixed center of ro-

tation. If the displacements of the lever's reference points are sufficiently small and the permanent weight of the lever remains unconsidered (ideal lever), it will be sufficient to consider only translational components of movements and forces. Then the lever operates as linear way transformer (see Sect. 3.1.3).

3.1.2 Coordinates

Within the scope of mechanics, the systems specified above are described by position coordinates of chosen reference points in a coordinate system. In addition, the forces affecting these reference points are taken into consideration. Thus, position coordinates and force vectors represent the mechanical coordinates.

As elaborated in Sect. 3.1.1, the junction points can move only in one direction in the arranged model. Forces or motions generated by source mechanisms can affect the junction points (reference points) of the lumped elements. Using the example of a spring, Fig. 3.3 a) shows how a compressive force F causes a decreasing in length ξ of the spring. For a positive force value, the arrow of the force vector of a compressive force points to the element (coordinate representation, see also Fig. 3.2 and Sect. 3.1.1). A positive value for ξ is defined as a decreasing in distance between the reference points of the component.

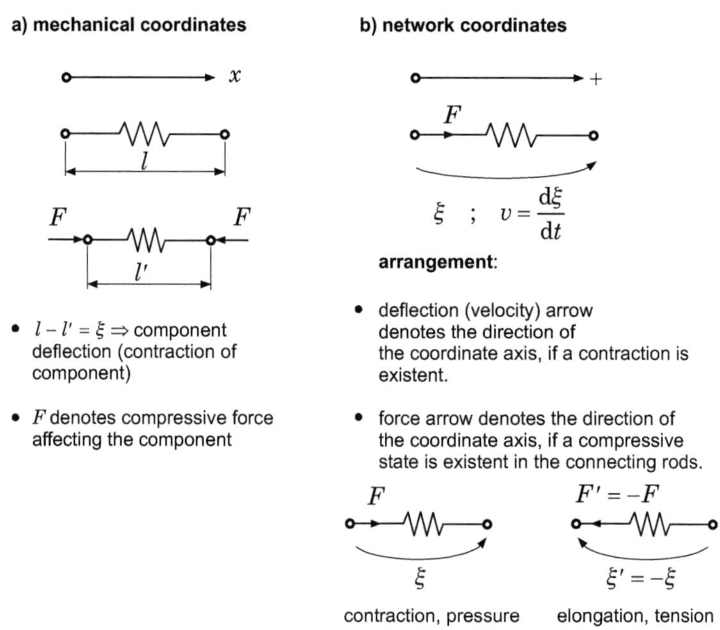

Fig. 3.3. Mechanical coordinates and network coordinates

Taking the aim of the next section to design global models on network level by means of electrical, mechanical and acoustic subsystems into consideration, it is reasonable to start the adaption to electrical systems already with the choice of coordinates. The set of equations which has to be set up, should be finally isomorphic to those of the electrical network shown in Table 2.2. That means that the mechanically derived system equations and their correlations with structures generating them are to differ from those of electrical networks only in terms of their variables and parameters.
In case of electrical networks, the product of coordinates results in the quantity power. Therefore, the coordinate of motion of the mechanical network should describe not the position of reference points on the translational axis but their time derivative, i.e. their *velocities*. In addition, with networks not the velocity of the individual reference points is considered but the difference of the velocities with respect to the end points of the particular component. That makes sense, because the force F affecting the component is causually associated with this velocity difference v. Therefore, following *network coordinates* are chosen:

- velocity difference v across the component
- cross-sectional force F in the imaginary rigid and massless connecting rods

Also in the network representation the directions of coordinates are represented by *arrows*. In Fig. 3.3 b), an example is shown. In order to define a distinct sign convention, further arrangements are necessary. For that purpose, one of the two possible directions of motion is specified as positive reference direction. This reference direction is represented as reference arrow (Fig. 3.3 above). Following direction arrangements refer to that:

- a velocity arrow whose direction is consistent with the direction of the reference arrow is expected to characterize a *time decreasing in distance* of the component for positive numerical value of velocity.
- a force arrow in a connecting rod whose direction is consistent with the reference direction is expected to characterize a *pressure state* within the connecting rod for positive numerical value of force.

These arrangements are structurally consistent with common representations in electrical networks. Within the scope of mechanics, it is usual to draw a force arrow at a reference point (see Fig. 3.3). That means that an external force F is expected to affect the reference point. In other words, a force generating source acts between the *point of origin* being located on the translational axis and the reference point. The source supplies the reference point with its force and rests thereby at the point of origin. For the network representation, this situation is represented by an external source (active one-port). Any point which is rigidly connected with the center of mass of the considered closed mechanical system or performs at most a *uniform motion* in this system, can be a point of origin. Thus, as a clue it is convenient to imagine an enclosing *rigid frame* around a considered configuration which is rigidly connected with an extremely large mass. Then all points being located on this frame are suitable as point of origin.

3.1.3 Components

Spring elements, friction elements, mass elements and lever elements represent the components of mechanical networks. At first, the transition from the real component to the generalized and idealized component is discussed using the example of spring elements.

Component Spring

In the left part of Fig. 3.4 three possible kinds of springs are presented. A square bar is compressed by force F which affects the upper and lower surface A.

The effect of force F on surface A is expected to happen in such a way that it uniformly affects the total surface, so that a location-independent mechanical longitudinal stress is generated in the compression bar. The application of force is effected according to an imaginary massless and rigid connecting rod. This rod with one force application point is symbolically added to both sides of the real component. In this way, two force application points are generated at the component that constitute the character of reference points. During action of force a distance change between the points can be observed.

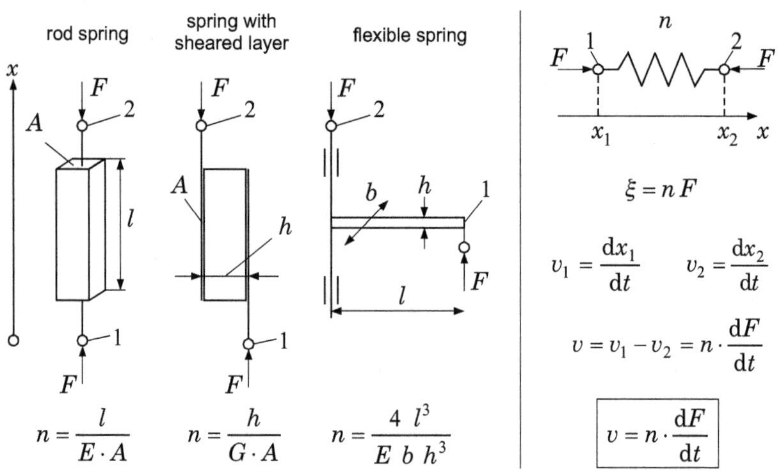

Fig. 3.4. Component spring – realization examples and model

In case of the rod spring with length l and YOUNG's modulus E, the decreasing in distance ξ during application of a longitudinal force F results in:

$$\xi = \frac{l}{EA}F \tag{3.1}$$

The ratio ξ/F denotes the *compliance* n. It is the reciprocal value of the commonly used spring constant c. The expediency for using the compliance n as spring characterizing parameter instead of the spring constant c results from the desired structural identity with electrical networks (see also Sect. 2.2).

In the second example illustrated in Fig. 3.4, a layer with thickness h is sheared at the surfaces A. The application of force F results in an angle change of the side surfaces of the sheared layer compared to the surfaces A. This angle change results in a decreasing of distance ξ between the reference points 1 and 2 which are to be defined by imaginary rigid and massless connecting rods at the ends in accordance to the example of the rod spring. The compliance n is defined by shear modulus G, layer thickness h and surface A.

A flexible spring element can also be considered in the same manner. As shown in Fig. 3.4 (center), the one end of a flexible spring with width b and bender thickness h is rigidly connected with a supported rigid rod being freely moveable only along the translational direction x. The reference point 2 is to be located at the rod. The reference point 1 is located at the other end of the bender. If a force F affects the reference points, their x-coordinates will change. As it can be seen in the examples, the center lines of the connecting rods do not need to be in one line. However, they must be in parallel and the motion may only be possible in the defined translational direction x.

In order that the chosen coordinates F and v can be utilized, it is still necessary to calculate the velocity of the reference points and subsequently the velocity difference by means of the change of the x-coordinates. Now the force can be connected with the velocity across the component spring.

Component Friction

Friction components are consciously built into some constructions in order to attenuate resonant effects. In other constructions they are unwanted, but due to the real behavior of spring components they are existent.

If e.g. a rigid body is moved slowly in an oil-immersed vessel, it will necessitate a force for this motion which is primarily determined by the viscous friction in the oil. The force is proportional to the velocity of motion. Such a configuration is symbolically illustrated among other possible configurations in the left upper part of Fig. 3.5. The friction element is located between the two reference points 1 and 2 which can move along the x-axis. The velocity difference $v = v_1 - v_2$ of both reference points 1 and 2 is proportional to force F affecting the reference points. The porportionality factor $r = F/v$ denotes the *frictional impedance*.

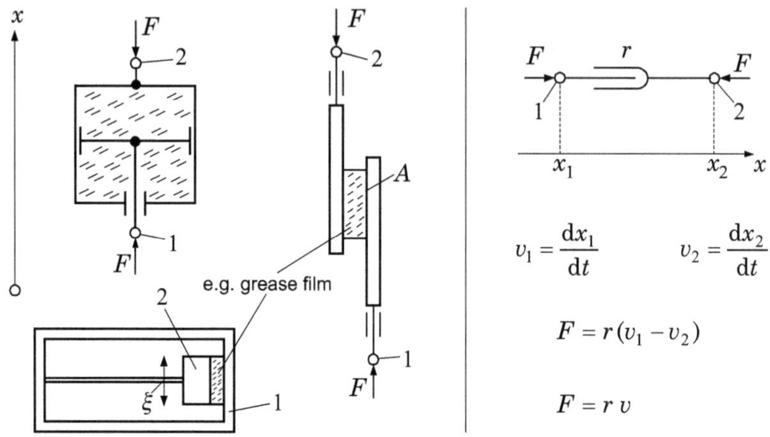

Fig. 3.5. Component friction – realization example and model

The symbol of the mechanical component is represented in the right part of Fig. 3.5. In reality, such components certainly feature also a mass. So the moved mass is represented as separate component. If the real component should comprise internal spring elements, also these elements must be represented separately. So the real friction element is an idealized element which represents only the frictional effect. In this book, only *viscous* frictional effects are modeled by means of linear models.
In Sect. 3.1.5, it is shown that analogy relations exist between a mechanical network representation and an electric circuit. Since it is usual in the field of electrical engineering to characterize the resistor by means of the parameter „electrical resistance" R and not by means of the electrical conductance, also the reciprocal value of the frictional impedance $1/r = h$ (the so-called frictional admittance h) is used as parameter due to representation analogy.

Component Mass

A movable and rigid body that is affected by a force will be accelerated. The proportionality factor between force and velocity change denotes the mass of the body.
If the mass of body is expected to be represented as component in a network with the coordinates F and v, two reference points will be necessary. The velocity difference between the reference points should be solely defined with correct sign by the mass property of the body and the application of force.
The body's center of mass which is affected by a force F represents the first reference point. In order to select the second reference point, it necessitates at first a distinction of cases.

- Assuming a compressive force affecting the center of mass along the x-direction (mechanical representation: force arrow points to center of mass along the x-direction), thus a point with *greater* x-value being located on the translational axis can be chosen as second reference point which is rigidly connected with the origin of the inertial system (see Sect. 3.1.2). Taking the requirement $\xi = l_0 - l$ into consideration, the velocity difference between center of mass and this reference point is defined with correct sign by the inertia relation (Fig. 3.6 left). The pressure state caused by the force affecting the mass results in a decreasing in distance between the reference points with time chosen in such a way. According to the network definition, the velocity arrow across the component and the force arrow within the connecting rod must therefore coincide with the direction of the reference arrow of the network (Fig. 3.6 right). The reference point is represented by a line at the mass symbol and by a connection point characterizing a reference point.

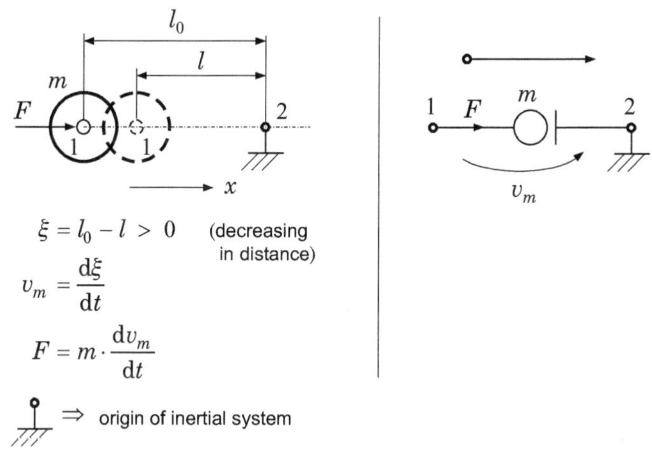

Fig. 3.6. Choice of reference point considering the component mass

- However, if a tensile force acting along the x-direction is assumed (mechanical representation: force arrow points away from center of mass along positive x-direction), a point of reference with *smaller* x-value must be chosen in order to achieve a representation with correct sign considering the same definition $\xi = l_0 - l$ (Fig. 3.7 left). The result is an increasing in distance between the reference points with time. Considering the network definition, the velocity arrow across the component and the force arrow within the connecting rod must therefore point against the direction of the reference arrow (Fig. 3.7 right).

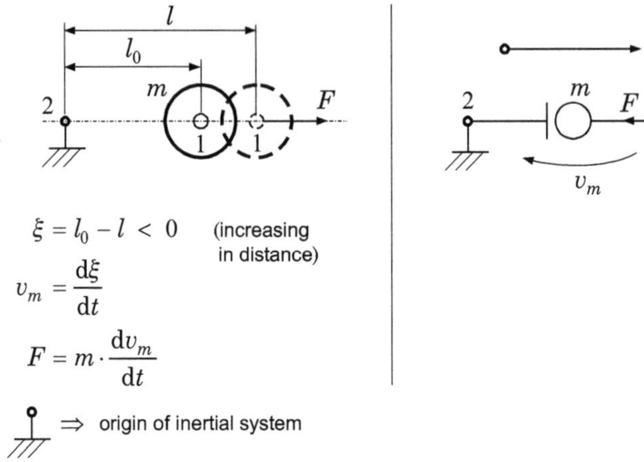

Fig. 3.7. Choice of reference point considering an altered force application side

Thus, in the network domain *both* considerations will result in the same representation, if the reference point is respectively supplemented on the side facing the direct ground connection. So, the general representation form of components results from that. The second connecting point of the mechanical mass *always* denotes either *the origin of the inertial system* or the center of mass of the closed mechanical system.
The result is a simple approach to the transformation of a designed real mechanical structure. A rigid mass block, whose reference point denotes its center of mass, can be represented by means of a component with two connections by extracting the *mass characteristic* of the block from the reference point. This is realized by leading a connection from the reference point to the circular part of a mass symbol and by connecting the other connection of the mass symbol (reference line) with the rigid frame which encloses the entire configuration. In this way, all mass elements are extracted from the present structure of mass, spring and friction components and are all connected to one side with the origin (rigid frame) (see also Sect. 3.1.7).

Interconnection of Mechanical Components

The network representation of a structure consisting of concentrated elements has been introduced in order to be able to analyze the dynamic behavior of linearly interconnected mechanical, acoustic and electrical systems by means of general calculation methods used in the field of electrical engineering. The component symbols which are used in the field of electrical engineering, are also used in network analysis programs (see Table 3.1). Therefore, it is convenient to approve the electrical symbols for mechanical components (as it will be shown later also for acoustic components).

Table 3.1. Network representation of mechanical components

	realization example	dimensioning or measurement rule	network element
spring	spring rod 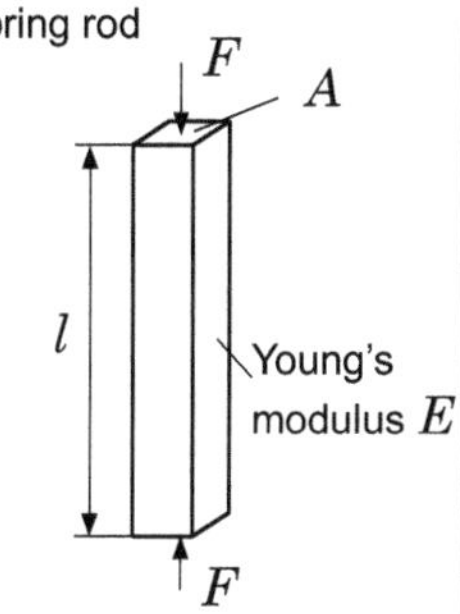	compliance $n = \frac{l}{E \cdot A}$	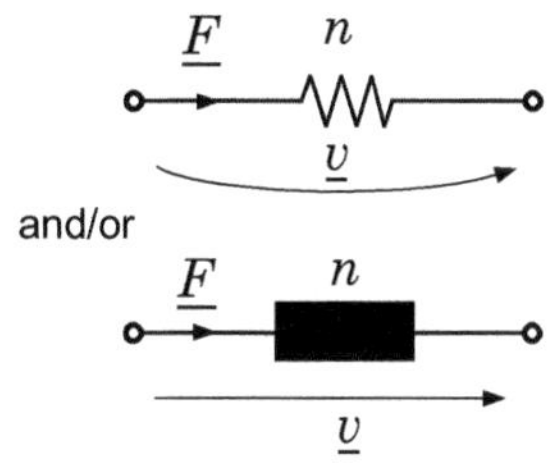 $\underline{v} = \mathrm{j}\omega n \cdot \underline{F}$
friction	attenuator 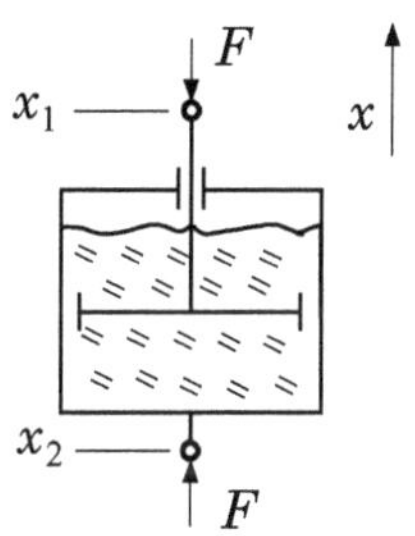	frictional admittance $h = \frac{\left(\frac{\mathrm{d}(x_1 - x_2)}{\mathrm{d}t}\right)}{F}$	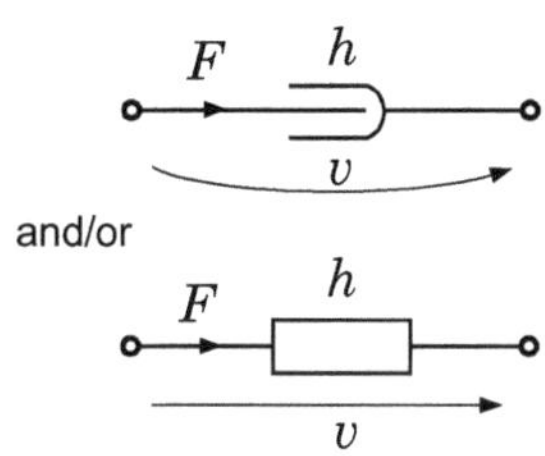 $v = h \cdot F$
mass	mass element 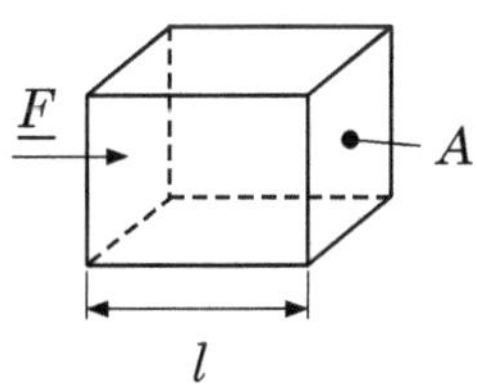	mass $m = \rho \cdot l \cdot A$	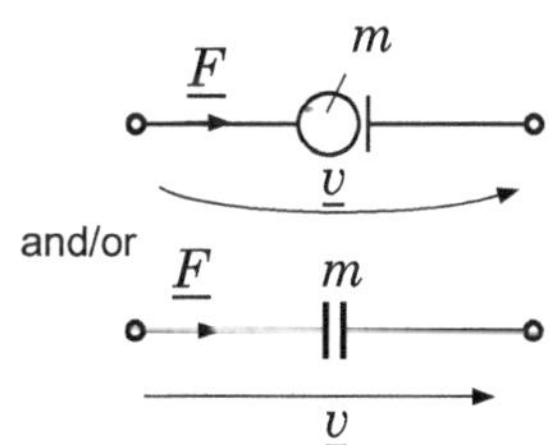 $\underline{v} = \frac{1}{\mathrm{j}\omega m} \cdot \underline{F}$

In addition, in the electrical engineering it is usual to assume sinusoidal source signals in order to analyze the frequency-dependent behavior. In Sect. 2.1, it has already been shown that this choice of time function does not constitute a loss of generality. With these definitions it is now also reasonable to apply the calculation method with complex amplitudes (see Sect. 2.1). Thus, on the part of the components the conditions are given to deduce a mechanical circuit from the real mechanical configuration by a first level of abstraction. The

mechanical circuit can be analyzed with general methods of linear network theory. In the following sections, the necessary considerations with respect to the sources, the interconnection rules and a further component, the so-called lever, are presented.
For the three mechanical (laterally moved) components spring, friction and mass, Table 3.1 respectively contains an realization example as well as the network element representation with the respective relation between the complex amplitudes of the coordinates concerning the component.

Sources of Force and Motion

As already described in the preceding sections, sources with sinusoidally variable source amplitude are used for system excitation. Sources of motion which force a definite oscillating amplitude independently of load and frequency, can be realized e.g. by means of a crank mechanism (source of deflection). In another way also sources of velocity and acceleration can be realized. The technical circuit representation is achieved by means of a source element, at which the source coordinates $\underline{F}_0$ or $\underline{v}_0$ are charted (Table 3.2).

Table 3.2. Sources of motion and force

	mechanical representation	circuit representation
source of motion	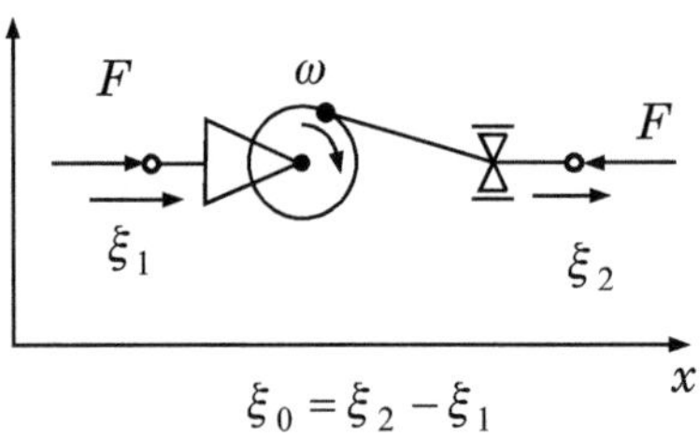	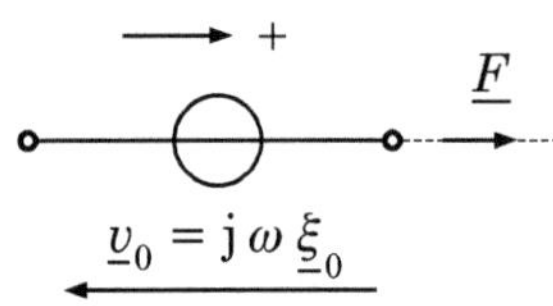
source of force	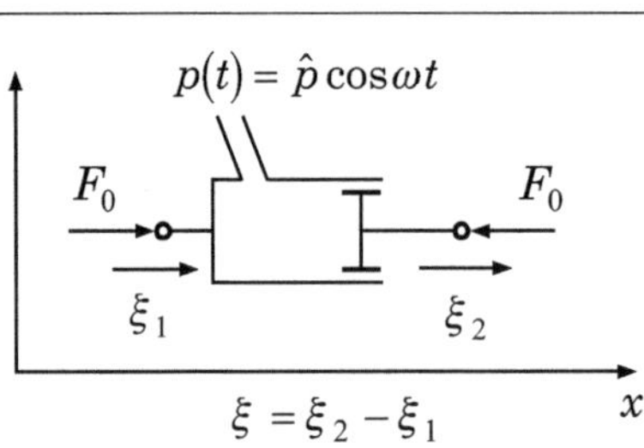	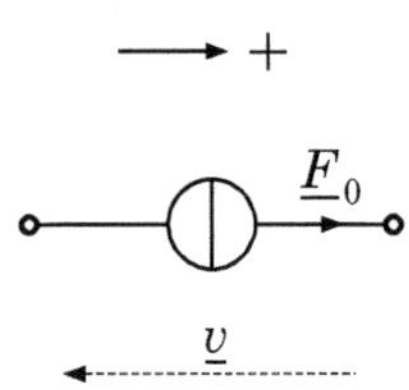

A positive value of v denotes an extension of the source element (direction of arrow in opposite direction to the directional arrow). If the velocity of the source is independent of frequency and load, then $\underline{v}_0$ will be constant. If the deflection and acceleration amplitudes of the sources are independent of frequency, then relations must be specified for $\underline{v}_0$ which characterize the frequency- and phase-dependent interrelation (as illustrated in Table 3.2). Then the force $\underline{F}$ within the rod enabling the connection to the network depends on the network.

However, a source of force always generates the same force amplitude $\underline{F}_0$ independently of the effective load. It can be realized with an impression cylinder which is loaded with a sinusoidally variable pressure of constant amplitude. In case of a connected network, a load-dependent velocity $\underline{v}$ will occur across the source.

In both cases this is only valid within the linearity limits of the configurations. For the source of motion they result from the maximum permitted forces in the crank mechanism and for the source of force they result from the maximum stroke of piston.

Component Lever

As transfer element of motions, the lever interconnects three reference points. In Fig. 3.8, it is shown that displacements ξ and forces F are defined at each reference point. All displacements are defined along the positive x-direction and all forces are defined as compressive forces within the connecting rods. For the following derivation, the lever length from point 0 to point 1 is defined as l_1 and the lever length from point 0 to point 2 is defined as l_2. In the right figure part, the displacements, the force and moment balances as well as the kinematic condition as a result of the rotation of the rigid rod are specified.

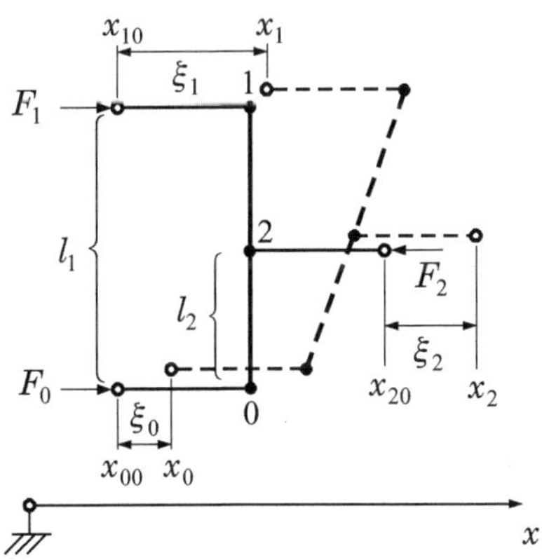

$$\xi_1 = x_1 - x_{10}$$
$$\xi_2 = x_2 - x_{20}$$
$$\xi_0 = x_0 - x_{00}$$

balance of forces: $F_1 - F_2 + F_0 = 0$

balance of moments: $F_1\, l_1 - F_2\, l_2 = 0$

kinematics: $\dfrac{\xi_1 - \xi_0}{l_1} = \dfrac{\xi_2 - \xi_0}{l_2}$

Fig. 3.8. Mechanical coordinates concerning a lever

If the reference points 1 and 2 are considered to be the main connections and point 0 is considered to be the point of reference, then a representation with respect to Fig. 3.9 can be achieved by means of the displacement differences

$$\xi_{10} = \xi_1 - \xi_0 \qquad \xi_{20} = \xi_2 - \xi_0$$

and after transition to the sinusoidal quantities with the complex velocities

$$\underline{v}_1 = \mathrm{j}\omega\underline{\xi}_{10} \qquad \underline{v}_2 = \mathrm{j}\omega\underline{\xi}_{20}.$$

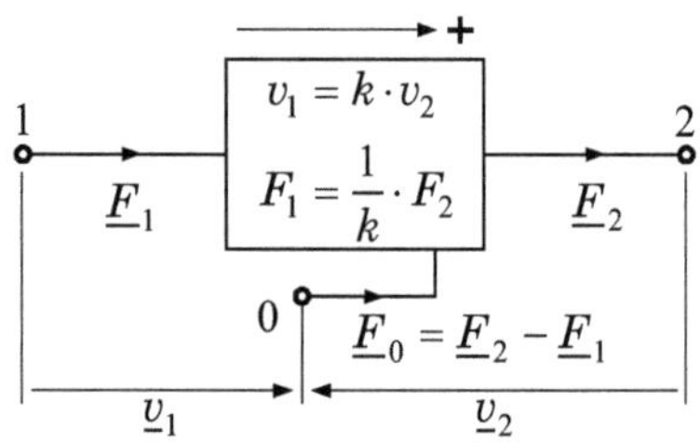

Fig. 3.9. Component lever with network coordinates

Furthermore, a special case being important in practise is achieved in combination with the presented figure. It is characterized by the fact that point 0 is rigidly connected with the origin. The relations shown in Fig. 3.8 provide also a basis for the derivation of the other special case (point 2 represents a fixed point). In order to achieve the *same* network representation also in this case, the denotations of reference points, the transmission ratio and the correlation of lever lengths are newly chosen. Both special cases are shown in Fig. 3.10. Due to reinterpretation of the initial situation, they result in the same network representation with respect to Fig. 3.9.

3.1.4 Rules of Interconnection

In mechanical networks KIRCHHOFF's *nodal and loop rules* are also valid. One can easily be convinced of that, if one locates three components between three reference points according to Fig. 3.11 and adds up the velocities $\underline{v}_i$ along a closed loop.
At first, it can be realized that the sum of extensions and contractions must be equal in both legs. By means of the mathematical method of *complete induction*, it can be proven that the derivative of a sum of finitely many differentiable functions equals the sum of the derivatives of the summands. From this follows that also the velocities must be equal in both legs.

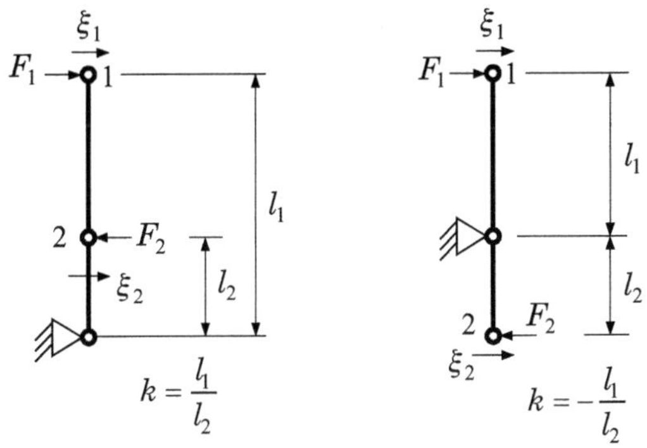

Fig. 3.10. Two forms of lever with fixed point of rotation

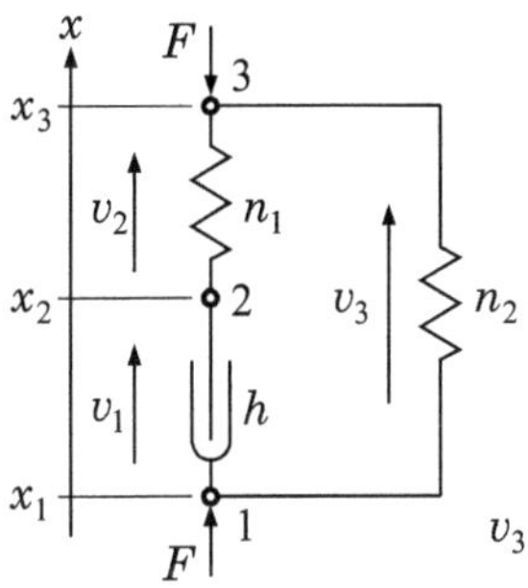

$$\xi_1(t) = x_2(t) - x_1(t)$$
$$\xi_2(t) = x_3(t) - x_2(t)$$
$$\xi_3(t) = x_3(t) - x_1(t)$$

$$\xi_1 + \xi_2 = x_2 - x_1 + x_3 - x_2 = x_3 - x_1 = \xi_3$$

$$v_3 = \frac{\mathrm{d}\xi_3(t)}{\mathrm{d}t} = \frac{\mathrm{d}(\xi_1 + \xi_2)}{\mathrm{d}t} = \frac{\mathrm{d}\xi_1}{\mathrm{d}t} + \frac{\mathrm{d}\xi_2}{\mathrm{d}t} = v_1 + v_2$$

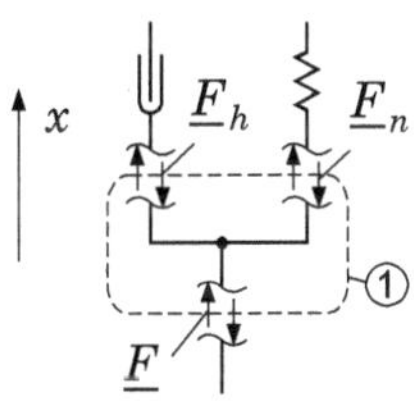

$$\sum_{①} \underline{F}_h + \underline{F}_n - \underline{F} = 0$$

Fig. 3.11. Explanation of validity of KIRCHHOFF's nodal and loop rules

This in turn results in KIRCHHOFF's *loop rule*:

$$\sum_{\text{loop}} \underline{v}_i = 0 \tag{3.2}$$

Here, the variables $\underline{v}_i$ denote the signed i velocities along the loop.
In the lower part of Fig. 3.11, the area of force splitting concerning the mechanical configuration of the upper figure part (reference point 1) is illustrated in more detail. The node is detached by means of the sectional drawing (1). 6 sectional forces are added as substitution. Since the force vectors run as defined parallel to the x-direction, it is sufficient to write signed values as al-

ready discussed at the beginning of Sect. 3.1. In order that the detached part remains in equilibrium, all force components must cancel each other. The sum (with correct sign) of all forces pointing to the node result thereby in zero. Each of the section forces can be considered as complex network coordinate $\underline{F}_i$. As a result, the sum of forces flowing toward the network node must equal zero.
This results in KIRCHHOFF's *nodal rule*:

$$\sum_{\text{node}} \underline{F}_i = 0 \tag{3.3}$$

With mechanical networks consisting of several elements, the question of the *impedance* or *admittance* between points being connected by several elements frequently arises. This task can be solved step by step, if the rules for calculation of impedances or admittances of elements connected in series and in parallel respectively are known.
The rules for interconnection of general complex impedances $\underline{z}_i$ and admittances $\underline{h}_i$ respectively are shown in Fig. 3.12. In the special case of series or parallel connection of spring or mass components, the rules shown in Fig. 3.13 are the result. The series connection of masses can be found, e.g. if a source of motion is loaded by a mass at both sides of connection.

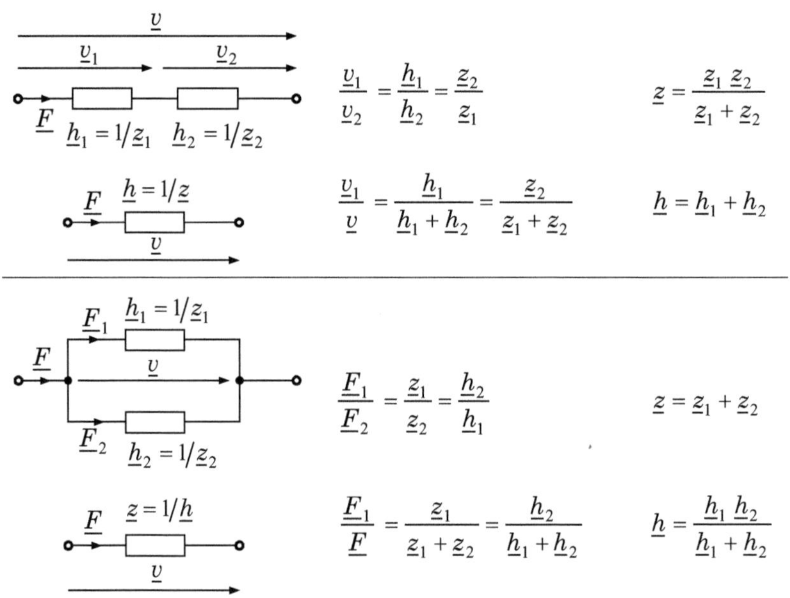

Fig. 3.12. Interconnection of two complex impedances and admittances

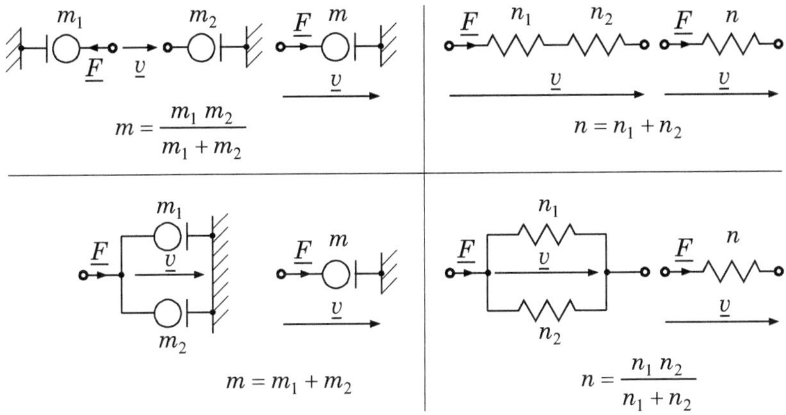

Fig. 3.13. Interconnection rules for mass components and spring components

3.1.5 Isomorphism between Mechanical and Electrical Circuits

Considering the mechanical network handling, the previous representations have substantial similarity to the handling of electrical circuits. This *isomorphism* is an intended aim of the approach. By selecting the coordinates (force summation in nodes, velocity loop) a complete isomorphism to electrical networks will be the result, if relations between coordinates and components as shown in Table 3.3 are chosen.

This isomorphic behavior of both physical structures is achieved by means of the relation between current and force and between voltage and velocity. Thus, topologically equal representations are generated. The application of abstract linear network theory (see also Chap. 4) finally equates both physical structures. Nevertheless, the existence of isomorphism has a very practical advantage. Without any rechecking and changes, network analysis and network synthesis programs which solely originate from the field of electrical engineering, can be consulted just like the entire knowledge of the signal analysis and signal processing of information engineering.

In order to carry out the relations also quantitatively, proportionality quantities must be defined between the coordinates. It is expected to be valid:

$$\underline{u} = G_1 \underline{v} \tag{3.4}$$

$$\underline{i} = \frac{1}{G_2} \underline{F} \tag{3.5}$$

Furthermore, it could be allowed that the frequencies between electrical and mechanical network differ by a constant factor. In practice, no use is made of this possibility. Therefore, it is not used in following considerations.

Table 3.3. Analogy between electrical and mechanical circuit

correlations between coordinates and components				
voltage	$\underline{u}$	o—o	$\underline{v}$	velocity
current	$\underline{i}$	o—o	$\underline{F}$	force
inductance	L	o—o	n	compliance
capacitance	C	o—o	m	mass
resistance	R	o—o	h	frictional admittance
conductance	G	o—o	$r = \frac{1}{h}$	frictional impedance
transformer	$k = \frac{W_1}{W_2}$	o—o	$k = \frac{l_1}{l_2}$	lever

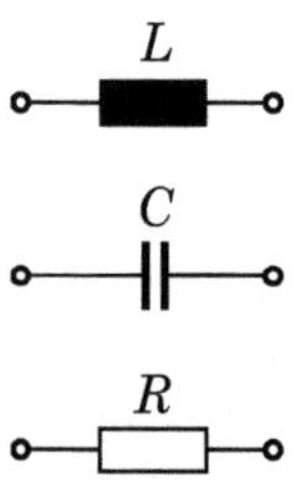

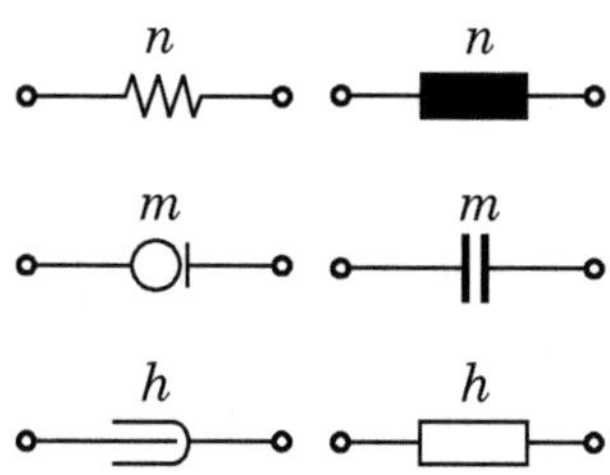

L	$\underline{u} = \mathrm{j}\omega L\underline{i}$	$\underline{v} = \mathrm{j}\omega n\underline{F}$	n
C	$\underline{u} = \frac{1}{\mathrm{j}\omega C}\underline{i}$	$\underline{v} = \frac{1}{\mathrm{j}\omega m}\underline{F}$	m
R	$\underline{u} = R\underline{i}$	$\underline{v} = h\underline{F}$	h
node of circuit structure	$\sum_{*} \underline{i}_\nu = 0$	$\sum_{*} \underline{F}_\nu = 0$	node of mechanical scheme
loop of circuit structure	$\sum_{\circlearrowright} \underline{u}_\nu = 0$	$\sum_{\circlearrowright} \underline{v}_\nu = 0$	loop of mechanical scheme

From this definition, it follows for the components:

$$C = \frac{m}{G_1 G_2} \tag{3.6}$$

$$L = G_1 G_2 n \tag{3.7}$$

$$R = G_1 G_2 h \tag{3.8}$$

The porportionality factors G_1 and G_2 are arbitrary with respect to their value. With regard to the character of network analysis programs (e.g. pSpice), the sequence of numbers, however, should be maintained. Thus, only powers of ten are available. In order to achieve a certain clearness of the electrical circuit for an experienced circuit practician, the factors 10^3 proved to be useful as power of ten for G_1 and G_2. Thus, the result is:

$$G_1 = 10^3\,\mathrm{V\,s\,m^{-1}} \qquad \text{and} \qquad G_2 = 10^3\,\mathrm{N\,A^{-1}}$$

If same numerical values are used for G_1 and G_2, then a representation with equal power rating will be achieved. It is often made use of this advantage. The generated relations 1 g to 1 nF, 1 N to 1 mA and 1 mm s^{-1} to 1 V result in conditions, for which experiences and imagination are often still useful.

3.1.6 Representation of Transient Characteristics of Mass Point Systems in the Frequency Domain (BODE diagram)

The *transfer function* $\underline{B}\,(\omega)$ of a time-invariant, linear electromechanical system was introduced in the Figs. 2.4, 2.5 and 2.9 of Sect. 2.1. The graphical representation of the transfer function (frequency response) with respect to absolute value and phase

$$\underline{B}\,(\omega) = \Re\left\{\underline{B}\,(\omega)\right\} + \mathrm{j}\Im\left\{\underline{B}\,(\omega)\right\} = \left|\underline{B}\,(\omega)\right| \mathrm{e}^{\mathrm{j}\varphi_B(\omega)} \tag{3.9}$$

is achieved by

$$\text{amplitude-frequency response} \qquad \left|\underline{B}\,(\omega)\right| \quad \text{and}$$

$$\text{phase-frequency response} \qquad \varphi_B\,(\omega).$$

In order to „compress" the scales and to allow for simple handling of the graphical representation of amplitude and phase responses, as it will be shown afterwards, BODE [42] introduced the double-logarithmic form of representation of amplitude response and the semi-logarithmic representation of phase response. This form of representation is called *BODE diagram*.
The logarithm is taken of the amplitude response (lg decade logarithm) and multiplied by 20. The decibel [dB] is introduced as unit:

$$\left|\underline{B}\,(\omega)\right| [\mathrm{dB}] = 20 \lg \left|\underline{B}\,(\omega)\right| \tag{3.10}$$

The phase response φ is linearly plotted against the logarithmic frequency axis. In order to assure dimensionless quantities concerning the logarithmic calculus, ω must be referred to a characteristic frequency ω_i and the amplitude must be referred to the constant reference value B_0 (*transfer factor*).
As example, Fig. 3.14 represents the amplitude-frequency response of a low-pass element of 1st order. Following is valid for the amplitude-frequency response:

$$\left|\underline{B}\,(\omega)\right| [\mathrm{dB}] = 20 \lg \left|\underline{B}\,(\omega)\right| = 20 \lg \sqrt{\Re^2\left\{\underline{B}\right\} + \Im^2\left\{\underline{B}\right\}}$$

$$20\lg\left|\frac{1}{1+\mathrm{j}\dfrac{\omega}{\omega_1}}\right| = \begin{cases} 0\,\mathrm{dB}, & \omega \ll \omega_1 \\ -3\,\mathrm{dB}, & \omega = \omega_1 \\ -20\,\mathrm{dB}\,/\mathrm{Decade}, & \omega \gg \omega_1 \end{cases}$$

In combination with

$$\varphi(\omega) = \arg\{\underline{B}(\omega)\} = \arctan\frac{\Im\{\underline{B}(\omega)\}}{\Re\{\underline{B}(\omega)\}},$$

$$\varphi(\omega) = \arg\left\{\frac{1}{1+\mathrm{j}\dfrac{\omega}{\omega_1}}\right\}\begin{cases} 0^\circ, & \omega \to 0 \\ -45^\circ, & \omega = \omega_1 \\ -90^\circ, & \omega \to \infty \end{cases},$$

the phase-frequency response illustrated in Fig. 3.15 is achieved.

In general, the transfer function can be specified by means of a rational function with zeros q_i and poles s_j in the form

$$\underline{B}(\omega) = \frac{(\mathrm{j}\omega - q_1)(\mathrm{j}\omega - q_2)\ldots(\mathrm{j}\omega - q_n)}{(\mathrm{j}\omega - s_1)(\mathrm{j}\omega - s_2)\ldots(\mathrm{j}\omega - s_m)}. \tag{3.11}$$

By introducing characteristic frequencies, basic functions can be generated and the transfer function can be represented as series connection – iterative network – of basic functions, e.g.

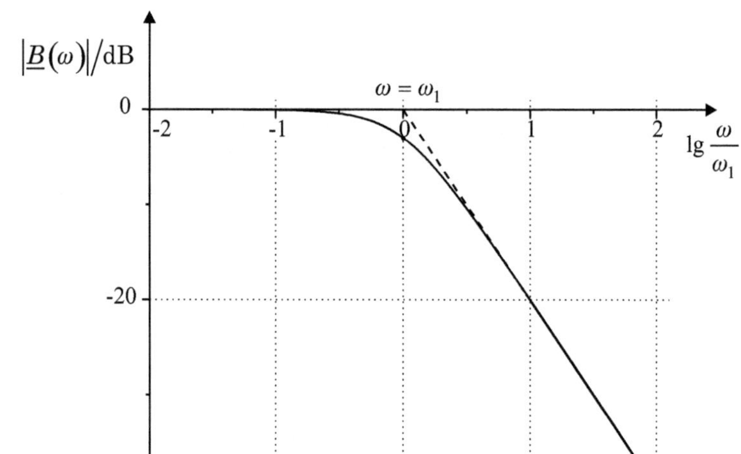

Fig. 3.14. Amplitude-frequency response of a low-pass element of 1st order in double-logarithmic representation

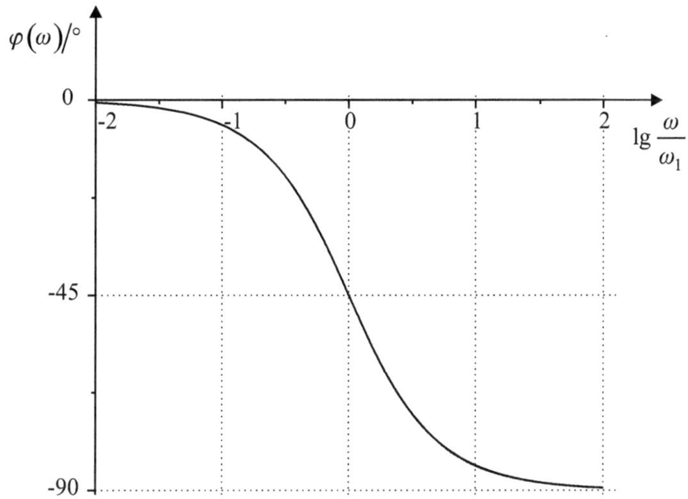

Fig. 3.15. Phase-frequency response of a low-pass element of 1st order in semi-logarithmic representation

$$\underline{B}(\omega) = B_0 \frac{\left(1 + \mathrm{j}\dfrac{\omega}{\omega_1}\right) \mathrm{j}\dfrac{\omega}{\omega_2} \ldots}{\mathrm{j}\dfrac{\omega}{\omega_3}\left(1 + \mathrm{j}\dfrac{\omega}{\omega_4}\right)\left(1 + \mathrm{j}\dfrac{\omega}{\omega_5}\right) \ldots}. \tag{3.12}$$

These basic functions have a descriptive meaning as

- derivative element: $\mathrm{j}\dfrac{\omega}{\omega'}$ $\quad \omega'$ reference frequency

- integrating element: $\dfrac{1}{\mathrm{j}\dfrac{\omega}{\omega'}}$

- high-pass of 1st order: $1 + \mathrm{j}\dfrac{\omega}{\omega'}$

- low-pass of 1st order: $\dfrac{1}{1 + \mathrm{j}\dfrac{\omega}{\omega'}}$

- resonant low-pass: $\dfrac{1}{1 + \mathrm{j}\dfrac{\omega}{\omega'}\dfrac{1}{Q} - \left(\dfrac{\omega}{\omega'}\right)^2}$, $\quad \omega' = \omega_0$ resonant frequency, $Q = \dfrac{1}{2D}$ Q-factor, D damping

The frequency-independent part represents the *transfer factor* B_0. The typical amplitude- and phase-frequency responses of selected basic elements are shown in Fig. 3.16. The iterative network of the basic elements corresponds to the multiplication of their individual transfer functions

$$\underline{B}(\omega) = \underline{B}_1(\omega) \cdot \underline{B}_2(\omega) \cdots \underline{B}_i(\omega)$$

and can be rapidly composed in the BODE diagram by means of addition of the individual amplitude- and phase-frequency responses.

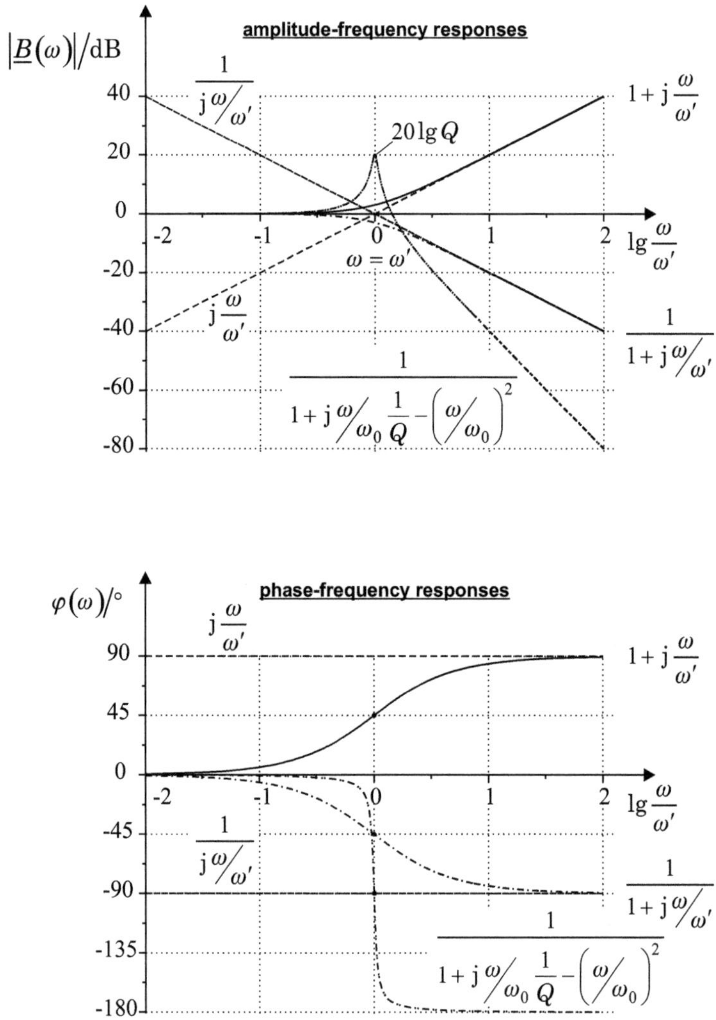

Fig. 3.16. Characteristic curve progressions of amplitude- and phase-frequency responses of selected basic functions

In the following, this main advantage of BODE representation is to be explored subsequently in more detail using the example of the acceleration sensor calculated in Sect. 9.2.7.

The basic design of a compression acceleration sensor is represented in Fig. 3.17. Its transient characteristic can be represented by means of an iterative network consisting of basic elements.

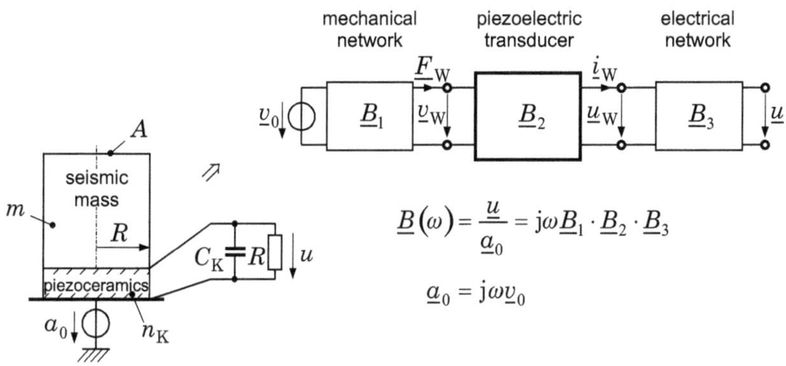

Fig. 3.17. Design principle and iterative network of a piezoelectric acceleration sensor

The transient characteristic of the mechanical network is described by a resonant low-pass element

$$\underline{B}_1 = \frac{\underline{F}_W}{\underline{v}_0} = B_{01}\frac{1}{1 + j\frac{\omega}{\omega_0}\frac{1}{Q} - \left(\frac{\omega}{\omega_0}\right)^2}, \qquad \omega_0^2 = \frac{1}{mn_K}, \qquad Q = \frac{1}{\omega_0 n_K r},$$

the transient characteristic of the piezoelectric transducer is described by the transducer constant Y

$$\underline{B}_2 = \frac{\underline{u}_W}{\underline{F}_W} = Y = B_{02}$$

and the transient characteristic of the following electrical data interpretation network is described by a low-pass element of 1st order

$$\underline{B}_3 = \frac{\underline{u}}{\underline{u}_W} = B_{03}\frac{1}{1 + j\frac{\omega}{\omega_1}}, \qquad \omega_1 = \frac{1}{R(C_K + C_n)}, \qquad C_n = \frac{1}{Y^2}n_K.$$

Thus, the total transfer function yields

$$\underline{B}(\omega) = \frac{\underline{u}}{\underline{a}_0} = B_{01} \cdot B_{02} \cdot B_{03}\frac{1}{1 + j\frac{\omega}{\omega_0}\frac{1}{Q} - \left(\frac{\omega}{\omega_0}\right)^2} \cdot \frac{1}{1 + j\frac{\omega}{\omega_1}} \cdot j\frac{\omega}{\omega_1}.$$

In BODE diagram representation, the addition of the individual transient characteristics enables the design of the total amplitude- and phase-frequency response of $\underline{B}(\omega)$. They are illustrated in Fig. 3.18.

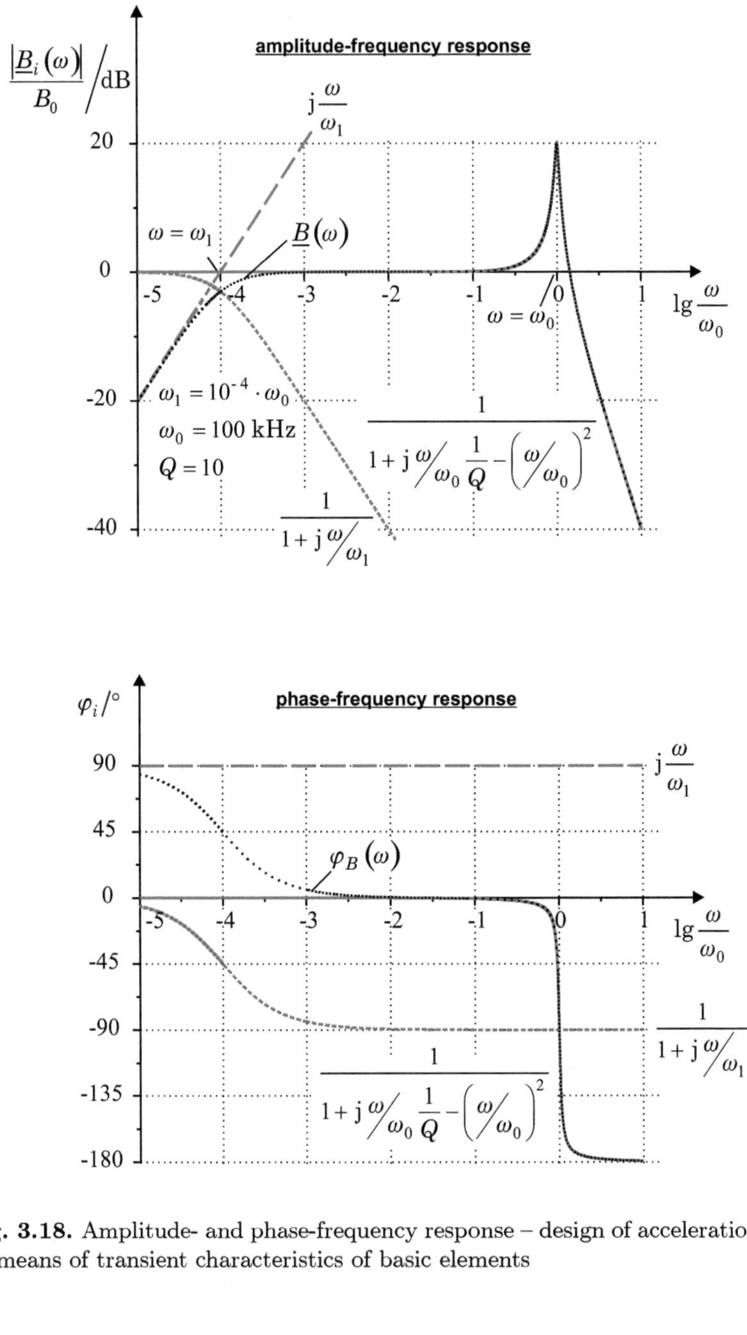

Fig. 3.18. Amplitude- and phase-frequency response – design of acceleration sensor by means of transient characteristics of basic elements

Due to the given advantages of BODE representation, it will be consistently utilized in the course of further chapters.

3.1.7 Network Representation of Mass Point Systems

With the preliminary considerations it must be noted that for the present only problems can be modeled which perform only directions of motion along one axis. Other directions of motion must be excluded by suitable bearings. In order to achieve a network model for a real technical configuration, it necessitates the determination of components being essential for the considered function. This is achieved by means of abstraction and simplification compared to the real technical configuration, e.g. with omitting nonessential elements. It makes sense to concentrate rigidly connected masses and to represent existing parallel connected spring elements as one combined component.
This is to be demonstrated by using the example of a simple foundation with a source of force. The foundation mass m rests on several spring and friction components which are connected to the firm ground with their other component sides. The motion of foundation in solely perpendicular direction is obtained by guides. The center of mass provides a reference point. Since all spring elements are connected to the reference point with the one side and to the origin with the other side, thus parallel, all spring elements can be concentrated in accordance with Fig. 3.13. Now they provide a combined spring with compliance n. The same considerations are valid for the friction elements. The new friction element is represented by the frictional impedance r. It is common practice to designate the component friction with the frictional admittance $h = 1/r$ as parameter. On the foundation stands a source of force which is expected to be supported at the origin. In the left part of Fig. 3.19, the mechanical representation of this technical problem is illustrated. Here, the source of force is symbolized by a force arrow pointing to the mass.

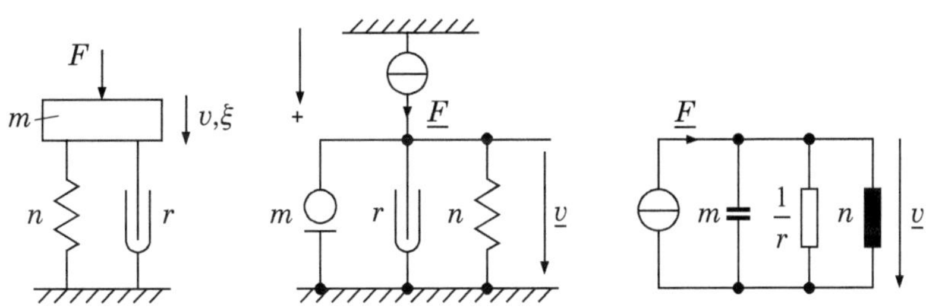

Fig. 3.19. Mechanical representation, mechanical scheme and network representation

A representation level between the mechanical representation and the desired network model provides the *second step* toward a network model. This intermediate step results in the „mechanical scheme". The mechanical scheme is characterized by maintenance of order of the mechanical representation, by replacement of mass elements by their components, by realization of all connections by means of massless and rigid rods and by application of network coordinates. Thus, velocity arrows are plotted across the components, force arrows are plotted into the rods. For that global designation, a positive directional arrow is necessary. If the direction of a force arrow within a connecting rod corresponds to the directional arrow, then, as already discussed before, a pressure state within the rod will be existent (see Sect. 3.1.2). In the middle of Fig. 3.19, the mechanical scheme is illustrated. On the basis of calculation results finally obtained from the network model, motion sequences of real mechanics can be analyzed by correct spatial allocation of the reference points. In the *third step*, reference points with same direction of motion (e.g. all points of reference at the rigid frame) can be combined. In addition, the sources are usually represented on the left side and the reference points which are connected to the origin, are drawn as bottom line. These rules originate from the representation practice of electrical circuits. The mechanical network representation achieved in this way is shown in Fig. 3.19. It proved to be necessary to proceed in these three steps in order to be able to assign surely the calculation results to the real configuration.

In case of excitation with a force $\underline{F}$, the dynamic behavior of the foundation can be described by calculating the frequency response of velocity $\underline{v}$ or oscillating deflection $\underline{\xi}$. By convention, a sinusoidal excitational force is assumed for this. But it is also possible to calculate system responses, e.g. the response to impulsive or stepwise force excitations. In order to calculate the frequency response, the impedance $\underline{z}$ of the parallel connection of mass, spring and friction is determined at first:

$$\underline{z} = \mathrm{j}\omega m + \frac{1}{\mathrm{j}\omega n} + r$$

$$\underline{z} = \frac{1}{\mathrm{j}\omega n}\left(1 - \omega^2 mn + \mathrm{j}\omega nr\right)$$

In combination with the characteristic values *assigned frequency* ω_0 (resonant frequency) and *quality factor* Q, the frequency response of the oscillating deflection and the oscillating velocity can be formulated in normalized form:

$$\omega_0 = \frac{1}{\sqrt{mn}}, \qquad \omega_0 nr = \frac{1}{Q}$$

$$\underline{z} = \frac{1}{\mathrm{j}\omega n}\left(1 - \left(\frac{\omega}{\omega_0}\right)^2 + \mathrm{j}\left(\frac{\omega}{\omega_0}\right)\omega_0 nr\right)$$

$$\frac{\underline{v}}{\underline{F}} = \underline{h} = \frac{1}{\underline{z}} = \omega_0 n \frac{\mathrm{j}\dfrac{\omega}{\omega_0}}{1 - \left(\dfrac{\omega}{\omega_0}\right)^2 + \mathrm{j}\left(\dfrac{\omega}{\omega_0}\right)\dfrac{1}{Q}} \begin{cases} \mathrm{j}\omega n & \omega \ll \omega_0 \\ Q\omega_0 n & \omega = \omega_0 \\ \dfrac{1}{\mathrm{j}\omega m} & \omega \gg \omega_0 \end{cases} \tag{3.13}$$

The frequency functions are represented in Fig. 3.20. In order that a logarithmic representation of amplitude-frequency responses is achieved, it is necessary to provide dimensionless quantities and to form the absolute values. Typed frequency responses are achieved by a suitable choice of reference quantities. The reduction of the alternating force which is led in the ground often represents the aim of the application of a foundation. Thus, perturbations of environment can be avoided extensively. Now it is easy to calculate the force passing through the springs and friction elements. The force $\underline{F}_\mathrm{B}$ leading in the ground results in:

$$\frac{\underline{F}_\mathrm{B}}{\underline{F}} = \frac{1 + \mathrm{j}\left(\dfrac{\omega}{\omega_0}\right)\dfrac{1}{Q}}{1 - \left(\dfrac{\omega}{\omega_0}\right)^2 + \mathrm{j}\left(\dfrac{\omega}{\omega_0}\right)\dfrac{1}{Q}} \tag{3.14}$$

For large mechanical quality factors Q and for frequencies considerably above the assigned frequency ω_0, the frequency response corresponds approximately to the frequency response of the oscillating deflection (see Fig. 3.20). At resonant frequency, an approximate force $\underline{F}_\mathrm{B} = \underline{F}Q$ affects the ground. Within the considered frequency range, it is considerably larger than without foundation. With foundation and with higher frequencies, the force is considerably smaller than without foundation.

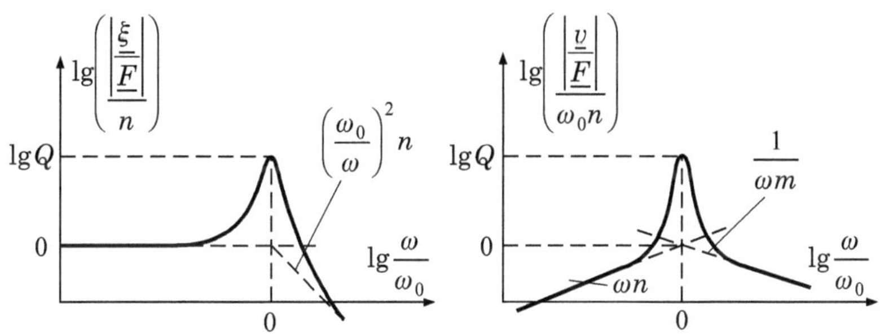

Fig. 3.20. Frequency response of deflection and velocity

3.1.8 Sample Applications

After having considered the quite simple translational system „foundation" in the previous section concerning particularly the aspect of methodical approach, now examples will follow, for which the result is not predictable by implication so easily.

Determination of the Dissipation Factor of a Spring

The relation between mechanical stress T and mechanical strain S within a rod in longitudinal direction and with respect to an allowed transversal contraction is denoted by YOUNG's *modulus* or *modulus of elasticity* E. Here, the relation $E = T/S$ is valid. The relations for a spring without loss presented in this book describe a proportionality between force and deflection. Thus, no phase shift exists between the complex amplitudes of force and deflection.
However, by measurements for nearly all *real* springs a phase shift can be observed. It originates from internal losses caused by deformation of the spring material. It is appropriate to describe these losses by a complex modulus of elasticity. It is expected to be valid

$$\underline{E} = E\,(1 + \mathrm{j}\eta)\,. \tag{3.15}$$

The real part of the complex modulus of elasticity corresponds to the modulus of elasticity mentioned above, whereas the factor η denotes the losses in the material. Generally, the dissipation factor η is frequency-dependent. Mostly, the frequency dependence of the real part of the modulus of elasticity is less than that of the dissipation factor.
The technical circuit interpretation of the complex modulus of elasticity succeeds e.g. by connecting a friction element in parallel to the actual spring element. If the dissipation factor η would be strictly proportional to frequency, the frictional impedance would be constant (frequency-independent). That does not usually apply in such a way. Within a narrow frequency range, e.g. in the neighborhood of the working frequency of a system, a constant frictional impedance can be approximately assumed.
Figure 3.21 shows the circuit representation of a lossy spring by means of a parallel connection. The components of the parallel connection result from following consideration:

$$\underline{z} = \frac{\underline{F}}{\underline{v}} = \frac{AT}{\mathrm{j}\omega l S} = \frac{AE\,(1+\mathrm{j}\eta)}{\mathrm{j}\omega l} = \frac{1}{\mathrm{j}\omega n} + \frac{1}{\left(\dfrac{\omega n}{\eta}\right)} = \frac{1}{\mathrm{j}\omega n} + r$$

Due to the generally quite low losses, for a constant frequency this parallel connection can be also converted to a series connection consisting of a spring and a friction element and showing approximately the same effect.

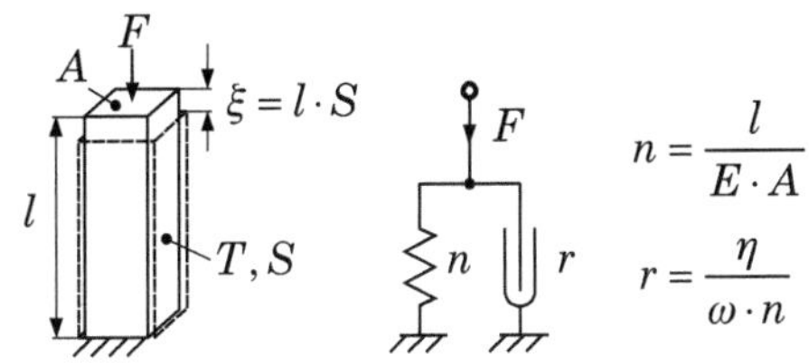

Fig. 3.21. Representation of a lossy spring rod

In practice, losses can often be considered only by specifying estimated resonator performances. Thus, it makes sense to place the component with loss in such a way in the circuit that these estimates can be easily converted to component parameters and that they remain constant within the operating frequency range. In a second step, corrections can be deduced from measured frequency responses or tests of free oscillations by means of a sample set-up providing the model verification.

Considering this described background, a measuring unit for the determination of modulus of elasticity and dissipation factor of a sample of material at a target frequency is of special interest. The sample of a material to be analyzed is set on an oscillating surface. Furhermore, a piece of mass covering all over is put on the sample. Preferably the oscillating surface is represented by the mounting surface of a shaker which can be used for generation of a frequency-variable sinusoidal oscillation. In addition to the sample, also an acceleration sensor is attached to the shaker's mounting surface. The sensor measures the acceleration amplitude of excitation a_0. By means of a second acceleration sensor attached to the mass, the amplitude a_2 is measured. The experimental set-up and the mechanical circuit are illustrated in Fig. 3.22.

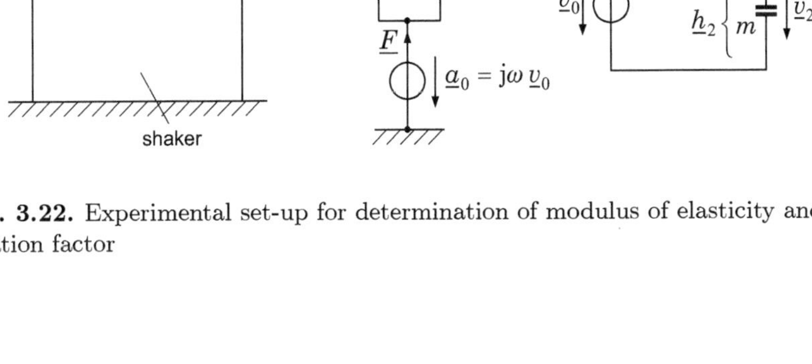

Fig. 3.22. Experimental set-up for determination of modulus of elasticity and dissipation factor

Instead of accelerations, the velocities are charted in the mechanical circuit. The ratio of the absolut amplitude values a_2/a_0 is plotted against the frequency.

The transfer function has the character of a resonant low-pass. By introducing the resonant frequency

$$\omega_0 = \frac{1}{\sqrt{mn}}$$

and the mechanical quality factor

$$Q = \frac{1}{\omega_0 nr},$$

the quantities to be determined result in

$$\eta(\omega_0) = \frac{1}{Q} \quad \text{and} \quad E(\omega_0) = \frac{lm\omega_0^2}{A}.$$

In addition to the mechanical dimensions sample height l, sample surface A and mass m (including mass of acceleration sensor), the resonant frequency and the mechanical quality factor have to be determined. If the samples of material do not represent high-loss materials, the resonant frequency can be determined by means of maximum position of the acceleration amplitudes' ratio. The quality factor Q is determined by means of frequency difference $\Delta\omega$ between those two points, at which the transfer function decreases by the factor $1/\sqrt{2}$ with respect to the resonant maximum value (see Fig. 3.23). Thus, the dissipation factor can be formulated in the following way:

$$\eta(\omega_0) = \frac{\Delta\omega}{\omega_0}$$

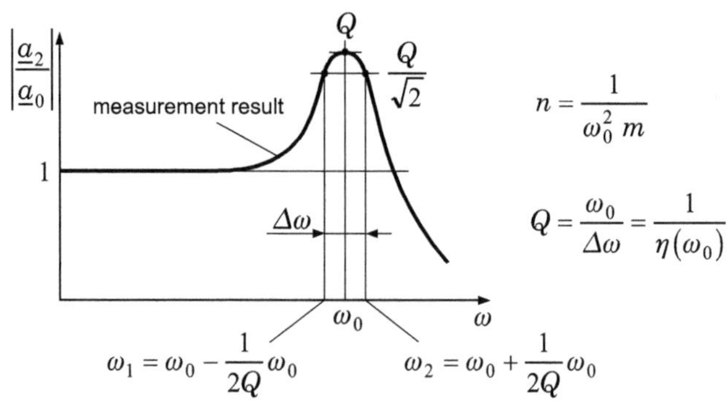

Fig. 3.23. Measurement of resonant frequency ω_0 and quality factor Q

In consideration of high quality factors, this method of determination of quality factor and resonant frequency is very accurate. Even if the quality factor amounts to $Q = 3$, the error of the characteristic values will be less than 3% and is accurate enough for most practical problems.

Vibration Isolation of a Machine

A machine with rotating components which are not completely balanced represents a source of force for that position, at which it is mounted. In order to avoid disturbances of environment, the forces induced by the machine into the mounting place should be as small as possible. For this purpose, usually an elastic element between the machine and the mounting place is attached. Now it is to be analyzed which reduction of force to be induced into the mouting place can be obtained in this way. The model of the machine comprises a rotor with mass m_1 rotating with an imbalance amplitude ϱ and angular velocity ω in a supporting stand with mass m_2. On the one hand, the supporting stand is directly installed on a surface assumed to be stationary (Fig. 3.24 a)), on the other hand, an elastic intermediate layer with internal absorption is installed between housing and surface (Fig. 3.24 b)). Now it is searched for both the force $\underline{F}_1$ which is applied to the mounting place without isolation and the improvement $\underline{F}_1'/\underline{F}_1$ resulting from mounting of the elastic intermediate layer. By means of the assembly, only vertical motions are permitted.
Due to the existing eccentricity ϱ, the shaft performs a motion with a vertical velocity component $\underline{v} = \mathrm{j}\omega\varrho$. The internal admittance $\underline{h}_i$ denotes the ratio of vertical force and velocity with respect to a non-rotating rotor. It is defined by the rotor mass m_1.
The force for short-circuit operation is formally obtained from $\underline{F}_0 = \underline{v}_0/\underline{h}_i = \mathrm{j}\omega m_1 \underline{v}_0$. It is the force which must affect the shaft of the rotating rotor in order to force a rotor velocity of $\underline{v} = 0$. The rotation of rotor with vertically blocked shaft can be replaced by a sinusoidal motion of a mass point with mass m_1 and

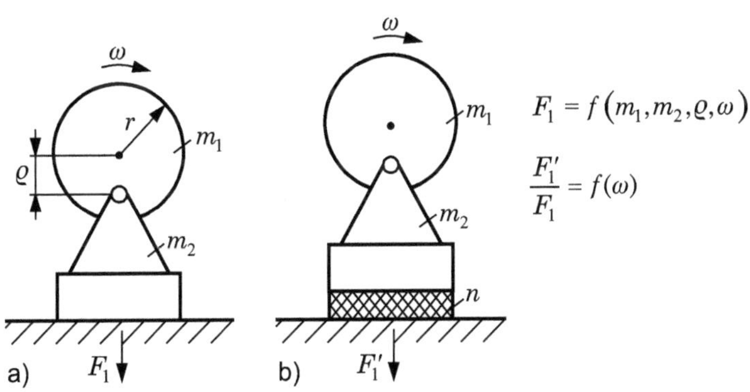

Fig. 3.24. Vibration isolation of a machine with imbalance

amplitude $\underline{\varrho}$. This motion generates exactly the force $\underline{F}_0 = m_1\underline{a} = m_1 \mathrm{j}\omega\underline{v}_0$. Thus, the parameters of the active one-port „rotor" are determined. Now there is the task to specify the mechanical scheme for the two cases represented in Fig. 3.24 and to deduce the mechanical circuits. That is accomplished in Fig. 3.25 and Fig. 3.26.

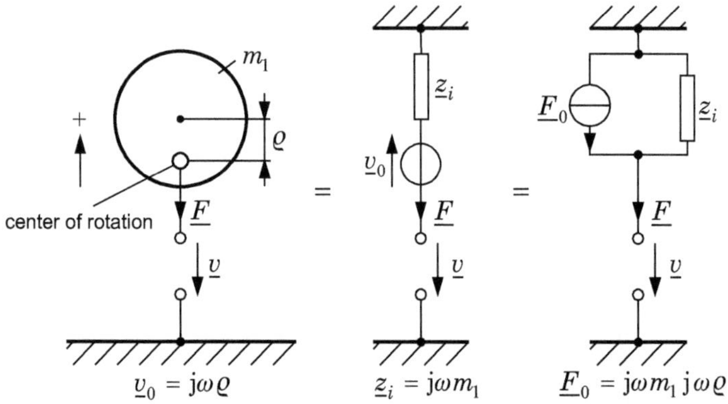

Fig. 3.25. Rotor with imbalance as active mechanical one-port

For case a) of the missing elastic support, the searched force $\underline{F}_1$ is identical to the force of source $\underline{F}_0$. Due to its velocity ($\underline{v} = 0$), the mass m_1 can not absorb forces. For case b), this is not so simple. Here, the force $\underline{F}_0$ must be divided up between both impedances $\underline{z}_1$ and $\underline{z}_2$.

It can be written:

$$\underline{F}_1' = \underline{F}_0 \frac{\dfrac{1}{\mathrm{j}\omega n} + r}{\dfrac{1}{\mathrm{j}\omega n} + r + \mathrm{j}\omega\underbrace{(m_1 + m_2)}_{m}} = \underline{F}_0 \frac{\dfrac{1}{\mathrm{j}\omega n}(1 + \mathrm{j}\eta)}{\dfrac{1}{\mathrm{j}\omega n}(1 + \mathrm{j}\eta) + \mathrm{j}\omega m}$$

$$\Rightarrow \left|\frac{\underline{F}_1}{\underline{F}_0}\right| = \left|\frac{\underline{F}_1'}{\underline{F}_1}\right| = \frac{\sqrt{1+\eta^2}}{\sqrt{\left(1 - \left(\dfrac{\omega}{\omega_0}\right)^2\right)^2 + \eta^2}}$$

with

$$\omega_0 = \frac{1}{\sqrt{mn}} \quad \text{and} \quad \eta = r\omega n = \eta\,(\omega_0)\,\frac{\omega}{\omega_0}.$$

The elastic support is assumed to be characterized by a frequency-independent friction r. The progression of ratio $\left|\underline{F}_1'/\underline{F}_1\right|$ as a function of frequency is represented in Fig. 3.27.

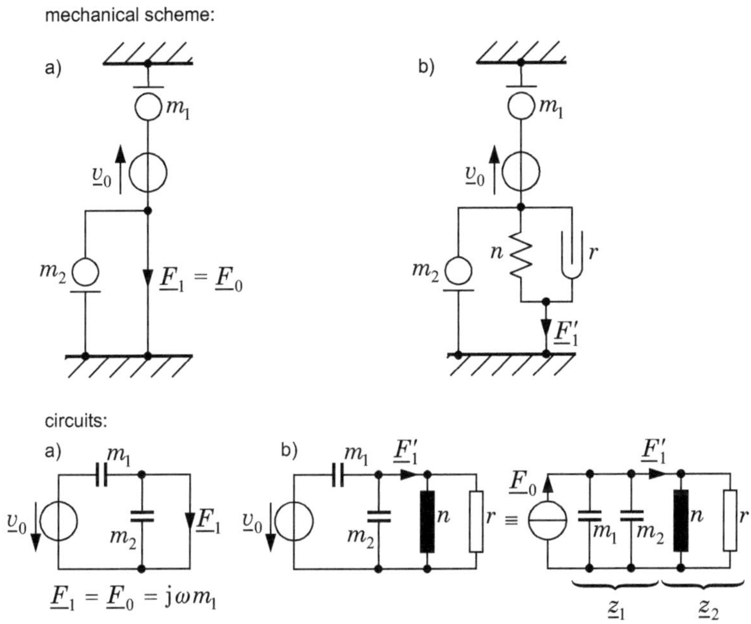

Fig. 3.26. Mechanical scheme and deduced circuits of the configurations illustrated in Fig. 3.24

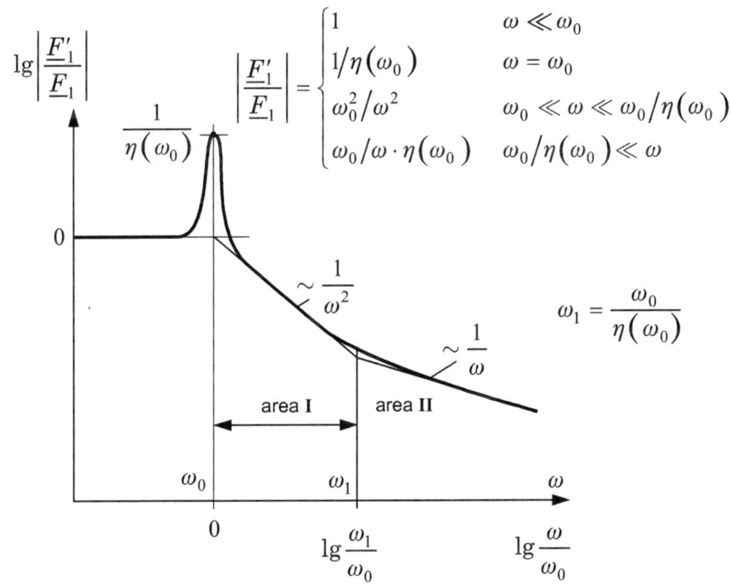

Fig. 3.27. Vibration damping of a vibration-isolated machine

For low frequencies both η^2 and $(\omega/\omega_0)^2$ are much less than 1 and $\underline{F}_1' = \underline{F}_1 = \underline{F}_0$. The elastic support has no influence. For $\omega = \omega_0$ the denominator has its smallest value. $\eta(\omega_0)$ is usually of the order of magnitude of 10^{-1}. The square root in the numerator can be considered to be still 1. The force $\underline{F}_1'$ generated in case of the elastic support is by a factor $1/\eta(\omega_0)$ higher than with permanent coupling. No attenuation but an amplification of the disturbing process is generated.
For $\omega > \omega_0$ the influence of the summand $(1-(\omega/\omega_0)^2)^2 \approx (\omega/\omega_0)^4$ preponderates in the square root of the denominator at first. The ratio $\left|\underline{F}_1'/\underline{F}_1\right|$ decreases with $1/\omega^2$. However, from a certain frequency of ω_1 also $\eta(\omega)$ in the numerator can not be neglected any longer. This frequency is defined by $\eta = 1$. Above ω_1 it is $\sqrt{1+\eta^2} \approx \eta$ and the ratio $\left|\underline{F}_1'/\underline{F}_1\right|$ decreases only with $1/\omega$.

A numerical example is to demonstrate the practically occurring orders of magnitude. The imbalance of a machine may be characterized by a rotor mass of $m_1 = 100\,\mathrm{kg}$ and an imbalance amplitude of $\varrho = 0.1\,\mathrm{mm}$. The operating speed frequency of the machine may be $n = 3000\,\mathrm{min}^{-1} = 50\,\mathrm{Hz}$. The support may be characterized by a resonant frequency of $f_0 = 10\,\mathrm{Hz}$ in combination with the total mass $m_1 + m_2$ of the machine. Its dissipation factor may be $\eta(\omega) = 0.1$ at $10\,\mathrm{Hz}$. So, the amplitude of the force $\underline{F}_0 = \underline{F}_1$ results in

$$\hat{F}_0 = \hat{F}_1 = \omega^2 \varrho m_1 = \begin{cases} 1000\,\mathrm{N} & f = 50\,\mathrm{Hz}, \\ 40\,\mathrm{N} & f = 10\,\mathrm{Hz}. \end{cases}$$

Due to $f_1 = 10 f_0$, for $f = 50\,\mathrm{Hz} = 5 f_0$ you are still in area I illustrated in Fig. 3.27. Therefore, the force $\underline{F}_1'$ results in

$$\text{for} \quad f = 50\,\mathrm{Hz}: \qquad \hat{F}_1' = \hat{F}_1\,(50\,\mathrm{Hz}) \left(\frac{\omega_0}{\omega}\right)^2 = 40\,\mathrm{N}$$

$$\text{for} \quad f = 10\,\mathrm{Hz}: \qquad \hat{F}_1' = \hat{F}_1\,(10\,\mathrm{Hz}) \frac{1}{\eta(\omega_0)} = 400\,\mathrm{N}$$

Thus, compared to the fixed assembly, at operating speed frequency a force reduction around the factor 25 is the result.

Passive Vibration Absorber

The considerations to the translational system „foundation" have shown that the force affecting the ground will only be reduced compared to the source force, if the operating frequency of the force generator is considerably above the resonant frequency. However, depending on the quality factor Q, in the neighborhood of the resonant frequency the ground force is larger. If the ground force ought to be reduced in resonance, a larger attenuation will be necessary. However, the ground force increases with higher operating frequency.

Then the flux of force flows through the damping element into the ground. By applying a second oscillating system, further possibilities are generated which can be utilized purposefully. This additional system is known under the designation „vibration absorber". Such a supplementary unit is appropriate for several fields of application. For the cases outlined in the following sections, a real mechanical system consisting of a simple resonant foundation in combination with an additional oscillating system is represented in Fig. 3.28. The frictional losses in the supplementary system are generated both in the spring and in the region of the guide rods of mass m_2. The guide rods are rigidly connected with the mass m_1. The viscous friction r_2^* may be freely adjustable in this example. Figure 3.28 shows the mechanical scheme of the system and the mechanical circuit, too.
At first, the case of an excitation of foundation with **constant frequency** ω **being effected in direct neighbourhood of the foundation's resonant frequency** ω_0 is to be considered. Without additional oscillating system, a

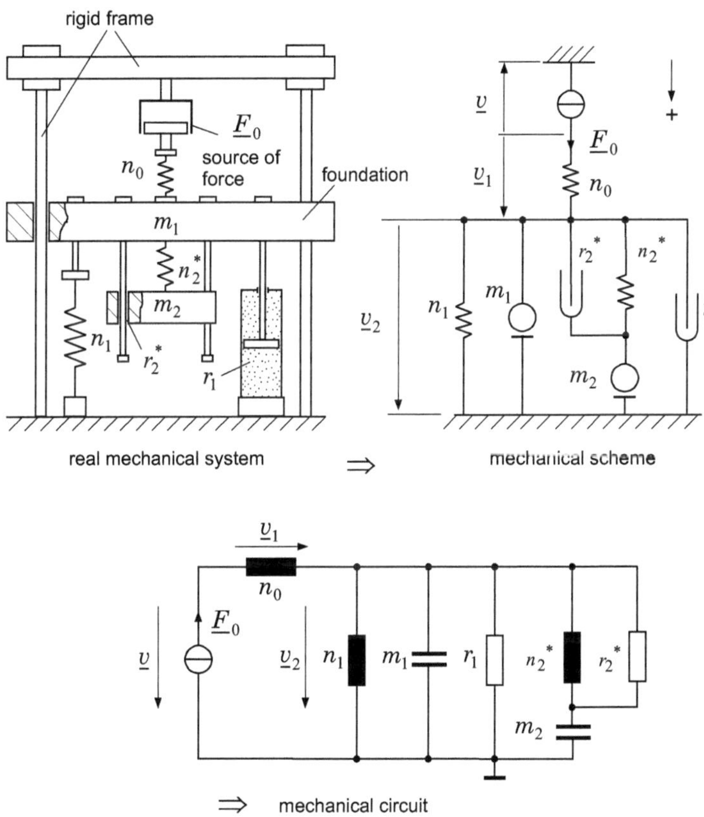

Fig. 3.28. Foundation with vibration absorber

high force amplitude affecting the ground compared to the excitational force amplitude and a high deflection amplitude on foundation mass would be generated in case of this operational mode. A considerable lowering of the resonant frequency ω_0 of the simple resonant foundation could be a solution. However, due to the often limited permitted deflection of foundation mass as a result of the gravity acceleration g

$$\xi_{\text{stat}} = \frac{g}{\omega_0^2}$$

a lowering of the resonant frequency will not be considered. Instead a second mechanical resonator is attached and dimensioned to the same resonant frequency ω_0. In the sample calculation, the mass m_2 of this system may amount to 20% of foundation mass only. In practice, this extension of foundation construction is applicable more easily than the considerable lowering of the resonant frequency specified above. In order to achieve a considerable decrease of the ground force in the neighborhood of resonant frequency, the quality factor of the additional oscillator must be sufficiently high (e.g. higher than 20). For an easy analytical calulation it is convenient to perform approximations and circuit simplifications. So e.g. the parallel connection of spring n_2^* and friction r_2^* can be transformed into a series connection (Fig. 3.29).

For the components it can be written

$$r_2 = \frac{r_2^*}{1 + \left(Q_2^* \frac{\omega_0}{\omega}\right)^2} \qquad n_2 = \frac{n_2^*}{1 + \left(\frac{\omega_0}{Q_2^* \omega}\right)^2} \qquad \text{with} \qquad Q_2^* = \frac{r_2^*}{\omega_0 n_2^*}.$$

For a high quality factor Q_2^* and for only minor deviation of frequency from the resonant frequency it can be written approximately:

$$\frac{\omega}{\omega_0} \approx 1, \qquad n_2 \approx n_2^* \qquad \text{and} \qquad r_2 \approx \frac{r_2^*}{(Q_2^*)^2}$$

Now the ratio of ground force $\underline{F}$ to excitational force $\underline{F}_0$ can be calculated very easily. The spring n_0 is of no importance for the mentioned problem. At first, the components are combined into complex impedances (Fig. 3.29).

The parallel connection of n_1, m_1 and r_1 provides the complex impedance $\underline{z}_1$. The components n_1 and r_1 are combined into $\underline{z}$. The series connection of n_2, r_2 and m_2 is described by the complex impedance $\underline{z}_2$. Now the force ratios can be described by the impedances $\underline{z}$, $\underline{z}_1$ and $\underline{z}_2$ according to:

$$\frac{\underline{F}_1}{\underline{F}_0} = \frac{\underline{z}_1}{\underline{z}_1 + \underline{z}_2} \qquad \text{and} \qquad \frac{\underline{F}}{\underline{F}_1} = \frac{\underline{z}}{\underline{z}_1}$$

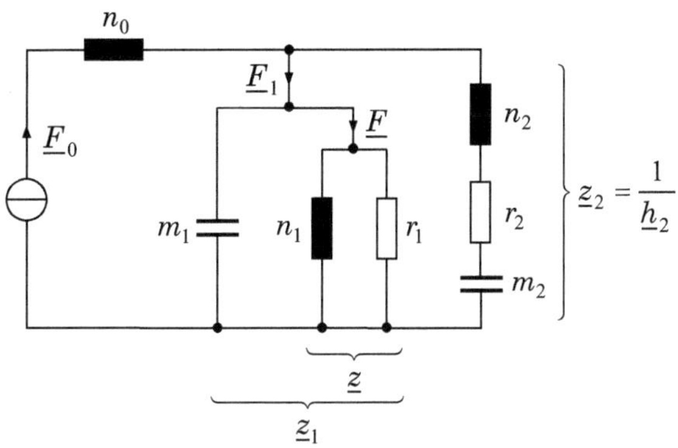

Fig. 3.29. Introduction of ground force $\underline{F}$ and impedances $\underline{z}$, $\underline{z}_1$ and $\underline{z}_2$

This results in:

$$\frac{\underline{F}}{\underline{F}_0} = \frac{\underline{z}}{\underline{z}_1 + \underline{z}_2} = \frac{\underline{z} \cdot \underline{h}_2}{1 + \underline{z}_1 \underline{h}_2} \qquad \text{with} \qquad \underline{h}_2 = \frac{1}{\underline{z}_2}$$

Insertion of the components yields

$$\frac{\underline{F}}{\underline{F}_0} = \frac{\left(\frac{1}{\mathrm{j}\omega n_1} + r_1\right)\left(\mathrm{j}\omega n_2 + \frac{1}{\mathrm{j}\omega m_2} + \frac{1}{r_2}\right)}{1 + \left(\mathrm{j}\omega m_1 + \frac{1}{\mathrm{j}\omega n_1} + r_1\right)\left(\mathrm{j}\omega n_2 + \frac{1}{\mathrm{j}\omega m_2} + \frac{1}{r_2}\right)}$$

and

$$\frac{\underline{F}}{\underline{F}_0} = \frac{(1 + \mathrm{j}\omega n_1 r_1)\left(1 - \omega^2 m_2 n_2 + \mathrm{j}\omega \frac{m_2}{r_2}\right)}{-\omega^2 m_2 n_1 + (1 - \omega^2 m_1 n_1 + \mathrm{j}\omega n_1 r_1)\left(1 - \omega^2 m_2 n_2 + \mathrm{j}\omega \frac{m_2}{r_2}\right)}, \tag{3.16}$$

respectively. With the appropriate *normalized quantities* (2 assigned frequencies and 2 quality factors)

$$\omega_1^2 = \frac{1}{m_1 n_1}, \qquad \omega_2^2 = \frac{1}{m_2 n_2}, \qquad \frac{1}{Q_1} = \omega_1 n_1 r_1, \qquad \frac{1}{Q_2} = \frac{\omega_2 m_2}{r_2}$$

it follows in normalized notation:

$$\frac{\underline{F}}{\underline{F}_0} = \frac{\left(1 + \mathrm{j}\frac{\omega}{\omega_1}\frac{1}{Q_1}\right)\left(1 - \frac{\omega^2}{\omega_2^2} + \mathrm{j}\frac{\omega}{\omega_2}\frac{1}{Q_2}\right)}{-\frac{\omega^2}{\omega_1\omega_2}\sqrt{\frac{m_2}{m_1}\frac{n_1}{n_2}} + \left(1 - \frac{\omega^2}{\omega_1^2} + \mathrm{j}\frac{\omega}{\omega_1}\frac{1}{Q_1}\right)\left(1 - \frac{\omega^2}{\omega_2^2} + \mathrm{j}\frac{\omega}{\omega_2}\frac{1}{Q_2}\right)} \tag{3.17}$$

According to the assumptions concerning frequency and quality factor, the expression can be simplified:

$$\left|\frac{\underline{F}}{\underline{F}_0}\right| \approx \frac{m_1}{m_2}\frac{1}{Q_2}$$

The smaller the additional mass is chosen, the higher the quality factor must be adjusted, so that an effective decrease of ground force is achieved.

In order to consider a broader frequency range, the mechanical circuit is calculated without using of approximations. Figure 3.30 shows the frequency response of ground force of the circuit illustrated in Fig. 3.28.

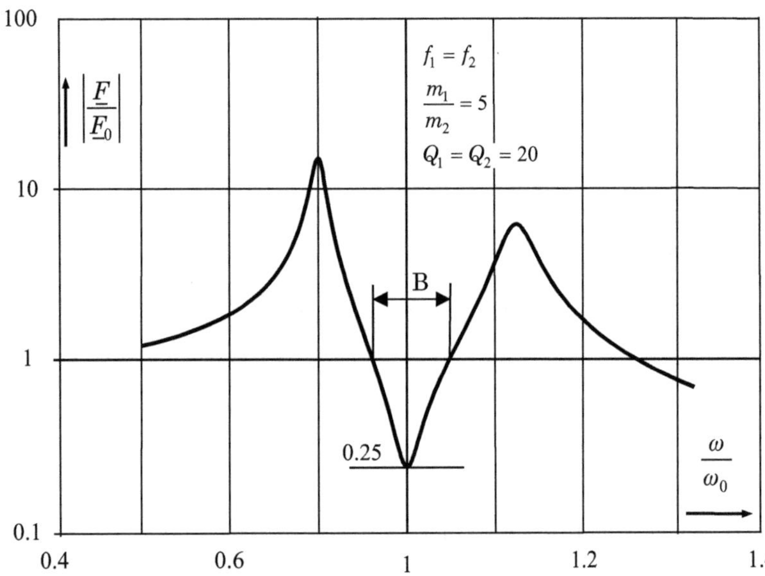

Fig. 3.30. Frequency response of ground force

If the deviation of the operating frequency from the common resonant frequency amounts to more than about 8% (with assumed mass ratio $m_1/m_2 = 5$ and quality factors $Q_1 = Q_2 = 20$), the advantage of the damping effect will be lost. For positive or negative deviations of 20% even new resonance areas are generated. With decreasing ratio m_1/m_2, the frequency area B in Fig. 3.30 increases, in which the operating frequency is allowed to deviate from the common resonant frequency. However, the complexity of the vibration absorber is much higher.
The new resonance areas which can be identified during powering the frequency up to operating frequency, will represent only a problem, if the change

of frequency (speed frequency) of a powering-up or powering-down drive system happens fast enough. The residence time in the critical resonance area must be sufficiently short in order to avoid high oscillation amplitudes.
The system must be dimensioned completely different, if it is to be utilized for the support of a fast decay after **impulsive or sudden loads**. The reduction of motion of a balance plate, on which always approximately same masses are weighed (e.g. in production control of food), represents a typical sample application. Compared to the direct damping of the simple spring-mass system, the application of the vibration absorber enables constructional advantages. For the additional resonator a quality factor of $Q_2^* = 2$ is advantageous. The decay time is longer for lower or higher quality factors. Only mass ratios of 10 or more can be mostly chosen for this application. If the mass ratio amounts to $m_1/m_2 = 10$ and the quality factor of the main oscillator amounts to $Q_1 = 20$, an equivalent quality factor of 5.3 of the total system will be achieved with respect to the additional resonator's quality factor of $Q_2^* = 2$ and thus the decay time will decrease considerably.
The same principle is also used successfully in systems comprising an electromechanical transducer in order to allow for damping the mechanical system electrically. The series connection of a suitably chosen resistor and a suitably chosen inductor in the electrical circuit of a piezoelectrically driven and composed longitudinal oscillator which is to work in pulsed mode, represents an example for that.

3.2 Mechanical Networks for Rotational Motions

For the reference points of mechanical networks concerning translational motions discussed in Sect. 3.1 only motions parallel to a straight line have been allowed. In the following section, only rotations around a fixed axis are allowed. The bearings of this axis are rigidly connected with the surrounding system. The surrounding system (represented by means of a „rigid frame" in the preceding section) is allowed to perform a uniform translational motion at most. Therefore, a surrounding rigid frame is provided with characteristics of an infinitely large mass and an infinitely large rotating mass Θ (see Sect. 3.2.2). With respect to this assumption, a physical structure is available again which is isomorphic in a mathematical sense to the systems discussed before. By means of a suitable coordinate choice, also the topology of the system is consistent with the model representation.
In practice, systems with rotational motions often interact with translational systems by means of transducer elements, e.g. rods connected with axis (see Sect. 5.1). But also torsional oscillation problems, e.g. generated by drive engines and the like, can be addressed by following below-mentioned model approaches. Figure 3.31 shows such a rotational system.
By means of two rods, forces affect an ideal shaft (massless and rigid) in such a way that a torsional moment is induced into the shaft. The torsional

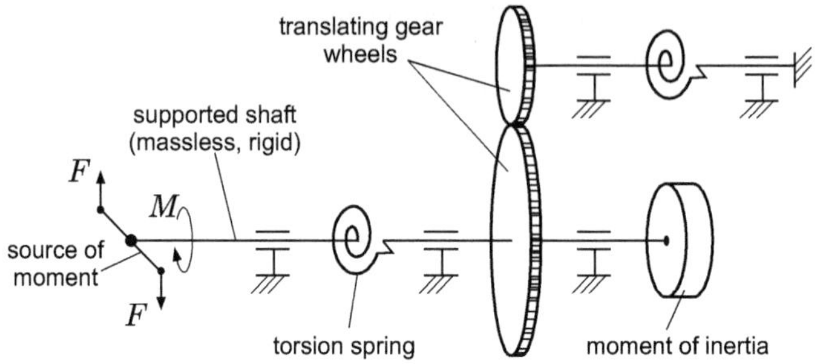

Fig. 3.31. Rotational system

moment results in a torsional twisting of a torsion spring which rests on a rotating mass. In addition, a torsional moment is transmitted to a further torsion spring by means of a gearing. Just like the shaft also the gearing is considered to be massless and rigid. The right connection of this spring is rigidly supported. The torsion angle of the torsion spring is proportional to the torsional moment. The proportionality constant is called *rotational compliance* n_R. The index R will be often omitted, if it concerns a purely rotational system (without transducer elements) and thus it is impossible to mistake it for a translational compliance.

3.2.1 Coordinates

For the description of a uniaxial rotational system it will suffice to use angle coordinates φ_i (cylindrical coordinate system) and torsional moment vectors $\mathbf{M}_i$ or, alternatively, the coordinate torsional moment M in combination with a directional arrow (often drawn as double-headed arrow [43]) and the right-hand rule. However, in order to achieve a system which provides isomorphism to electrical networks while maintaining the topology, it is necessary to apply the torsional moment M and the angular velocity difference

$$\Omega = \frac{\mathrm{d}\varphi_1}{\mathrm{d}t} - \frac{\mathrm{d}\varphi_2}{\mathrm{d}t}$$

(angular velocity across the component) as pair of coordinates. As in case of electrical and mechanical-translational systems, the product of coordinates results in a power quantity. Figure 3.32 shows the complex network coordinates torsional moment $\underline{M}$ and angular velocity $\underline{\Omega}$ using the example of a rotational component. As in case of translational systems, a *directional system arrow* is introduced indicating the positive direction and in order to achieve a uniqueness of signs for a technical circuit representation. The angular velocity $\underline{\Omega}$ will be positive, if both the $\underline{\Omega}$-network arrow and the directional system arrow point to the same direction. A torsional moment along the direction

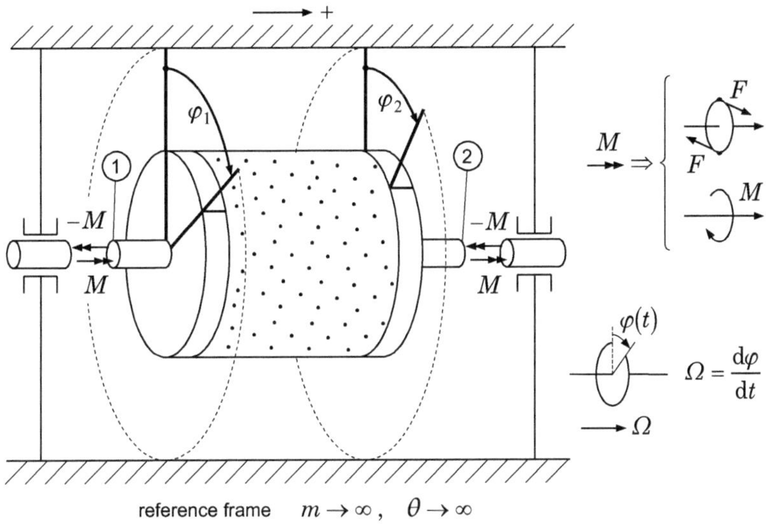

Fig. 3.32. Mechanical coordinates at a general rotational component

of the torsional moment arrow is effective at that side of section, the directional system arrow points to (reference point 1, positive component surface, direction of rotation according to right-hand rule).

3.2.2 Components and System Equations

According to the type of coupling of angular velocity and torsional moment, three different kinds of components can be distinguished within the scope of linear processes. It is referred to the components as torsion spring, torsional friction and rotating mass (moment of inertia). Table 3.4 shows descriptive and schematic representations of rotational components and their system equations. In order to achieve a technical circuit representation, the velocity differences across the elements must be introduced and the rotating mass must get a second reference point. The transition to a technical circuit representation is performed in Table 3.5. In addition to the three types of components, the gearing (analogously to the lever) is specified as „transducer" in the two possible gear wheel configurations.
Analogously to the consideration of translational systems, two source mechanisms are introduced. With the first source mechanism, an angle change is forced by the source (angle source). With the second source mechanism, the source generates a torsional moment. Its value does not depend on the occurring angle change. Table 3.6 shows the mechanical and technical circuit representation of these two source types. Active rotational one-ports (source with source impedance) can be represented by means of these sources. In Sect.

Table 3.4. Rotational components - descriptive and schematic representation

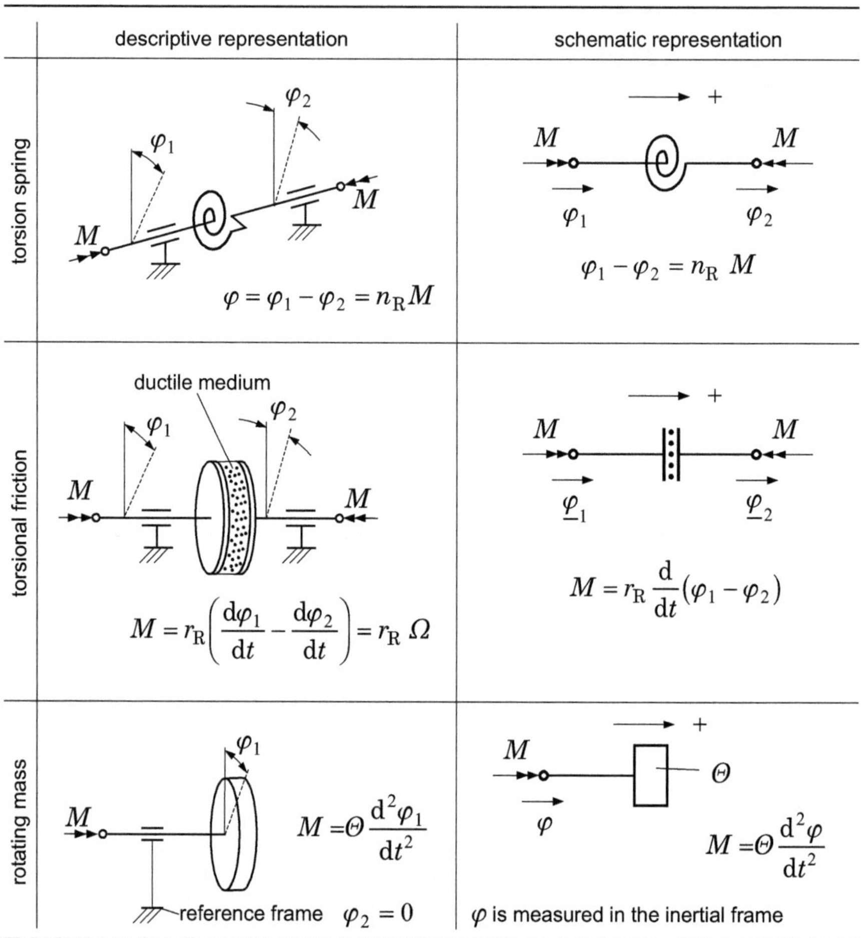

3.2.4, it is exemplified that KIRCHHOFF's laws can be formed analogously to translational systems. The interconnection rules of Sect. 3.1.4 can also be applied formally and conceptually to rotational networks.

3.2.3 Isomorphism between Mechanical and Electrical Circuits

By comparing the equations describing completely rotational networks with appropriate equations of electrical networks, the isomorphism of both physical structures existing in a mathematical sense can be distinguished. The identity will also relate to the topological structure, if the electric voltage is assigned to the angular velocity and the electric current is assigned to the torsional

Table 3.5. Technical circuit representation of rotational components

schematic representation	technical circuit representation
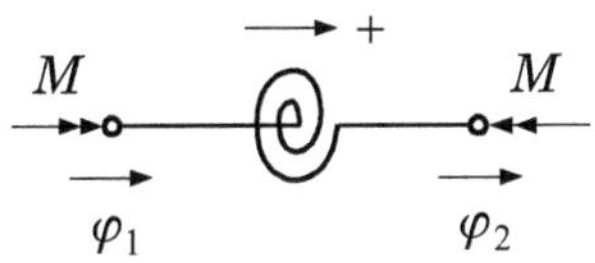 $\varphi_1 - \varphi_2 = n_\mathrm{R}\ M$	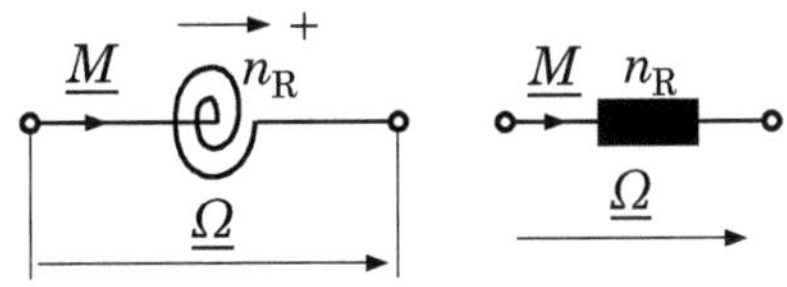 $\underline{\Omega} = \mathrm{j}\omega\ n_\mathrm{R}\ \underline{M}$
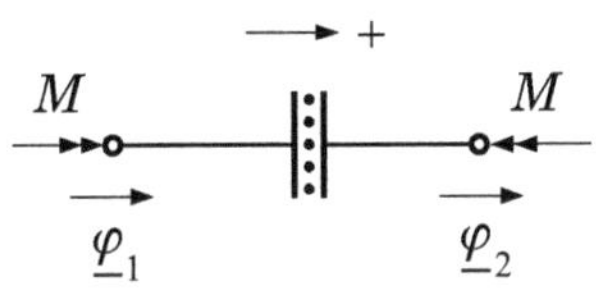 $M = r_\mathrm{R}\dfrac{\mathrm{d}}{\mathrm{d}t}(\varphi_1 - \varphi_2)$	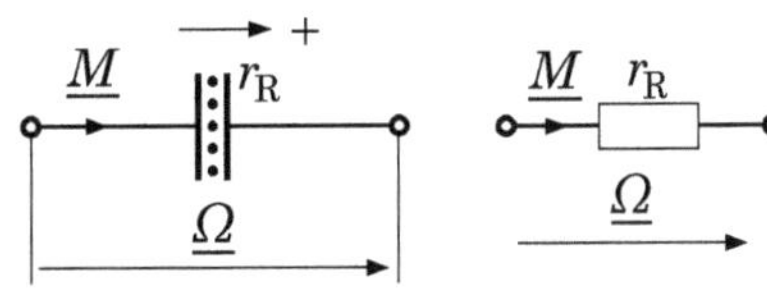 $\underline{M} = r_\mathrm{R}\ \underline{\Omega}$ $\underline{M} = \mathrm{j}\omega\,(\varphi_1 - \varphi_2)\cdot r_\mathrm{R}$
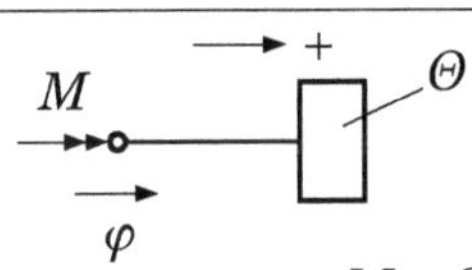 $M = \Theta\dfrac{\mathrm{d}^2\varphi}{\mathrm{d}t^2}$ φ is measured in an inertial frame	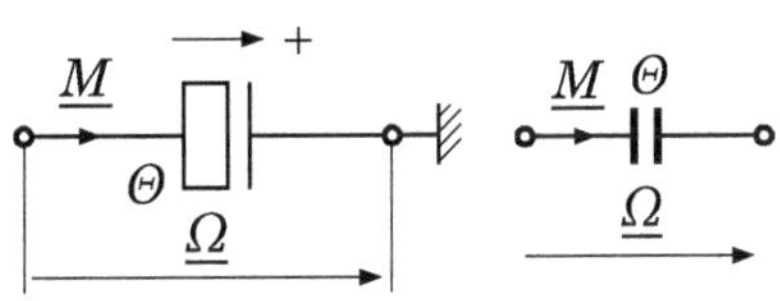 $\underline{M} = \mathrm{j}\omega\ \Theta\ \underline{\Omega}$
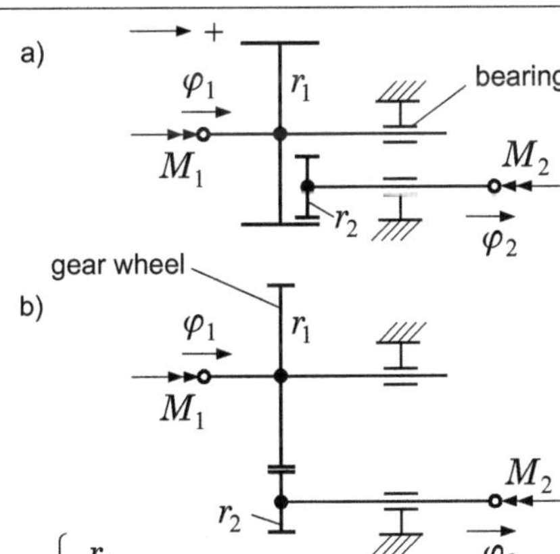 $k = \begin{cases} \dfrac{r_2}{r_1} & \text{for a)} \\ -\dfrac{r_2}{r_1} & \text{for b)} \end{cases}$ $\varphi_1 = k\ \varphi_2$ $M_1 = \dfrac{1}{k} M_2$	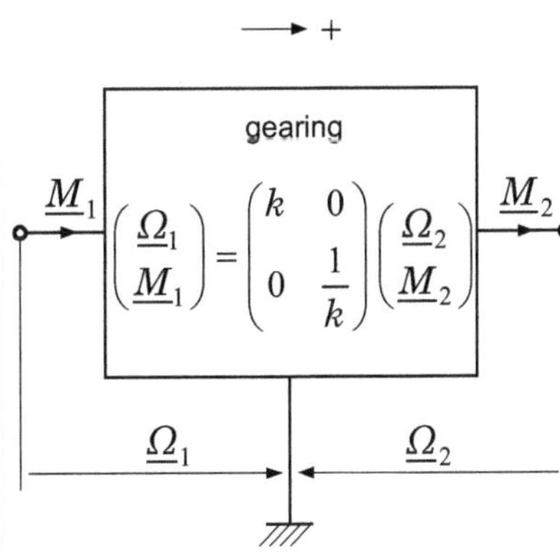 $\begin{pmatrix}\underline{\Omega}_1 \\ \underline{M}_1\end{pmatrix} = \begin{pmatrix} k & 0 \\ 0 & \dfrac{1}{k}\end{pmatrix}\begin{pmatrix}\underline{\Omega}_2 \\ \underline{M}_2\end{pmatrix}$

Table 3.6. Rotational sources

	mechanical representation	technical circuit representation
angle source	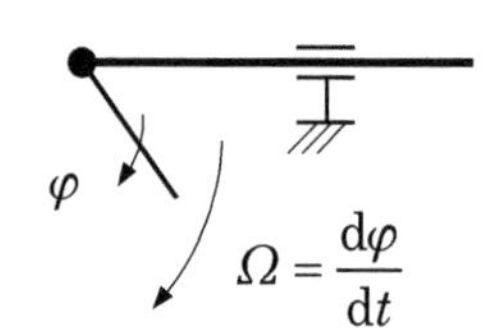	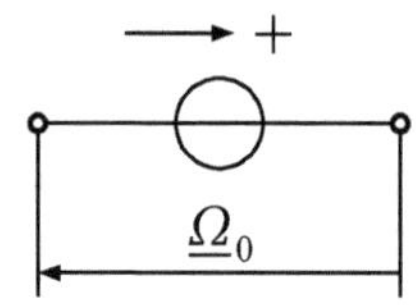
moment source	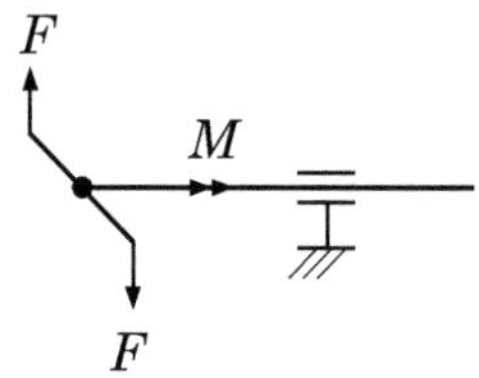	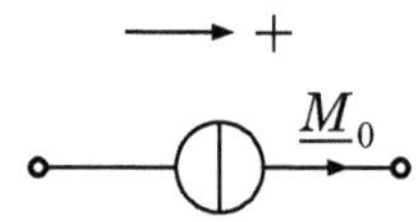

moment. The appropriate relations between electrical and rotational system are summarized in Table 3.7.
In order to provide also quantitative relations, proportionality quantities must be defined between the coordinates. It is to be valid:

$$\underline{u} = G_3 \underline{\Omega} \qquad \text{and} \qquad \underline{i} = \frac{1}{G_4}\underline{M} \tag{3.18}$$

In addition, it could be allowed that the frequencies between the electrical and the mechanical network differ in a constant factor. In practice, no use is made of this possibility. Therefore, it is not used in the following. By means of this definition, it can be written for the components:

$$C = \frac{\Theta}{G_3 G_4}, \qquad L = G_3 G_4 n_{\mathrm{R}} \qquad \text{and} \qquad R = G_3 G_4 h_{\mathrm{R}} \tag{3.19}$$

At first, the porportionality factors G_3 and G_4 are arbitrary with respect to their absolute values. However, with regard to the characteristics of network analysis programs (e.g. pSpice), the sequences of numbers should be maintained. Thus, only the powers of ten are available. If same numerical values are used for G_3 and G_4, a representation with equal power rating will be generated. It is usually made use of this advantage.

Table 3.7. Isomorphic correlations between electrical and rotational network

correlations between coordinates and components				
voltage	$\underline{u}$	o——o	$\underline{\Omega}$	angular velocity
current	$\underline{i}$	o——o	$\underline{M}$	torsional moment
inductance	L	o——o	n_R	rotational compliance
capacitance	C	o——o	Θ	rotating mass (moment of inertia)
resistance	R	o——o	h_R	torsional friction admittance
transformer	$\frac{w_1}{w_2}$	o——o	k	see figure gearing

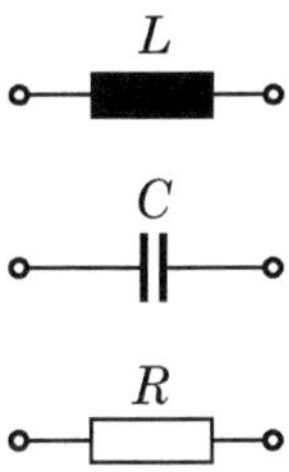

$\underline{u} = \mathrm{j}\omega L \underline{i}$

$\underline{u} = \frac{1}{\mathrm{j}\omega C}\underline{i}$

$\underline{u} = R\underline{i}$

$\underline{\Omega} = \mathrm{j}\omega n_\mathrm{R} \underline{M}$

$\underline{\Omega} = \frac{1}{\mathrm{j}\omega \Theta}\underline{M}$

$\underline{\Omega} = h_\mathrm{R} \underline{M}$

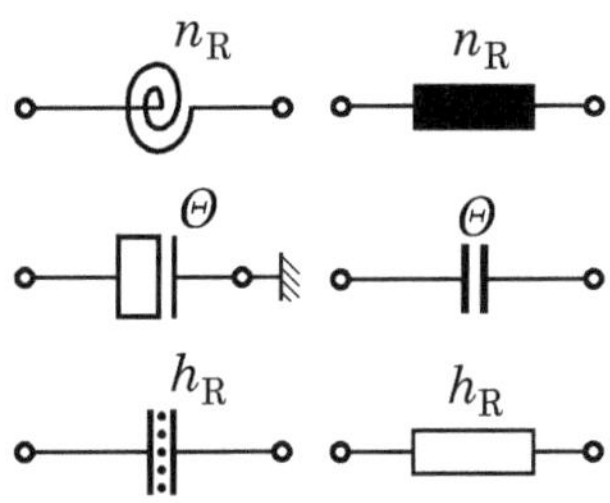

node of electrical circuit structure	$\sum_{*} \underline{i}_\nu = 0$	$\sum_{*} \underline{M}_\nu = 0$	node of mechanical scheme
loop of electrical circuit structure	$\sum_{\circlearrowright} \underline{u}_\nu = 0$	$\sum_{\circlearrowright} \underline{\Omega}_\nu = 0$	loop of mechanical scheme

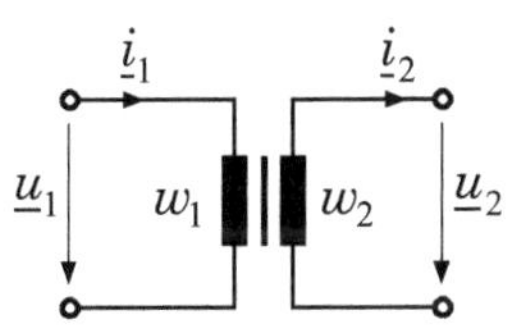

$$\underline{u}_2 = \frac{w_2}{w_1}\underline{u}_1$$

$$\underline{i}_2 = \frac{w_1}{w_2}\underline{i}_1$$

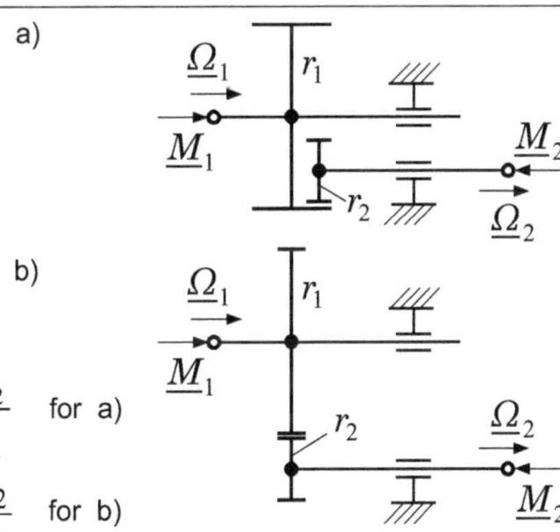

$$k = \begin{cases} \frac{r_2}{r_1} & \text{for a)} \\ -\frac{r_2}{r_1} & \text{for b)} \end{cases}$$

$$\underline{\Omega}_2 = \frac{1}{k}\underline{\Omega}_1 , \quad \underline{M}_2 = k\,\underline{M}_1$$

3.2.4 Sample Application for a Rotational Network

The deduction of a technical circuit representation from a real rotational configuration should happen again in two steps. While maintaining the geometrical configuration, at first it is appropriate to sketch a representation in which the branching of the torsional moment is distinguishable. One part of the drive torque applied to an axis is absorbed by bearing friction, another part is branched off for acceleration of rotating masses. In a schematic representation this should be demonstrated by means of a moment node. The branch point can be considered to be a gearing fitted on the axis with a gear ratio of one (Fig. 3.33). Then the new output axis supplies the rotational friction or the rotating mass. By means of this separation of the moment flows into several imaginary parallel axis, the separation into network legs is easy to imagine.

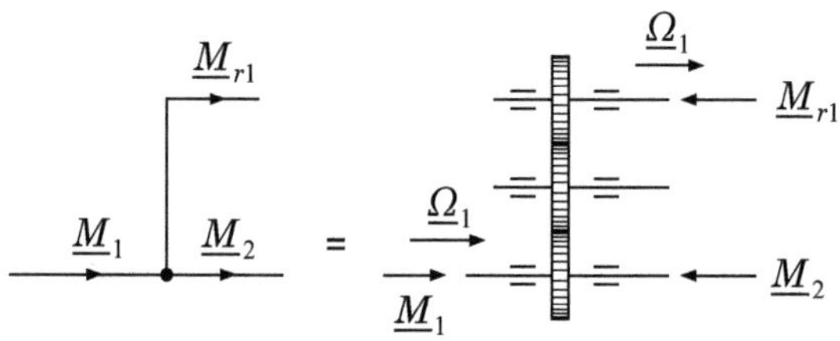

Fig. 3.33. Model of a torsional moment branching (node)

Using the example of a gas turbine, in Fig. 3.34 a system consisting of a source of moment, frictional elements, flexible clutch element and a vibration absorber is represented. A rotational load can be connected to the system output. The moment branchings into the frictional elements and rotating masses specified before are shown in the mechanical scheme. A torsion spring represents the flexible coupling. By means of this spring, the torsional moment is transmitted to the output side. A difference of angular velocity is generated across the clutch. But the flexible clutch comprises also parts with mass. They are represented by the moments of inertia Θ_2 and Θ_3. The vibration absorber consists of a ring with the moment of inertia Θ_4 which is connected to the main axis by means of 4 flexible springs. As a whole the flexible springs represent the torsion spring n_{R2}. The mechanical scheme comprises only rotational components. Thus, a mix-up with translational components is excluded. In this case, the index R can be omitted. Here, a connected load is indicated as a general component with resistance symbol r_L, since the character of this load is unknown.

The mechanical circuit represents the third representation stage. The references to direction and position can be canceled. The representation rules for networks are valid. Now the representation as active source one-port with connected general load is completed.

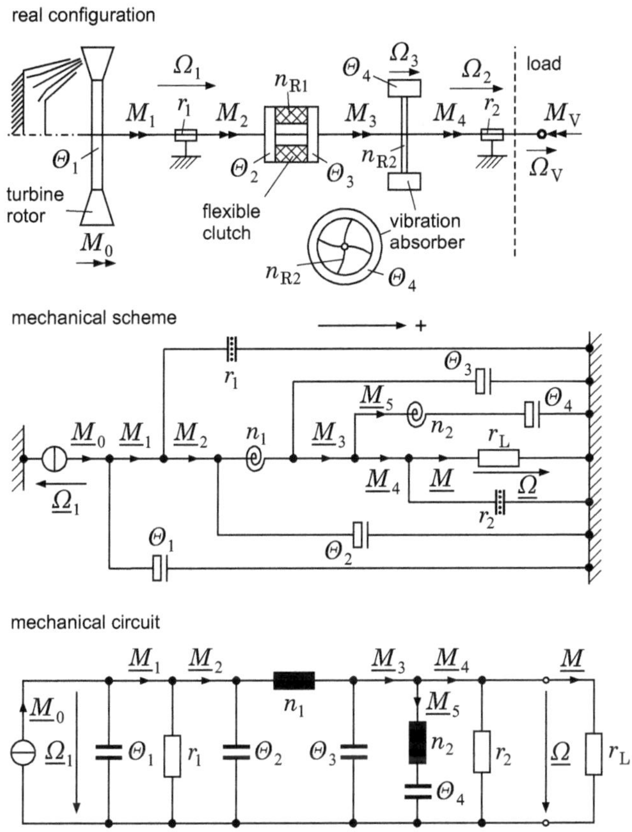

Fig. 3.34. Model of a rotational system in three representation stages

3.3 Acoustic Networks

If several gas-filled volumetric cavities are interconnected by canals, tubes or simply only by holes, then this structure can be regarded as an acoustic network on the assumption of some restrictions. Also a single cavity provided with a hole or a short tube section represents a frequently occurring acoustic resonator (HELMHOLTZ resonator) which can be represented as acoustic network.

Volumetric cavities and *canal-like cavities* represent the components of the acoustic network. The modeling calls for the embedding of the system consisting of cavities into an environment of same medium in which a constant pressure p_0 is existent. Its value equals the average pressure in the cavities. Furthermore, the linear dimensions of these cavities must be much smaller than the wavelength in the compressible medium the cavities are filled with. The latter can be always fulfilled by choice of a sufficiently low upper frequency

limit. Whether the technical question can be still reasonably answered or not must be analyzed in individual cases.
Real configurations will be suitable for a modeling by means of an acoustic network, if the structure of the configuration allows for a discretization in such a way that almost no compression but only motion is existent in canal-like elements and only a compression but almost no motion arises in volumetric elements. Thus, the same medium quantity streams out simultaneously from ideal canal-like components which streams in the component. Due to the missing motion, in volumetric components the pressure is position-independent. As a consequence of these restrictions outlined above, the upper frequency limit of a dynamic process to be considered must be chosen so low that the acoustic wavelength in the compressible medium is large compared to the linear dimensions of the components. In Sect. 6.2 it is shown how to handle canal-like cavities in order to provide a modeling for higher frequencies.
In order to be able to neglect the effect of nonlinearities of the general differential equations of gas dynamics, general approximations in the field of *linear acoustics* are used. For that, it is assumed that the medium elements move around a rest position with sufficiently low medium velocity and that the pressure change remains sufficiently small compared to the average pressure.
For mechanical networks, the reference points were defined at the connecting points of the mechanical components. The connection was realized by means of massless and rigid rods. Here, *virtual canal elements* are defined analogously which are filled with a massless and incompressible medium. They are to provide the connection of defined acoustic components. No pressure difference is generated between the ends of these connecting elements and no density changes arise within these elements. The streaming in volume flows out of the end instantaneously. Any local discretization of the real configuration will succeed by means of the theoretical introduction of these ideal canal elements.
The modeling by means of acoustic networks proved to be useful for microphones, headphones, bass reflex boxes, resonance sound absorbers, hydraulic coupling systems, vibration dampers with fluids, resonance systems for space acoustic corrections and similar configurations.

3.3.1 Coordinates

Figure 3.35 shows a detail of a canal-like cavity at two different points of time. It can be realized that the marked medium elements moved on a bit further during time difference Δt, thus a volume displacement ΔV occurred. The medium elements can have different velocities. The calculation shown in Fig. 3.35 exemplifies that the volume displacement ΔV related to the time difference Δt equals the product of cross-sectional area A and average velocity of the medium elements. This product is called *volumetric flow rate* q and represents a coordinate of acoustic networks. The second coordinate is the *pressure difference* p across the canal-like component. The reference points are characterized by ⓐ and ⓑ. By analogy with the coordinate force, the volumetric flow rate q is charted as arrow pointing to the element into the connecting element. The product of coordinates represents a power quantity again.

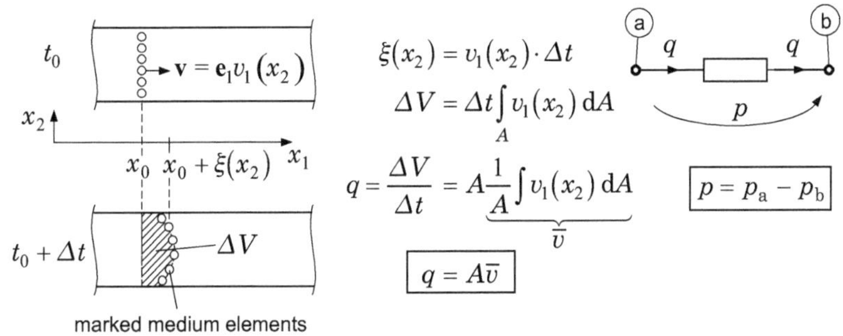

Fig. 3.35. Coordinates volumetric flow rate q and pressure difference p in the acoustic network

A compression is generated in volumetric components by means of the streaming in volumetric flow rate. The pressure change is effective toward the constant ambient pressure. Thus, volumetric and canal-like cavities can be represented as general component of similar type.

3.3.2 Acoustic Components

Volumetric cavities show a spring character. That can be easily understood, if a closed gas volume is compressed by means of a piston. If this compression is performed so fast that no heat exchange with the container wall is possible ($\Delta Q = 0$), an *adiabatic change of state* will exist. If the volume changes are very small, a linear relation between volume change and pressure change can be assumed. On the assumption of a constant mass m in the volume V evaluated at the point (p_0, V_0, T_0), the thermal equation of state of the ideal gas

$$pV = mRT \tag{3.20}$$

with *gas constant* R and absolute temperature T can be written in the following way:

$$p_0 \Delta V + V_0 \Delta p = mR\Delta T \tag{3.21}$$

On the assumption of $\Delta Q = 0$ at the evaluation point,

$$\Delta Q = mc_V \Delta T + p_0 \Delta V = 0 \tag{3.22}$$

results from the *first law of thermodynamics* with c_V denoting the *specific heat capacity* at constant volume. The insertion of (3.22) into (3.21) results in

$$-\frac{\Delta V}{\Delta p} = \frac{V_0}{p_0 \left(1 + \dfrac{R}{c_V}\right)}. \tag{3.23}$$

In combination with the *adiabatic exponent*

$$\kappa = 1 + \frac{R}{c_V} = \frac{c_p}{c_V} \tag{3.24}$$

(for air, the adiabatic exponent amounts to $\kappa = 1.4$) and the definition of the *acoustic compliance*

$$N_\mathrm{a} = -\frac{\Delta V}{\Delta p}, \tag{3.25}$$

it can be written:

$$N_\mathrm{a} = \frac{V_0}{\kappa p_0} \quad \text{(for adiabatic changes of state)} \tag{3.26}$$

The transition to complex amplitudes is completed in Table 3.8 b). The representation as acoustic compliance is shown in part c) of same Table.
The acoustic compliance is defined by means of *complex coordinates* of the acoustic network and by means of graphical network representation of the component. Concerning their component behavior, *canal-like components* show two limiting cases. In the first limiting case, the moved gas acts solely as mass. In the second limiting case, the friction prevails. The first case is approximately existent for very high frequencies in a canal only filled with gas. In order to provide the case of predominant friction, for low frequencies the canal can be considered to be filled with a porous medium additionally.
Figure 3.36 shows components with dominant acoustic friction $Z_{\mathrm{a},r}$. Figure 3.37 shows the conditions for an ideal acoustic mass M_a. *Real acoustic canal elements* show a behavior comprising both characteristics. A representation is possible by series connection of both elements. On the basis of components with simple geometries, in Sect. 3.3.4 the frequency-dependent characteristics of real elements are discussed.

3.3.3 Network Representation of Acoustic Systems

If acoustic elements are connected with tube elements (imaginary connecting tubes filled with a massless and incompressible substance), then reference points can be defined again at the connecting points. Figure 3.38 shows a

Table 3.8. Volumetric cavity, acoustic spring

	graphical representation	adiabatic	isothermal
a)	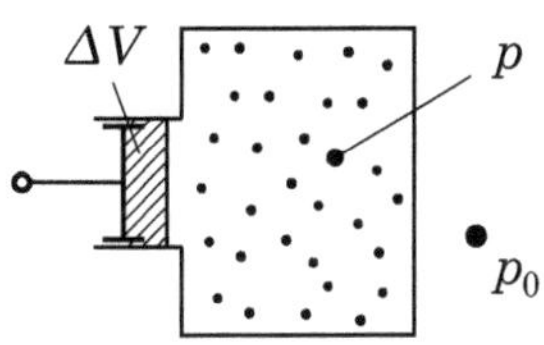	$\Delta V = V - V_0$ $\Delta p = p - p_0$ $-\frac{\Delta V}{\Delta p} = \frac{1}{\kappa} \cdot \frac{V_0}{p_0}$ with $\kappa = \frac{c_p}{c_V}$	$\Delta V = V - V_0$ $\Delta p = p - p_0$ $-\frac{\Delta V}{\Delta p} = \frac{V_0}{p_0}$
b)	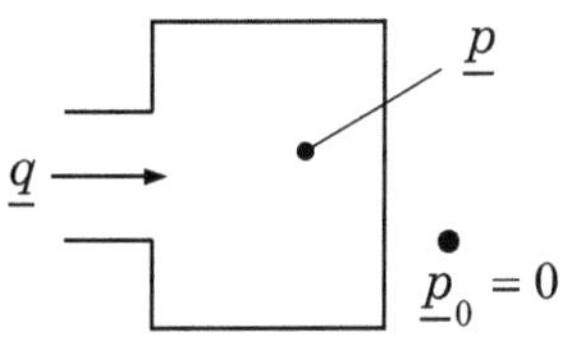	$-\Delta V \rightarrow \underline{V};\quad \Delta p \rightarrow \underline{p}$ $\frac{\underline{V}}{\underline{p}} = \frac{1}{\kappa} \cdot \frac{V_0}{p_0} = N_\mathrm{a}$	$-\Delta V \rightarrow \underline{V};\quad \Delta p \rightarrow \underline{p}$ $\frac{\underline{V}}{\underline{p}} = \frac{V_0}{p_0} = N_\mathrm{a,iso}$
c)	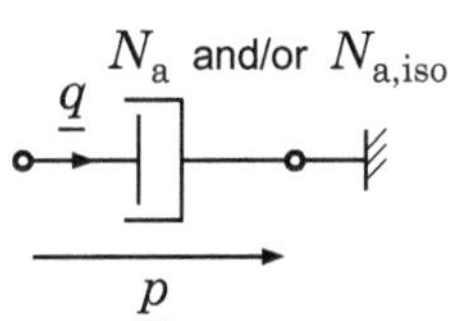	due to $\underline{q} = \mathrm{j}\omega\underline{V}$: $\frac{\underline{q}}{\underline{p}} = \mathrm{j}\omega N_\mathrm{a}$ with $N_\mathrm{a} = \frac{1}{\kappa} \cdot \frac{V_0}{p_0}$	due to $\underline{q} = \mathrm{j}\omega\underline{V}$: $\frac{\underline{q}}{\underline{p}} = \mathrm{j}\omega N_\mathrm{a,iso}$ with $N_\mathrm{a,iso} = \frac{V_0}{p_0}$

possible network structure consisting of three volumetric cavities (acoustic compliances) and four canal-like cavities. The cavity being located in the leg with q_1 may have a mass-like character. The other three canal-like components may solely represent friction components. The source of flow represented in the left figure part generates the volumetric flow rate $q_0 = v_0 A$.
The more abstract acoustic scheme can be deduced from this still very realistic represented structure (Fig. 3.39). This representation provides a basis for the acoustic circuit (Fig. 3.40) of the structure. In order to calculate the network, KIRCHHOFF's laws are required again.
For the flows flowing toward a node, KIRCHHOFF's nodal rule is valid:

$$\sum_\nu q_\nu = 0 \tag{3.27}$$

The sum of all pressures around the loop fulfills KIRCHHOFF's loop rule:

$$\sum_\nu p_\nu = 0 \tag{3.28}$$

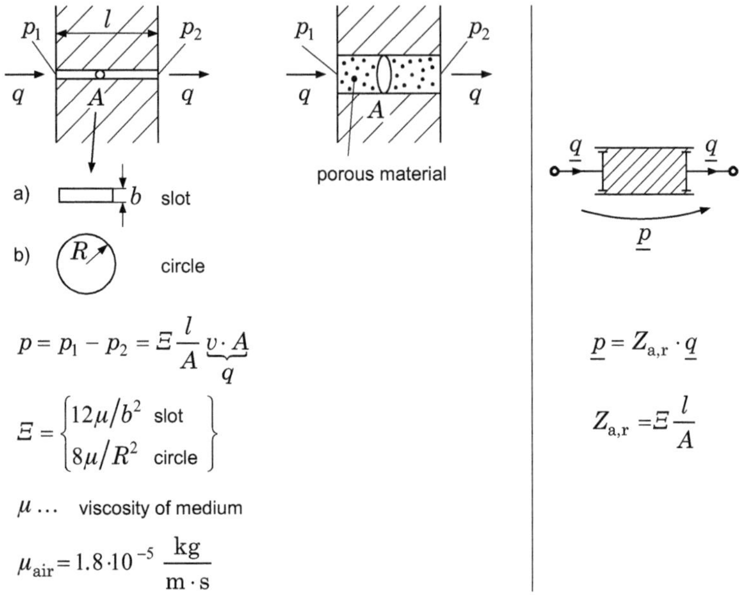

Fig. 3.36. Component *acoustic friction*

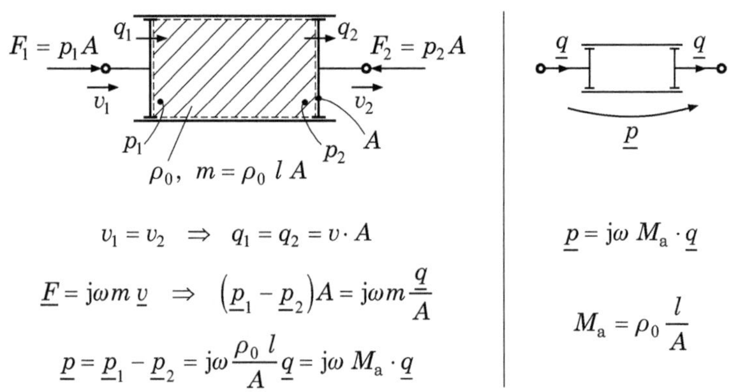

Fig. 3.37. Component *acoustic mass*

The characteristics of acoustic networks represented so far show again an isomorphism to electrical networks. The acoustic coordinates specified in Sect. 3.2.2 result in an analogy of topology. Compared to mechanical networks, the flow coordinate q represents a coordinate of motion.
Also acoustic networks possess a transformation mechanism. It consists of two piston transducers connected in series with different area and rigid connection between the pistons. Figure 3.41 illustrates the structure.

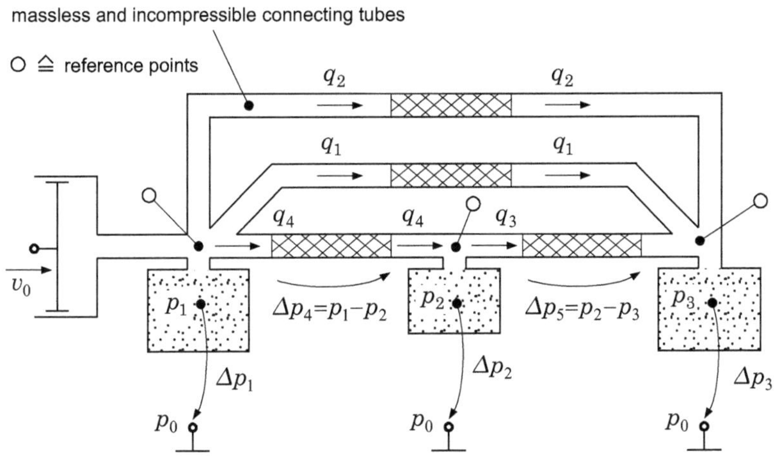

Fig. 3.38. Volumetric and canal-like cavities (connected with tube elements)

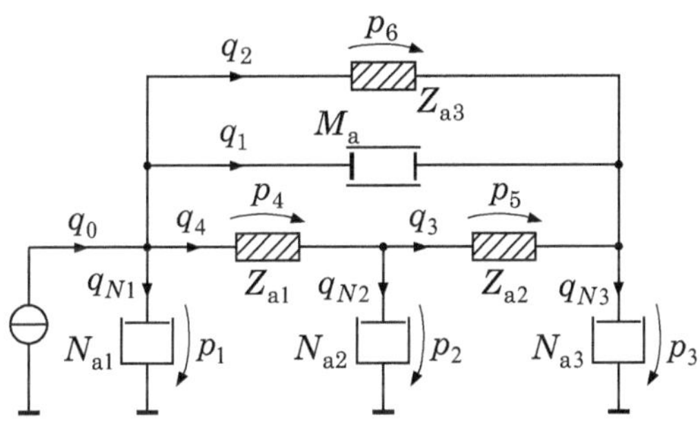

Fig. 3.39. Acoustic scheme

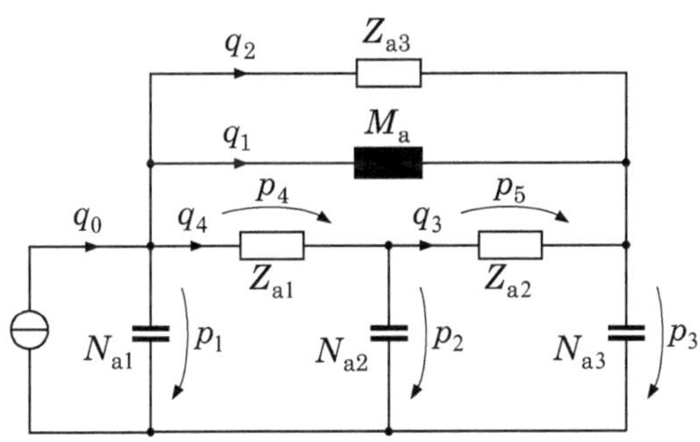

Fig. 3.40. Acoustic circuit

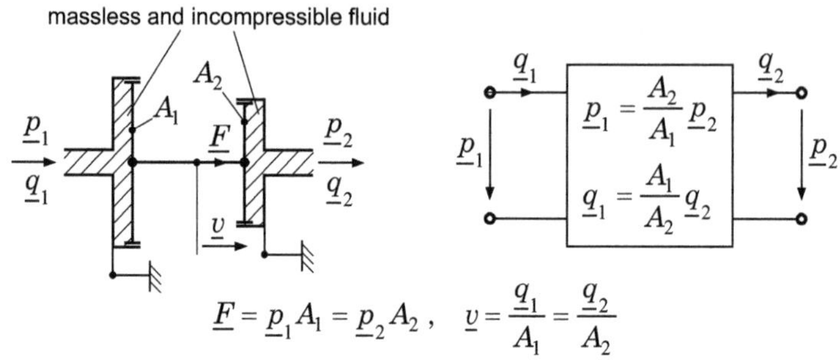

Fig. 3.41. Acoustic transformer

3.3.4 Real Acoustic Components

The components discussed in Sect. 3.3.2 are based on the assumption of ideal limiting cases. Possible energy losses were neglected. With these assumptions, in the limited frequency range a sufficiently exact representation of reality can be really achieved for many acoustic structures. In case of too large deviations from the ideal models, the limiting cases are not suitable for the basic principle of calculation any longer. Too large deviations arise between model approach and reality.
The thermal conductivity and viscous friction of the fluid result in irreversible energy transformations resulting in the fact that the components do not have a pure reactance behavior. Therefore, the representation of components must be supplemented with dissipative elements. It can be shown that a series connection consisting of reactive and dissipative elements results in a particularly good representation of reality. Reason for that is the characteristic of acoustic structures. For decreasing frequencies, the acoustic frictional impedance connected in parallel approaches a frequency-independent limit $Z_{a,0}$ [44].
For simple geometrical forms such as slot, circular cylinder and ball, complete analytical solutions for real acoustic elements comprised by a series connection of storage and loss elements are available [44]. However, the solutions comprise cylindrical functions which make practical work concerning structure optimization even more difficult. In order to keep the complexity for representation of these non-idealizable but simply formed structures smaller, the results of well-manageable approximate representations are compiled in the following.

Real Acoustic Volume Components

Table 3.9 shows the relations for representation of real acoustic compliances for the two configurations slot-like volume (smallest dimension of area A is much larger than gap height d) and spherical volume.

Table 3.9. Real acoustic volume components

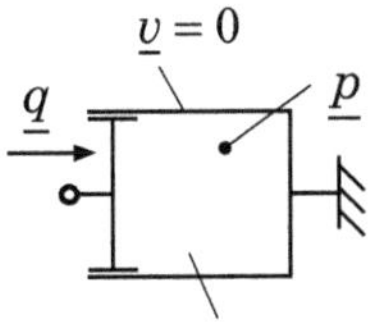

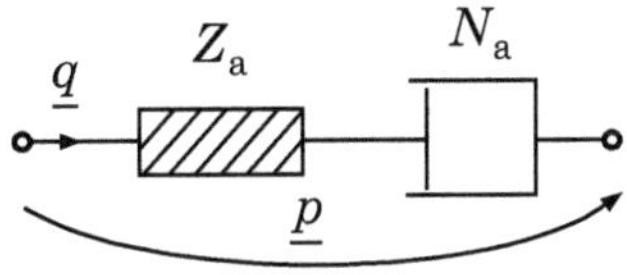

$$\underline{Z}_{\mathrm{a}} = \frac{1}{\mathrm{j}\omega\, N_{\mathrm{a}}} + Z_{\mathrm{a}}$$

	density ρ_0	heat conductivity λ	specific heat capacity c_p	adiabatic exponent $\kappa = c_p / c_V$
air	$1.3 \frac{\mathrm{kg}}{\mathrm{m}^3}$	$0.026 \frac{\mathrm{W}}{\mathrm{mK}}$	$1007 \frac{\mathrm{J}}{\mathrm{kg \cdot K}}$	1.4

slot

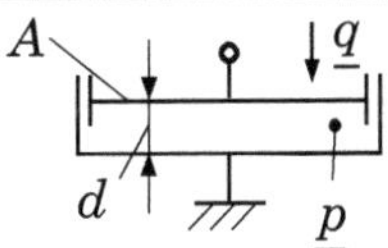

smallest dimension of $A \gg d$

$$V_0 = A d$$

$$N_{\mathrm{a0}} = \frac{V_0}{\kappa\, p_0}$$

$$Z_{\mathrm{a0}} = \frac{1}{12}\left(1 - \frac{1}{\kappa}\right)\frac{\rho_0\, c_p\, p_0}{\lambda}\frac{d}{A}$$

$$\omega_{\mathrm{g}} = \frac{1}{Z_{\mathrm{a0}}\, N_{\mathrm{a0}}} = 12 \frac{\lambda}{\rho_0\, c_p}\frac{1}{d^2}\frac{\kappa^2}{\kappa - 1}$$

low frequencies $\left(\omega < 0.1\,\omega_{\mathrm{g}}\right)$:

$$Z_{\mathrm{a}} = Z_{\mathrm{a0}} \quad , \quad N_{\mathrm{a}} = \kappa\, N_{\mathrm{a0}}$$

high frequencies $\left(\omega > 0.3\,\omega_{\mathrm{g}}\right)$:

$$\frac{Z_{\mathrm{a}}}{Z_0} = \frac{a}{\left(\frac{\omega}{\omega_{\mathrm{g}}}\right)^{\frac{3}{2}}}\left(1 + \frac{2a}{\sqrt{\frac{\omega}{\omega_{\mathrm{g}}}}}\right)$$

$$\frac{N_{\mathrm{a}}}{N_0} = 1 + \frac{a}{\sqrt{\frac{\omega}{\omega_{\mathrm{g}}}}}$$

$$a = \sqrt{\frac{(\kappa - 1)^3}{6\,\kappa^2}} = 0.0738 \quad (\text{air})$$

ball

$$V_0 = \frac{4}{3}\pi R^3$$

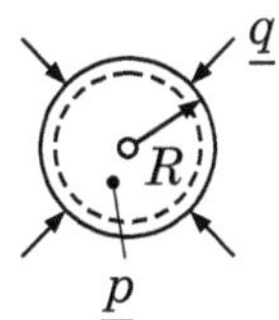

$$N_{\mathrm{a0}} = \frac{V_0}{\kappa\, p_0}$$

$$Z_{\mathrm{a0}} = \frac{1}{20\pi}\left(1 - \frac{1}{\kappa}\right)\frac{\rho_0\, c_p\, p_0}{\lambda}\frac{1}{R}$$

$$\omega_{\mathrm{g}} = \frac{1}{Z_{\mathrm{a0}}\, N_{\mathrm{a0}}} = 15 \frac{\lambda}{\rho_0\, c_p}\frac{1}{R^2}\frac{\kappa^2}{\kappa - 1}$$

low frequencies $\left(\omega < 0.1\omega_{\mathrm{g}}\right)$:

$$Z_{\mathrm{a}} = Z_{\mathrm{a0}} \quad , \quad N_{\mathrm{a}} = \kappa\, N_{\mathrm{a0}}$$

high frequencies $\left(\omega > 0.3\,\omega_{\mathrm{g}}\right)$:

$$\frac{Z_{\mathrm{a}}}{Z_0} = \frac{a}{\left(\frac{\omega}{\omega_{\mathrm{g}}}\right)^{\frac{3}{2}}}\left(1 + \frac{2a}{\sqrt{\frac{\omega}{\omega_{\mathrm{g}}}}}\right)$$

$$\frac{N_{\mathrm{a}}}{N_0} = 1 + \frac{a}{\sqrt{\frac{\omega}{\omega_{\mathrm{g}}}}}$$

$$a = \sqrt{\frac{3}{10}\frac{(\kappa - 1)^3}{\kappa^2}} = 0.0990 \quad (\text{air})$$

The walls of the volumes which are filled with ideal gas were assumed to represent a sufficiently large thermal storage system, so that the wall temperature remains constant independently of processes running in the volume. For low frequencies, the components Z_a and N_a are frequency-independent. For high frequencies, the compliance N_a remains constant on adiabatic level $N_a = N_{a,0}$, whereas the loss impedance Z_a further decreases.
Graphical representations of the functions are charted in Figs. 3.42 and 3.43. The stated approximation functions were achieved by intuitive approximation of theoretical solutions [45]. Maximum deviations of a few percent from original function arise. It is remarkable that the phase angle between volumetric flow rate $\underline{q}$ and pressure $\underline{p}$ arising with sinusoidal source quantity is considerably smaller than $\pi/2$ in the transition region between isothermal and adiabatic compression. Within this region losses arise. If the acoustic volume is a component part of an acoustic resonator and the technical task requires a quality factor as high as possible or small losses respectively, then the mentioned transient region should be avoided by a suitable dimensioning.

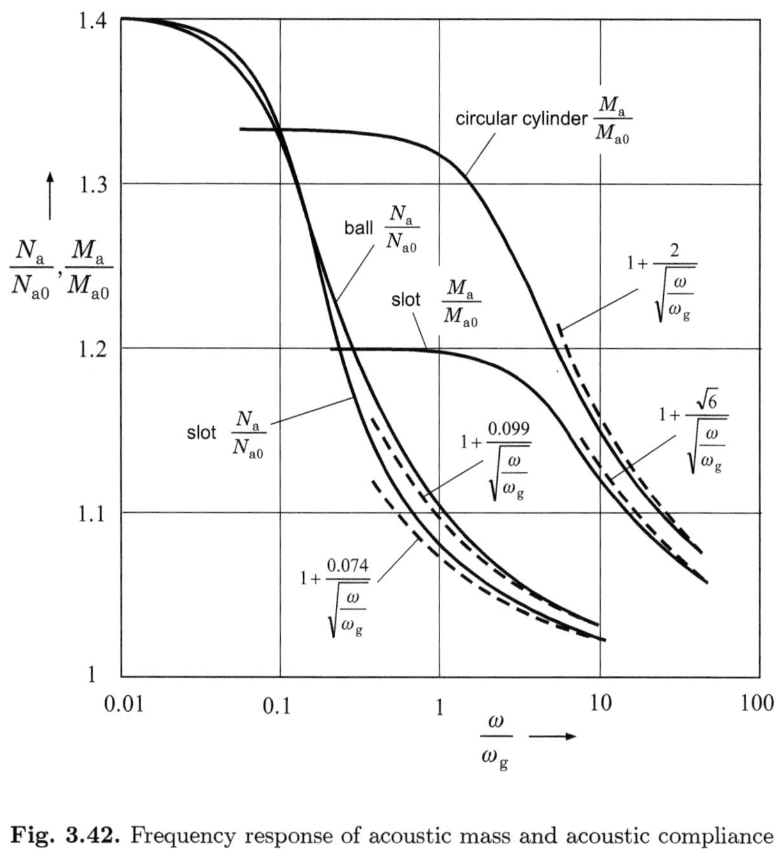

Fig. 3.42. Frequency response of acoustic mass and acoustic compliance

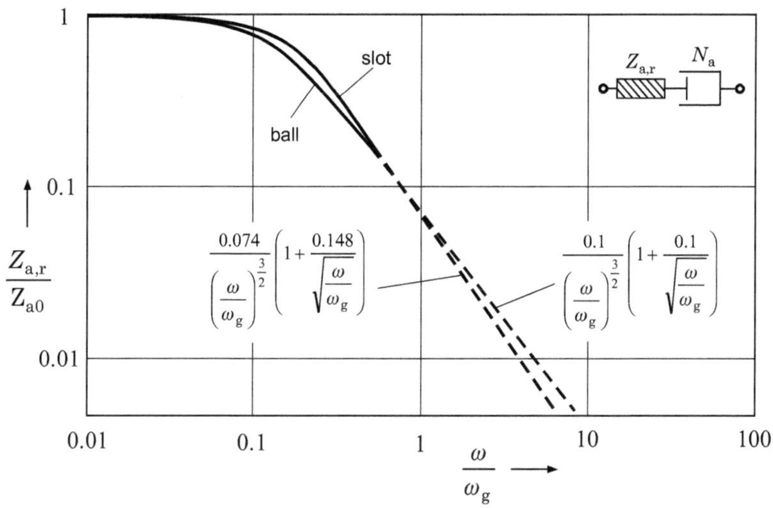

Fig. 3.43. Frequency response of frictional impedance according to Table 3.9 (volume component)

For real canal elements both mass effects and frictional effects are present. Also here, a formal representation is possible by means of a series connection of an ideal mass element and an ideal fluid friction.

Real Acoustic Canal Components

For a slot-like and a circular canal cross-section which are both filled with an ideal gas of known density ρ_0 and known viscosity μ (determined by assuming a stationary laminar flow), Table 3.10 shows the impedances of the components.

Also here, the components are frequency-independent for low frequencies. The frequency responses for higher frequencies specified in Table 3.10 are illustrated in Figs. 3.42 and 3.44.

The frequency-dependent behavior of component parameters is disturbing for a wideband simulation in network analysis programs, since the user must program a special component formation. In addition, not any network analysis program have the ability to integrate freely programmable component macros. Nevertheless, in order to provide for a wideband simulation in a simple way, a further approximation step can be applied which results in representation errors less than 10% for reactive components shown in Table 3.11 and up to approximately 50% for dissipative components. This approximation stage will suffice for first estimations. If it proves to be important as a result of rough calculation for certain frequencies to achieve more exact results, the

Table 3.10. Real acoustic canal components

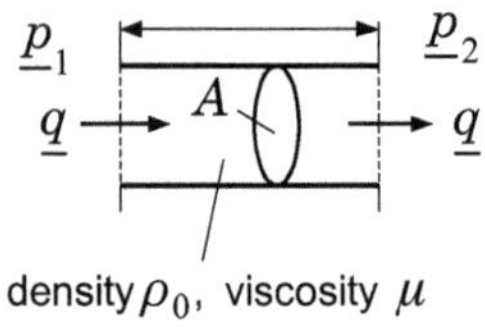

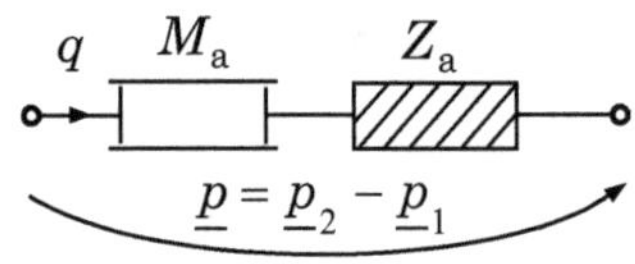

$$\underline{Z}_a = j\omega M_a + Z_a$$

slot-like cross-section	circular cross-section
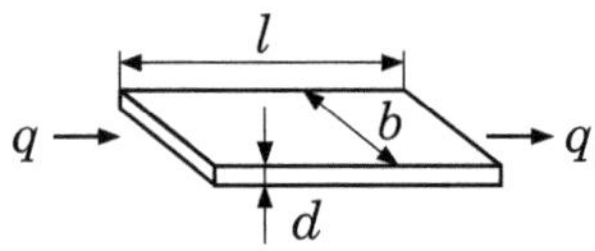 $A = b\,d$	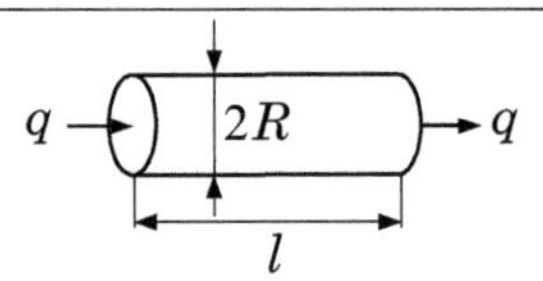 $A = \pi R^2$
$M_{a0} = \rho_0 \frac{l}{A}$, $Z_{a0} = \frac{12\mu}{d^2}\frac{l}{A}$	$M_{a0} = \rho_0 \frac{l}{A}$, $Z_{a0} = \frac{8\mu}{R^2}\frac{l}{A}$
$\omega_g = \frac{Z_{a0}}{M_{a0}} = \frac{12\mu}{\rho_0 d^2}$	$\omega_g = \frac{Z_{a0}}{M_{a0}} = \frac{8\mu}{\rho_0 R^2}$
low frequencies $(\omega < \omega_g)$:	low frequencies $(\omega < \omega_g)$:
$Z_a = Z_{a0}$, $M_a = \frac{6}{5} M_{a0}$	$Z_a = Z_{a0}$, $M_a = \frac{4}{3} M_{a0}$
high frequencies $(\omega > 3\,\omega_g)$:	high frequencies $(\omega > 3\,\omega_g)$:
$\frac{Z_a}{Z_{a0}} = \sqrt{\frac{\omega}{6\,\omega_g}}\left(1 + \frac{\sqrt{3/2}}{\sqrt{\frac{\omega}{\omega_g}}}\right)$	$\frac{Z_a}{Z_{a0}} = \sqrt{\frac{\omega}{4\,\omega_g}}\left(1 + \frac{1}{\sqrt{\frac{\omega}{\omega_g}}}\right)$
$\frac{M_a}{M_{a0}} = 1 + \frac{1}{\sqrt{\frac{\omega}{6\,\omega_g}}}$	$\frac{M_a}{M_{a0}} = 1 + \frac{1}{\sqrt{\frac{\omega}{4\,\omega_g}}}$

circuits according to Table 3.9 and Table 3.10 can be used for selected discrete frequencies. For special problems, it will be made also an effort for the complete solution. The method of approximation consists in defining a circuit consisting of several ideal acoustic components in such a way, that a correct representation is generated for low frequencies. The dimensioning of free parameters is effected by means of error minimization criteria. While accepting the errors mentioned above, these defined equivalent elements can be

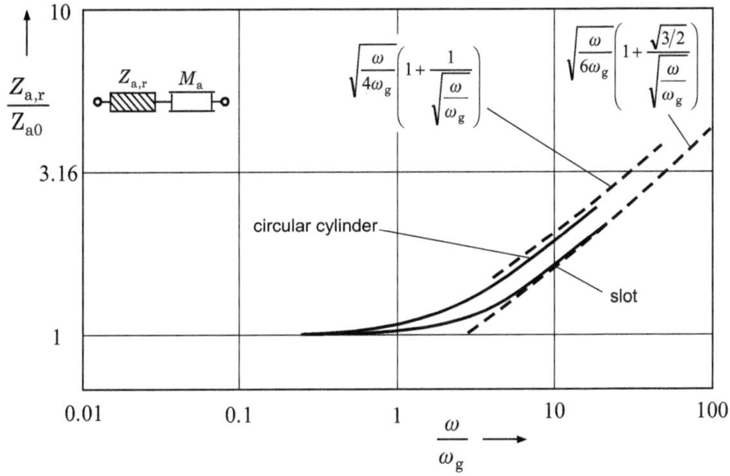

Fig. 3.44. Frequency response of frictional impedance according to Table 3.10 (canal component)

inserted into the network circuit of the total structure to be analyzed in order to perform a wideband analysis. Table 3.11 represents the circuits with frequency-independent component parameters.

Table 3.11. Circuit representation of real acoustic components

volume components according to Table 3.9	canal components according to Table 3.10
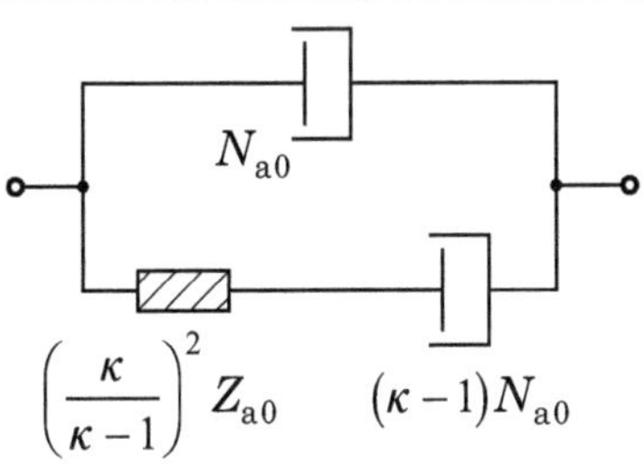	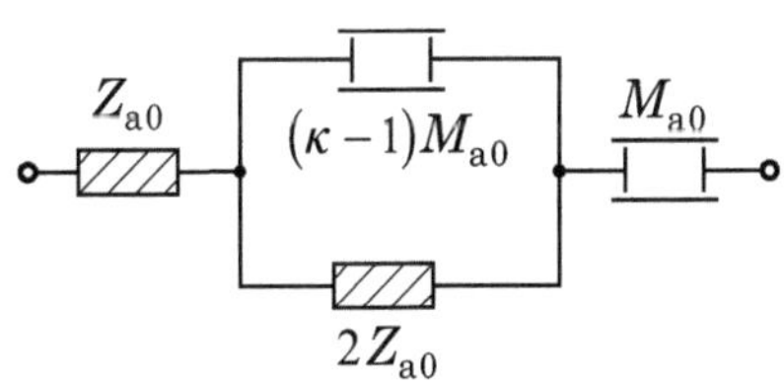
valid for: $\frac{\omega}{\omega_g} < 10$	valid for: $\frac{\omega}{\omega_g} < 100$
N_{a0}, Z_{a0}, ω_g see Table 3.9	M_{a0}, Z_{a0}, ω_g see Table 3.10

3.3.5 Isomorphism between Acoustic and Electrical Circuits

Due to existent isomorphism and same topology, analogous relations between the acoustic and electrical circuit can be set up again. The relations constituted with coordinate definition are summarized in Table 3.12.
These relations allow for the application of network analysis programs utilized in the field of electrical engineering (e.g. pSpice) for the dynamic analysis of acoustic and hydraulic systems. In order to provide also quantitative relations, proportionality quantities must be defined between the coordinates. It is expected to be valid:

$$\underline{u} = G_5 \underline{p} \qquad \text{and} \qquad \underline{i} = \frac{1}{G_6} \underline{q} \tag{3.29}$$

In addition, it could be allowed that frequencies between the electrical and mechanical network differ by a constant factor. In practice, no use is made of this possibility. Therefore, it is not used in the following. By means of this definition, it can be written for the components:

$$C = \frac{N_\mathrm{a}}{G_5 G_6}, \qquad L = G_5 G_6 M_\mathrm{a} \qquad \text{and} \qquad R = G_5 G_6 Z_\mathrm{a} \tag{3.30}$$

At first, the porportionality factors G_5 and G_6 are arbitrary with respect to their absolute values. However, with regard to characteristics of network analysis programs (e.g. pSpice), the sequences of numbers should be maintained. Thus, only powers of ten are available. In order to achieve a certain clearness of the electrical circuit, the factor 1 is used for G_5 and G_6 as power of ten. Thus, it can be written:

$$G_5 = 1 \frac{\mathrm{Vm}^2}{\mathrm{N}}, \qquad G_6 = 1 \frac{\mathrm{m}^3}{\mathrm{As}}$$

If same numerical values are used for G_5 and G_6, a representation with equal power rating will be generated. It is usually called on this advantage.

3.3.6 Sample Applications

Small cavity systems, whose information transferring characteristics must be optimized, can be found particularly in the field of sound receiver technology. In addition, acoustic networks can be used for modeling and optimization also in sound generating systems (loudspeaker boxes, headphones and earphones, telephone earpieces, calibrators, piezo phones, horns) and for mufflers in exhaust pipes. Using the example of a *pistonphone*, a *microphone* and an *exhaust gas muffler*, the modeling by means of acoustic networks is demonstrated in this section.

Table 3.12. Isomorphism between electrical and acoustic network

correlation between coordinates and components			
voltage	$\underline{u}$ ⟷ $\underline{p}$		pressure
current	$\underline{i}$ ⟷ $\underline{q}$		volumetric flow rate
inductance	L ⟷ M_a		acoustic mass
capacitance	C ⟷ N_a		acoustic compliance
resistance	R ⟷ $Z_\mathrm{a,r}$		acoustic friction

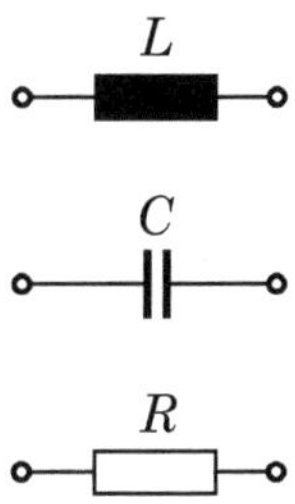

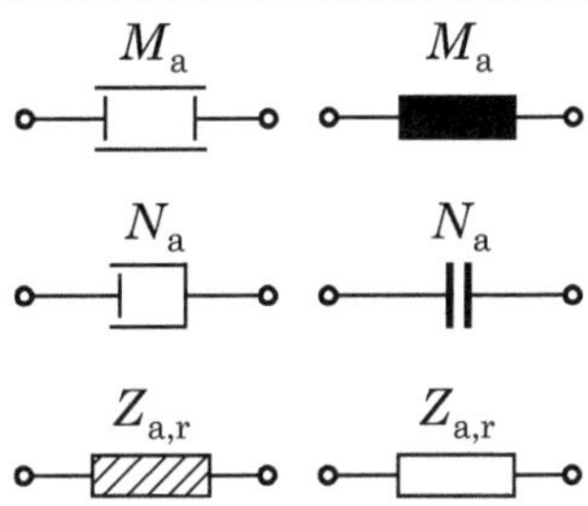

	$\underline{u} = \mathrm{j}\omega L \underline{i}$	$\underline{p} = \mathrm{j}\omega M_\mathrm{a} \underline{q}$	
	$\underline{u} = \frac{1}{\mathrm{j}\omega C} \underline{i}$	$\underline{p} = \frac{1}{\mathrm{j}\omega N_\mathrm{a}} \underline{q}$	
	$\underline{u} = R \underline{i}$	$\underline{p} = Z_\mathrm{a,r} \underline{q}$	
node of electrical circuit structure	$\sum_{*} \underline{i}_\nu = 0$	$\sum_{*} \underline{q}_\nu = 0$	node of acoustic network
loop of electrical circuit structure	$\sum_{\circlearrowright} \underline{u}_\nu = 0$	$\sum_{\circlearrowright} \underline{p}_\nu = 0$	loop of acoustic network

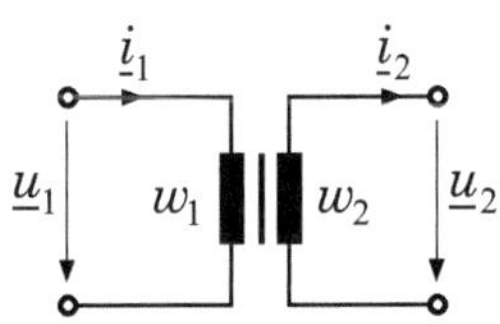

$$\underline{u}_2 = \frac{w_2}{w_1} \underline{u}_1$$

$$\underline{i}_2 = \frac{w_1}{w_2} \underline{i}_1$$

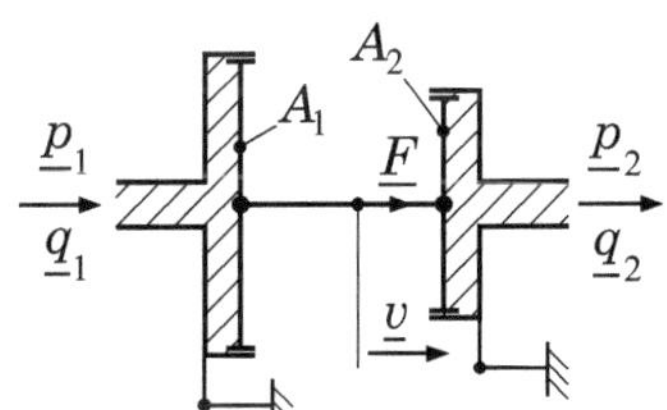

$$p_2 = \frac{A_1}{A_2} p_1$$

$$q_2 = \frac{A_2}{A_1} q_1$$

Pistonphone

For the calibration or inspection of microphones or a complete acoustic pressure measurement chain it necessitates a simple, small, lightweight and reliable testing unit. The simplest embodiment consists of a cavity, an oscillating piston and a microphone connection. The cavity is supplied with an alternating volumetric flow rate $\underline{q}_0$ by means of a sinusoidally moved piston with area A_{K} (Fig. 3.45). The microphone is extended into the volume in a spatially defined way. It is sealed with respect to the surrounding area. A volumetric flow rate forced by the piston motion will generate an alternating pressure within the volume. The dimensions of the volume are so small that the same pressure prevails at all positions of volume. If this pressure amplitude is exactly known, then the pressure response of the microphone can be controlled at the operating frequency of the pistonphone.

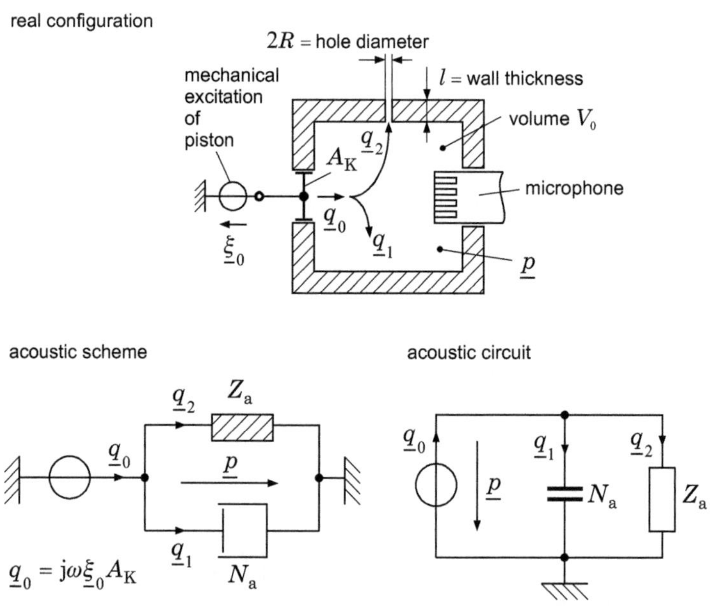

Fig. 3.45. Real configuration, acoustic scheme and acoustic circuit of a pistonphone

In order to provide for identity of the average internal pressure in the cavity and the external pressure, it is necessary to establish a connection between the internal volume and the external space. This connecting canal may have the character of a frictional impedance. The normal function of pistonphone may not be influenced by this canal. For this purpose, the frequency response of the pressure must be analyzed in consideration of a deflection excitation with constant amplitude. The ratio of pressure to volumetric flow rate results in:

$$\frac{\underline{p}}{\underline{q}_0} = \frac{1}{\mathrm{j}\omega N_{\mathrm{a}} + \dfrac{1}{Z_{\mathrm{a}}}} = \frac{1}{\mathrm{j}\omega N_{\mathrm{a}}} \frac{1}{1 + \dfrac{1}{\mathrm{j}\omega N_{\mathrm{a}} Z_{\mathrm{a}}}} \tag{3.31}$$

Thus, for the pressure amplitude related to deflection, it can be written

$$\frac{\underline{p}}{\underline{\xi}_0} = \mathrm{j}\omega A_\mathrm{K} \cdot \frac{\underline{p}}{\underline{q}_0} = \frac{A_\mathrm{K}}{N_\mathrm{a}} \frac{1}{1 - \dfrac{1}{\omega N_\mathrm{a} Z_\mathrm{a}}} = \frac{A_\mathrm{K}}{N_\mathrm{a}} \frac{1}{1 - \frac{\omega_\mathrm{g}}{\omega}} \tag{3.32}$$

with $\omega_\mathrm{g} = 1/\left(N_\mathrm{a} Z_\mathrm{a}\right)$. The consideration of the absolute values is sufficient for an evaluation. Thus, it follows:

$$\left|\frac{\underline{p}}{\underline{\xi}_0}\right| = \frac{A_\mathrm{K}}{N_\mathrm{a}} \frac{1}{\sqrt{1 + \left(\frac{\omega_\mathrm{g}}{\omega}\right)^2}} = \begin{cases} \dfrac{\omega}{\omega_\mathrm{g}} \cdot \dfrac{A_\mathrm{K}}{N_\mathrm{a}}, & \dfrac{\omega}{\omega_\mathrm{g}} \ll 1 \\[2ex] \dfrac{1}{\sqrt{2}} \cdot \dfrac{A_\mathrm{K}}{N_\mathrm{a}}, & \dfrac{\omega}{\omega_\mathrm{g}} = 1 \\[2ex] \dfrac{A_\mathrm{K}}{N_\mathrm{a}}, & \dfrac{\omega}{\omega_\mathrm{g}} \gg 1 \end{cases} \tag{3.33}$$

The operating frequency ω of the pistonphone should be at least one decade over the cut-off frequency ω_g. The dimensions of volume limit the application for higher frequencies. A volume reduction is limited at least in one direction by the microphone diameter . All individual dimensions must be considerably smaller than a fourth of the wavelength in air.

Condenser Microphone with Equalization of Pressure

A thin plate with an area A is resiliently mounted on a front surface of a cylindrical cavity (Fig. 3.46).

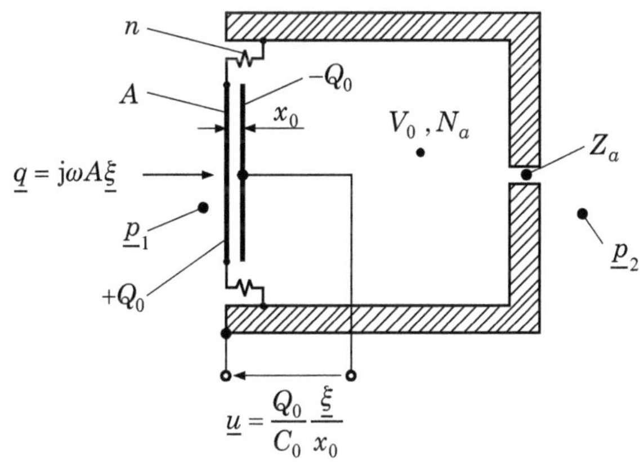

Fig. 3.46. Simple condenser microphone

The spring suspension may have the mechanical compliance n. The chosen compliance is so rigid that no resonant effects develop in the given transfer frequency range. Therefore, the mass of plate can be neglected for modeling. In a very short distance x_0, an electrode charged with an electric charge Q_0 is attached behind the plate. The electrical capacitance between the electrodes amounts to C_0. The charge Q_0 is kept constant by technical measures. In case of a plate displacement ξ, a voltage change is generated between the electrode and the housing.

Thus, an alternating voltage $\underline{u}$ can be observed at the electrical output of the microphone in case of excitation with an alternating pressure $\underline{p}$. The voltage is proportional to the deflection $\underline{\xi}$ and thus to the pressure affecting the plate. The cavity behind the plate represents a volume element. On the right side of Fig. 3.46, a small hole is visible. This hole connects the volume with the ambient pressure. The hole represents a canal element. The canal element is filled with a porous material and may have the character of a flow resistance. The filled hole is necessary, so that slow changes of the ambient pressure (caused by fluctuations of the barometric pressure or heating of the internal volume) do not generate a basic plate displacement. After modeling, the necessary dimensioning of flow resistance Z_a can be specified for a given transfer frequency range (lower frequency limit).

Two limiting cases are possible for the application of the microphone. Concerning the first limiting case, the microphone can be implemented into a housing in such a way that the acoustic pressure can not affect the external connection of the flow resistance ($\underline{p}_2 = 0$). Concerning the second limiting case, the microphone capsule is exposed to the sound field. Thus, for low frequencies, the acoustic pressure can affect plate and flow resistance ($\underline{p}_1 = \underline{p}_2$). In the following, the transfer function is calculated for both cases. In order to represent the effect of mechanical compliance in the acoustic circuit diagram, another additional consideration is necessary yet. With respect to the plate area A, the pressure difference $p_1 - p_V$ between plate and volume is transformed into a force affecting the spring with compliance n. This results in a displacement $\xi = np_1A$. At the same time a volume shift $\Delta V = \xi A$ is generated. The flow rate generated due to the pressure difference across the plate amounts to:

$$q = \frac{\mathrm{d}V}{\mathrm{d}t} = \frac{\mathrm{d}}{\mathrm{d}t}\left(n\left(p_1 - p_V\right)A^2\right)$$

Thus, the plate provides an acoustic compliance $N_{\mathrm{a,P}} = nA^2$. In case of insertion of the plate into an acoustic circuit diagram, it has to be noticed that the flow rate $\underline{q}$ flows through the plate into the internal volume. The pressure difference $p_1 - p_V$ occurs across the plate. Thus, it is in series with the external pressure source.

The flow resistance is connected in series with the pressure source $\underline{p}_2$. The circuit shown in the left part of Fig. 3.47 results from these facts. If the second limiting case mentioned above applies (microphone capsule in free sound field, thus $\underline{p}_1 = \underline{p}_2$), then the circuit can be simplified like shown in the right

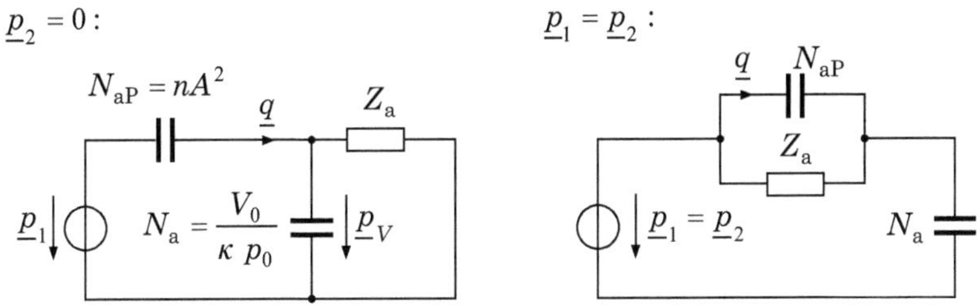

Fig. 3.47. Acoustic circuits of condenser microphone

part of Fig. 3.47. Now the transfer function can be easily calculated for both limiting cases. In case of the freely arranged microphone ($\underline{p}_1 = \underline{p}_2$), it can be written:

$$\frac{\underline{\xi}}{\underline{p}_1} = \frac{N_{a,P} N_a}{A\,(N_{a,P} + N_a)} \frac{1}{1 - j\dfrac{\omega_1}{\omega}} \qquad \text{with} \qquad \omega_1 = \frac{1}{(N_{a,P} + N_a)\, Z_a}$$

In case of the implemented microphone ($\underline{p}_2 = 0$) follows:

$$\frac{\underline{\xi}}{\underline{p}_1} = \frac{N_{a,P}}{A} \frac{1 + j\dfrac{\omega}{\omega_2}}{1 + j\dfrac{\omega}{\omega_1}} \qquad \text{with} \qquad \omega_2 = \frac{1}{N_a Z_a}$$

Figure 3.48 shows the calculated transfer functions for both mentioned cases. Above the cut-off frequency ω_1, in both cases the transfer characteristic is frequency-independent. In the first case (capsule in housing), the amplification factor is higher for frequencies below the cut-off frequency ω_1.
The example shows the influence of the installation environment of the respective system on the total frequency response.

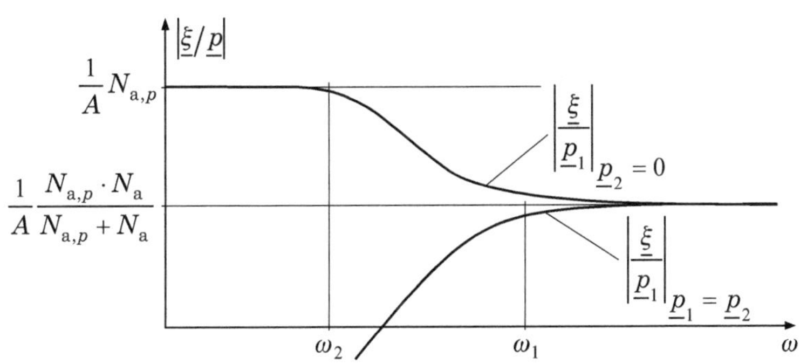

Fig. 3.48. Frequency response of plate displacement for two limit cases

Exhaust Gas Muffler

A simple exhaust gas muffler of a motor vehicle with two volumetric cavities is able to reduce the acoustic pressure by 90% in the far sound field.
Figure 3.49 shows an example of a muffler. Due to the simple structure, it is suitable here for demonstration. Two separated cavities are connected with the continuous main pipe by means of two tube sockets. In order to achieve a realistic transfer function, estimated frictional components are added to all acoustic masses.

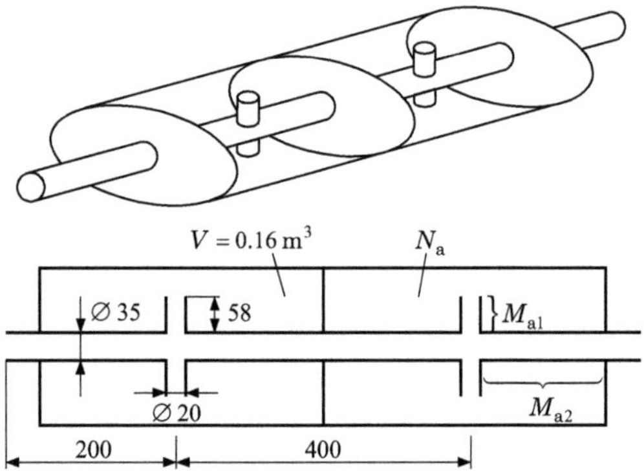

Fig. 3.49. Exhaust gas muffler

Figure 3.50 shows the acoustic circuit of the assumed general arrangement. Assuming a spherical wave-like radiation, the acoustic current appearing at the tube outlet and the acoustic pressure in the far sound field are interconnected by the relation specified in Fig. 3.51. The impedance $Z_{a,r5}$ caused by sound field is assumed to be frequency-independent and real-valued. The caused representation errors are in the range of the remaining errors generated by means of damping estimations. However, for the reader the calculation is comprehensible more easily. With the network analysis program pSpice frequency-dependent real resistors can be programmed, thus it succeeds a more exact representation with little effort.
In case of „without muffler", the acoustic current $\underline{q}_1$ acts as a source quantity of a spherical wave. Both the muffler and the exhaust pipe are not provided. The case „pipe as muffler" already causes a damping of the acoustic pressure level in the far sound field by means of the acoustic mass and friction in the pipe. By means of adding resonator cavities, further damping effects are generated in the far sound field depending on the number of cavities. Using a network analysis program, the calculation was made by means of direct input of circuit according to Fig. 3.50 using the transformation factors G_3 and G_4 between acoustic and electrical circuit (see Sect. 3.3.5).

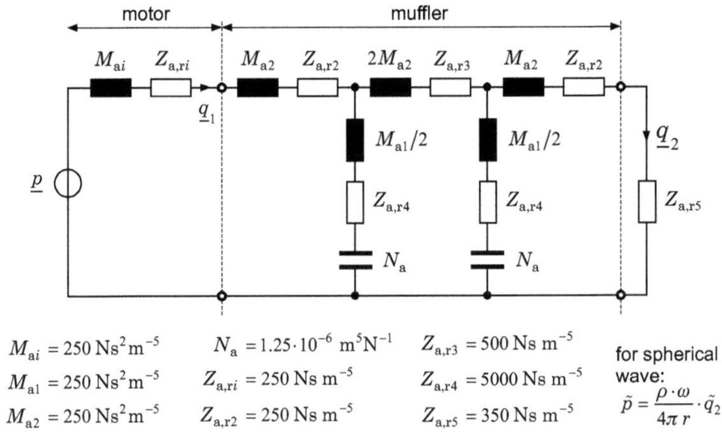

Fig. 3.50. Acoustic circuit of exhaust gas muffler according to Fig. 3.49

For the cases mentioned above, Fig. 3.51 shows the variations of the acoustic pressure level in the far field at a specified position.

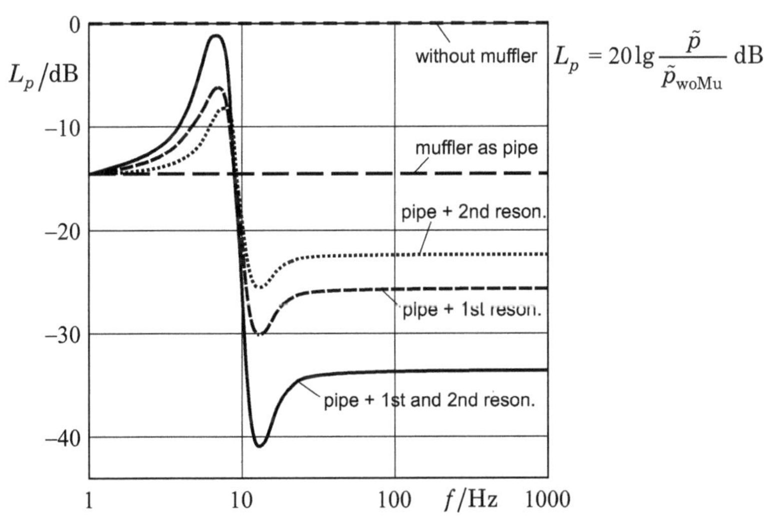

Fig. 3.51. Damping of the acoustic pressure level for different configurations

4

Abstract Linear Network

Four physical structures resulted from the considerations in the preceding sections which are isomorphic, concerning both the relations between coordinates which describe the system status and the topology of schemes which can be assigned to the structures.
This fact suggests the introduction of an abstract linear network which comprises the characteristics of all structures previously discussed. The importance of such an abstract network is based on the fact that all results of electrical network theory can be transfered to this abstract network due to the definite relations to electrical networks. Thus, the validity of these relations is ensured not only for the physical structures previously discussed, but also for all other occurring structures which can be described by equations summarized in the following. Among these structures are e.g. *thermal* and *magnetic networks*. However, they are not discussed within the context of this book.
Instead of the imaginary frequency $\mathrm{j}\omega$ previously used, in these equations the complex frequency $p = \mathrm{j}\omega + \sigma$ is written now. That means an analytical continuation of the frequency function from imaginary axis on complex plane or the extension of allowed time functions from stationary sinusoidal processes previously discussed on exponentially increasing and decreasing sinusoidal oscillations (see Chap. 2).

4.1 Coordinates

In the following, the state of the abstract network is described by means of two different kinds of coordinates.

1. Differential coordinates have the meaning of the difference of two scalar quantities between the endpoints of an element:

A. Lenk et al., *Electromechanical Systems in Microtechnology and Mechatronics*,
Microtechnology and MEMS, DOI 10.1007/978-3-642-10806-8_4,

$$\underline{\mu} = \begin{cases} \underline{u} & \text{voltage} \\ \underline{v} & \text{velocity} \\ \underline{p} & \text{pressure} \\ \underline{\Omega} & \text{angular velocity} \end{cases}$$

2. Flow-like coordinates invariably permeate an element, they have the same value at both endpoints of the element:

$$\underline{\lambda} = \begin{cases} \underline{i} & \text{current} \\ \underline{F} & \text{force} \\ \underline{q} & \text{volumetric flow rate} \\ \underline{M} & \text{torsional moment} \end{cases}$$

The product of a linked pair of coordinates in each case results in a power quantity which can not be negative in real networks without sources:

$$P = \frac{1}{4}\left(\underline{\mu}\underline{\lambda}^* + \underline{\mu}^*\underline{\lambda}\right) \tag{4.1}$$

4.2 Components

The abstract network comprises three different kinds of components (α, β, γ), one transformation mechanism and two kinds of sources which correspond to both kinds of coordinates.
The three components are represented in Table 4.1. In each case there are two energy storage elements and one element, in which an irreversible process runs in a thermodynamic sense.
It should be noticed that the general coordinates $\underline{\mu}$ and $\underline{\lambda}$ concerning the elements α, β and γ shown in Table 4.1 are interconnected by the complex frequency $p = \mathrm{j}\omega + \sigma$ according to

$$\underline{\mu} = p\alpha\underline{\lambda}, \tag{4.2}$$

$$\underline{\mu} = \frac{1}{p\beta}\underline{\lambda} \tag{4.3}$$

and

$$\underline{\mu} = \gamma\underline{\lambda}. \tag{4.4}$$

Table 4.1. Coordinates and components of abstract networks

$$\underline{\mu} = p\,\alpha\,\underline{\lambda}$$

$$\underline{\mu} = \frac{1}{p\,\beta}\,\underline{\lambda}$$

$$\underline{\mu} = \gamma\,\underline{\lambda}$$

$\alpha =$	L	inductance
	n	compliance
	M_a	acoustic mass
	n_R	rotational compliance
$\beta =$	C	capacitance
	m	mass
	N_a	acoustic compliance
	Θ	moment of inertia
$\gamma =$	R	resistance
	$1/r$	frictional admittance
	Z_a	acoustic friction
	$1/r_R$	rotational friction admittance

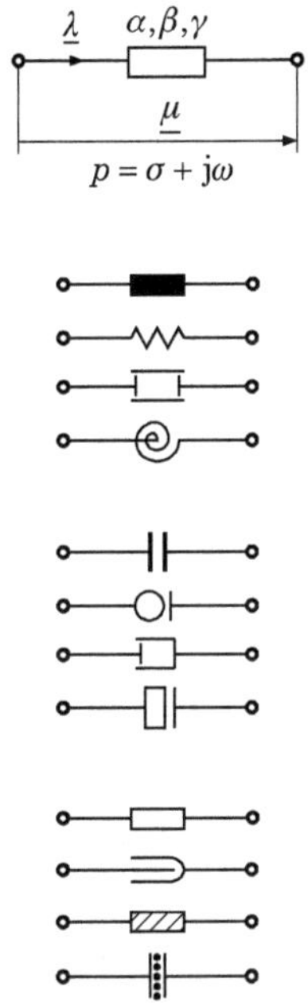

Both source mechanisms and transformation mechanism are summarized in Fig. 4.1.

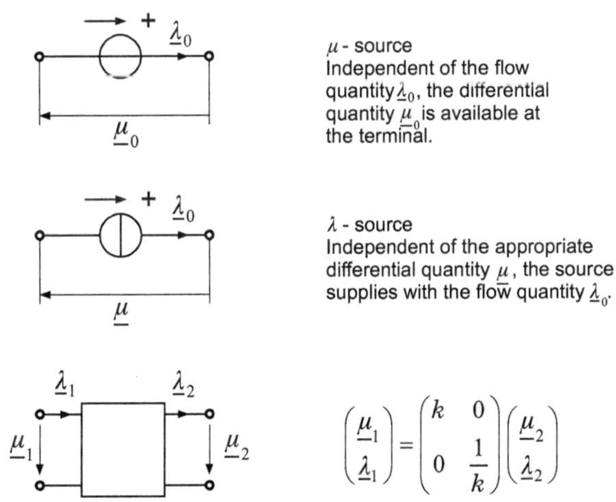

$$\begin{pmatrix} \underline{\mu}_1 \\ \underline{\lambda}_1 \end{pmatrix} = \begin{pmatrix} k & 0 \\ 0 & \frac{1}{k} \end{pmatrix} \begin{pmatrix} \underline{\mu}_2 \\ \underline{\lambda}_2 \end{pmatrix}$$

Fig. 4.1. Sources and coupling mechanisms of abstract linear networks

4.3 Nodal and Loop Rules

The systems consisting of the elements specified in Sect. 4.2 can be represented by schemes comprising legs and nodes (Fig. 4.2). Then the equations (4.5) and (4.6) are valid for the differential coordinates $\underline{\mu}_n$ being contained in each case in a mesh and for the flow coordinates $\underline{\lambda}_m$ converging at a node. They are referred to as nodal and loop rules:

$$\text{loop rule:} \qquad \sum_{\text{loop}} \underline{\mu}_n = 0 \tag{4.5}$$

$$\text{nodal rule:} \qquad \sum_{\text{node}} \underline{\lambda}_m = 0 \tag{4.6}$$

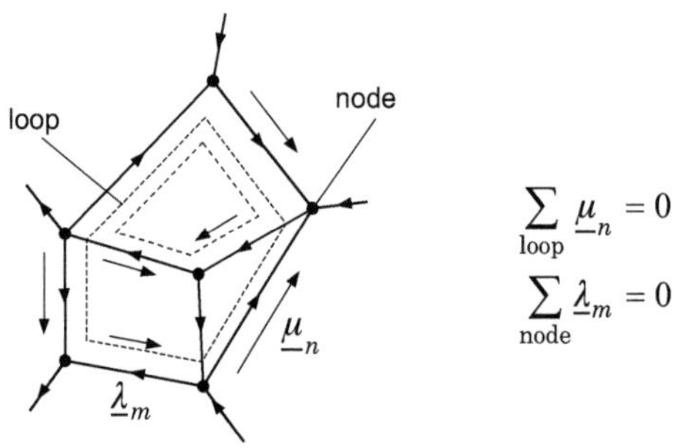

Fig. 4.2. Loop and nodal rules

4.4 Characteristics of the Abstract Linear Network

The equations (4.1) up to (4.6) describe the abstract network completely. In order to solve all tasks which can occur in combination with such a network, no further assumptions are necessary. By means of these equations, the state of a network can be particularly calculated as a function of operating sources. If a network is able to interact energetically with the environment by means of N pairs of coordinates, then it results from the basic equations (4.2) up to (4.6) that N flow coordinates are well-defined in each case by N differential coordinates and vice versa (Fig. 4.3 and Sect. 2.2).
The coefficients of the systems of equations

$$\underline{\mu}_n = \sum_{m=1}^{\mathrm{N}} \underline{\rho}_{nm} \underline{\lambda}_m \qquad n = 1, 2, \ldots, \mathrm{N} \tag{4.7}$$

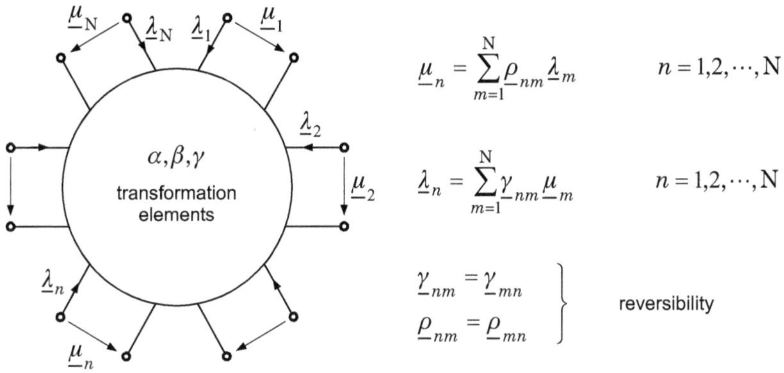

Fig. 4.3. N-port in the abstract linear network

$$\underline{\lambda}_n = \sum_{m=1}^{N} \underline{\gamma}_{nm}\underline{\mu}_m \qquad n = 1, 2, \ldots, N \tag{4.8}$$

denote generalized impedances and admittances. However, for historical reasons the linguistic usage is different for different physical structures. For *electrical and acoustic networks*, it is referred to the quotient

$$\frac{\text{differential coordinate } \underline{\mu}}{\text{flow coordinate } \underline{\lambda}} \Rightarrow \frac{\underline{u}}{\underline{i}} \Rightarrow \frac{\underline{p}}{\underline{q}}$$

as impedance and to its reciprocal value as admittance.
However, the reverse definition is usual for *mechanical networks*. It is referred to the quotient

$$\frac{\text{flow coordinate } \underline{\lambda}}{\text{differential coordinate } \underline{\mu}} \Rightarrow \frac{\underline{F}}{\underline{v}} \Rightarrow \frac{\underline{M}}{\underline{\Omega}}$$

as impedance and to its reciprocal value as admittance. Thus, for the impedance always a „force-like" quantity (F, M, p, u) can be found in the numerator.

Concerning network problems, three kinds of typical quotients occur between the coordinates which differ in their function theoretical character as functions of the complex variable $p = \sigma + \mathrm{j}\omega$. In the following, the essential characteristics of these quotients are summarized. Using the example of an electrical network, their derivation from the basic equations is contained in [46].
It is referred to the elements of the main diagonals of the matrices in (4.7) and (4.8) as impedances and admittances, respectively (Fig. 4.3). They denote the quotients of flow coordinates and differential coordinates at each pair of points of the system.

These quotients have following characteristics:

$$\left.\begin{matrix}\underline{\rho}_{nn}\\ \underline{\gamma}_{nn}\end{matrix}\right\} = \begin{Bmatrix}\rho_0\\ \gamma_0\end{Bmatrix} \frac{(p-p_1')(p-p_2')\cdots(p-p_k')}{(p-p_1'')(p-p_2'')\cdots(p-p_l'')} = \frac{Z(p)}{N(p)} = \frac{\sum\limits_{\nu=1}^{k} a_v p^{\nu}}{\sum\limits_{\nu=1}^{l} b_v p^{\nu}} \tag{4.9}$$

$$\Re\begin{Bmatrix}\underline{\rho}_{nn}\\ \underline{\gamma}_{nn}\end{Bmatrix} > 0 \quad \text{for} \quad \Re\{p\} > 0. \tag{4.10}$$

$$k = \begin{Bmatrix} l+1\\ l\\ l-1\end{Bmatrix} \tag{4.11}$$

$$a_\nu, b_v > 0, \quad \text{real-valued} \tag{4.12}$$

$$\Re\{p_\lambda', p_\lambda''\} \le 0 \tag{4.13}$$

$$\underline{\rho}_{nn}(p^*) = \underline{\rho}_{nn}^*(p) \quad \text{and} \quad \underline{\gamma}_{nn}(p^*) = \underline{\gamma}_{nn}^*(p) \tag{4.14}$$

The form of (4.9) results from the fact that the relations for the components (4.2) up to (4.4) can only result in rational functions of variable p in combination with the nodal and loop rules. The a_ν and b_ν are real-valued and > 0, because this was postulated for the α, β, γ in (4.2) up to (4.4) (real components). Equation (4.10) implies the conclusion that a real component consumes power at most on average for sinusoidal processes with constant ($\sigma = 0$) or exponentially decreasing ($\sigma > 0$) amplitude. But the component is not able to dissipate power. Equation (4.13) represents an equivalent conclusion to that fact. The real parts of the zeros of the polynomial functions in the numerator and denominator appear in form $e^{\sigma t}$ as factors in the time functions of the possible free oscillations in network.
At most, the amplitudes of the free oscillations can be constant (loss-free passive system $\sigma_\lambda = 0$), but they can not increase. In real physical systems always an – even though very small – damping is existent ($\sigma_\lambda < 0$), thus the free oscillations disappear for $t \to \infty$. Equation (4.11) results from the fact that for high frequencies a network between two points is only able to take the character of one of the three basic components, i.e. for $p \to \infty$ the $\underline{\rho}_{nn}$ or $\underline{\gamma}_{nn}$ can only be constant, $\sim p$ or $\sim 1/p$. The characteristic of (4.14) finally results from the easily verifiable fact that this relation is valid for each of the three basic components.
It is referred to the quotients of congeneric coordinates at different positions of network as *transfer factors*.

It applies:

$$\left.\begin{aligned} \underline{T}_\mu &= \frac{\underline{\mu}_n}{\underline{\mu}_m} \\ \underline{T}_\lambda &= \frac{\underline{\lambda}_n}{\underline{\lambda}_m} \end{aligned}\right\} = \begin{Bmatrix} \rho_0 \\ \gamma_0 \end{Bmatrix} \frac{(p-p_1')(p-p_2')\cdots(p-p_k')}{(p-p_1'')(p-p_2'')\cdots(p-p_l'')} = \frac{Z(p)}{N(p)} = \frac{\sum\limits_{\nu=1}^{k} a_\upsilon p^\nu}{\sum\limits_{\nu=1}^{l} b_\upsilon p^\nu}$$

$$\underline{\mu}_m, \underline{\lambda}_m \text{ excitation,} \qquad \underline{\mu}_n, \underline{\lambda}_n \text{ response} \tag{4.15}$$

$$a_\nu, b_\upsilon = \text{real-valued} \tag{4.16}$$

$$l \geq k \tag{4.17}$$

$$\Re\{p_\nu''\} \leq 0 \tag{4.18}$$

Just like (4.12), equation (4.16) results from the fact that α, β, γ are real-valued. If it is considered that a phase opposition of response and excitation can occur, it can be easily understood that the a_ν and b_ν need not to be necessarily positive. For energetic reasons that was impossible for impedances. For very high frequencies ($p \to \infty$), (4.17) implies the conclusion that a transfer factor converges toward a constant value at most. But it is not able to increase indefinitely with increasing p. However, compared to impedances and admittances, for $p \to \infty$ transfer factors are able to converge toward zero at any power of p. Both predications can be easily verified by means of special examples. Equation (4.18) finally originates from the same fact as (4.13). The real parts of the p_ν'' occur again with the factor $\mathrm{e}^{\sigma t}$ for free oscillations of the network which can be observed at the positions ν. Therefore, they must be positive or equal zero at most.

It is referred to the quotients of flow coordinates and differential coordinates at two different network positions as *core impedances* or *core admittances* (Fig. 4.4). If it is considered that a core admittance or core impedance can always be defined as a product of an admittance and an impedance with a transfer factor respectively, the characteristics of these quotients directly result from those of the two groups mentioned before.

For example it applies:

$$\underline{\varrho}_{nm} = \left(\frac{\underline{\mu}_n}{\underline{\lambda}_m}\right)_{(\cdots)} = \left(\frac{\underline{\mu}_n}{\underline{\mu}_m}\right)_{(\cdots)} \left(\frac{\underline{\mu}_m}{\underline{\lambda}_m}\right)_{(\cdots)} \tag{4.19}$$

$$\underline{\varrho}_{nm} = \left(\underline{T}_\mu\right)_{mn} \underline{\varrho}_{mm} \tag{4.20}$$

It is remarkable that the $\underline{\varrho}_{nm}$ and $\underline{\gamma}_{nm}$ increase proportionally at most with p for high frequencies and are able to converge toward zero at any high power of p.

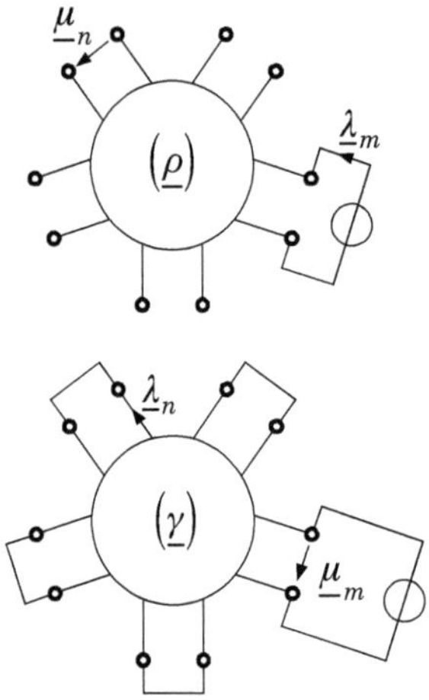

$$\underline{\rho}_{nm} = \left(\frac{\underline{\mu}_n}{\underline{\lambda}_m} \right)_{\underline{\lambda}_1, \underline{\lambda}_2 \cdots \underline{\lambda}_{m-1}, \underline{\lambda}_{m+1}, \cdots \underline{\lambda}_\mathrm{N} = 0}$$

$\underline{\lambda}_m$ excitation, $\underline{\mu}_n$ response

$$\underline{\gamma}_{nm} = \left(\frac{\underline{\lambda}_n}{\underline{\mu}_m} \right)_{\underline{\mu}_1, \underline{\mu}_2 \cdots \underline{\mu}_{m-1}, \underline{\mu}_{m+1}, \cdots \underline{\mu}_\mathrm{N}}$$

$\underline{\mu}_m$ excitation, $\underline{\lambda}_n$ response

$$\left\{ \begin{matrix} \underline{\rho}_{nm} \\ \underline{\gamma}_{nm} \end{matrix} \right\} = \left\{ \begin{matrix} \rho_0 \\ \gamma_0 \end{matrix} \right\} \frac{\left(p - p_1'\right)\left(p - p_2'\right) \cdots \left(p - p_k'\right)}{\left(p - p_1''\right)\left(p - p_2''\right) \cdots \left(p - p_l''\right)} = \frac{\sum_{\nu=1}^{k} a_\nu p^\nu}{\sum_{\nu=1}^{l} b_\nu p^\nu}$$

a_ν, b_ν = real-valued $\qquad k \le l + 1 \qquad \mathrm{Re}\left\{ p_\nu'' \right\} \le 0$

Fig. 4.4. Impedances and admittances in the abstract linear network

5

Mechanical Transducers

The still separated mechanical-translational, mechanical-rotational and acoustic networks represented in Chap. 3 are often interconnected by means of transducer elements. The component *lever* discussed in Sect. 3.1.3 represents such a transducer element between two mechanical-translational networks. The interconnection between translational and rotational networks is generated by the component type *rod* (rigid rod and bending rod). The interconnection between mechanical and acoustic networks is realized by means of the component type *area element* (piston transducer and bending plate).

5.1 Translational-Rotational Transducer

Rigid and flexible rods are frequently applied elements which realize the objective of coupling between networks or which are primarily necessary in the field of construction and thus require a modeling with both network types. The application of bending rods as guidance components for translationally moved assemblies represents a typical case for the latter consideration (e.g. parallel guide by means of flat springs). Here, the rotational part of the network occurs only in the representation domain of bending rods. Also an assembly consisting of two ideal rigid rods is able to act solely as lever and does not have to use any rotational components to the outside. At first, the relations are represented at the rigid rod and, in the following, the relations are demonstrated using the example of bending rod.

5.1.1 Rigid Rod

If a rigid and massless rod is rigidly connected at one side with an axis as it is shown in Fig. 5.1, the rotational coordinates affecting the axis and the translational coordinates affecting the free end of rod will be only interconnected by the rod length l. Since also in this case only small deflections and

A. Lenk et al., *Electromechanical Systems in Microtechnology and Mechatronics*,
Microtechnology and MEMS, DOI 10.1007/978-3-642-10806-8_5,

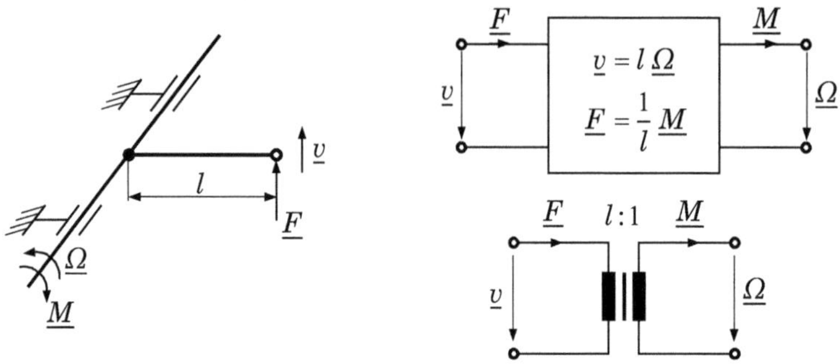

Fig. 5.1. One side pivoted rod as translational-rotational transducer

thus small angles are considered to be valid, thus following linear relations between the coordinates are generated:

$$\underline{\Omega} = \frac{\underline{v}}{l} \qquad \text{and} \qquad \underline{M} = \underline{F}l \tag{5.1}$$

Assuming the axis to be loaded with a rotational impedance, an affecting velocity results in a torsional moment at the free end of rod. The torsional moment is reflected in a force at the free rod end. Thus, a transformed impedance is effective at the end of rod. The transformation factor is represented by the rod length l. In the same manner, this consideration can also be done from the rotational reference point. Also in this case, the rod length l represents the transformation factor.
However, in general case translational and rotational coordinates should be allowed on both sides. This can be achieved by attaching an axis to the free end and by the fact that the previously pivoted side becomes a translationally flexible reference point. Now both pairs of coordinates can be defined at both sides. The two-port network representation of the one side pivoted rod must now be replaced by a four-port network representation (Fig. 5.2).

The equilibrium of moments and forces affecting the massless rod and the kinematic conditions resulting from the stiffness of rod result into following equations:

$$\begin{array}{ll} \underline{F}_1 = \underline{F}_2 = \underline{F} & \underline{\Omega}_1 = \underline{\Omega}_2 = \underline{\Omega} \\ \underline{v}_2 - \underline{v}_1 = l\underline{\Omega} & \underline{M}_1 - \underline{M}_2 = l\underline{F} \end{array} \tag{5.2}$$

Figure 5.2 shows a four-port representation of these interconnections and thus a circuit representation of the component „ideal rod". The internal coupling two-port network is an ideal transformer with the transmission ratio *rod length* l.

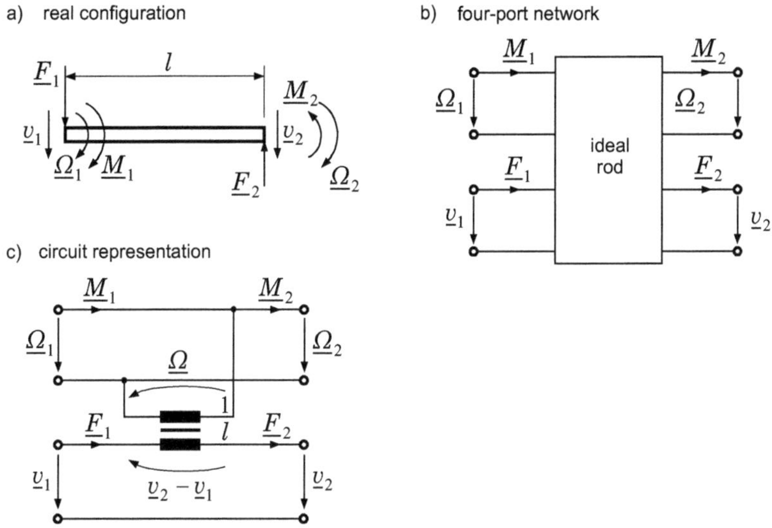

Fig. 5.2. Ideal rod a) real configuration, b) coupling four-port network, c) circuit representation

Figure 5.3 illustrates how the one side pivoted rod model shown in Fig. 5.1 results from the model of ideal rod shown in Fig. 5.2 by setting the boundary conditions $\underline{v}_1 = 0$ (short-circuit operation) and $\underline{M}_2 = 0$ (open-circuit operation).

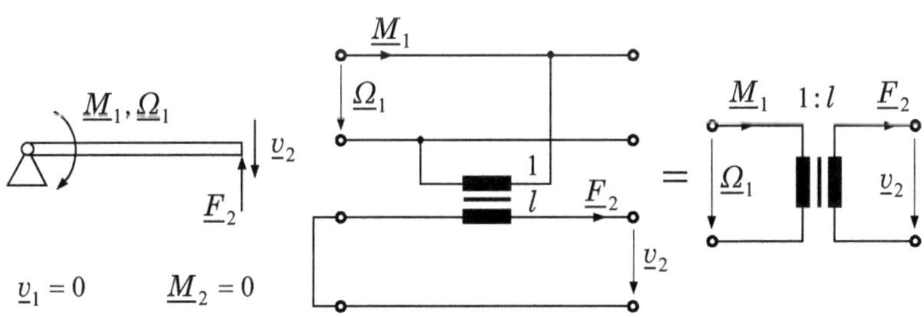

Fig. 5.3. One side pivoted rod modeled by means of ideal rod

In addition, also the derivation of the relations of the internally supported lever represented in Sect. 3.1.3 succeeds by interconnection of two ideal rods. The procedural method is represented in Fig. 5.4. The coupling conditions for two ideal rods of length l_1 and l_2 result from the characteristics of ideal support which can absorb forces but no moments.

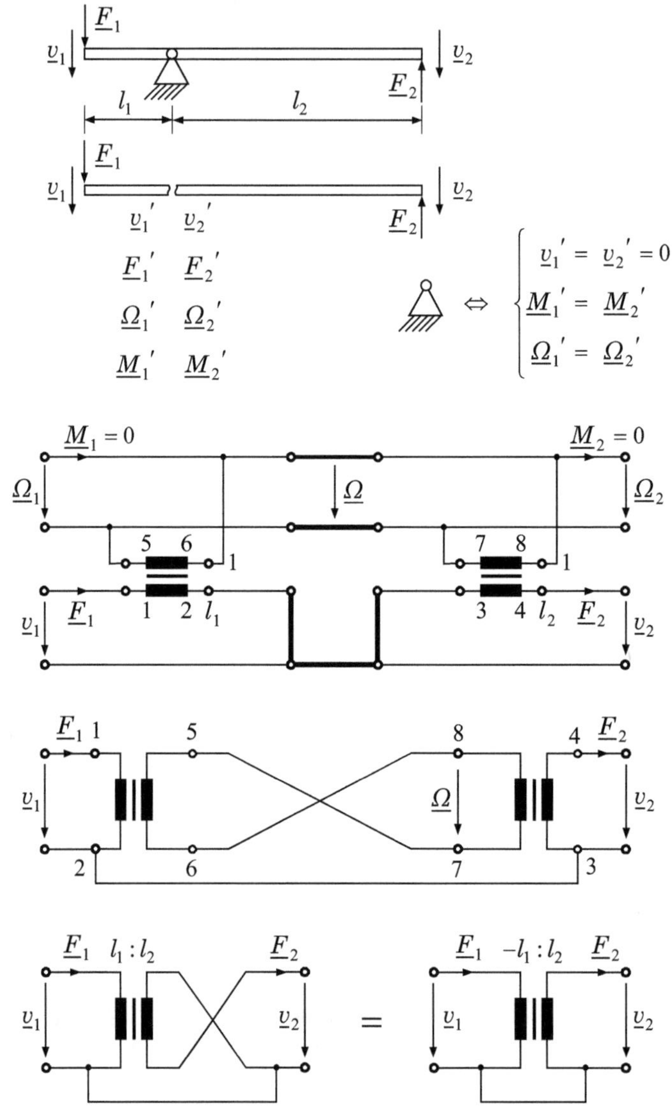

Fig. 5.4. Two-port network representation of lever modeled by means of two rods

In Sect. 5.1.2 it is demonstrated that the model of bending rod can be composed of ideal rods and torsion springs. In this context, the four-port model of ideal rod is of prime importance.

5.1.2 Bending Rod

Bending rods and circular bending plates (see Sect. 5.2) are often found as components for springs, guiding devices, mechanical-acoustic transducers and in connection with piezoelectric transducers. However, a representation of the bending theory (e.g. in [43]) of rod and plate goes beyond the scope of this book. Therefore, only a short introduction to rod bending with the special objective of derivation of a four-port circuit of the finite bending rod and the corresponding chain matrix is given.

Connecting Equations at the Differential Bending Element

It is to be referred to such deformations as rod bending which are caused by moments being parallel to each other and acting perpendicularly to the rod axis. It is assumed that the rod length is large compared to its lateral dimensions. Only small bending angles are expected to be allowed. Furthermore, only a longitudinal stress T_1 is assumed to be existent within the rod which is linearly dependent on the coordinate x_2 (Fig. 5.5). The position x_2, where the longitudinal stress amounts to $T_1 = 0$, is defined as $x_2 = 0$ (*neutral axis position*). At first, a rectangular cross-sectional area of the bending rod with the width b and thickness h is assumed. For following consideration, a short rod segment with length Δx is cut out of this rod and cross-sectional forces are added. By means of the analysis of deformation and cross-sectional forces, a finite bending element can be deduced which consists of two short rods interconnected by a *rotational compliance*. In the following, this will be demonstrated.

Figure 5.5 shows the conditions at the loaded bending element. The strain on the bender's upper surface $S_1(h/2)$ and the compression strain on the bender's lower surface are equal according to amount. Due to the predefinition of very small bending angles $\Delta\varphi$, the trigonometric functions can be approximately represented by their arguments. For isotropic materials, also the dependence of the mechanical longitudinal stress on x_2 results from Fig. 5.6, the total torsional moment M results from integration with respect to the cross-sectional forces. Thereby, the quantity I denotes the *area moment of inertia* around the neutral axis.

In combination with the modulus of elasticity E $(T_1 = ES_1)$ and the relation between moment and mechanical stress illustrated in Fig. 5.6

$$M = T_1(h/2) \cdot \frac{1}{h/2} \int\limits_{-h/2}^{h/2} x_2^2 b \mathrm{d}x = T_1(h/2) \cdot \frac{1}{h/2} \cdot I \tag{5.3}$$

follows

$$M = E \cdot S_1(h/2) \cdot \frac{I}{h/2}. \tag{5.4}$$

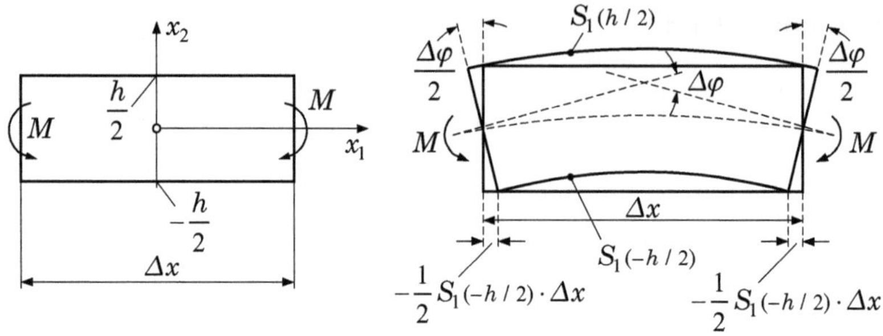

$$\frac{\Delta\varphi}{2} = \frac{\frac{1}{2}S_1(h/2)\cdot\Delta x}{\frac{h}{2}} \quad\Rightarrow\quad \frac{\mathrm{d}\varphi}{\mathrm{d}x_1} = \frac{1}{\frac{h}{2}}S_1(h/2)$$

$$S_1(x_2) = S_1(h/2)\cdot\frac{x_2}{\frac{h}{2}}$$

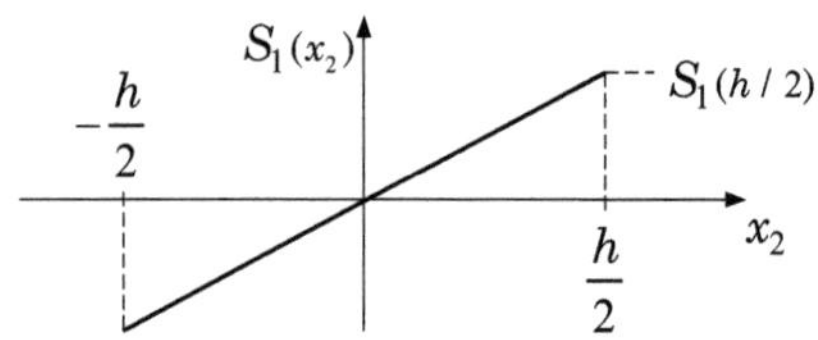

Fig. 5.5. Deformation of a short bending element

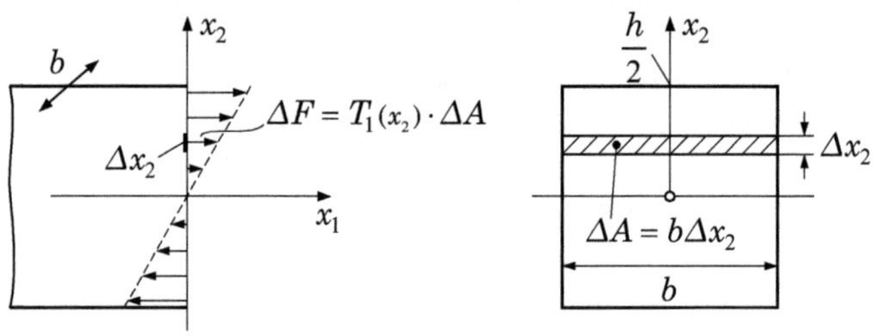

from $S_1(x_2) = S_1(h/2)\cdot\frac{x_2}{\frac{h}{2}}$ follows $T_1(x_2) = T_1(h/2)\cdot\frac{x_2}{\frac{h}{2}}$

$$\Delta M = \Delta F\cdot x_2 = T_1(x_2)\cdot x_2 b\Delta x_2 = T_1(h/2)\cdot\frac{x_2}{\frac{h}{2}}\cdot x_2 b\Delta x_2$$

$$M = T_1(h/2)\cdot\frac{1}{\frac{h}{2}}\int_{-h/2}^{+h/2} x_2^2\, b\,\mathrm{d}x_2$$

for bending rod with rectangular cross section: $I = \frac{b\cdot h^3}{12}$

Fig. 5.6. Correlation between torsional moment and mechanical stress

By using Fig. 5.5, the following expression is achieved:

$$S_1\,(h/2)\cdot\frac{1}{h/2}=\frac{\Delta\varphi}{\Delta x} \tag{5.5}$$

In combination with (5.4) follows:

$$\Delta\varphi=\frac{\Delta x}{EI}M=\Delta n_{\mathrm{R}}M \tag{5.6}$$

Thus, the differential bending element can be recreated by means of a rotational compliance Δn_{R} and two rods with length $\Delta x/2$ (Fig. 5.7). If the rod degenerates into a plate by extension of width along the x_3 direction, it can be generally assumed that no deformations are possible in this direction ($S_3=0$). For this case, it applies:

$$\Delta n_{\mathrm{R}}=\frac{1-\nu^2}{EI}\Delta x \tag{5.7}$$

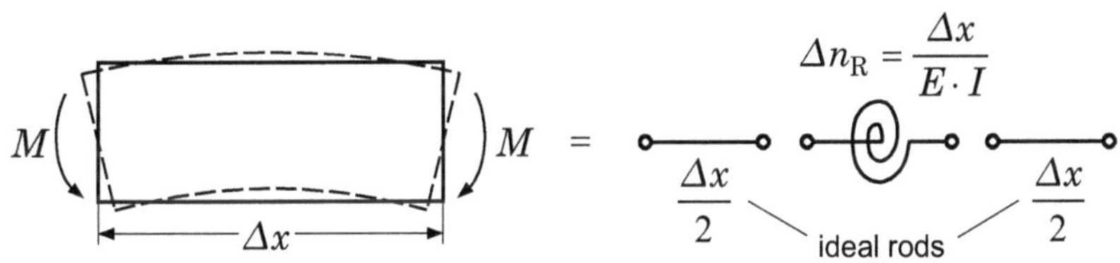

Fig. 5.7. Model of differential bending element

Finite Bending Rod and its Four-Port Representations

In order to apply the bender model universally, it is necessary to be able to let forces act along the x_2 direction in addition to the two torsional moments shown in Fig. 5.5. Then the resulting shear load causes additional displacements which are, however, significantly smaller than the displacements caused by bending. In the following, the resulting shear strains are neglected. Thus, the model represented in Fig. 5.5 is also valid for forces and moments affecting both sides.

In order to model the finitely long bending rod, a complex model must be composed of the model for ideal rods (Sect. 5.1.1) and the rotational compliance mentioned above.

Figure 5.8 shows the circuit representation of the finite bending element and a cascade connection consisting of finite bending elements for representation of the long bender. By means of this representation, it is also possible to represent benders with position-variable area moments of inertia, with local changes in density or with external pressure loads by technically correct circuits. The cascade connection of many bender four-ports shouldn't be a substantial problem for usually applied network analysis programs.

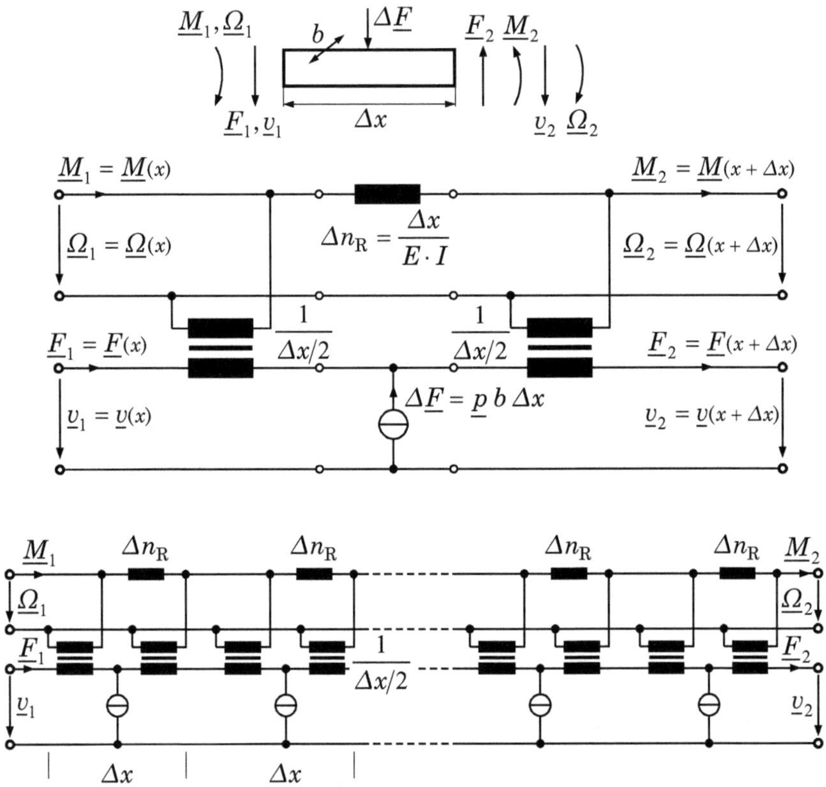

Fig. 5.8. Bending element and cascade connection

In the case of quasi-static processes without pressure load, also long benders can be exactly represented by means of a modified, considerably smaller four-port circuit, as long as only interconnections between the coordinates at both bender ends are of interest (Fig. 5.9).
This circuit was achieved by means of n-pole theory. It comprises a negative compliance. Most network analysis programs permit the input of negative values for component parameters. However, for all circuitry summarizing of components up to a real interface, in combination with other compliances always positive characteristic spring values can be generated.

Thus, also a representation of finitely long bending rods concerning their coordinates at the rod ends is possible with network analysis programs without any difficulty.

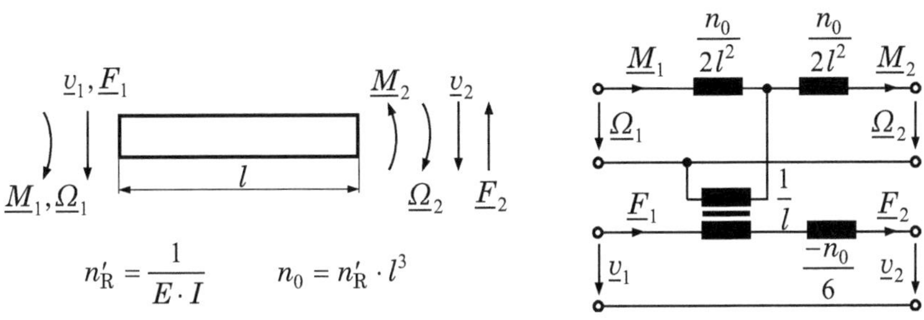

Fig. 5.9. Modified four-port network of finite bender

Compared to the cascade connection, a significantly less programming effort is necessary. The chain matrix is needed for analytical calculations. It is defined in following way:

$$\begin{pmatrix} \underline{v}_2 \\ \underline{\Omega}_2 \\ \underline{M}_2 \\ \underline{F}_2 \end{pmatrix} = \begin{pmatrix} 1 & l & -\mathrm{j}\omega n_0/(2l) & \mathrm{j}\omega n_0/6 \\ 0 & 1 & -\mathrm{j}\omega n_0/l^2 & \mathrm{j}\omega n_0/(2l) \\ 0 & 0 & 1 & -l \\ 0 & 0 & 0 & 1 \end{pmatrix} \begin{pmatrix} \underline{v}_1 \\ \underline{\Omega}_1 \\ \underline{M}_1 \\ \underline{F}_1 \end{pmatrix}, \quad n_0 = \frac{l^3}{EI} \tag{5.8}$$

Sample Application: Bending Rod as Spring Element

The application of the modified four-port circuit will be demonstrated by using a simple example. The free end of a clamped-free bender is loaded by a force. Here, the deflection of the force-loaded reference point and the bender's angle of inclination under the force application point are of interest. Figure 5.10 shows the bender being rigidly clamped on the right side and being loaded by a force F on the left side. The modified bender circuit connected with the boundary conditions is represented in the middle of Fig. 5.10. Due to the clamp, the velocity and the angular velocity are set at zero (jumpers) on the right side of the central circuit. The force source operates on the left side of the circuit. The velocity $\underline{v}_1$ and angular velocity $\underline{\Omega}_1$ are available as observation quantities. Thus, also deflection and angle of inclination are available. The result of circuit summary and circuit simplification is illustrated on the right side of Fig. 5.10.

The total compliance yields

$$n = \frac{\xi_1}{F_1} = \frac{l^3}{3EI}. \tag{5.9}$$

This result corresponds to the specification of bending rods listed in formularies.

The angle of inclination $\underline{\varphi}_1$ under load can be deduced from the velocity

$$\underline{v}_\mathrm{R} = -l\underline{\Omega}_1 = \frac{\mathrm{j}\omega n_0/2}{\mathrm{j}\omega n_0/2 - \mathrm{j}\omega n_0/6}\underline{v}_1 = \frac{3}{2}\underline{v}_1. \tag{5.10}$$

The angle of inclination is calculated according to:

$$\underline{\varphi}_1 = -\frac{\underline{v}_\mathrm{R}}{\mathrm{j}\omega l} = -\frac{3}{2}\frac{\underline{v}_1}{\mathrm{j}\omega l} = -\frac{n_0}{2l}\underline{F}_1 = -\frac{l^2}{2EI}\underline{F}_1 \tag{5.11}$$

With this method also much more complicated problems can be calculated fast, clearly and reliably.

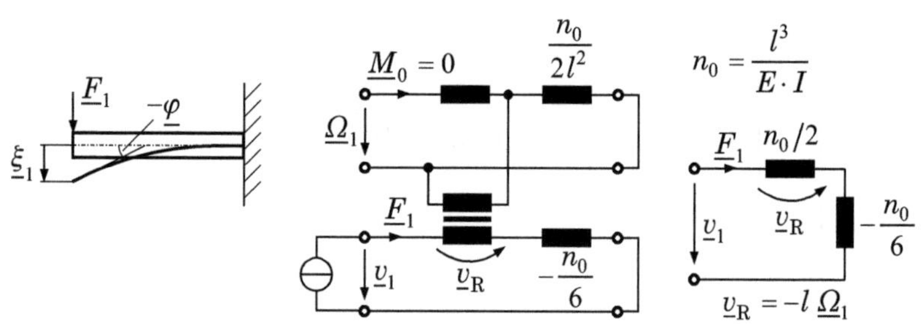

Fig. 5.10. Clamped-free bender with source of force

5.2 Mechanical-Acoustic Transducer

The coupling of mechanical and acoustic domains can particularly be found with electroacoustic, pneumatic and hydraulic systems. Forces and motions are to generate acoustic pressures and volumetric flow rates and vice versa. In the simplest case this transformation occurs with a rigid and massless piston being arranged in a tube. In practice, usually diaphragms and bending plates are used for this task, since here the problem of edge sealing is already solved.

5.2.1 Ideal and Real Mechanical-Acoustic Piston Transducers

The basic element of an ideal piston transducer is represented in Fig. 5.11. A massless and rigid piston is supported axially shiftable in a tube without tolerance and without friction and is connected with an also rigid and massless coupling rod.

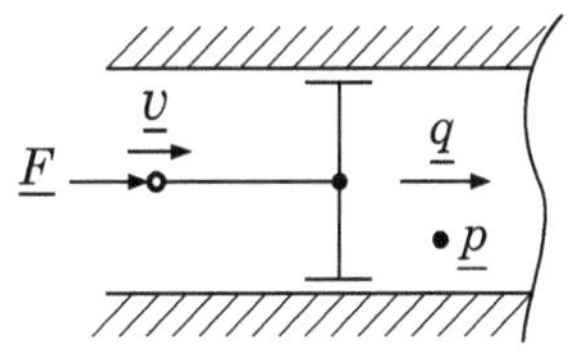

Fig. 5.11. Ideal piston transducer

At first, it is assumed that the tube remains open at the mechanical side, so that a piston shift is not able to generate a pressure there. However, if the acoustic side is terminated with an acoustic impedance, then a forced velocity $\underline{v}$ will result in a volumetric flow rate $\underline{q} = \underline{v}A$ and a pressure $\underline{p}$ which counteracts with a force $\underline{F} = \underline{p}A$ the source of motion at the mechanical side. Thus, the flow coordinate force is interconnected with the differential coordinate pressure.
Furthermore, the differential coordinate velocity is interconnected with the flow coordinate volumetric flow rate. These coupling systems are called *gyrators*. The mechanical-acoustic coupling two-port network and its gyrator-like transducer characteristics are summarized in Table 5.1. For a better understanding, the characteristics of electrical gyrators are also specified. It can be realized that e.g. a capacitor connected with one side of the electric gyrator acts as an inductor at the other gyrator side. A parallel connection of two impedances at one gyrator side appears as series connection of separately transformed impedances on the other gyrator side.
For the mechanical-acoustic transducer, the gyrator-like characteristic results in the fact that an acoustic compliance (corresponding to the mentioned electrical capacitance) operates as mechanical compliance at the other gyrator side (corresponding to the mentioned electrical inductance). Similarly, acoustic masses result in mechanical masses after application of the transducer.
With the realization of piston transducers it is difficult to obtain a sufficiently leakproof and frictionless arrangement of the piston in the tube. Therefore, for constructions e.g. an elastic sealing element is added between piston and tube which keeps the piston slidable within the scope of motional limits of the sealing element.
In addition, it must be generally assumed that an acoustic impedance can

Table 5.1. Gyrator-like transducers

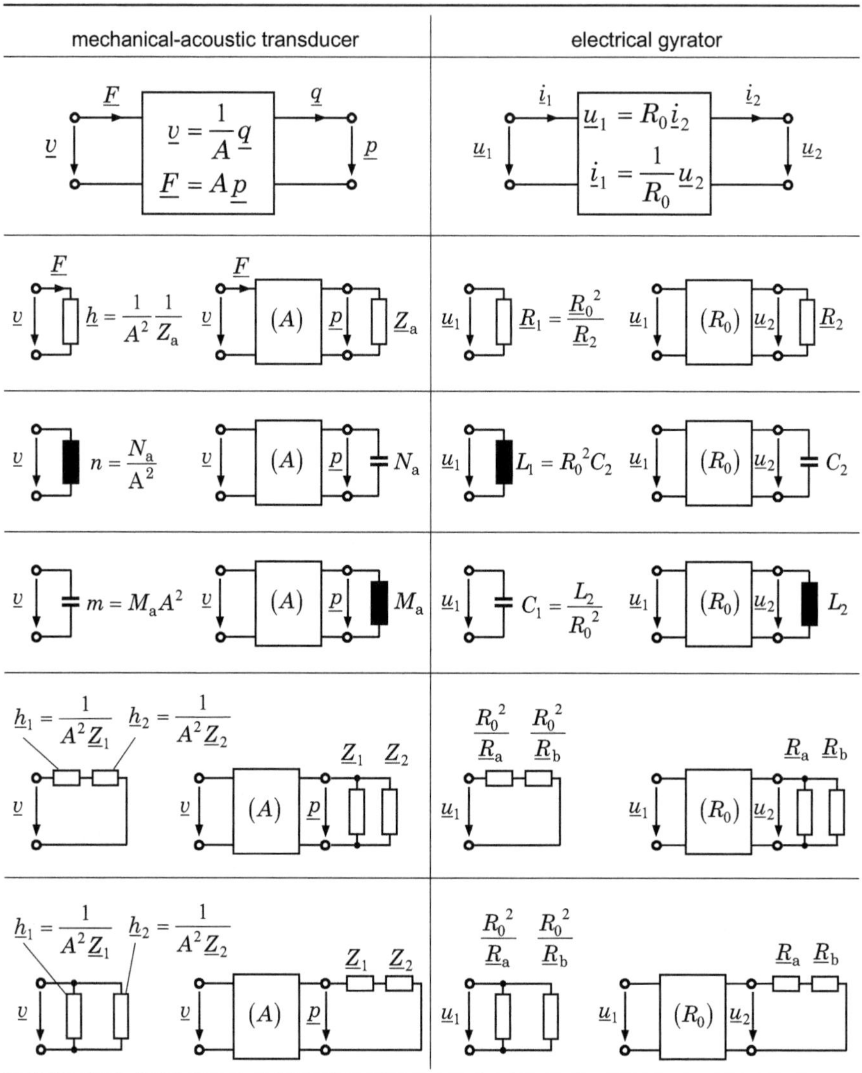

also affect the second tube side. This fact results in a three-port configuration. Figure 5.12 shows such a variant of the mechanical-acoustic transducer.
The force at the mechanical port of the transducer is generated both by the resilience of the sealing element during the forced shift of piston and by the pressure difference between right and left piston side affecting the piston area. The volumetric flow rate is equal at both acoustic ports and is only defined by the piston area and its velocity.

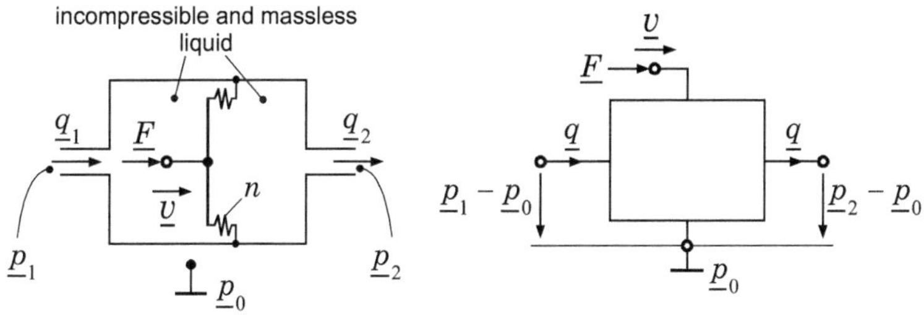

Fig. 5.12. Real piston transducer with two acoustic ports

This results in:

$$\left.\begin{aligned} F &= \frac{1}{n}\xi + \underbrace{(p_2 - p_1)}_{p_W}A \\ v &= \frac{q}{A} \end{aligned}\right\} \Rightarrow \begin{cases} \underline{v} = \frac{1}{A}\underline{q} \\ \underline{F} - \frac{1}{j\omega n}\underline{v} = A\underline{p}_W \end{cases} \tag{5.12}$$

After performed transformation, (5.12) enables a circuit design interpretation which is represented in Fig. 5.13. The ambient pressure $\underline{p}_0$ affects the common connection of both acoustic ports.

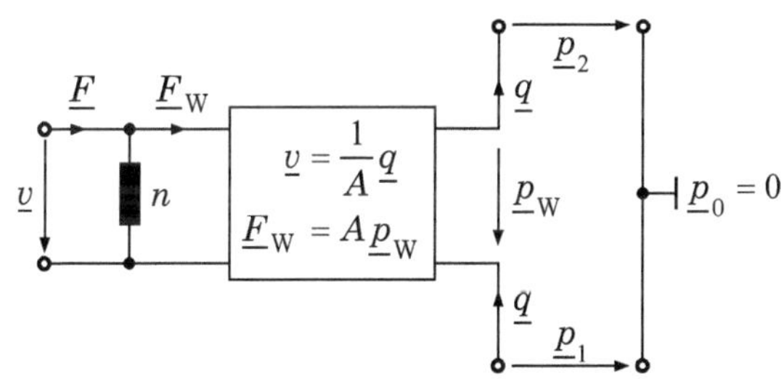

Fig. 5.13. Circuit of piston transducer according to Fig. 5.12

5.2.2 General Elastomechanical-Acoustic Plate Transducer

If the rigid piston is replaced by an elastomechanical plate element, a transducer circuit will be derivable (as shown in the following) which can be described also by only one transducer coefficient and two components. Concerning the following derivation, at first *two* transducer coefficients will be found. However, their identity is shown at the end.

An elastomechanical plate element is embedded into a mechanical-acoustic transducer. The plate element is deformed by action of force and pressure. The shifted volume results from occurring deflection (Fig. 5.14). Since the deflection ξ is position-dependent, it has to be integrated over the whole plate surface in order to determine the shifted volume V. The reference deflection ξ_0 is to occur at the application point of force F.

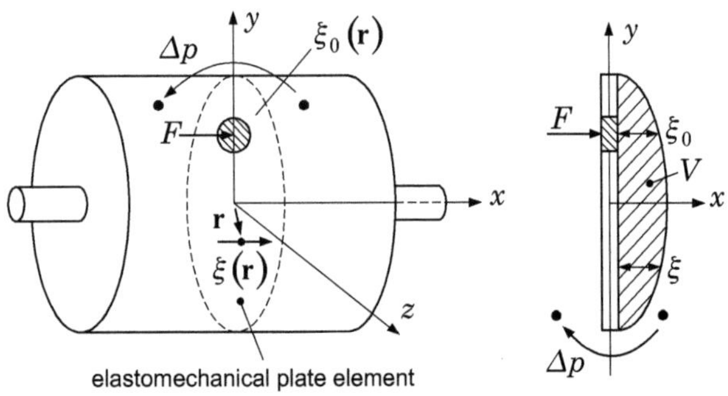

Fig. 5.14. General mechanical-acoustic plate transducer

For the following modeling and parameter determination, a restriction on special geometries and boundary conditions is done. Circular and lamellar diaphragms with pivoting and rigid restraint are chosen as base forms (Fig. 5.15).

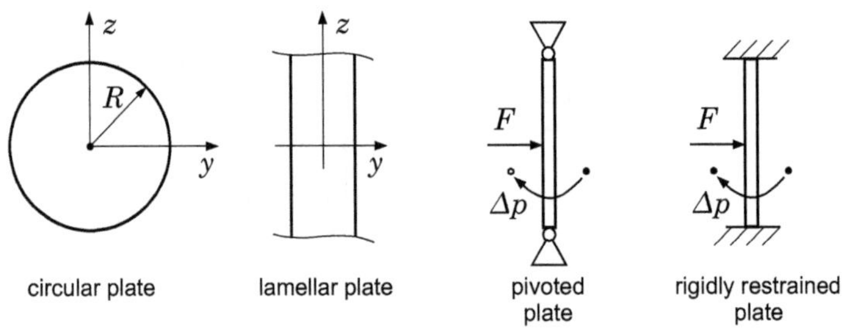

Fig. 5.15. Chosen geometries and boundary conditions for plate transducers

Concerning these elastic components, deflection functions for a centered force load and deflection functions for a pressure load can be gleaned from text books of elasticity theory (e.g. [43]).

By means of the source-referred position functions

$$\frac{\xi_F(x)}{F} = \left.\frac{\xi(x)}{F}\right|_{p=0} \quad \text{and} \quad \frac{\xi_p(x)}{p} = \left.\frac{\xi(x)}{p}\right|_{F=0}, \tag{5.13}$$

both the volumes (Fig. 5.16)

$$V_F = \int \xi_F(x)\,\mathrm{d}A \quad \text{and} \quad V_p = \int \xi_p(x)\,\mathrm{d}A \tag{5.14}$$

and the volumetric transfer factors

$$\frac{V_F}{F} = \left.\frac{V}{F}\right|_{p=0} \quad \text{and} \quad \frac{V_p}{p} = \left.\frac{V}{p}\right|_{F=0} \tag{5.15}$$

and the point transfer factors at the force application point

$$\frac{\xi_{F0}}{F} \quad \text{and} \quad \frac{\xi_{p0}}{p} \tag{5.16}$$

can be determined. Now the force response A_F and pressure response A_p

$$A_F = \frac{V_F}{\xi_{F0}} \quad \text{and} \quad A_p = \frac{V_p}{\xi_{p0}} \tag{5.17}$$

can be calculated by means of the available quotients.

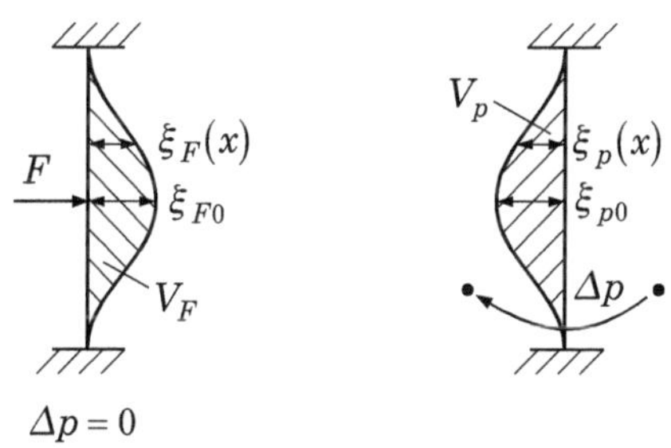

Fig. 5.16. Displaced volume in case of force and pressure load

Due to different deflection functions for different load (force or pressure), numerical values for A_F and A_p will be generally different. In order to facilitate the modeling considerations, following two relations are provided:

$$\frac{V_F}{F} = \frac{V_F}{\xi_{F0}}\frac{\xi_{F0}}{F} = A_F\frac{\xi_{F0}}{F} \tag{5.18}$$

$$\frac{V_p}{p} = \frac{V_p}{\xi_{p0}}\frac{\xi_{p0}}{p} = A_p\frac{\xi_{p0}}{p} \tag{5.19}$$

With a mechanical-acoustic transducer, both a force and pressure load of the elastomechanical plate element occurs simultaneously. Assuming linear relations being existent between cause and effect, both sources have a share in plate deformation independently. Thus, due to the required linearity, no reaction of an occurred deformation on a further deformation exists induced by the same or another cause. Thus, the position-dependent displacement results in $\xi(x) = \xi_F(x) - \xi_p(x)$ due to the specified direction of pressure. Thus, the position of force application and the displaced volume result in:

$$\xi_0 = \xi_{F0} - \xi_{p0} \quad \text{and} \quad V = V_F - V_p \tag{5.20}$$

By means of fraction expansion follows:

$$\xi_0 = \frac{\xi_{F0}}{F}F - \frac{\xi_{p0}}{p}p \quad \text{and} \quad V = \frac{V_F}{F}F - \frac{V_p}{p}p \tag{5.21}$$

The quotients in (5.21) can be physically interpreted. According to the definition, the ratio ξ_{F0}/F denotes the mechanical compliance of the force application point in absence of pressure. This compliance is called n_K. According to the definition, the ratio V_p/p denotes an acoustic compliance in absence of a force. This compliance is called $N_\mathrm{a,L}$. Thus, it follows:

$$n_\mathrm{K} = \frac{\xi_{F0}}{F} \quad \text{and} \quad N_\mathrm{a,L} = \frac{V_p}{p} \tag{5.22}$$

Equation (5.22) and the equations (5.18) and (5.19) result in:

$$\frac{V_F}{F} = A_F n_\mathrm{K} \quad \text{and} \quad \frac{\xi_{p0}}{p} = \frac{N_\mathrm{a,L}}{A_p} \tag{5.23}$$

In combination with (5.22) and (5.23), the equations in (5.21) can now be represented as system equations in the following way:

$$\xi_0 = n_\mathrm{K} F - \frac{N_\mathrm{a,L}}{A_p} p \tag{5.24}$$

$$V = A_F n_\mathrm{K} F - N_\mathrm{a,L} p \tag{5.25}$$

In order to provide a circuit interpretation, the linear system of equations (5.24) and (5.25) must be formulated according to $F, V = f(\xi, p)$. Rearranging yields:

$$F = \frac{1}{n_\mathrm{K}}\xi_0 + \underbrace{\frac{N_\mathrm{a,L}}{n_\mathrm{K}}\frac{1}{A_p}}_{1/A'} p \tag{5.26}$$

$$V = A_F \xi_0 - \underbrace{N_\mathrm{a,L}\left(1 - \frac{A_F}{A_p}\right)}_{N_\mathrm{a,K}} p \tag{5.27}$$

The physical meaning of the factors $1/A'$ and $N_{\mathrm{a,K}}$ in (5.26) and (5.27) is shown in Fig. 5.17. For a fixed force application point ($\xi_0 = 0$), the quantity A' denotes the effective area for pressure p, in order to generate the force F_{K} in the rigid rod for short-circuit operation. The acoustic compliance $N_{\mathrm{a,K}}$ denotes the displaced pressure-related volume V_{K} at $\xi_0 = 0$.

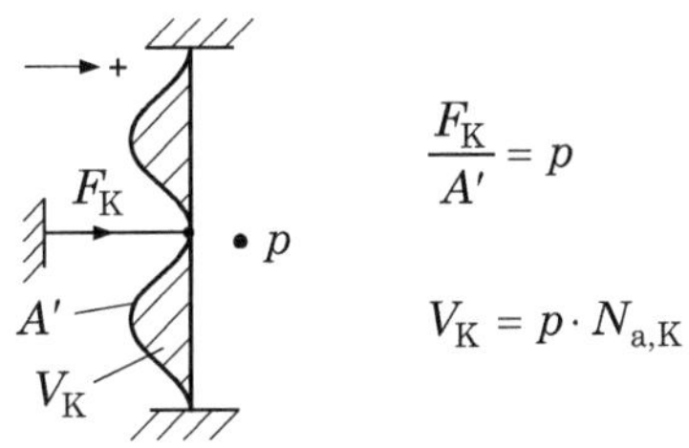

Fig. 5.17. To the explanation of the parameters $N_{\mathrm{a,K}}$ and A'

As already mentioned at the beginning, the quantities A_F and A' are identical. That is a consequence of the reciprocity of mechanical point systems. It results either directly from the associated differential equations under the assumption of loss-free elastomechanical systems or more clearly from the conception that an elastomechanical continuum can be described by a mechanical network according to Sect. 2.3, if a sufficiently fine discretization is assumed. For that purpose, according to (2.103) up to (2.105) for a certain quantity of point coordinates a pair of generalized coordinates pressure p and volume displacement V is introduced into this network.
The relation between the pair of coordinates (p, V) and the pair of coordinates (F, ξ) of a single reference point can be represented in the form

$$F = \alpha_{11}\xi + \alpha_{12}V \tag{5.28}$$

$$p = \alpha_{21}\xi + \alpha_{22}V \tag{5.29}$$

in combination with the reciprocity relation

$$\alpha_{21} = \alpha_{12}$$

resulting from (2.104). By rearranging the system of equations $F, V = f(\xi, V)$ into the form of (5.26) and (5.27), the identity

$$A' = A_F \tag{5.30}$$

follows. For the assumed case of a purely elastomechanical loss-free system, equation (5.30) is associated with the existence of a state function *internal energy*, as it is explained in detail in Sect. 2.4.2 and in Chap. 7. This fact enables an even more simple interpretation of (5.30). For that purpose, complex amplitudes are introduced into (5.26) and (5.27). Thus, the following equations are generated after rearranging and multiplication with and division by $\mathrm{j}\omega$, respectively:

$$\underline{F} - \frac{1}{\mathrm{j}\omega n_{\mathrm{K}}}\underline{v} = \frac{1}{A'}\underline{p} \tag{5.31}$$

$$\underline{q} + \mathrm{j}\omega N_{\mathrm{a,K}}\underline{p} = A_F\underline{v} \tag{5.32}$$

These two equations can be immediately interpreted by means of a circuit. The circuit is illustrated in Fig. 5.18.

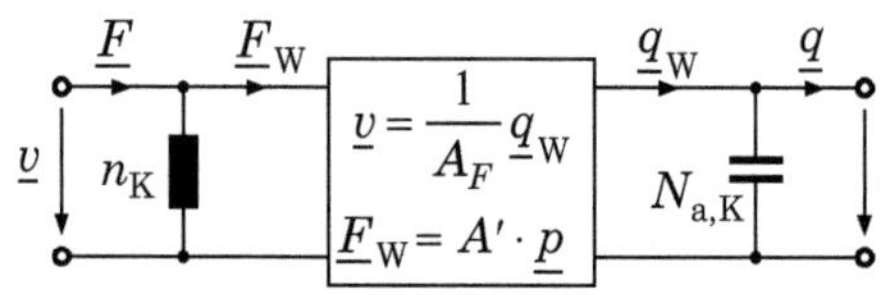

Fig. 5.18. Plate transducer circuit derived from (5.31) and (5.32)

Assuming that no internal losses occur in the transducer, the input power of the transducer must equal its output power. From this follows:

$$\hat{v}\hat{F} = \hat{p}\hat{q} \qquad \text{and} \qquad A' = A_F \tag{5.33}$$

Like in case of the piston transducer, for the coupling mechanism in the elastomechanical plate transducer only one quantity can be found as constituted above. For a selected plate shape and a certain surround, this quantity A_F is an inherent part of the effective plate area A. Thus, it makes sense to formulate this transducer constant by means of the plate A and a factor φ which is always less than 1. Thus, it applies:

$$A_F = A \cdot \varphi \tag{5.34}$$

The mechanical compliance n_{K} shown in the circuit diagram (Fig. 5.18) is indexed with the symbol K denoting a „short-circuited" output of the acoustic transducer ($\underline{p} = 0$). The acoustic compliance $N_{\mathrm{a,K}}$ which is also indexed with K due to the boundary condition „short-circuit" ($\underline{v} = 0$), can also be transformed to the mechanical side. In the following, the two-port network is also used in this form of representation.
Thus, it did succeed also in specifying a network representation for the elastomechanical plate transducer.

5.3 Characteristics of Selected Mechanical-Acoustic Transducers

In this section, the transducer parameters of the mechanical-acoustic plate and diaphragm transducer for chosen forms and boundary conditions initially specified according to Fig. 5.15 are summarized in tables.
The tables for the bending rod (Table 5.4), the circular bending plate (Table 5.7), the circular diaphragm (Table 5.8) and the lamellar diaphragm (Table 5.9) in combination with both selected supports (pivoted or clamped) include the transducer parameters n_{K}, $N_{\mathrm{a,K}}$, $N_{\mathrm{a,L}}$ and φ according to the circuit models illustrated in each case.
In order to complete the overview, for both load cases and both boundary conditions the quasi-static deflection functions, the displaced volume and the reference quantities are specified in each case for bending rod (Tables 5.2 and 5.3) and bending plate (Tables 5.5 and 5.6).
For bending rods (Tables 5.2 and 5.3) the results are represented in such a way that they are also valid for a clamped-free rod with free or angled force application. Therefore, the effective force amounts to $2F$ for rods supported on both sides.

Tables for Bending Rods

Force Load

Table 5.2. Characteristic functions and characteristic values of bending rod under force load

	pivoted	clamped
	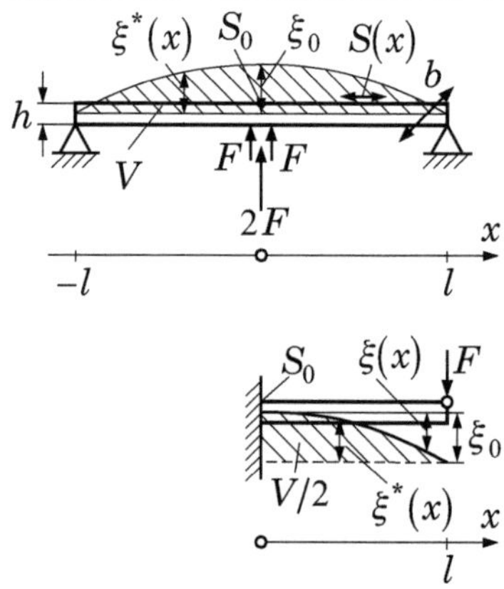	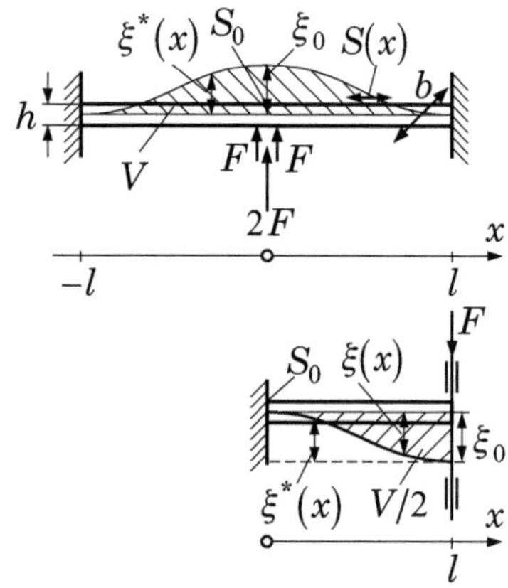
$\frac{\xi(x)}{\xi_0}$	$\frac{\xi(x)}{\xi_0} = \frac{1}{2}\left(\frac{x}{l}\right)^2\left(3 - \frac{x}{l}\right)$	$\frac{\xi(x)}{\xi_0} = \left(\frac{x}{l}\right)^2\left(3 - 2\frac{x}{l}\right)$
	$\frac{\xi^*(x)}{\xi_0} = 1 - \frac{3}{2}\left(\frac{x}{l}\right)^2 + \frac{1}{2}\left(\frac{x}{l}\right)^3$	$\frac{\xi^*(x)}{\xi_0} = 1 - 3\left(\frac{x}{l}\right)^2 + 2\left(\frac{x}{l}\right)^3$
$\frac{S(x)}{S_0}$	$\frac{S(x)}{S_0} = 1 - \frac{x}{l}$	$\frac{S(x)}{S_0} = 1 - 2\frac{x}{l}$ $(x \geq 0)$
ξ_0	$\frac{\xi_0}{F} = \frac{1}{3} n_0$	$\frac{\xi_0}{F} = \frac{1}{12} n_0$
S_0	$\frac{S_0}{\xi_0} = \frac{3}{2}\frac{h}{l^2}$	$\frac{S_0}{\xi_0} = 3\frac{h}{l^2}$
$\frac{V}{\xi_0 A}$	$\frac{V}{\xi_0 A} = \frac{5}{8}$ $A = 2l\,b$	$\frac{V}{\xi_0 A} = \frac{1}{2}$ $A = 2l\,b$
	$n_0 = \frac{l^3}{E \cdot I} = \frac{12\, l^3}{E\, b\, h^3}$	

Pressure Load

Table 5.3. Characteristic functions and characteristic values of bending rod under pressure load

	pivoted	clamped
	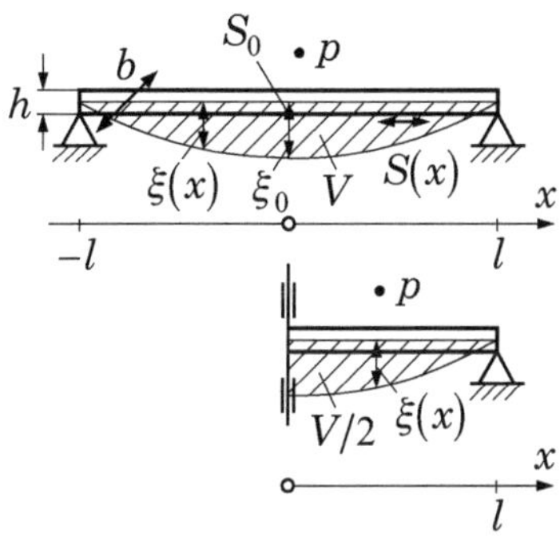	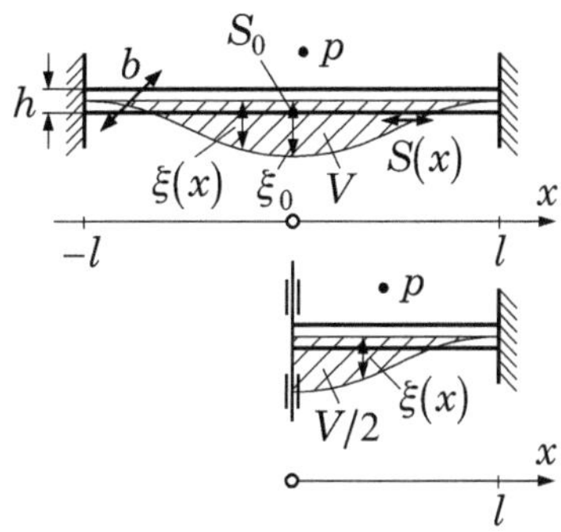
$\frac{\xi(x)}{\xi_0}$ $\frac{S(x)}{S_0}$	$\frac{\xi(x)}{\xi_0} = 1 - \frac{6}{5}\left(\frac{x}{l}\right)^2 + \frac{1}{5}\left(\frac{x}{l}\right)^4$ $\frac{S(x)}{S_0} = 1 - \left(\frac{x}{l}\right)^2$	$\frac{\xi(x)}{\xi_0} = \left(1 - \left(\frac{x}{l}\right)^2\right)^2$ $\frac{S(x)}{S_0} = 1 - 3\left(\frac{x}{l}\right)^2$
ξ_0	$\frac{\xi_0}{p\,A} = \frac{5}{48} n_0$ $A = 2l\,b$ $\frac{\xi_0}{p} = \frac{5}{2}\frac{l^4}{h^3}\frac{1}{E}$	$\frac{\xi_0}{p\,A} = \frac{1}{48} n_0$ $A = 2l\,b$ $\frac{\xi_0}{p} = \frac{1}{2}\frac{l^4}{h^3}\frac{1}{E}$
S_0	$\frac{S_0}{\xi_0} = \frac{6}{5}\frac{h}{l^2}$	$\frac{S_0}{\xi_0} = 2\frac{h}{l^2}$
$\frac{V}{\xi_0 A}$	$\frac{V}{\xi_0 A} = \frac{16}{25}$ $A = 2l\,b$	$\frac{V}{\xi_0 A} = \frac{8}{15}$ $A = 2l\,b$
	$n_0 = \frac{l^3}{E \cdot I} = \frac{12\,l^3}{E\,b\,h^3}$	

Table 5.4. Bending rods as mechanical-acoustic transducers

	clamped	pivoted
	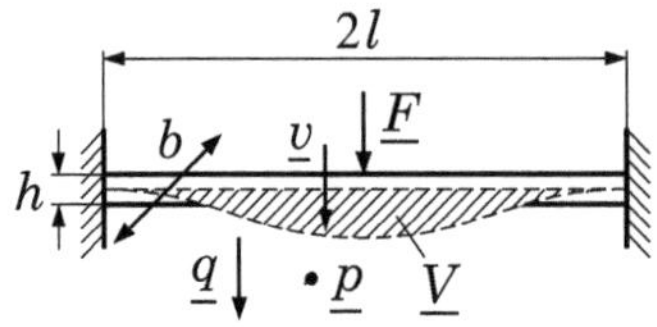	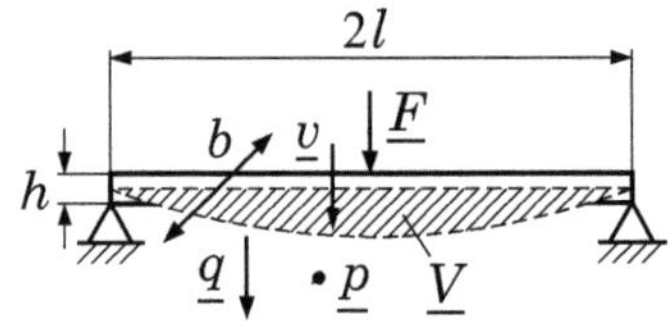
n_{K}	$\frac{1}{24} n_0 = \frac{1}{2} \frac{l^3}{bh^3} \frac{1}{E}$	$\frac{1}{6} n_0 = 2 \frac{l^3}{bh^3} \frac{1}{E}$
$N_{\mathrm{a,K}}$	$N_{\mathrm{a,K}} = \frac{1}{15} (A\varphi)^2 n_{\mathrm{K}} = \frac{1}{30} \frac{bl^5}{h^3} \frac{1}{E}$	$N_{\mathrm{a,K}} = \frac{3}{125} (A\varphi)^2 n_{\mathrm{K}} = \frac{3}{40} \frac{bl^5}{h^3} \frac{1}{E}$
$N_{\mathrm{a,L}}$	$N_{\mathrm{a,L}} = \frac{8}{15} \frac{bl^5}{h^3} \frac{1}{E}$	$N_{\mathrm{a,L}} = 3.2 \frac{bl^5}{h^3} \frac{1}{E}$
φ	$\frac{1}{2}$	$\frac{5}{8}$

$$n_0 = \frac{l^3}{E\,I} = \frac{12}{E} \frac{l^3}{b\,h^3}$$

$$A = 2l\,b$$

$$N_{\mathrm{a,L}} = \left(\frac{\underline{V}}{\underline{p}} \right)_{\underline{F}=0} = N_{\mathrm{a,K}} + (A\varphi)^2 n_{\mathrm{K}}$$

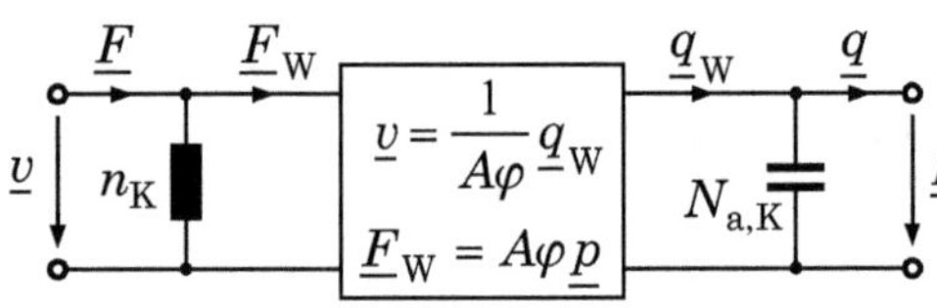

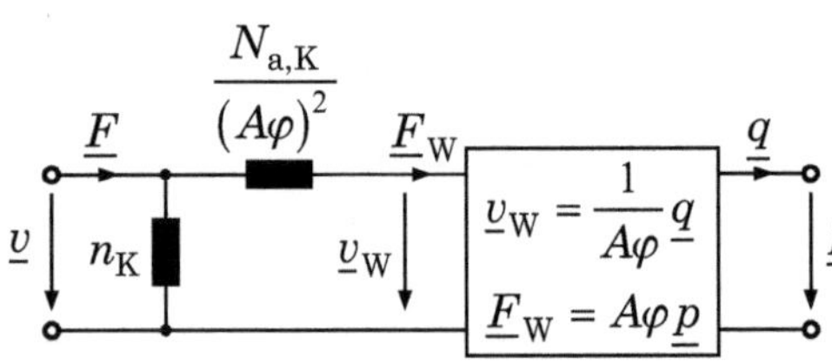

Tables for Bending Plates

Force Load

Table 5.5. Characteristic functions and characteristic values of bending plates under force load

	clamped	pivoted
	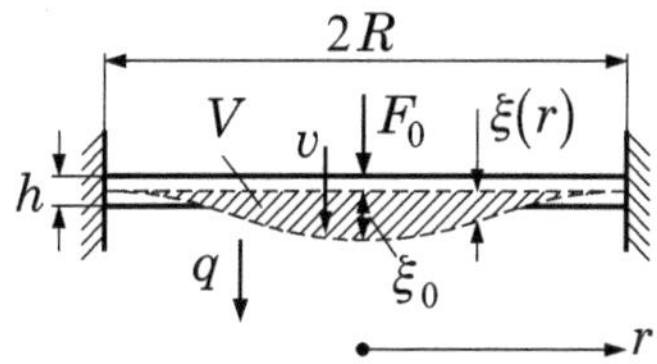	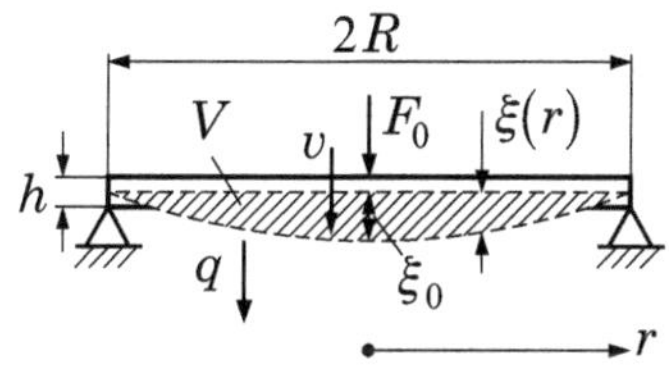
$\dfrac{\xi(r)}{\xi_0}$	$1-\left(\dfrac{r}{R}\right)^2-2\left(\dfrac{r}{R}\right)^2\ln\dfrac{R}{r}$	$1-\left(\dfrac{r}{R}\right)^2-2\left(\dfrac{r}{R}\right)^2\dfrac{1+\nu}{3+\nu}\ln\dfrac{R}{r}$
$\dfrac{S_r(r)}{S_0}$	$\ln\dfrac{R}{r}-1$	$\dfrac{\nu}{3+\nu}\left[\dfrac{1+\nu}{\nu}\left(\ln\dfrac{R}{r}\right)-1\right]$
$\dfrac{S_\vartheta(r)}{S_0}$	$\ln\dfrac{R}{r}$	$\dfrac{1}{3+\nu}\left[(1+\nu)\left(\ln\dfrac{R}{r}\right)+1\right]$
$\dfrac{T_r(r)}{T_0}$	$(1+\nu)\left(\ln\dfrac{R}{r}\right)-1$	$(1+\nu)\left(\ln\dfrac{R}{r}\right)$
$\dfrac{T_\vartheta(r)}{T_0}$	$\nu\left[\dfrac{1+\nu}{\nu}\left(\ln\dfrac{R}{r}\right)-1\right]$	$(1-\nu)\left[1+\dfrac{1+\nu}{1-\nu}\ln\dfrac{R}{r}\right]$
ξ_0	$\dfrac{R^2}{16\pi K}F_0$	$\dfrac{R^2}{16\pi K}\dfrac{3+\nu}{1+\nu}F_0$
$\dfrac{V}{\pi R^2\xi_0}$	$\dfrac{1}{4}$	$\dfrac{1}{2}-\dfrac{1}{4}\dfrac{1+\nu}{3+\nu}=\dfrac{5+\nu}{4(3+\nu)}$
S_0	$S_0=\dfrac{2h}{R^2}\xi_0$	
T_0	$T_0=\dfrac{3}{2\pi}\dfrac{1}{h^2}F_0=\dfrac{3}{2}\left(\dfrac{R}{h}\right)^2\dfrac{F_0}{\pi R^2}$	
	$K=\dfrac{h^3}{12}\dfrac{1}{s_{11}(1-\nu^2)}$ $\quad s_{11}=\dfrac{1}{E}$	Poisson ratio $\nu=-\dfrac{s_{12}}{s_{11}}$

Pressure Load

Table 5.6. Characteristic functions and characteristic values of bending plates under pressure load

	clamped	pivoted
	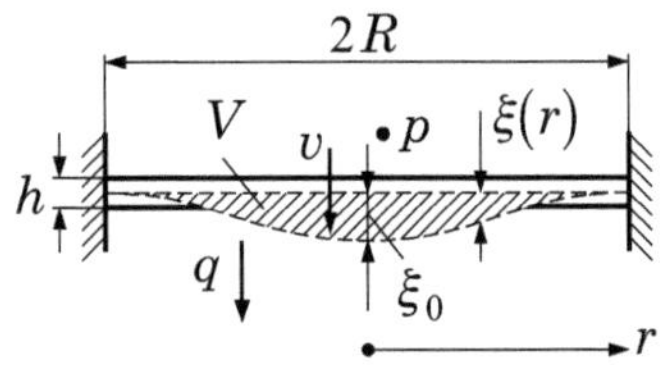	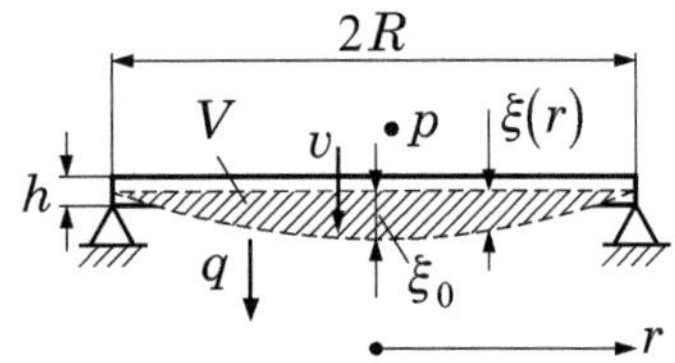
$\frac{\xi(r)}{\xi_0}$	$\left(1-\left(\frac{r}{R}\right)^2\right)^2$	$\frac{1+\nu}{5+\nu}\left[1-\left(\frac{r}{R}\right)^2\right]\left[\frac{5+\nu}{1+\nu}-\left(\frac{r}{R}\right)^2\right]$
$\frac{S_r(r)}{S_0}$	$1-3\left(\frac{r}{R}\right)^2$	$\frac{3+\nu}{5+\nu}\left[1-3\frac{1+\nu}{3+\nu}\left(\frac{r}{R}\right)^2\right]$
$\frac{S_\vartheta(r)}{S_0}$	$1-\left(\frac{r}{R}\right)^2$	$\frac{3+\nu}{5+\nu}\left[1-\frac{1+\nu}{3+\nu}\left(\frac{r}{R}\right)^2\right]$
$\frac{T_r(r)}{T_0}$	$1-\frac{3+\nu}{1+\nu}\left(\frac{r}{R}\right)^2$	$\frac{3+\nu}{1+\nu}\left[1-\left(\frac{r}{R}\right)^2\right]$
$\frac{T_\vartheta(r)}{T_0}$	$1-\frac{3\nu+1}{1+\nu}\left(\frac{r}{R}\right)^2$	$\frac{3+\nu}{1+\nu}\left[1-\frac{3\nu+1}{3+\nu}\left(\frac{r}{R}\right)^2\right]$
ξ_0	$\frac{R^4}{64\,K}p$	$\frac{R^4}{64\,K}\frac{5+\nu}{1+\nu}p$
$\frac{V}{\pi R^2\xi_0}$	$\frac{1}{3}$	$\frac{1}{2}-\frac{1}{6}\frac{1+\nu}{5+\nu}=\frac{7+\nu}{3(5+\nu)}$
S_0	$S_0=\frac{2h}{R^2}\xi_0$	
T_0	$T_0=\frac{3}{8}\left(\frac{R}{h}\right)^2(1+\nu)\,p$	
	$K=\frac{h^3}{12}\frac{1}{s_{11}\left(1-\nu^2\right)}$ $\quad s_{11}=\frac{1}{E}$	Poisson ratio $\nu=-\frac{s_{12}}{s_{11}}$

Table 5.7. Bending plates as mechanical-acoustic transducers

	clamped	pivoted
	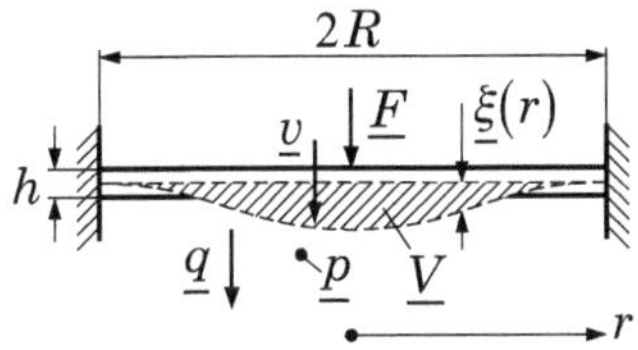	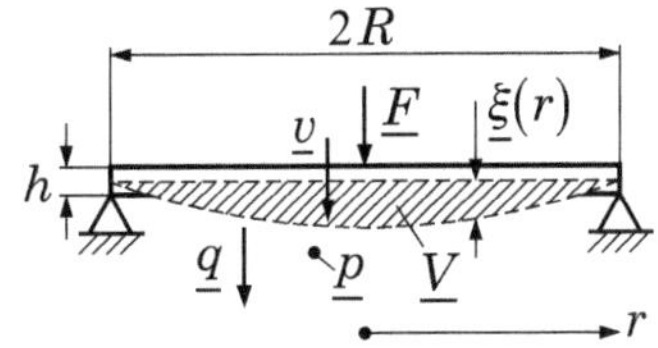
n_K	$\frac{R^2}{16\pi K} = \frac{3}{4\pi}\frac{R^2}{E\,h^3}\left(1-\nu^2\right)$ $\approx 0.22\frac{R^2}{E\,h^3}$ *)	$\frac{R^2}{16\pi K}\frac{3+\nu}{1+\nu} = \frac{3}{4\pi}\frac{R^2}{E\,h^3}(1-\nu)(3+\nu)$ $\approx 0.55\frac{R^2}{E\,h^3}$ *)
$N_{a,K}$	$\frac{1}{3}(A\varphi)^2 n_K = \frac{\pi}{64}\frac{R^6}{E\,h^3}\left(1-\nu^2\right)$ $\approx 0.045\frac{R^6}{E\,h^3}$ *)	$\frac{4}{3}\frac{(7+\nu)(3+\nu)}{(5+\nu)^2}(A\varphi)^2 n_K$ $\frac{\pi}{16}\frac{R^6}{E\,h^3}\frac{(7+\nu)^2-\frac{3}{4}(5+\nu)^2}{(7+\nu)(3+\nu)}(1-\nu)$ $\approx 0.184\frac{R^6}{E\,h^3}$ *)
$N_{a,L}$	$\frac{\pi}{16}\frac{R^6}{E\,h^3}\left(1-\nu^2\right) \approx 0.18\frac{R^6}{E\,h^3}$ *)	$\frac{\pi}{16}\frac{R^6}{E\,h^3}(1-\nu)(7+\nu) \approx 1.06\frac{R^6}{E\,h^3}$ *)
φ	$\frac{1}{4}$	$\frac{1}{2}-\frac{1}{4}\frac{1+\nu}{3+\nu} \approx 0.4$ *)

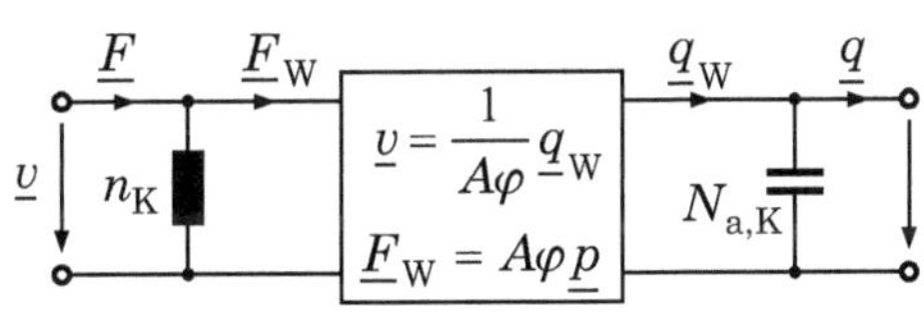

$$N_{a,L} = N_{a,K} + (A\varphi)^2 n_K$$

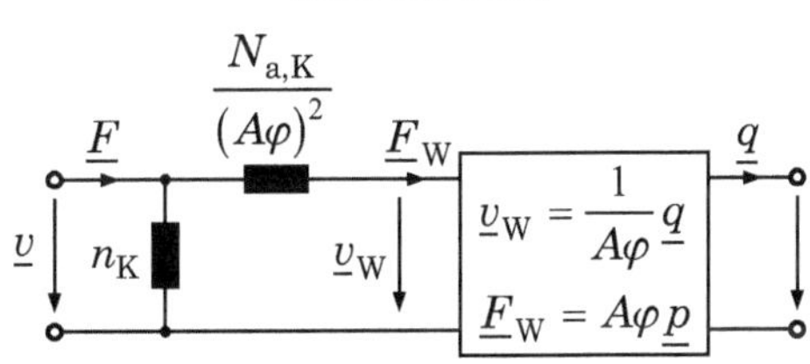

*) $\nu = 0.3$

Tables for Diaphragm Transducers

Table 5.8. Circular diaphragms as mechanical-acoustic transducers

	circular diaphragm with ring force	circular diaphragm with area force
	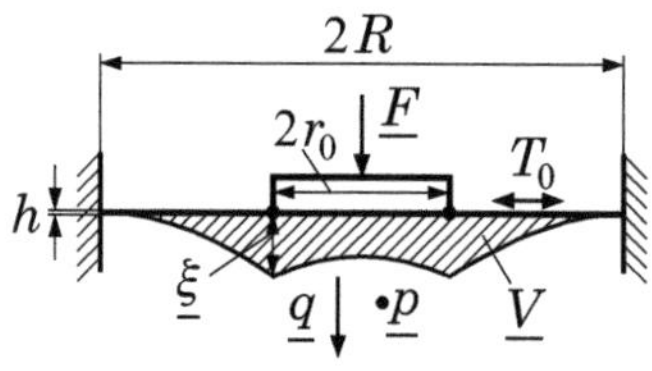	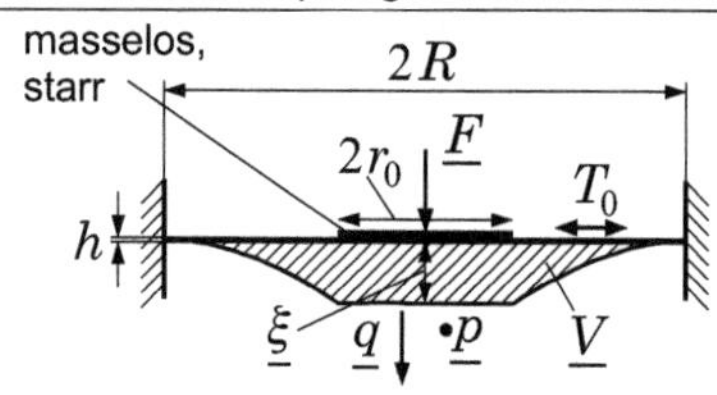
n_K	$\dfrac{\ln(R/r_0)}{2\pi T_0 h}$	$\dfrac{\ln(R/r_0)}{2\pi T_0 h}$
$N_{a,K}$	$\left(\dfrac{\ln(R/r_0)}{\left(1-(r_0/R)^2\right)^2}-1\right) n_K (A\varphi)^2$	$\left(\ln\dfrac{R}{r_0}\dfrac{1+(r_0/R)^2}{1-(r_0/R)^2}-1\right) n_K (A\varphi)^2$
$N_{a,L}$	$\dfrac{\pi}{8}\dfrac{R^4}{T_0\, h}$	$\dfrac{\pi}{8}\dfrac{R^4}{T_0\, h}\left(1-\left(\dfrac{r_0}{R}\right)^4\right)$
φ	$\dfrac{1}{2}\dfrac{1-(r_0/R)^2}{\ln(R/r_0)}$	$\dfrac{1}{2}\dfrac{1-(r_0/R)^2}{\ln(R/r_0)}$

$$N_{a,L} = N_{a,K} + (A\varphi)^2 n_K$$

$$A = \pi R^2$$

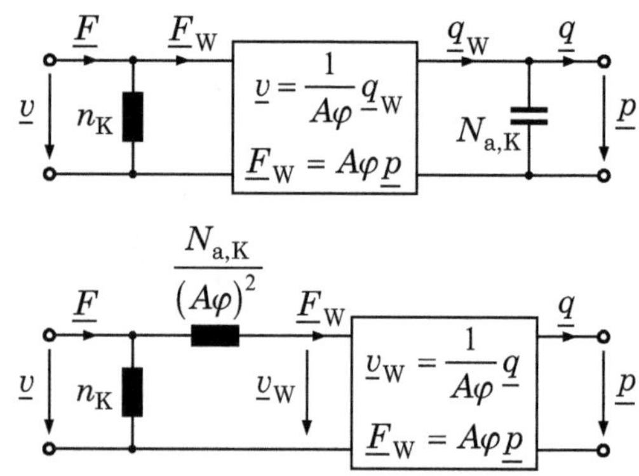

Table 5.9. Lamellar diaphragm as mechanical-acoustic transducer

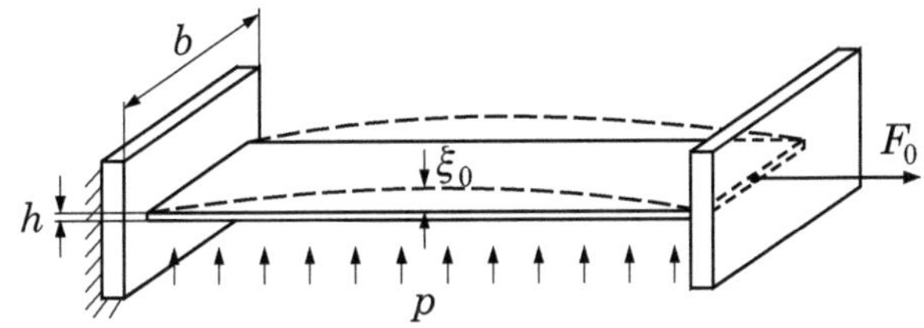

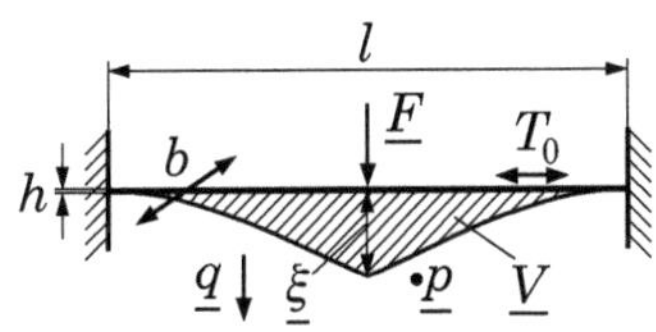

n_{K}	$\dfrac{l}{4T_0\, b\, h}$
$N_{\mathrm{a,K}}$	$\dfrac{1}{3}\, n_{\mathrm{K}}(A\varphi)^2 = \dfrac{1}{48}\, \dfrac{l^3 b}{T_0 h}$
$N_{\mathrm{a,L}}$	$\dfrac{1}{12}\, \dfrac{l^3 b}{T_0\, h}$
φ	$\dfrac{1}{2}$

6

Mechanical and Acoustic Networks with Distributed Parameters

The previous network analysis was solely carried out for systems with lumped components. Now an extension to systems with distributed, thus position-dependent parameters is realized. If the wavelength of extensional waves at higher frequencies reaches the order of magnitude of component dimensions, this extension will be inevitable. In the Sects. 6.1 and 6.2, these mechanical and acoustic systems are described as one-dimensional waveguides. The discretized representation in the form of finite network elements takes place in Sect. 6.3. Finally, a method of combined simulation of FEM calculation and network description is presented in Sect. 6.4.

6.1 Representation of Mechanical Systems as one-dimensional Waveguides

For many technical applications there are mechanical components, whose longest dimension is not small any longer compared to the wavelength within the component for the interesting frequency component. For rod-shaped structures (also with variable cross-section) a network representation can be specified for the component. The two cross-sectional dimensions must be small compared to the wavelength. According to terms of transmission line theory, it is referred to such components as one-dimensional waveguides. For selected frequency ranges it even succeeds to specify a substitute of waveguide by means of a few components. Thus, clarity is improved and it is provided for a fast assessability concerning preliminary drafts.
In the following, a rod is represented by means of a multiplicity of short rod pieces (Fig. 6.1). The relation between the coordinates at the one rod end depending on the coordinates at the other rod end can be determined by means of this model. In the circuit domain, this can be perceived as two-port network which can be represented by a T- or Π-circuit. This circuit comprises frequency-dependent components that can not be represented by simple spring, mass or frictional elements any longer. However, with these

A. Lenk et al., *Electromechanical Systems in Microtechnology and Mechatronics*,
Microtechnology and MEMS, DOI 10.1007/978-3-642-10806-8_6,

mathematically described impedances it can be calculated without any problems. That does not pose a problem for network analysis programs, since both loss-free and dissipative conductors are contained as standard components. For both limiting cases – low frequencies and in the neighborhood of resonances – an approximate representation works with usual components. Thus, for these cases a simplification of circuit design is possible which enables a constriction of the technical problem to the operation of few components as approximate representation. That allows for concentration on basic effects. As a result, the calculations with exact solutions can be checked for plausibility much better.

6.1.1 Extensional Waves within a Rod

Figure 6.1 shows a rod with length l, cross-sectional area A and YOUNG's modulus E. The force $\underline{F}$ is expected to affect both rod ends and an observation of the reference points x_1 and x_2 should be possible. The mass of rod can remain unconsidered for quasi-static forces. The result in network domain is a two-port network with a spring component n. The velocities amount to $\underline{v}_1$ and $\underline{v}_2$ at both rod ends, the force affecting both rod ends represents a compressive force. The direction of arrow in the network domain corresponds with the reference direction of arrow pointing to the right. The decrease in length $\underline{\xi}$ of the spring amounts to $(\underline{v}_1 - \underline{v}_2)/\mathrm{j}\omega$.
The rod can be split into many small rod parts of length l_i. Each of these rod elements will represent a two-port network of similar type (Fig. 6.2).
The two-port networks are interconnected in a cascade connection. These considerations provide for a position-dependent variation of the rod's cross-sectional area. Now each element features a different spring compliance.
If the mass inertia can not be neglected any longer with higher frequencies and if the rod is thin, there will be the possibility to branch off a force component at each connecting node between the partial springs in order to overcome the

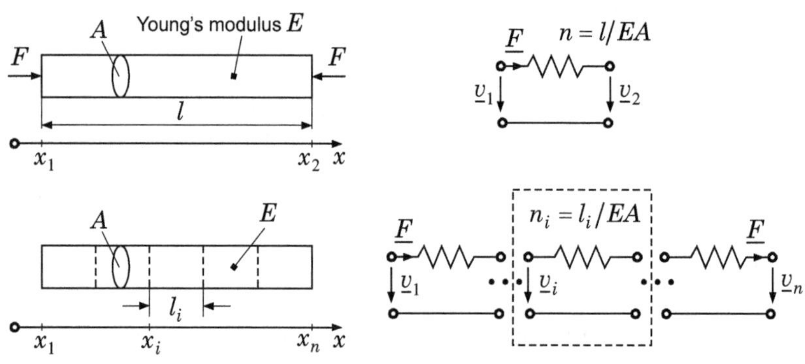

Fig. 6.1. Finite element decomposition of a rod

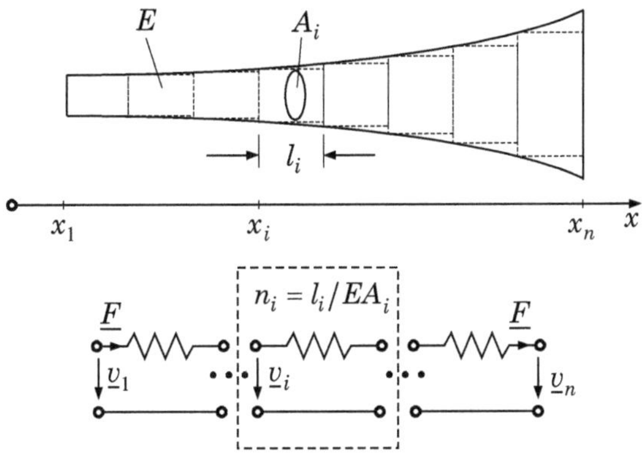

Fig. 6.2. Rod with variable cross-section

inertia of the partial mass (Fig. 6.3). The small cross-sectional dimension of rod was assumed, in order that the inertia forces for transverse contraction are so small that they may remain unconsidered. Thus, only one longitudinal state of stress but states of strain in all directions are existent within the rod. In case of a sufficient fine splitting of the rod, it is of no practical importance whether the node in front or behind the spring element is used for mass coupling. In order to achieve a symmetric representation, the first spring element is divided up into two half elements and the output is loaded with the second half of the rod spring element.

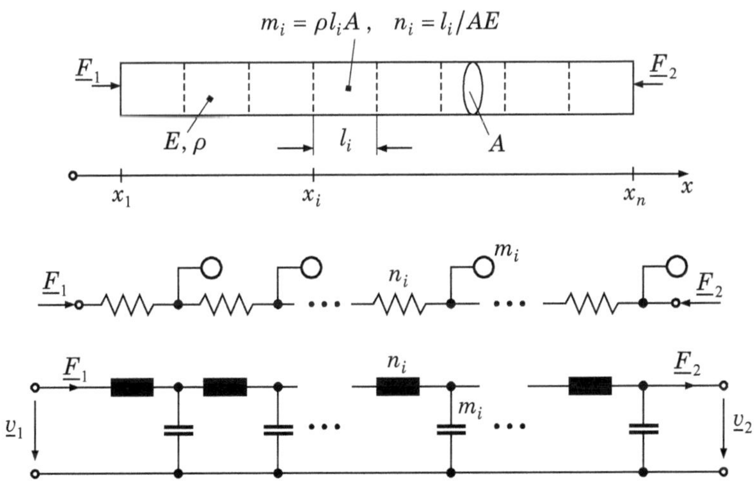

Fig. 6.3. Rod with considered mass inertia

If it is passed on to very short elements and in the limiting case to infinitely short elements of length Δx, KIRCHOFF's loop rule will yield according to Fig. 6.4:

$$\underline{v}(x) = \mathrm{j}\omega \Delta n \underline{F}(x) + \underline{v}(x + \Delta x)$$

$$\frac{\underline{v}(x + \Delta x) - \underline{v}(x)}{\Delta x} = -\mathrm{j}\omega n' \underline{F}(x) \Rightarrow \frac{\mathrm{d}\underline{v}}{\mathrm{d}x} = -\mathrm{j}\omega n' \underline{F},$$

where $n' = 1/AE$ denotes the compliance per unit length. For the node of the upper mass point it follows in combination with KIRCHHOFF's nodal rule:

$$\underline{F}(x) = \underline{F}(x + \Delta x) + \mathrm{j}\omega \Delta m \underline{v}(x + \Delta x)$$

$$\frac{\underline{F}(x + \Delta x) - \underline{F}(x)}{\Delta x} = -\mathrm{j}\omega m' \underline{v}(x + \Delta x) \Rightarrow \frac{\mathrm{d}\underline{F}}{\mathrm{d}x} = -\mathrm{j}\omega m' \underline{v},$$

where $m' = \rho A$ denotes the mass per unit length. Here, the quantity ρ denotes the density of material.

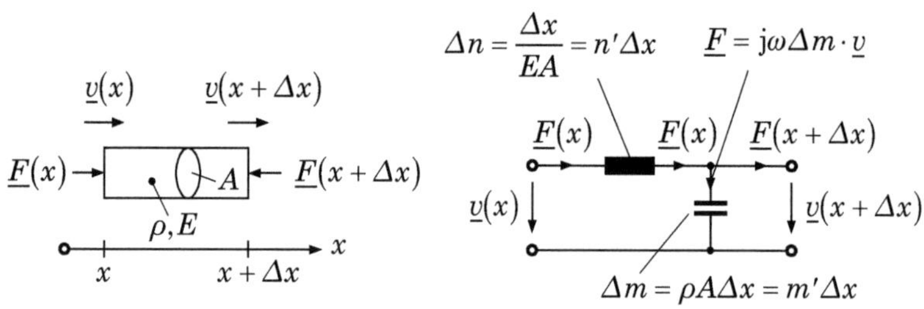

Fig. 6.4. Rod element

In the following, a position independence of both the cross-sectional area A and the material properties is to be assumed, in order that m' and n' can be considered as constants. By means of differentiation and insertion of both equations, two differential equations for $\underline{v}(x)$ and $\underline{F}(x)$ can be derived:

$$\left.\begin{aligned} \frac{\mathrm{d}\underline{F}}{\mathrm{d}x} &= -\mathrm{j}\omega m' \underline{v} \\ \frac{\mathrm{d}\underline{v}}{\mathrm{d}x} &= -\mathrm{j}\omega n' \underline{F} \end{aligned}\right\} \Rightarrow \left\{\begin{aligned} \frac{\mathrm{d}^2\underline{F}}{\mathrm{d}x^2} + \omega^2 m' n' \underline{F} = 0 \\ \frac{\mathrm{d}^2\underline{v}}{\mathrm{d}x^2} + \omega^2 m' n' \underline{v} = 0 \end{aligned}\right. \tag{6.1}$$

By means of the short form $\omega^2 m' n' = \frac{\omega^2 \rho}{E} = \beta^2$ and β denoting the position-independent *wave number*, the general solution for the velocity results in

$$\underline{v}(x) = \underline{C}_1 \, e^{j\beta x} + \underline{C}_2 \, e^{-j\beta x} \, .$$

If this solution is inserted into (6.1), a general solution for the force will be achieved:

$$\underline{F}(x) = -\frac{j\beta}{j\omega n'} \left(\underline{C}_1 \, e^{j\beta x} - \underline{C}_2 \, e^{-j\beta x} \right) \tag{6.2}$$

It is referred to the expression $\frac{j\beta}{j\omega n'} = \sqrt{\frac{m'}{n'}} = A\sqrt{\rho E}$ as wave impedance z_W. The complex amplitudes $\underline{C}_1$ and $\underline{C}_2$ still to be determined are specified by boundary conditions.
For $x = 0$ follows:

$$\underline{v}_1 = \underline{C}_1 + \underline{C}_2 \quad \text{and} \quad \underline{F}_1 = z_W \left(\underline{C}_2 - \underline{C}_1 \right)$$

Thus, the constants result in:

$$\underline{C}_1 = \frac{1}{2} \left(\underline{v}_1 - \frac{\underline{F}_1}{z_W} \right) \quad \text{and} \quad \underline{C}_2 = \frac{1}{2} \left(\underline{v}_1 + \frac{\underline{F}_1}{z_W} \right)$$

At the rod end ($x = l$), $\underline{v}(l) = \underline{v}_2$ and $\underline{F}(l) = \underline{F}_2$ are valid. Thus, the relation between input and output coordinates at the rod can be determined.
It applies:

$$\underline{v}_2 = \cos(\beta l) \, \underline{v}_1 - j \frac{1}{z_W} \sin(\beta l) \, \underline{F}_1 \tag{6.3}$$

and

$$\underline{F}_2 = -j z_W \sin(\beta l) \, \underline{v}_1 + \cos(\beta l) \, \underline{F}_1 \tag{6.4}$$

For the following network considerations it is desirable to represent the input coordinates as function of the output coordinates. As a result of coordinate transformation it can be written:

$$\begin{pmatrix} \underline{v}_1 \\ \underline{F}_1 \end{pmatrix} = \begin{pmatrix} \cos(\beta l) & j \frac{1}{z_W} \sin(\beta l) \\ j z_W \sin(\beta l) & \cos(\beta l) \end{pmatrix} \begin{pmatrix} \underline{v}_2 \\ \underline{F}_2 \end{pmatrix} \tag{6.5}$$

In order to improve the physical understanding for the introduced quantities, two thought experiments are made. If the output of rod is loaded with the real-valued impedance z_W, exactly this impedance takes effect also at the input of rod independently of rod length. Thus, the wave impedance has a descriptive meaning. If this rod loaded in such a way is excited with the frequency ω and if the position dependence of the velocity $\underline{v}(x)$ is considered, the length of one period λ (local repetition of same oscillating state, also called wavelength) is achieved according to the relation $\beta\lambda = 2\pi$. Thus, the constant $\beta = 2\pi/\lambda$ has a meaning corresponding to the reciprocal value of wavelength. The propagation velocity c of the wave image on the rod results from length λ of the local period divided by duration T of the temporal period.

Thus, it follows:

$$c = \frac{\lambda}{T} = f\lambda = \frac{\omega}{\beta} = \sqrt{\frac{E}{\rho}} \tag{6.6}$$

Thus, the wave velocity c depends only on the material of rod. In combination with (6.5), the two-port network solutions of the one-dimensional waveguide are available.
Additionally, it must be noted that by means of the described methods also exponential dependences of the components m' and n' on x can result in an analytical expression of the system equations.

In order to achieve a network-compatible solution, now the equivalent T- and Π-circuits (Fig. 6.5) are calculated.

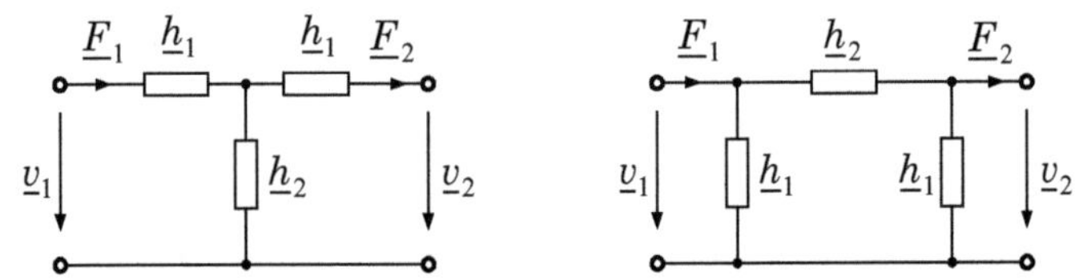

Fig. 6.5. Equivalent T- and Π-circuit of finite waveguide

The general admittances $\underline{h}$ are charted in the circuits, since the electrical resistances (and not their conductances) are usually used in two-port network analysis. The admittances can be determined by variation of the boundary conditions at the two-port network. $\underline{F}_2 = 0$ is chosen as boundary condition for the T-circuit. The admittances result in

$$\underline{h}_2 = \left(\frac{\underline{v}_2}{\underline{F}_1}\right)_{\underline{F}_2=0} \quad \text{and} \quad \underline{h}_1 + \underline{h}_2 = \left(\frac{\underline{v}_1}{\underline{F}_1}\right)_{\underline{F}_2=0} .$$

For the Π-circuit $\underline{v}_2 = 0$ is chosen advantageously. Now the admittances result in

$$\frac{1}{\underline{h}_2} = \left(\frac{\underline{F}_2}{\underline{v}_1}\right)_{\underline{v}_2=0} \quad \text{and} \quad \frac{1}{\underline{h}_1} + \frac{1}{\underline{h}_2} = \left(\frac{\underline{F}_1}{\underline{v}_1}\right)_{\underline{v}_2=0} .$$

Table 6.1 shows the components of the circuits, the approximations deduced for frequencies $\beta l \ll 1$ and their circuit. Thus, it is e.g. immediately possible to specify the mechanical input impedance $\underline{z}_\mathrm{e}$ of a rod with an open end at the other side ($\underline{F}_2 = 0$) (Fig. 6.6).
It applies

$$\underline{z}_\mathrm{e} = \frac{1}{\underline{h}_1 + \underline{h}_2} = \mathrm{j} z_\mathrm{W} \tan(\beta l) . \tag{6.7}$$

Table 6.1. Components of the T- und Π-circuit of one-dimensional waveguides

T -circuit	Π - circuit
$\boxed{\underline{h}_2 = \frac{1}{z_\mathrm{W}} \frac{1}{\mathrm{j}\sin\beta l}}$	$\boxed{\frac{1}{\underline{h}_2} = \frac{z_\mathrm{W}}{\mathrm{j}\sin\beta l}}$
$\underline{h}_1 + \underline{h}_2 = \frac{1}{z_\mathrm{W}\ \mathrm{j}\tan\beta l}$	$\frac{1}{\underline{h}_1} + \frac{1}{\underline{h}_2} = \frac{z_\mathrm{W}}{\mathrm{j}\tan\beta l}$
$\underline{h}_1 = \mathrm{j}\frac{1}{z_\mathrm{W}}\left(\frac{1}{\sin\beta l} - \frac{1}{\tan\beta l}\right)$	$\frac{1}{\underline{h}_1} = z_\mathrm{W}\left(\frac{1}{\mathrm{j}\tan\beta l} - \frac{1}{\mathrm{j}\sin\beta l}\right)$
$\boxed{\underline{h}_1 = \mathrm{j}\frac{1}{z_\mathrm{W}}\tan\beta\frac{l}{2}}$	$\boxed{\frac{1}{\underline{h}_1} = z_\mathrm{W}\ \mathrm{j}\tan\beta\frac{l}{2}}$
approximation for low frequencies $\beta l << 1$	
$\underline{h}_1 \approx \mathrm{j}\sqrt{\frac{n'}{m'}}\frac{1}{2}\omega\sqrt{m'n'}\cdot l$	$\frac{1}{\underline{h}_1} \approx \mathrm{j}\sqrt{\frac{m'}{n'}}\frac{1}{2}\omega\sqrt{m'n'}\cdot l$
$\underline{h}_1 \approx \mathrm{j}\omega\frac{n}{2}$	$\frac{1}{\underline{h}_1} \approx \mathrm{j}\omega\frac{m}{2}$
$\underline{h}_2 \approx \sqrt{\frac{n'}{m'}}\frac{1}{\mathrm{j}}\frac{1}{\omega\sqrt{m'n'}\cdot l}$	$\frac{1}{\underline{h}_2} \approx \sqrt{\frac{m'}{n'}}\frac{1}{\mathrm{j}\omega\sqrt{m'n'}\cdot l}$
$\underline{h}_2 \approx \frac{1}{\mathrm{j}\omega m}$	$\underline{h}_2 \approx \mathrm{j}\omega n$

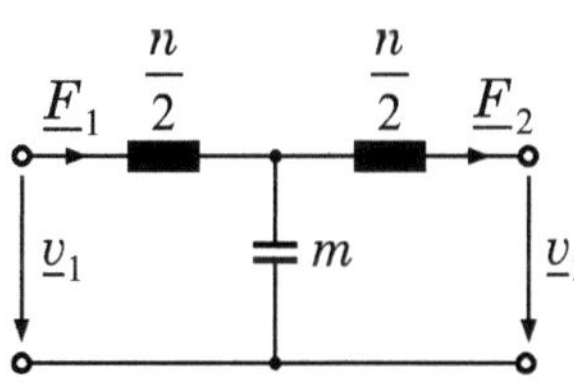

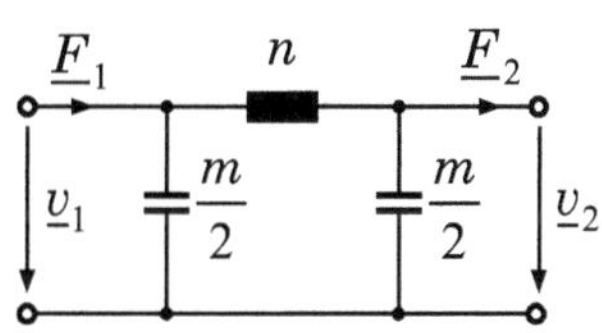

$m = m'l$, $n = n'l$ $l\ldots$ rod length

The input impedance $\underline{z}_\mathrm{e}$ into the rod is that one of a mass, whereby the absolute value of mass is increasing constantly with frequency up to the resonant frequency of rod.

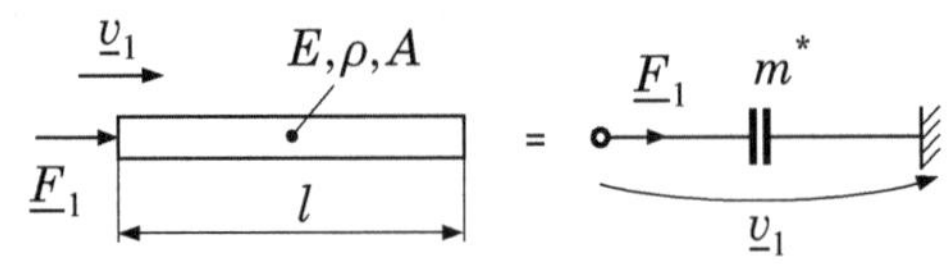

Fig. 6.6. Rod as mass load

From (6.7) follows

$$\omega m^* = z_{\mathrm{W}} \tan(\beta l)$$

and as a result of double-sided multiplication with $l\sqrt{m'n'}$ it can be written

$$m^* = m \frac{\tan(\beta l)}{\beta l}. \tag{6.8}$$

The pole is at $\beta l = \pi/2$. At the same time the rod length amounts to $l = \lambda/4$. This leads to a consequence being important for practice. The form of mass load is often used, in order to prevent the motion of a reference point (which can not be connected with a fixed point (e.g. point of origin) due to the present design) in steady state at a fixed operating frequency by means of an almost infinitely large mass m^*. The secondary effects which occur during the transient time must be considered separately.

6.1.2 Approximate Calculation of the Input Impedance

For practical tasks, the impedance $\underline{z}_0(\omega)$ at a selected point of a more complicated system or continuum is often of interest. In the following, it is assumed that this system is characterized by only very low internal losses and that no active power can be transfered to the environment. An estimate of frequency of the expected first pole or zero of this impedance will often suffice, in order to be able to assess to what extent it is permitted to calculate with the quasi-static solution. Often a quasi-static description of system can be specified faster than the frequency function of impedance. In the following, it is demonstrated how an estimate for the first pole or zero from a well-known quasi-static system description and how a simple model of the searched impedance can be specified.

As a first approximation, the searched impedance of the system can be represented by a parallel or a series connection of a mass and spring. A parallel connection will be necessary, if the reference point, for which the impedance function has to be identified, shows a resilent character in case of very low frequencies. If the reference point shows the character of a mass for low frequencies, then the approximate representation of the system can be realized by means of a series connection of a spring and mass. The values of m_{ers} and n_{ers} represent the equivalent parameters of the system (Fig. 6.7). On the supposition of negligible losses it can be assumed that the total kinetic and potential energy of the system can be found in the mass and spring, respectively. By means of restriction to sinusoidal processes, these conclusions can now be referred to the power ratings. In the following, this will now be demonstrated for parallel connection using the example of a mass.

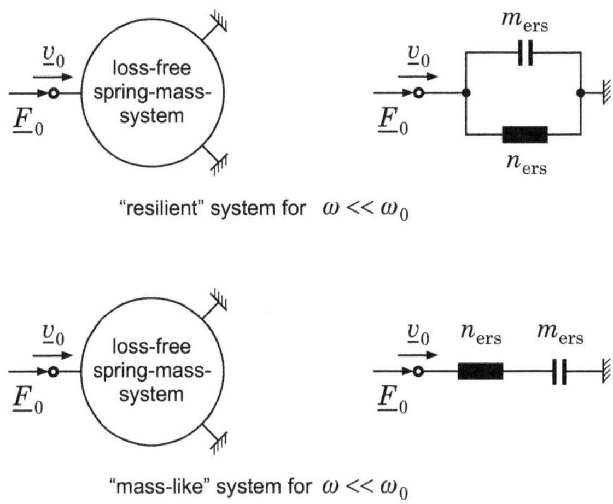

Fig. 6.7. Approximate representation of input impedance

The kinetic energy of the equivalent mass amounts to

$$W_{\text{kin}} = m_{\text{ers}} \frac{v^2}{2}. \tag{6.9}$$

With $v = \hat{v}_0 \sin(\omega t)$ and $Q(t) = \mathrm{d}W_{\text{kin}}/\mathrm{d}t$ it can be written:

$$Q(t) = \frac{1}{2} \hat{v}_0^2 \omega m_{\text{ers}} \sin(2\omega t)$$

$$\hat{Q} = \frac{1}{2} \hat{v}_0^2 \omega m_{\text{ers}} = \tilde{v}_0^2 \omega m_{\text{ers}}$$

A similar consideration for the spring yields

$$\hat{Q}_{\text{pot}} = \frac{\tilde{v}_0^2}{\omega n_{\text{ers}}}. \tag{6.10}$$

$\hat{Q}$ denotes the peak value of the oscillating reactive power. Figure 6.8 shows the relations between equivalent components and reactive power for both the series and parallel connection.

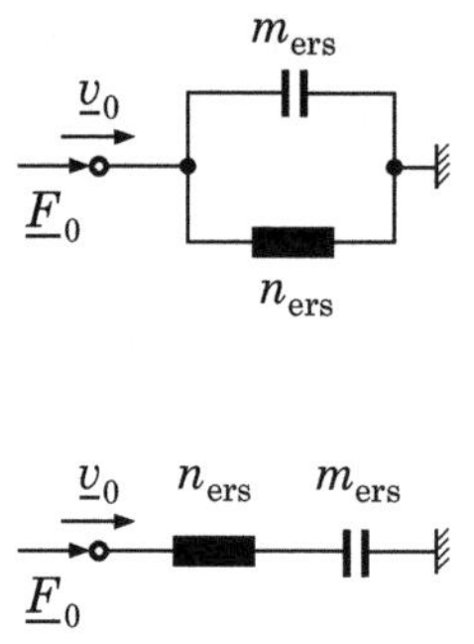

$$\hat{Q}_{kin} = \tilde{v}_0^{\,2} \cdot \omega \cdot m_{ers} \qquad \hat{Q}_{pot} = \frac{\tilde{v}_0^{\,2}}{\omega \cdot n_{ers}}$$

$$\hat{Q}_{kin} = \frac{\tilde{F}_0^{\,2}}{\omega \cdot m_{ers}} \qquad \hat{Q}_{pot} = \tilde{F}_0^{\,2} \cdot \omega \cdot n_{ers}$$

Fig. 6.8. Determination of equivalent components

If the calculation of the kinetic or potential reactive power succeeds for a real system, the equivalent parameters of the circuit can be specified. A particularly easy access consists in a direct determination of one of both equivalent parameters from the quasi-static behavior and in the calculation of the other parameter from the reactive power which the system receives at low frequencies (thus quasi-static). The equivalent parameters are frequency-independent. For low frequencies, the circuit represents the impedance very well. As a suitable approximation, the calculated first resonant frequency corresponds with the real resonant frequency.
Since in the preceding section exact solutions were calculated for the input impedance of a homogeneous rod with constant cross-sectional area, in the following, the approximation method is demonstrated using the example of the system *rod with rigid load.* Generally, velocity and force along the rod are dependent on the position. The rod is split again into volume elements. In order to calculate the power, it has to be integrated over their volumes.
The kinetic power being contained in a rod element of width $\mathrm{d}x$ amounts to

$$\mathrm{d}Q_{\mathrm{kin}} = \tilde{v}^2\,(x)\,\omega\rho A \mathrm{d}x.$$

The total power in the rod amounts to

$$Q_{\mathrm{kin}} = \int_0^l \mathrm{d}Q_{\mathrm{kin}} = \omega \frac{m}{l} \tilde{v}_0^2 \int_0^l \frac{\tilde{v}^2\,(x)}{\tilde{v}_0^2} \mathrm{d}x.$$

This power must also be found in the equivalent mass. Thus, it applies:

$$\omega m \tilde{v}_0^2 \frac{1}{l} \int_0^l \frac{\tilde{v}^2\,(x)}{\tilde{v}_0^2} \mathrm{d}x = \omega m_{\mathrm{ers}} \tilde{v}_0^2$$

Thus, the calculation rule for the equivalent mass results in

$$m_{\mathrm{ers}} = m\frac{1}{l}\int\limits_0^l \frac{\tilde{v}^2(x)}{\tilde{v}_0^2}\mathrm{d}x. \tag{6.11}$$

In combination with

$$\mathrm{d}Q_{\mathrm{pot}} = \tilde{F}^2(x)\,\omega\frac{1}{EA}\mathrm{d}x,$$

the calculation of potential power results in the relation enabling the calculation of the equivalent compliance

$$n_{\mathrm{ers}} = n\frac{1}{l}\int\limits_0^l \frac{\tilde{F}^2(x)}{\tilde{F}_0^2}\mathrm{d}x. \tag{6.12}$$

Now the application of the considerations to the example *rod with rigid load* is possible in an easy way. Assuming quasi-static dependences

$$\underline{v}(x) = \mathrm{j}\omega\underline{\xi}(x) = \underline{v}_0\left(1-\frac{x}{l}\right) \qquad \text{(constant strain)} \qquad \text{and}$$

$$\underline{F}(x) = \underline{F}_0 \qquad \text{(position-independent force)},$$

it follows

$$m_{\mathrm{ers}} = m\frac{1}{l}\int\limits_0^l \left(1-\frac{x}{l}\right)^2\mathrm{d}x = \frac{m}{3} \quad \text{and} \quad n_{\mathrm{ers}} = n\frac{1}{l}\int\limits_0^l \mathrm{d}x = n. \tag{6.13}$$

The rod with respect to the boundary condition „rigid load" and the equivalent solution are represented in Fig. 6.9. The resonant frequency of the equivalent system amounts to:

$$\omega_{\mathrm{r,ers}} = \frac{1}{\sqrt{\frac{m}{3}n}} = \frac{\sqrt{3}}{l}\sqrt{\frac{E}{\rho}} = \sqrt{3}\frac{c}{l} \approx 1.73\frac{c}{l}$$

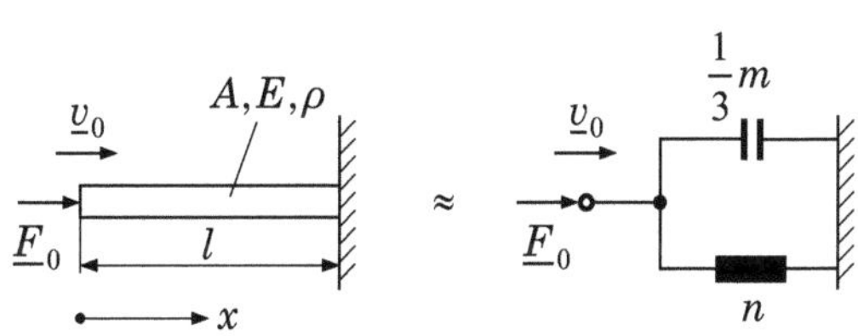

Fig. 6.9. Sample calculation of input impedance of a rigidly loaded rod

The real resonant frequency of the system amounts to:

$$\omega_{\mathrm{r}} = 2\pi \frac{1}{4} \frac{c}{l} = 1.57 \frac{c}{l}$$

This relation is sufficient for a rough estimate.

The same numerical values result for the rod with „no-load". The allocation to compliance and mass is only interchanged. This results in $m_{\mathrm{ers}} = m$ and $n_{\mathrm{ers}} = n/3$ for the rod with no-load.
If the solution for low frequencies is taken from Table 6.1 (T-circuit) for the rod with no-load (Fig. 6.6), the result will be a somewhat too low resonant frequency in combination with $n_{\mathrm{ers}} = n/2$. However, its deviation is nearly of same size. The mass m^* being effective at the front rod end according to Fig. 6.6 is represented in Fig.6.10 as a function of frequency for both cases „exact solution according to (6.8)" and „approximate solution resulting from $n/2$ and m". The quality of approximation can be assessed from the position of poles.

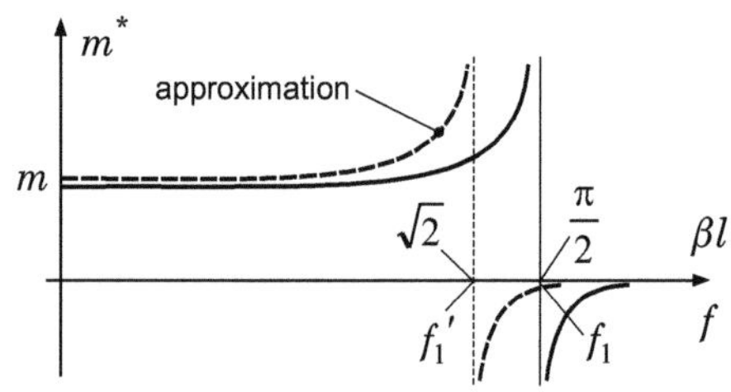

Fig. 6.10. Frequency response of mass being effective at the input
Comparison of approximation and exact solution

In case of a detailed knowledge of the dynamic ratios, the calculation of the kinetic or potential powers and thus the determination of the equivalent components represent another possibility. In this case, the result will be a good mapping in the neighborhood of frequency used for the calculation, but not for low frequencies.
However, the most important advantage of the reactive power method consists in the fact that the method enables an estimate of the first resonant frequency with structures, for which the *quasi-static deformation* can be taken from reference books (e.g. bending of beams and plates). In case of a clamped and homogeneous rod with constant cross-sectional area and subjected to a bending stress by means of a force affecting the free end, the method yields an equivalent mass $m_{\mathrm{ers}} = 0.235m$ on the assumption of a quasi-static compliance. The exact value amounts to $m_{\mathrm{ers}} = 0.25m$.

6.1.3 Approximate Representation of an Impedance at Resonance

In particular, filter systems (e.g. signal filters, structure-borne sound filters) and monofrequent sound radiators (e.g. ultrasonic transducers) operate only in the neighborhood of an operating frequency. They usually comprise components which can not be defined as lumped components any longer. In these cases, the task is to substitute the components being available at first as one-dimensional waveguides by simple mass-spring-systems generating a good representation in the neighborhood of operating frequency. The methods of approximation are also of importance for acoustic and rotational systems and therefore they are presented somewhat in more detail. The advantage of applying the approximate solution consists in the possibility that different circuit designs can be combined (see Sect. 3.1.4). Thus, it succeeds to represent the conditions being effective at operating frequency by means of basic structures. These basic structures can be interpreted by engineers very well and the effects of changes can be grasped quickly. Using the complete waveguide structure, independent of that, the exact calculation is made for the finally selected array of parameters by means of network analysis programs. However, the functional understanding primarily results from the deduced basic structure. It is developed for this objective.

The method represented subsequently in Fig. 6.11 is based on the assumption that in the operating frequency range zeros of the admittance or impedance are existent, whose neighborhood is to be simulated by components. That is mostly fulfilled with the initially specified assignments of tasks. An analytically well-known frequency function of a complex admittance $\underline{h}(\omega)$ is presupposed for the first of the two following considerations. For this function, it is sought for a series connection consisting of a spring and mass component which results in a correct representation at $\omega = \omega_\nu$ in the neighborhood of a selected zero of the function $\underline{h}(\omega)$. For that purpose, the frequency function of admittance is formulated as a Taylor series expansion around the point of the νth zero:

$$\underline{h}(\omega) = \underline{h}(\omega_\nu) + \left(\frac{\mathrm{d}\underline{h}}{\mathrm{d}\omega}\right)_{\omega_\nu} (\omega - \omega_v) + \ldots$$

Using the zero $\underline{h}(\omega_\nu) = 0$ and neglecting the higher terms of series yields

$$\underline{h}(\omega) = \left(\frac{\mathrm{d}\underline{h}}{\mathrm{d}\omega}\right)_{\omega_\nu} (\omega - \omega_v) .$$

The series connection of the equivalent components m_ν and n_ν results in the admittance

$$\underline{h}_\nu(\omega) = \mathrm{j}\omega n_\nu + \frac{1}{\mathrm{j}\omega m_\nu} = \mathrm{j}\omega_\nu n_\nu \left(\frac{\omega}{\omega_\nu} - \frac{\omega_\nu}{\omega}\right) \approx \mathrm{j}\omega_\nu n_\nu \frac{2(\omega - \omega_\nu)}{\omega_\nu} .$$

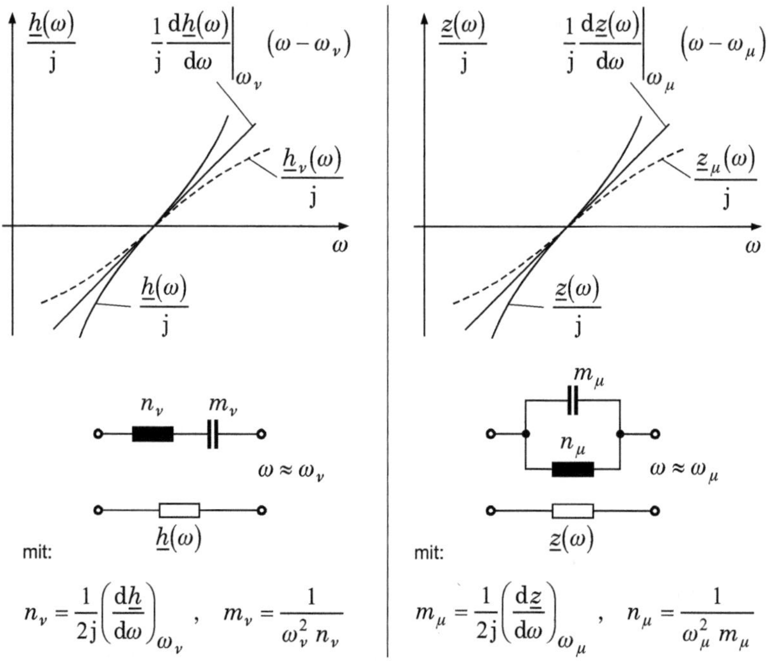

Fig. 6.11. Simulation of admittance/impedance in the neighborhood of a zero

By equating both admittance approximations, the equivalent components can be defined according to

$$n_\nu = \frac{1}{2\mathrm{j}}\left(\frac{\mathrm{d}\underline{h}}{\mathrm{d}\omega}\right)_{\omega_\nu} \tag{6.14}$$

and

$$m_\nu = \frac{1}{\omega_\nu^2 n_\nu}. \tag{6.15}$$

The second consideration is based on zeros of the impedance at $\omega = \omega_\mu$. Using the same calculation method, similar results are achieved for a parallel connection of a spring and mass component. A summary of the approaches is represented in Fig. 6.11. Often required results for zeros and poles of typical admittance functions are summarized in Table 6.2.

6.1.4 Approximated two-port Network Representation at Resonance

In addition to the possibility of representing mechanical and acoustic one-dimensional waveguides as T- or Π-circuit, models with transformer-like and

Table 6.2. Approximate solutions for the input admittance of one-dimensional waveguides

$\underline{h}$	m_1	n_1	ω_1	l/λ_1	f_1	
$\mathrm{j}h_z \tan \beta l$	$\frac{m}{2}$	$\frac{8}{\pi^2} n$	$\frac{\pi}{2} \frac{1}{\sqrt{mn}}$	$\frac{1}{4}$	$\frac{1}{4} \frac{c}{l}$	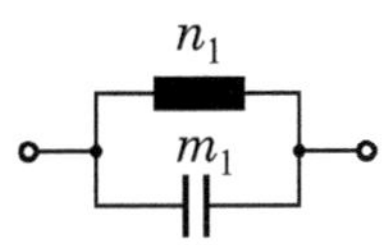
$\frac{h_z}{\mathrm{j} \tan \beta l}$	$\frac{m}{2}$	$\frac{2}{\pi^2} n$	$\frac{\pi}{\sqrt{mn}}$	$\frac{1}{2}$	$\frac{1}{2} \frac{c}{l}$	$\omega \approx \omega_1$
$\mathrm{j}h_z \tan \beta l$	$\frac{2}{\pi^2} m$	$\frac{n}{2}$	$\frac{\pi}{\sqrt{mn}}$	$\frac{1}{2}$	$\frac{1}{2} \frac{c}{l}$	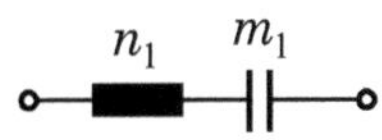
$\frac{h_z}{\mathrm{j} \tan \beta l}$	$\frac{8}{\pi^2} m$	$\frac{n}{2}$	$\frac{\pi}{2} \frac{1}{\sqrt{mn}}$	$\frac{1}{4}$	$\frac{1}{4} \frac{c}{l}$	$\omega \approx \omega_1$

gyrator-like coupling two-port networks can be found. In principle, linear reversible two-port networks can be represented by means of one of three possible circuit structures (Fig. 6.12). In order to determine the new circuit representation, exactly the same method is used as with the T- or Π-circuit (Fig. 6.5).

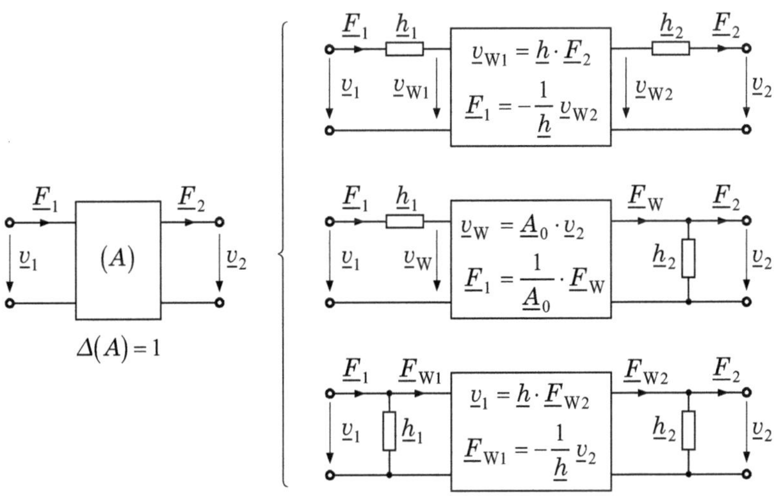

Fig. 6.12. Representation possibilities of linear reversible two-port networks
$\underline{h}$ *transformation admittance,* $\underline{A}_0$ *transformation ratio*

Based on the chain matrix (6.5) of the waveguide, the parameters of a circuit as illustrated in Fig. 6.12 are defined by the wave parameters $z_W = 1/h_W$ and βl. A further consideration is required for the selection of the three possible circuits and for checking the suitability.
The separation of a transformer or gyrator respectively is only reasonable, if their *transformation ratio* $\underline{A}_0$ or their *transformation admittance* $\underline{h}$ hardly depends on frequency, respectively. If the attempt of obtaining a representation by this circuit results in a sufficiently low frequency dependence for $\underline{A}_0$ or $\underline{h}$, then this circuit representation can be considered to be suitable. Thus, it would be possible to assume approximately a frequency independence for a limited frequency range. Therefore, all advantages of circuit representation would be available for the representation. In Fig. 6.13, this procedural method is demonstrated in the following using the example of a $\lambda/2$-waveguide.
At first, it is assumed on trial that a loss-free waveguide can be represented by a transformer-like two-port network in the neighborhood of $\lambda/2$-resonance (Fig. 6.12 center). The formation rule for the parameters $\underline{h}_{1K}$, $\underline{A}_0$ and $\underline{h}_{2L}$

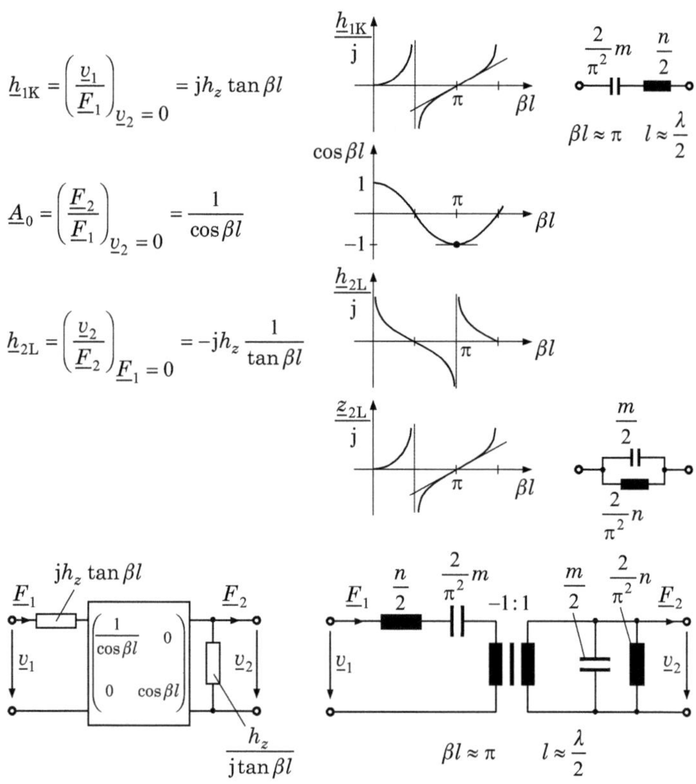

Fig. 6.13. $\lambda/2$-waveguide with transformer-like coupling element

shown in Fig. 6.13 is applied to (6.5) and thus yields the results specified in same figure.

The analytical expressions for the admittances can be represented as circuit by means of the results presented in Table 6.2. In the neighborhood of $\beta l \approx \pi$, the expression $\underline{A}_0 = 1/\cos(\beta l)$ yields the value -1. Therefore, it can be represented by the component *transformer*. The transformation ratio is hardly frequency-dependent within a certain range around resonance. Thus, it is reasonable to represent the waveguide as simple circuit within this frequency range. Circuit solutions for the frequency area around $\beta l \approx \pi/2$ according to the waveguide length $\lambda/4$ are achieved by similar considerations. Here, the other two circuit variants result equivalently in almost frequency-independent transformation admittances within the range of signal frequency. The choice of solution to be applied advantageously depends on further circuit environment (sources and loads of waveguide). Figure 6.14 shows the results.

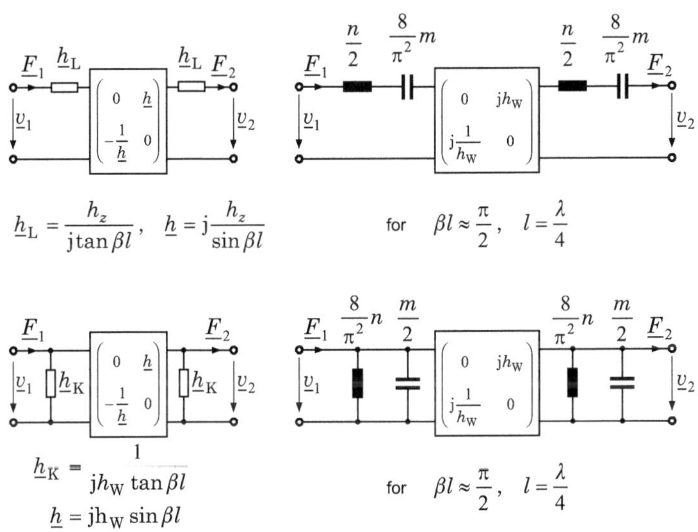

Fig. 6.14. $\lambda/4$-waveguide with gyrator-like coupling element

The capability of the presented methods is now demonstrated using the example of a *mechanical filtering device*. The series connection of three waveguides $\lambda/2 + \lambda/4 + \lambda/2$ according to Fig. 6.15 will result in a band filter structure, if the excitation is achieved by means of a source of force $\underline{F}_1$ and the output signal is generated by the velocity $\underline{v}_2$ of the unloaded third piece of waveguide. Now the waveguide pieces can be represented as cascade connection. Figure 6.16 shows the interconnection and parameter definitions.

It is referred to the wave admittance of the central piece as h_z. By means of interconnection of the elements connected in parallel in each case and by

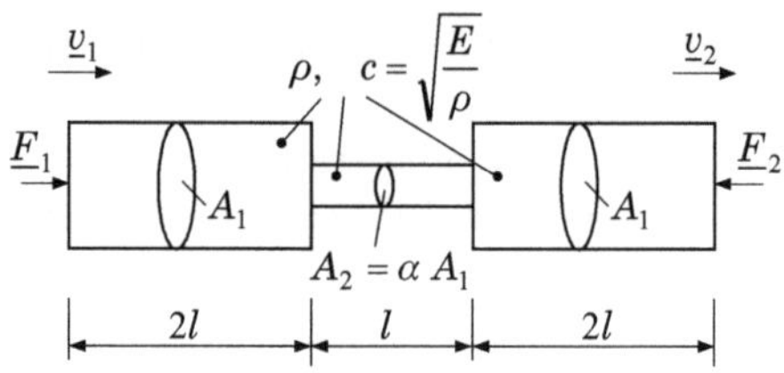

Fig. 6.15. Filter structure

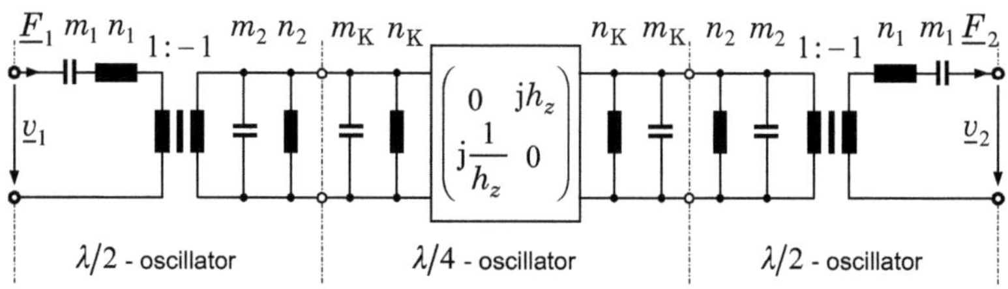

$$m_1 = \frac{4}{\pi^2}\rho l A_1 \qquad m_2 = \rho l A_1 \qquad m_K = \frac{1}{2}\rho l A_2 \qquad h_z = \frac{1}{A_2 \rho c}$$

$$n_1 = \frac{l}{\rho c^2 A_1} \qquad n_2 = \frac{4}{\pi^2}\frac{l}{\rho c^2 A_1} \qquad n_K = \frac{8}{\pi^2}\frac{l}{\rho c^2 A_2}$$

Fig. 6.16. Circuit of mechanical filter

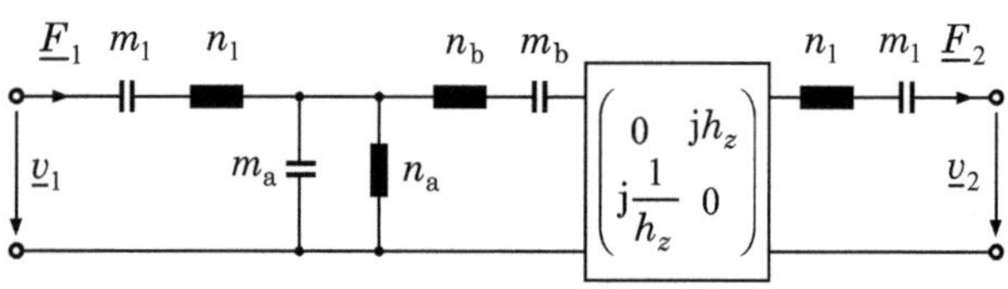

$$\alpha = \frac{A_2}{A_1}$$

$$\frac{m_1}{m_a} = \frac{1}{\pi^2}\frac{4}{1+\frac{\alpha}{2}} \approx \frac{4}{\pi^2}\bigg|_{\alpha \ll 1} \qquad \frac{n_1}{n_a} = \frac{\pi^2}{4}\left(1+\frac{\alpha}{2}\right) \approx \frac{\pi^2}{4}\bigg|_{\alpha \ll 1}$$

$$\frac{m_b}{m_a} = \frac{4}{\pi^2}\frac{\alpha^2}{\left(1+\frac{\alpha}{2}\right)^2} \approx \frac{4}{\pi^2}\alpha^2\bigg|_{\alpha \ll 1} \qquad \frac{n_b}{n_a} = \frac{\pi^2}{4}\frac{1}{\alpha^2}\left(1+\frac{\alpha}{2}\right)^2 \approx \frac{\pi^2}{4}\frac{1}{\alpha^2}\bigg|_{\alpha \ll 1}$$

Fig. 6.17. Summarized circuit of Fig. 6.16

means of transformation of the right parallel connection by the gyrator, the structure according to Fig. 6.17 is achieved. The area ratio was abbreviated with α. The new components m_a, m_b, n_a and n_b are generated. Due to the missing load at the filter output and due to the force feeding at the input side, the external series connections can remain unconsidered and the gyrator can be replaced by a short-circuit. Now the force in this short-circuit jumper denotes the output quantity (Fig. 6.18).

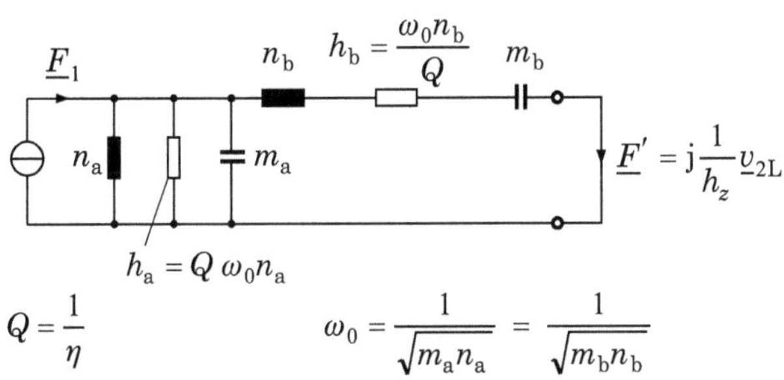

Fig. 6.18. Basic structure of mechanical filter

If now all the losses (being necessary for a reasonable function) by series and parallel connections of frictional admittances with selectable value for Q are considered, the desired basic structure of the filter component will be available. Figure 6.19 shows the filter frequency response in principle assuming same and suitably chosen values for quality factor Q. Section 9.1.2 comprises a practical example of a micromechanical filter.

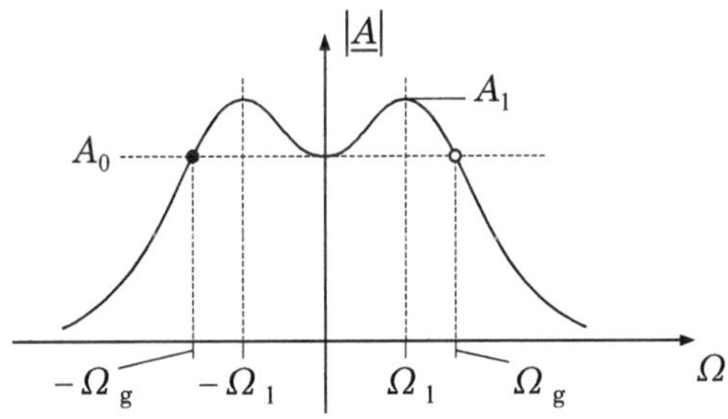

Fig. 6.19. Filter frequency response in principle

6.1.5 Flexural Vibrations within a Rod

The waveguide described in Sect. 6.1.1 is specified by the fact that a one-dimensional wave is defined by **one** pair of coordinates.

In this section, now a finite bending rod (see Sect. 5.1.2) is considered, whose ends are affected both by vertical forces and velocities and torsional moments

and angular velocities (Fig. 6.20 above). Thus, for bending waves discussed in the following the wave process is defined by **two** pairs of coordinates.
The deformations being generated within the rod are defined by following assumptions

- no shear deformation and
- strain $S_1(x)$ and mechanical stress $T_1(x)$ proportional to distance from neutral axis

For thin bending rods these simplifying assumptions result only in insignificant deviations from reality.
In order to handle dynamic processes, the mass of rod must be taken into consideration. Therefore, as already presented in Sect. 5.1.2, a circuit of a differential rod element is assumed, whose associated mass is now concentrated centrally. The structure of the differential rod element with length Δx consists of two ideal rods with length $\Delta x/2$ which are interconnected by a torsion spring $\Delta n_{\mathrm{R}} \sim \Delta x$ and a torque-free joint. In order to take the influence of mass into consideration, the mass $\Delta m = \varrho A \Delta x$ is attached to this joint. Figure 6.21 illustrates the associated circuit.

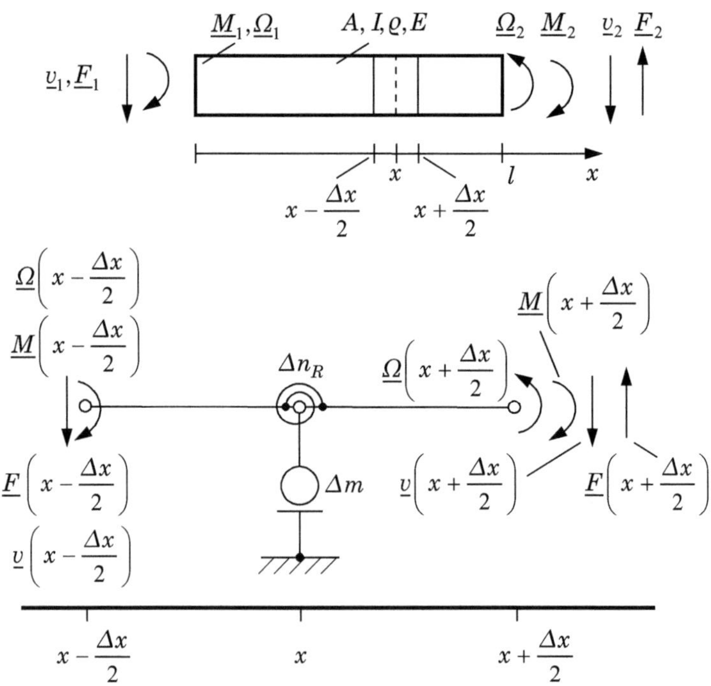

Fig. 6.20. Coordinates of bending rod and rod element

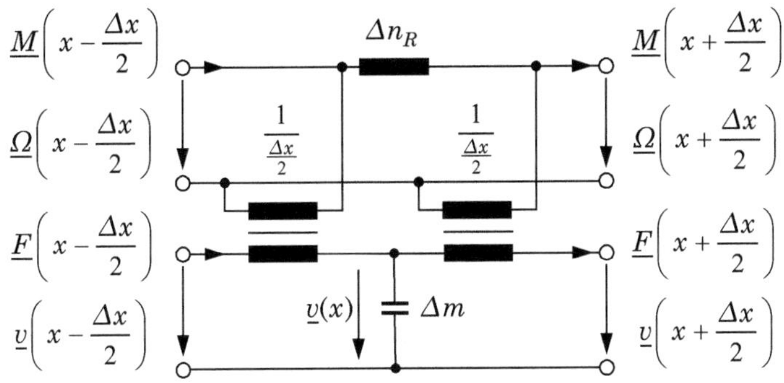

Fig. 6.21. Circuit of a rod element with concentrated mass

As already demonstrated in the preceding sections, loop and nodal equations can be derived from the circuit shown above in an analogous manner. They finally pass into differential equations for $\Delta x \to 0$. The loop and nodal equations yield:

$$\begin{aligned}&\underline{M}\,(x-\Delta x/2)-\underline{M}\,(x+\Delta x/2)\\&=\Delta x/2\,[\underline{F}\,(x-\Delta x/2)+\underline{F}\,(x+\Delta x/2)]\approx\Delta x\underline{F}\,(x)\,,\end{aligned}$$

$$\begin{aligned}&\underline{\Omega}\,(x-\Delta x/2)-\underline{\Omega}\,(x+\Delta x/2)\\&=\mathrm{j}\omega n'_{\mathrm{R}}\Delta x\,[\underline{M}\,(x-\Delta x/2)-\underline{F}\,(x-\Delta x/2)\,\Delta x/2]\,,\end{aligned}$$

$$\begin{aligned}&\underline{F}\,(x-\Delta x/2)-\underline{F}\,(x+\Delta x/2)=\mathrm{j}\omega m'\Delta x\underline{v}\,(x)\\&=\mathrm{j}\omega m'\Delta x\,[\underline{v}\,(x-\Delta x/2)-\underline{\Omega}\,(x-\Delta x/2)\,\Delta x/2]\,,\end{aligned}$$

$$\begin{aligned}&\underline{v}\,(x-\Delta x/2)-\underline{v}\,(x+\Delta x/2)\\&=-\,[\underline{\Omega}\,(x-\Delta x/2)+\underline{\Omega}\,(x+\Delta x/2)]\,\Delta x/2\approx-\Delta x\underline{\Omega}\,(x)\,.\end{aligned}$$

Similarly to the preceding sections it is written $\Delta n_{\mathrm{R}} = n'_{\mathrm{R}}\Delta x$ and $\Delta m = m'\Delta x$.

For $\Delta x \to 0$ following expressions are achieved:

$$\frac{\mathrm{d}\underline{M}}{\mathrm{d}x}=-\underline{F},\qquad\frac{\mathrm{d}\underline{v}}{\mathrm{d}x}=\underline{\Omega},$$

$$\frac{\mathrm{d}\underline{\Omega}}{\mathrm{d}x}=-\mathrm{j}\omega n'_{\mathrm{R}}\underline{M},\qquad\frac{\mathrm{d}\underline{F}}{\mathrm{d}x}=-\mathrm{j}\omega m'\underline{v},$$

$$n'_{\mathrm{R}}=\frac{1}{EI},\qquad m'=\varrho A.$$

These equations result in a differential equation for $\underline{v}(x)$

$$\frac{\mathrm{d}^4\underline{v}}{\mathrm{d}x^4} - \omega^2 m' n'_{\mathrm{R}} \underline{v} = 0.$$

The solution of this differential equation amounts to

$$\underline{v}(x) = A_1 \exp(\beta x) + A_2 \exp(-\beta x) + A_3 \exp(\mathrm{j}\beta x) + A_4 \exp(-\mathrm{j}\beta x)$$

$$\beta l = l\sqrt{\omega\sqrt{m' n'_{\mathrm{R}}}} = \sqrt{\omega\sqrt{m' l n'_{\mathrm{R}} l^3}} = \sqrt{\omega\sqrt{n_0 m}} = \sqrt{\frac{\omega}{\omega_0}}.$$

In the following, the reference quantities ω_0, n_{R}, z_0 and the total mass m of a rod with length l are introduced:

$$n_0 = \frac{l^3}{EI}, \qquad m = \varrho l A,$$

$$\omega_0 = \frac{1}{\sqrt{m n_0}}, \qquad z_0 = \sqrt{\frac{m}{n_0}} = \omega_0 m = \frac{1}{\omega_0 n_0}.$$

In the following, the solution is not formulated with the solution components $\exp(\pm\beta x)$ and $\exp(\pm\mathrm{j}\beta x)$ but with *Rayleigh functions* which also form a complete solution system as independent linear combinations of the solution components $\exp(\pm\beta x)$ and $\exp(\pm\mathrm{j}\beta x)$, respectively:

$$S(x) = \tfrac{1}{2}(\sinh x + \sin x), \qquad C(x) = \tfrac{1}{2}(\cosh x + \cos x),$$

$$s(x) = \tfrac{1}{2}(\sinh x - \sin x), \qquad c(x) = \tfrac{1}{2}(\cosh x - \cos x).$$

The functions S, C, s, c merge in the mentioned order by differentiation.

$$\underline{v}(x) = \underline{\alpha}_1 S(\beta x) + \underline{\alpha}_2 C(\beta x) + \underline{\alpha}_3 s(\beta x) + \underline{\alpha}_4 c(\beta x) \tag{6.16}$$

The remaining coordinates $\underline{\Omega}$, $\underline{M}$ and $\underline{F}$ also emerge from differentiation:

$$\underline{\Omega}(x) = \beta\left[\underline{\alpha}_1 C(\beta x) + \underline{\alpha}_2 s(\beta x) + \underline{\alpha}_3 c(\beta x) + \underline{\alpha}_4 S(\beta x)\right] \tag{6.17}$$

$$\underline{M}(x) = -\frac{\beta^2}{j\omega n'_{\mathrm{R}}}\left[\underline{\alpha}_1 s(\beta x) + \underline{\alpha}_2 c(\beta x) + \underline{\alpha}_3 S(\beta x) + \underline{\alpha}_4 C(\beta x)\right] \tag{6.18}$$

$$\underline{F}(x) = -\frac{\beta^3}{j\omega n'_{\mathrm{R}}}\left[\underline{\alpha}_1 c(\beta x) + \underline{\alpha}_2 S(\beta x) + \underline{\alpha}_3 C(\beta x) + \underline{\alpha}_4 s(\beta x)\right] \tag{6.19}$$

The constants $\underline{\alpha}_1$ up to $\underline{\alpha}_4$ can be determined by means of the two boundary conditions $(\underline{v}, \underline{F}, \underline{\Omega}, \underline{M})_{x=0} = (\underline{v}_1, \underline{F}_1, \underline{\Omega}_1, \underline{M}_1)$ and $(\underline{v}, \underline{F}, \underline{\Omega}, \underline{M})_{x=l} =$

$(\underline{v}_2, \underline{F}_2, \underline{\Omega}_2, \underline{M}_2)$ and thus the coordinates at position $x = 0$ can be expressed by those at position $x = l$. The resultant chain, admittance and impedance matrices are summarized in Table 6.3.
Following relations with $\eta = \sqrt{\omega/\omega_0}$ are advantageous for practical calculations:

$$\left.\begin{aligned}
&c(\eta)\,C(\eta) - s^2(\eta) = S^2(\eta) - c(\eta)\,C(\eta) = \tfrac{1}{2}\sinh\eta\sin\eta,\\
&S(\eta)\,C(\eta) - s(\eta)\,c(\eta) = \tfrac{1}{2}(\sinh\eta\cos\eta + \cosh\eta\sin\eta),\\
&s(\eta)\,C(\eta) - c(\eta)\,S(\eta) = \tfrac{1}{2}(\sinh\eta\cos\eta - \cosh\eta\sin\eta),\\
&C^2(\eta) - s(\eta)\,S(\eta) = \tfrac{1}{2}(1 + \cosh\eta\cos\eta)\\
&N = c^2(\eta) - s(\eta)\,S(\eta) = \tfrac{1}{2}(1 - \cosh\eta\cos\eta)
\end{aligned}\right\} \tag{6.20}$$

For low frequencies ($\eta \ll 1$) the following approximations including the fourth powers of η are achieved:

$$\left.\begin{aligned}
&S(\eta) = \eta && N = \eta^4/12\\
&c(\eta) = \eta^2/2 && S(\eta)\,C(\eta) - s(\eta)\,c(\eta) = \eta\\
&s(\eta) = \eta^3/6 && c(\eta)\,C(\eta) - s^2(\eta) = \eta^2/2\\
&C(\eta) = 1 + \eta^4/24 && c(\eta)\,S(\eta) - s(\eta)\,C(\eta) = \eta^3/3
\end{aligned}\right\} \tag{6.21}$$

In order to demonstrate the meaning of the quantities β and z_0, a solution $\underline{v}(x)$

$$\underline{v}(x) = \underline{v}_0 \exp(-\mathrm{j}\beta x) \qquad \text{with} \qquad \beta = \sqrt{\omega\sqrt{m' n'_{\mathrm{R}}}} = \frac{1}{l}\sqrt{\frac{\omega}{\omega_0}}$$

is considered which corresponds to a wave traveling in positive x-direction. Thus, the remaining coordinates also proceed just like velocity:

$$\underline{F}(x) = \underline{F}_0 \exp(-\mathrm{j}\beta x), \quad \underline{\Omega}(x) = \underline{\Omega}_0 \exp(-\mathrm{j}\beta x), \quad \underline{M}(x) = \underline{M}_0 \exp(-\mathrm{j}\beta x).$$

The quotients $\underline{\Omega}_0/\underline{M}_0$ and $\underline{v}_0/\underline{F}_0$ have the meaning of wave admittances in the rotational and translational leg, respectively:

$$(h_z)_{\mathrm{T}} = \frac{\underline{v}_0}{\underline{F}_0} = \frac{\beta}{\omega m'} = \frac{1}{z_0}\frac{1}{\sqrt{\dfrac{\omega}{\omega_0}}} = l\sqrt{\frac{n'_{\mathrm{R}}}{m'}}\frac{1}{\sqrt{\dfrac{\omega}{\omega_0}}},$$

$$(h_z)_{\mathrm{R}} = \frac{\underline{\Omega}_0}{\underline{M}_0} = \frac{\omega n'_{\mathrm{R}}}{\beta} = \frac{1}{l^2 z_0}\sqrt{\frac{\omega}{\omega_0}} = \frac{1}{l}\sqrt{\frac{n'_{\mathrm{R}}}{m'}}\sqrt{\frac{\omega}{\omega_0}}.$$

Table 6.3. Four-port matrices of flexural iterative network

Note: The Rayleigh functions are considered to be functions of $\eta = \sqrt{\omega/\omega_0}$.

chain matrix

$$\begin{pmatrix} \underline{v}_1 \\ \underline{\Omega}_1 \\ \underline{M}_1 \\ \underline{F}_1 \end{pmatrix} = \begin{pmatrix} C & -\frac{l}{\eta}S & \frac{1}{\mathrm{j}z_0 l}c & \frac{1}{\mathrm{j}z_0\eta}s \\ -\frac{\eta}{l}s & C & -\frac{\eta}{\mathrm{j}z_0 l^2}S & -\frac{1}{\mathrm{j}z_0 l}c \\ \mathrm{j}z_0 lc & -\frac{\mathrm{j}z_0 l^2}{\eta}s & C & \frac{l}{\eta}S \\ \mathrm{j}z_0\eta S & -\mathrm{j}z_0 lc & \frac{\eta}{l}s & C \end{pmatrix} \begin{pmatrix} \underline{v}_2 \\ \underline{\Omega}_2 \\ \underline{M}_2 \\ \underline{F}_2 \end{pmatrix}$$

impedance matrix

$$\begin{pmatrix} \underline{F}_1 \\ \underline{M}_1 \\ \underline{F}_2 \\ \underline{M}_2 \end{pmatrix} = \frac{\mathrm{j}z_0}{N} \begin{pmatrix} -\eta(SC-sc) & -l\left(cC-s^2\right) & \eta S & -lc \\ -l\left(cC-s^2\right) & -\frac{l^2}{\eta}(cS-sC) & lc & -\frac{l^2}{\eta}s \\ -\eta S & -lc & \eta(SC-sc) & -l\left(cC-s^2\right) \\ lc & \frac{l^2}{\eta}s & -l\left(cC-s^2\right) & -\frac{l^2}{\eta}(sC-cS) \end{pmatrix} \begin{pmatrix} \underline{v}_1 \\ \underline{\Omega}_1 \\ \underline{v}_2 \\ \underline{\Omega}_2 \end{pmatrix}$$

admittance matrix

$$\begin{pmatrix} \underline{v}_1 \\ \underline{\Omega}_1 \\ \underline{v}_2 \\ \underline{\Omega}_2 \end{pmatrix} = \frac{\mathrm{j}}{z_0 N} \begin{pmatrix} \frac{1}{\eta}(Cs-Sc) & \frac{1}{l}\left(cC-s^2\right) & -\frac{1}{\eta}s & -c \\ \frac{1}{l}\left(cC-s^2\right) & \frac{\eta}{l^2}(cs-CS) & \frac{1}{l}c & \frac{\eta}{l^2}S \\ \frac{1}{\eta}s & -\frac{\eta}{l}c & \frac{1}{\eta}(Sc-Cs) & \left(cC-s^2\right) \\ \frac{1}{l}c & -\frac{\eta}{l^2}S & \frac{1}{l}\left(cC-s^2\right) & \frac{\eta}{l^2}(CS-cs) \end{pmatrix} \begin{pmatrix} \underline{F}_1 \\ \underline{M}_1 \\ \underline{F}_2 \\ \underline{M}_2 \end{pmatrix}$$

A phase velocity c_{B} and a wavelength λ_{B} can be assigned to the wave number β. Both wave quantities are frequency-dependent:

$$c_{\mathrm{B}} = c_l \sqrt{\frac{\omega}{\omega_0} \frac{r}{l}} = \sqrt{c_l r \omega},$$

$$\lambda_{\mathrm{B}} = 2\pi \sqrt{\frac{c_l r}{\omega}},$$

$$c_l = \sqrt{\frac{E}{\varrho}}, \qquad r = \sqrt{\frac{I}{A}}.$$

Here, the quantity c_l denotes the wave velocity for extensional waves on rods, the quantity I denotes the areal moment of inertia of the rod's cross-sectional area A.

In the following, the application of the presented relations is exemplified by two simple examples:

As a *first example* a cantilevered rod is considered and it is asked for the translational impedance at the free end in case of an excitation not caused by a torsional moment (see Fig. 6.22). In the specified relation between force and motion vectors it is referred to the elements of the impedance matrix shown in Table 6.3 as $\underline{z}_{ik}$. Thus, the boundary conditions $\underline{M}_1 = \underline{\Omega}_2 = 0$ result in both following equations:

$$\underline{F}_1 = \underline{z}_{11}\underline{v}_1 + \underline{z}_{12}\underline{\Omega}_1,$$

$$\underline{M}_1 = \underline{z}_{21}\underline{v}_1 + \underline{z}_{22}\underline{\Omega}_1 = 0,$$

$$\frac{\underline{F}_1}{\underline{v}_1} = \underline{z} = \frac{\underline{z}_{11}\underline{z}_{22} - \underline{z}_{12}^2}{\underline{z}_{22}}.$$

After an intermediate calculation, the application of the relations (6.20) results in the impedance $\underline{z}$ as a combination of circular and hyperbolic functions with $\eta = \sqrt{\omega/\omega_0}$:

$$\underline{z} = \mathrm{j} z_0 \eta \frac{1 + \cosh\eta \cos\eta}{\sinh\eta\cos\eta - \cosh\eta\sin\eta} \tag{6.22}$$

In case of $\eta \ll 1$, it follows in combination with the relations (6.21)

$$\underline{z} = \frac{1}{\mathrm{j}\omega \dfrac{n_0}{3}}.$$

The achieved result corresponds with the result presented in Sect. 5.1.2. The first singularity denotes a zero of the impedance $\underline{z}$ and a pole of the admittance

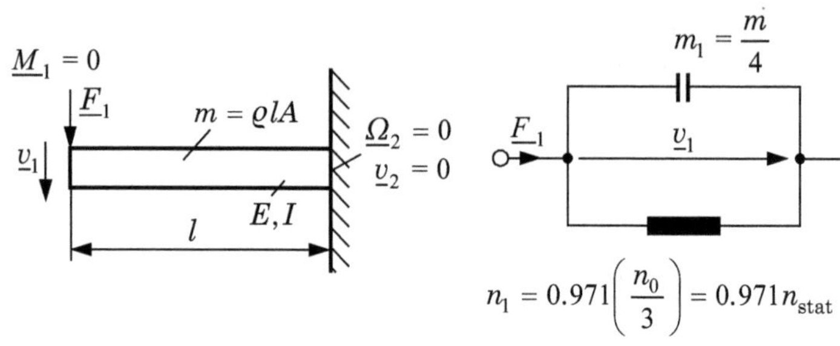

Fig. 6.22. Circuit representation of translational impedance at the free end of a cantilevered bending rod

$\underline{h} = 1/\underline{z}$, respectively. The frequency ω_1 is defined by the zero $\eta_1 = \sqrt{\omega_1/\omega_0}$ of the numerator of (6.22):

$$\cosh\eta_1 \cos\eta_1 = -1$$

The smallest solution of this equation yields $\eta_1 = 1.875$. In order to achieve a circuit representation of impedance $\underline{z}$, the results achieved by considerations made in Sect. 6.1.2 are used. Due to the zero of impedance $\underline{z}$, a representation as parallel resonant circuit must be possible. According to Sect. 6.1.2, the elements m_1 and n_1 result from the derivative of the function $\underline{z}(\omega)$ at the zero ω_1:

$$\left(\frac{\mathrm{d}\underline{z}}{\mathrm{d}\omega}\right)_{\omega_1} = \left(\frac{\mathrm{d}\underline{z}}{\mathrm{d}\eta}\right)_{\eta_1} \left(\frac{\mathrm{d}\eta}{\mathrm{d}\omega}\right)_{\omega_1} = \mathrm{j}\eta_1 z_0 \frac{1}{2\eta_1}\frac{1}{\omega_0} = \mathrm{j}\frac{1}{2}\frac{z_0}{\omega_0} = \mathrm{j}\frac{1}{2}m,$$

$$m_1 = \frac{1}{2\mathrm{j}}\left(\frac{\mathrm{d}\underline{z}}{\mathrm{d}\omega}\right)_{\omega_1} = \frac{m}{4},$$

$$n_1 = \frac{4}{\omega_1^2 m} = \frac{4\omega_0^2}{\omega_1^2}\frac{1}{\omega_0^2 m} = \frac{4}{\eta_1^4}n_0 = \frac{12}{\eta_1^4}\frac{n_0}{3},$$

$$n_1 = 0.971\frac{n_0}{3} = 0.971 n_{\mathrm{stat}}.$$

Thus, it is conceivable that almost the static compliance is effective in the neighborhood of resonance and that a fourth of total mass of rod can be thought to be concentrated at the end. The total canonical circuit can be formed by means of calculation of the equivalent parameters at further zeros of (6.22). The demonstrated calculation shows that the equivalent mass of all resonators amounts to $m/4$. The compliances can be calculated from the associated resonant frequencies.

As a *second example*, the translational admittance is determined. It is measurable at the clamping position of a bending rod (Fig. 6.23). Here, the admittance matrix is taken expediently as starting point.

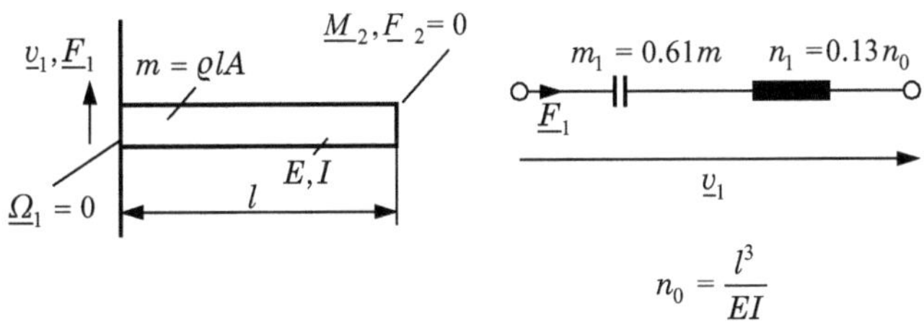

Fig. 6.23. Circuit representation of translational admittance at the clamped side of a cantilevered bending rod

In combination with the specified boundary conditions, the two following equations are achieved:

$$\underline{v}_1 = \underline{h}_{11}\underline{F}_1 + \underline{h}_{12}\underline{M}_1,$$

$$\underline{\Omega}_1 = \underline{h}_{21}\underline{F}_1 + \underline{h}_{22}\underline{M}_1 = 0.$$

From this follows the admittance according to

$$\underline{h} = \frac{\underline{v}_1}{\underline{F}_1} = \frac{\underline{h}_{11}\underline{h}_{22} - \underline{h}_{12}\underline{h}_{21}}{\underline{h}_{22}}.$$

In combination with the elements of admittance matrix presented in Table 6.3 and in combination with the relations (6.20), $\underline{h}$ can be formulated by means of circular and hyperbolic functions:

$$\underline{h} = \frac{1}{\mathrm{j} z_0 \eta} \frac{1 + \cosh\eta\cos\eta}{\sinh\eta\cos\eta - \cosh\eta\sin\eta}$$

In case of $\eta \ll 1$, for very low frequencies the admittance results in combination with the relations (6.21) in:

$$\underline{h} = \frac{1}{\mathrm{j}\omega m}$$

The first singularity results again from the zero of the numerator. Due to the zero of admittance, a representation as series resonant circuit must be possible. By analogy with the preceding example the equivalent elements n_1 and m_1 are achieved by means of the derivative $\mathrm{d}\underline{h}/\mathrm{d}\omega$:

$$\left(\frac{\mathrm{d}\underline{h}}{\mathrm{d}\omega}\right)_{\omega_1} = \left(\frac{\mathrm{d}\underline{h}}{\mathrm{d}\eta}\right)_{\eta_1} \left(\frac{\mathrm{d}\eta}{\mathrm{d}\omega}\right)_{\omega_1} = \frac{1}{\mathrm{j} z_0 \eta_1} \frac{\sinh\eta_1\cos\eta_1 - \cosh\eta_1\sin\eta_1}{\sinh\eta_1\cos\eta_1 + \cosh\eta_1\sin\eta_1} \cdot \frac{1}{2\eta_1}\frac{1}{\omega_0}$$

Taking $\cosh\eta_1 \cos\eta_1 = -1$ into account, the factor consisting of circular and hyperbolic functions can be simplified:

$$\frac{\sinh\eta_1 \cos\eta_1 - \cosh\eta_1 \sin\eta_1}{\sinh\eta_1 \cos\eta_1 + \cosh\eta_1 \sin\eta_1} = -\frac{\cosh\eta_1 + 1}{\cosh\eta_1 - 1} = -1.856,$$

$$\left(\frac{\mathrm{d}\underline{h}}{\mathrm{d}\omega}\right)_{\omega_1} = \mathrm{j}\frac{1.856}{2\eta_1^2}\frac{1}{\omega_0 z_0} = \mathrm{j}\,0.264 n_0 = \mathrm{j}2n_1,$$

$$n_1 = 0.132 n_0,$$

$$m_1 = \frac{1}{\omega_1^2 n_1} = \frac{\omega_0^2}{\omega_1^2}\frac{1}{\omega_0^2 n_1} = \frac{1}{\eta_1^4}\frac{1}{0.132}\frac{1}{\omega_0^2 n_0} = \frac{1}{\eta_1^4\, 0.132} m,$$

$$m_1 = 0.613 m.$$

6.2 Network Representation of Acoustic Systems as Linear Waveguides

It is often the case with acoustic problems that a representation with concentrated elements works only for a limited frequency range, since the dimensions of components are not sufficiently small compared to the wavelength within the liquid. For canal-like elements, this limitation can be avoided in canal direction by representing the canal as one-dimensional waveguide, if the two lateral dimensions remain sufficiently small compared to wavelength. That is often the case. With volumetric elements (elements being effective as acoustic compliance at low frequencies) also a representation as waveguide in case of one-dimensional extension is possible. However, the configuration of the closed-end pipe being now available is rarely found in practice. In principle, an extension of network representation in the three space directions is indeed possible without difficulties, however, nearly all advantages of a representation by concentrated elements with network methods are lost, so that such tasks should be directly processed with special solution methods of finite element (FEM) and boundary element method (BEM), respectively. Also a mix of techniques is possible and is usually applied nowadays. The fact that one-dimensional waveguides can be additionally analyzed with significant advantages in time in combination with network representations is based on the fact that the dissipative or loss-free conductor is integrated in network analysis programs as standard component. Thus, it can be addressed as separate component.

In order to attain a one-dimensional waveguide, the canal element is divided up into finite acoustic elements in longitudinal direction. For that purpose, a volumetric section of width h is regarded to consist of a differential volume and two differential mass elements (Fig. 6.24). The elements defined in such a way are referred to width h and are distinguished by a superscript asterisk. Now the complete pipe can be assembled from these elements, since the condition of small dimensions compared to wavelength is fulfilled for each element. However, often this complex representation is not required. A consideration of the interfaces at the beginning and end of pipe and the specification of a coupling chain matrix yield the desired results usually faster. Even if position-dependent cross-sectional areas are existent, solutions can be found in this way. At first, the solution for a constant cross-sectional area A is presented.

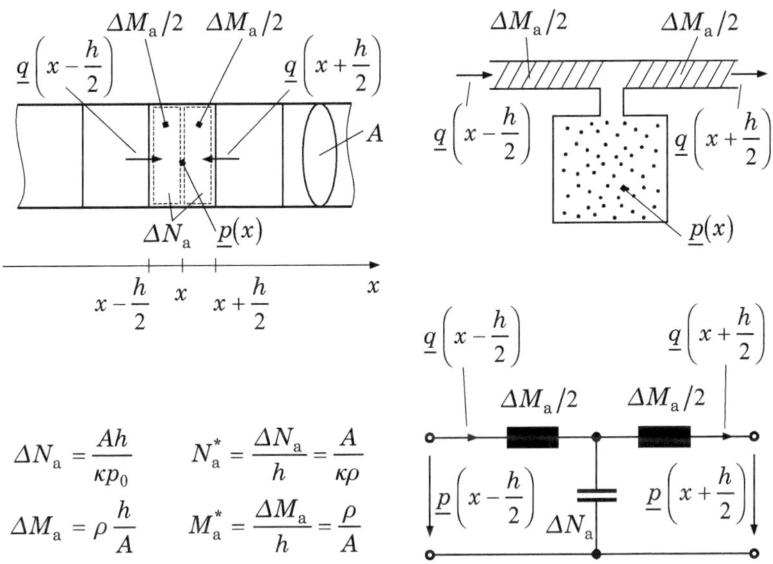

Fig. 6.24. Finite element of a one-dimensional acoustic waveguide

As already demonstrated in Sect. 6.1.1, the differential equation for pressure or volumetric flow rate can be extracted from the finite element. The solution of the differential equation and the determination of the constants from boundary conditions result in the chain matrix of the one-dimensional waveguide according to Fig. 6.25.

For a pipe with variable cross-sectional area according to

$$A(x) = A_1\,\mathrm{e}^{2mx} \tag{6.23}$$

also a chain matrix for the two-port network can be defined. The related parameters M_a^* and N_a^* are described as position function

$$\frac{\xi_F(x)}{F} = \left.\frac{\xi(x)}{F}\right|_{p=0} \quad \text{and} \quad \frac{\xi_p(x)}{p} = \left.\frac{\xi(x)}{p}\right|_{F=0}. \tag{6.24}$$

$$\left.\begin{aligned}\frac{\mathrm{d}\underline{p}}{dx} &= -\mathrm{j}\omega M_{\mathrm{a}}^{*}\underline{q}\\ \frac{\mathrm{d}\underline{q}}{dx} &= -\mathrm{j}\omega N_{\mathrm{a}}^{*}\underline{p}\end{aligned}\right\} \quad \begin{aligned}&\frac{\mathrm{d}^2\underline{p}}{\mathrm{d}x^2}+\omega^2 M_{\mathrm{a}}^{*}N_{\mathrm{a}}^{*}\underline{p}=0\\ &\underline{p}(x)=C_1\mathrm{e}^{\mathrm{j}\beta x}+C_2\mathrm{e}^{-\mathrm{j}\beta x}\end{aligned}$$

boundary conditions:

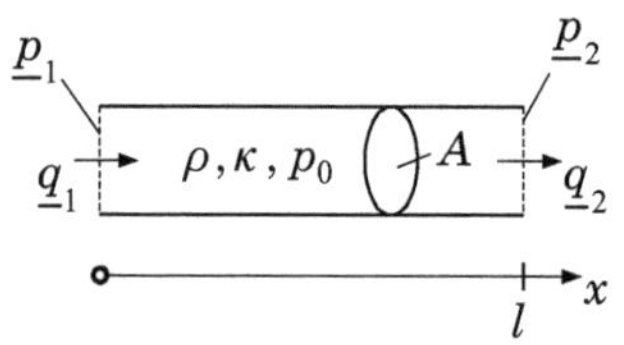

$$\begin{pmatrix}\underline{p}_1\\ \underline{q}_1\end{pmatrix}=\begin{pmatrix}\cos\beta l & \mathrm{j}Z_{\mathrm{a}0}\sin\beta l\\ \mathrm{j}\dfrac{1}{Z_{\mathrm{a}0}}\sin\beta l & \cos\beta l\end{pmatrix}\begin{pmatrix}\underline{p}_2\\ \underline{q}_2\end{pmatrix}$$

$$\beta=\omega\sqrt{M_{\mathrm{a}}^{*}N_{\mathrm{a}}^{*}}=\frac{\omega}{c}\ ,\quad Z_{\mathrm{a}0}=\sqrt{\frac{M_{\mathrm{a}}^{*}}{N_{\mathrm{a}}^{*}}}\ ,\quad c=\frac{1}{\sqrt{M_{\mathrm{a}}^{*}N_{\mathrm{a}}^{*}}}=\sqrt{\frac{\kappa p_0}{\rho}}$$

Fig. 6.25. Differential equation and chain matrix of one-dimensional acoustic waveguide

It applies:

$$M_{\mathrm{a}}^{*}(x)=\frac{\rho}{A_1}\,\mathrm{e}^{-2mx}\qquad\text{and}\qquad N_{\mathrm{a}}^{*}(x)=\frac{A_1}{\kappa p_0}\,\mathrm{e}^{2mx}$$

After substituting, it follows in combination with the differential equations of the acoustic waveguide specified in Fig. 6.25

$$\frac{\mathrm{d}^2\underline{p}}{\mathrm{d}x^2}+\frac{\omega^2}{c^2}\underline{p}+\frac{1}{A(x)}\frac{\mathrm{d}A}{\mathrm{d}x}\frac{\mathrm{d}\underline{p}}{\mathrm{d}x}=0 \qquad (6.25)$$

with

$$c^2=\frac{\kappa p_0}{\rho}\qquad\text{and}\qquad\frac{1}{A(x)}\frac{\mathrm{d}A}{\mathrm{d}x}=2m.$$

Solving this equation by means of an exponential approach and applying the boundary conditions results in the two-port network equation which interconnects the beginning and end of this pipe with variable cross-sectional area. This results in:

$$\begin{pmatrix}\underline{p}_1\\ \\ \underline{q}_1\end{pmatrix}=\mathrm{e}^{ml}\begin{pmatrix}\cos\beta l-\dfrac{m}{\beta}\sin\beta l & \dfrac{\rho c}{A_2}\dfrac{\omega}{\beta c}\mathrm{j}\sin\beta l\\ \\ \dfrac{A_1}{\rho c}\dfrac{\omega}{\beta c}\mathrm{j}\sin\beta l & \dfrac{A_1}{A_2}\left(\cos\beta l+\dfrac{m}{\beta}\sin\beta l\right)\end{pmatrix}\begin{pmatrix}\underline{p}_2\\ \\ \underline{q}_2\end{pmatrix}. \qquad (6.26)$$

With this solution, also acoustic interfacing devices (cones) can be integrated now in network solutions, if it is aimed for an analytical solution. If cones are to be handled with network analysis programs, the cone must be divided up into several pipe components of different cross-sectional areas. Depending on change of cross-sectional area, 10 up to 20 elements are sufficient. In order to avoid disturbances by discontinuities, dissipative pipe elements should be used.

The methods of approximate calculation of the input impedance of waveguides, the methods of simplified representation of the impedance in the neighborhood of a resonant frequency and the approximate two-port network representation is possible by analogy with the Sects. 6.1.2 up to 6.1.4. By analogy with Fig. 6.14, it can be demonstrated with these methods that also simple $\lambda/4$-pipes can be applied as impedance transformers for monofrequent applications.

By applying these methods of approximation, circuit representations with usually narrow frequency ranges being of interest can be specified which provide an insight into the basics of solution due to their simplicity. This insight allows for an improved checking of numerical solutions with respect to errors and mistakes.

6.3 Modeling of Transducer Structures with Finite Network Elements

Also with the description of transducer systems states arise which show a dependence of the interconnections on position. Furthermore, it can happen that the transition to summarily acting quantities (e.g. pressure and acoustic volume velocity or current and voltage, respectively) succeeds only at the transducer's input and output. As already demonstrated in the Sects. 6.1 and 6.2, in these cases small regions are defined, in which the simplification of position independence can be assumed.

Thus, the complete transducer is generated by means of a structurally correct interconnection of finite network elements. This procedure is demonstrated with two examples.

6.3.1 Ultrasonic Microactuator with Capacitive Diaphragm Transducer

Capacitive transducers showing a metallized synthetic diaphragm tensioned over an electrode shape according to Fig. 6.26 are suitable for applications which require a directional ultrasonic beam into a gaseous medium. The excitation is realized by means of an AC voltage superposed by a DC voltage.

In [47] and [48] this structure was modeled, calculated and experimentally verified with network methods and with new not previously used superposition solutions for the first time.

The diaphragm is attached to strips and electrostatically fixed by an electrical bias voltage. In the region between the strips, the diaphragm is able to perform

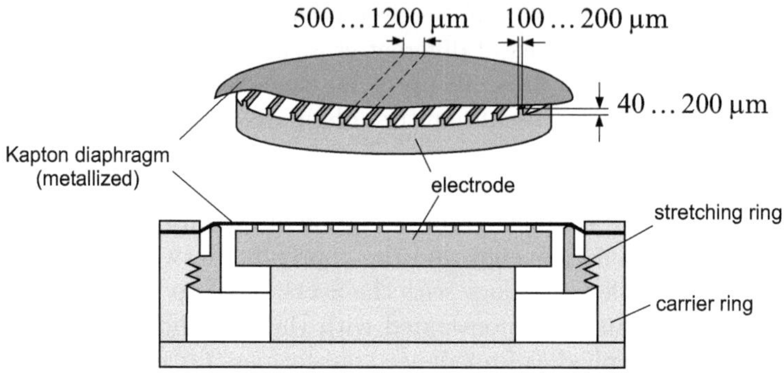

Fig. 6.26. Ultrasonic transducer with strip diaphragm elements [47]

free oscillations. The exciting action of force on the diaphragm results from the time-variant electric field between diaphragm and electrode.

At first, a model of a strip diaphragm must developed which results in a correct deflection for every position of action of force. Figure 6.27 shows the solution.

In order to achieve a closed model for the single strip, the reaction of sound field to the radiating transducer area and the action of cavity between diaphragm and strip electrode on the inward area must be represented correctly. This is realized by means of loading the acoustic transducer output with a complex impedance $\underline{Z}_a$.

Due to the strip structure (inhomogenous electric field), in the example illustrated in Fig. 6.26 a strong position dependence of the exciting force by the electric field exists. In addition, the mass of diaphragm must be considered in order to achieve a correct dynamic representation. Thus, a closed modeling succeeds only with finite network elements.

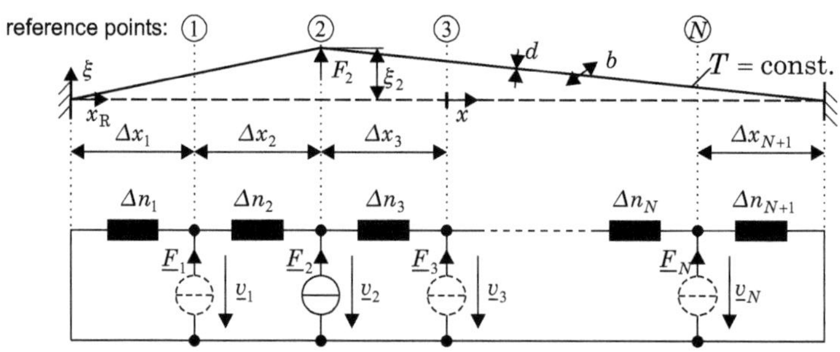

$$\Delta n_k = \frac{\Delta x_k}{F_T} \quad \text{with} \quad F_T = Tbd \quad \text{and} \quad k = 1 \dots N+1$$

Fig. 6.27. Strained strip diaphragm with network model

A further difficulty exists in the fact that the idealization of a diaphragm (no flexural rigidity) is hardly feasible technically. Therefore, the modeling is to comprise also the case of plate (sheet with diaphragm stress and flexural rigidity). Due to force addition at the model nodes, the superposition of the models of diaphragm and bending plate is possible. The choice of force as flow quantity works also here. Figure 6.28 shows the network representation of stretched strip plate achieved by interconnection.

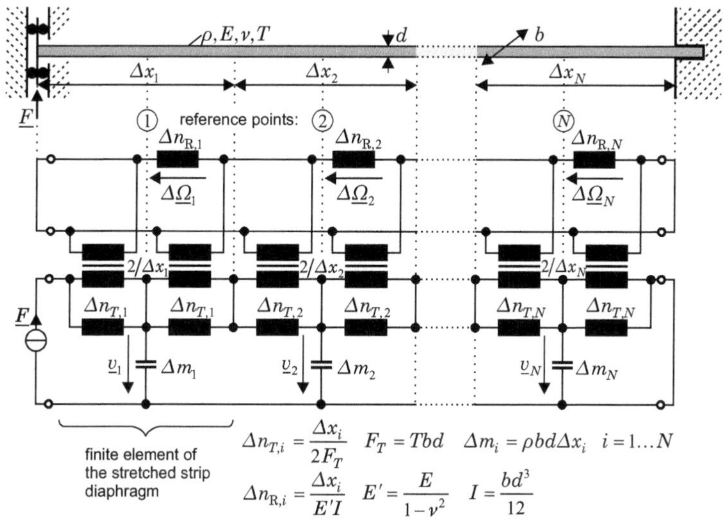

Fig. 6.28. Dynamic model of stretched strip plate

Investigations to representation errors caused by discretization show that 11 sections (reference points) already yield good results. Therefore, a calculation with network analysis programs is not really a temporal problem.

Now the mechanical-acoustic transducer (Y) must be put on each plate section. The acoustic transducers combine their acoustical volume velocity for position-independent pressure. The complex load $\underline{Z}_a$ caused by the sound field and the load caused by the dissipative acoustic compliance of the back volume affect the acoustic output. Figure 6.29 is achieved after combining the circuit components.

On the basis of geometry of the multistriped transducer, the calculation of acoustic pressure in the sound field can be done by means of common acoustic solution possibilities [47]. For the user of the ultrasonic transducer it was appropriate to define an equivalent piston acoustic radiator generating the same effects in the far field.

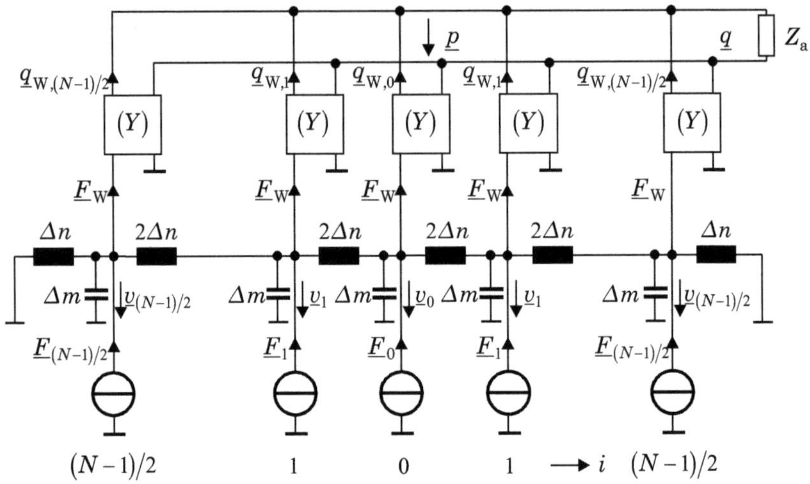

Fig. 6.29. Model according to Fig. 6.28 with acoustic load

6.3.2 Fluid-filled Pressure Transmission System of a Differential Pressure Sensor

Within the scope of research studies [49, 50], miniaturized pressure and differential pressure sensors with piezoresistive silicon measuring element were developed (Fig. 6.30). The pressure transmission is realized by means of a separating diaphragm electroplated on a high-grade steel body and liquid-filled capillaries.

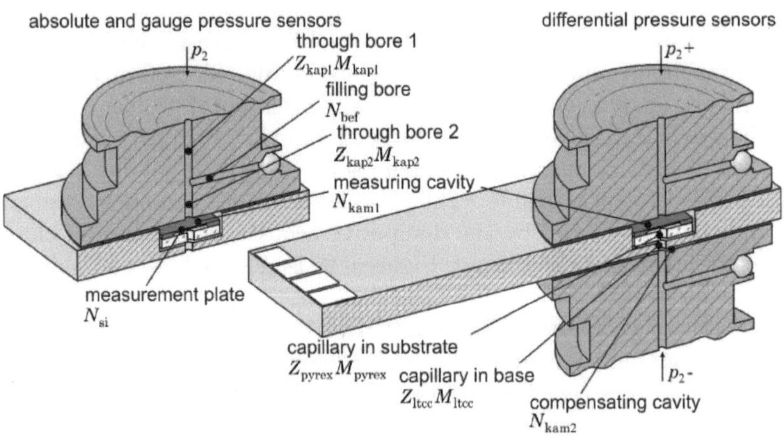

Fig. 6.30. Structure of pressure and differential pressure sensors and classification of acoustic components

In [49], the dimensioning of the acoustic low-pass system consisting of metallic separating diaphragm and oil filling represented the subject of design. On the one hand, a sufficient operating frequency range, on the other hand, also the required attenuation of overload pressure peaks which would result in a destruction of the silicon measuring element, were to be provided. The parameters being necessary for network design are summarized in Table 6.4.

Table 6.4. Parameters of differential pressure sensor

component	parameter	value
measuring element	edge dimensions	(3 x 3) mm^2
	height	1.2 mm
	silicon on borosilicate glas	
	silicon pressure measurement plate	(2.4 x 2.4 x 0.04) mm^3
	measuring range	100 mbar
base	edge dimensions	(25 x 12.5) mm^2
	height	1.2 mm
	material	LTCC
diaphragm carrier	diameter	12 mm
	height	5 mm
	through bore	∅ 0.8 mm
	material	high-grade steel, DIN1.4404
separating diaphragm	diameter	9 mm
	thickness	0.02 mm
	height above diaphragm bed	0.035 mm
	material	electroplated nickel
filling medium	silicone oil	Baysilone M100
	kinematic viscosity	0.093 Pa·s
	dynamic viscosity	10^{-4} m^2s^{-1}

Figure 6.31 illustrates the network model of the capillary systems for both sides of pressure feed to the silicon differential pressure measuring element. The classification of the components to the structural parts and cavities can be seen in Fig. 6.30.

The calculation of the acoustic components – acoustic compliance N, acoustic mass M, acoustic friction Z – is made on the basis of values specified in Table 6.4. Figure 6.32 shows the comprehensive model of the liquid-filled differential

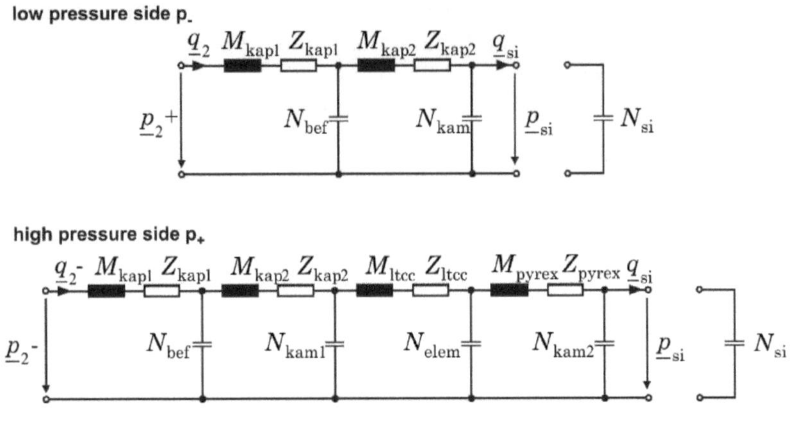

Fig. 6.31. Network models of capillary system for the p_+-and p_--side of differential pressure sensor

pressure sensor. The separating diaphragms were represented as concentrated components by series connection of acoustic compliance N_{tm}, mass M_{tm} and friction Z_{tm}. The filling oil bed volume is included as horizontal compliance N_{bett} connected in parallel.

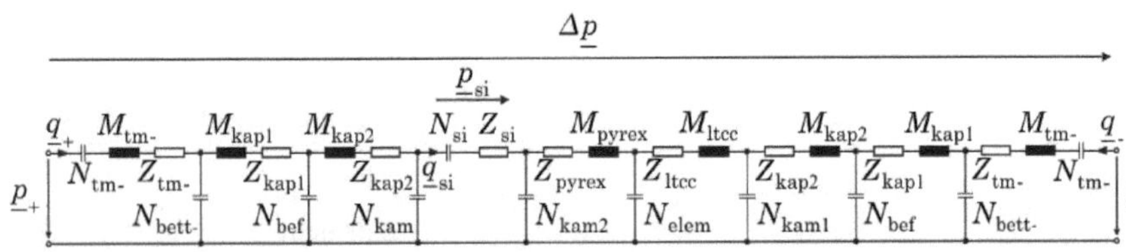

Fig. 6.32. Comprehensive model of fluid system with separating diaphragm and silicon measuring element

In combination with this network model, the pressure transfer function

$$\underline{B}_{\mathrm{diff}} = \frac{\underline{p}_{\mathrm{Si}}}{\Delta \underline{p}}$$

can be calculated using the numerical circuit calculation program pSpice. Figure 6.33 shows its frequency-response characteristic in comparison with the measured curve. Indeed the qualitative curve shape coincides roughly, however, the quantitative curve shape shows very large deviations up to 60 dB. These deviations are caused by representing the compliance of oil volume between separating diaphragm and diaphragm bed as concentrated component and by neglecting frictional effects in the slot-like pressure transmission canal between the outside radius of diaphragm and diaphragm center. Thus, it is necessary to describe the system bed volume and separating diaphragm with distributed parameters. Finite network elements are used as solution approach.

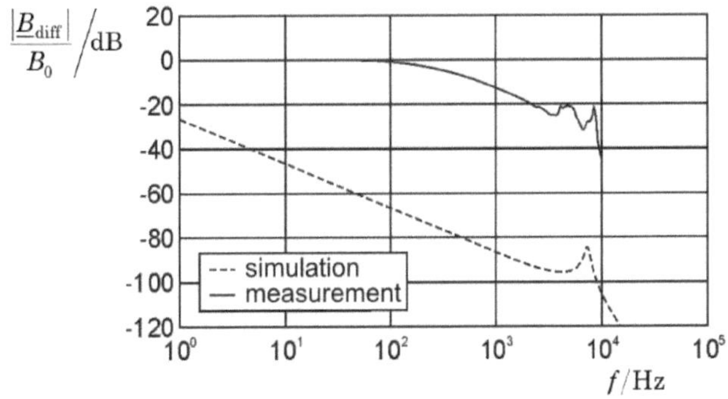

Fig. 6.33. Transfer function $|\underline{B}_{diff}| = f(\omega)$ calculated with concentrated acoustic components and obtained experimentally

In Fig. 6.34, the segmentation of separating diaphragm and oil is represented for five elements. The separating diaphragm and bed volume are divided up into five circular ring-shaped strips. In spite of a uniform pressure excitation $\underline{p}$, different pressures $\underline{p}_{11} \cdots \underline{p}_{15}$ are generated in the bed volume strips. The different diaphragm strip compliances are the reason for this. The pressure differences between two neighboring volume strips generate a volumetric flow rate $\underline{q}_{12} \cdots \underline{q}_{45}$ in each case. The pressure $\underline{p}_{15} = \underline{p}_2$ in the middle segment represents the output quantity providing the input quantity for following capillary system.

Figure 6.35 represents the network model of the separating diaphragm with bed volume for five elements. It consists of the parallel connection of the diaphragm segments and the series connection of the bed volume elements.

The combination of this finite network model for separating diaphragm and bed volume with the network model of the capillary system and silicon measuring element results in a comprehensive model which allows for the calculation of the transfer function of the differential pressure sensor $\underline{B}_{diff}$. Figure 6.36 shows the result for 70 elements. It corresponds very well to the experimentally determined curve shape. Now the desired frequency-response characteristic, as it is illustrated in Fig. 6.37, can be adjusted by variation of bed volume and capillary dimensions.

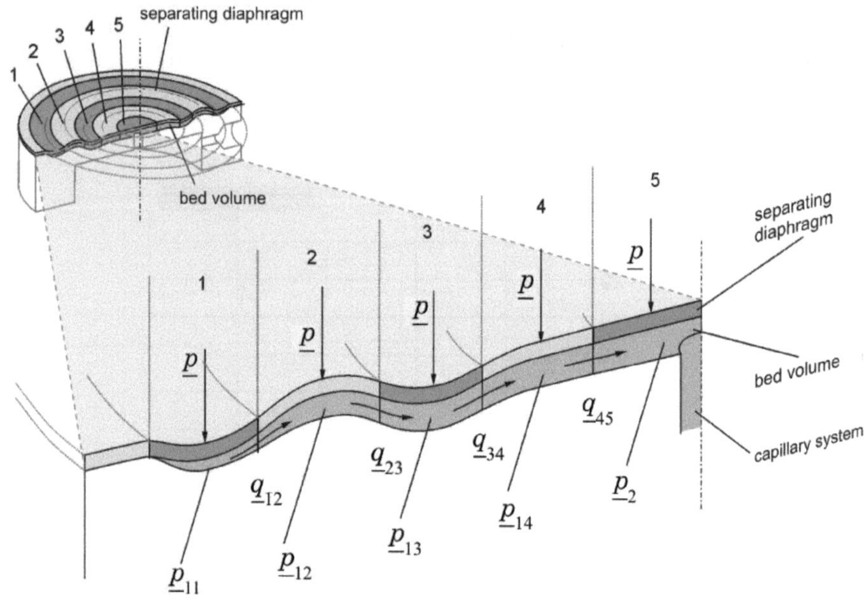

Fig. 6.34. Segmentation of separating diaphragm and bed volume into five equivalent circular ring-shaped elements of same width

6.4 Combined Simulation with Network and Finite Element Methods

Expanding scientific computing methods have resulted in growing requirements to the simulation being necessary for design and accompanying design

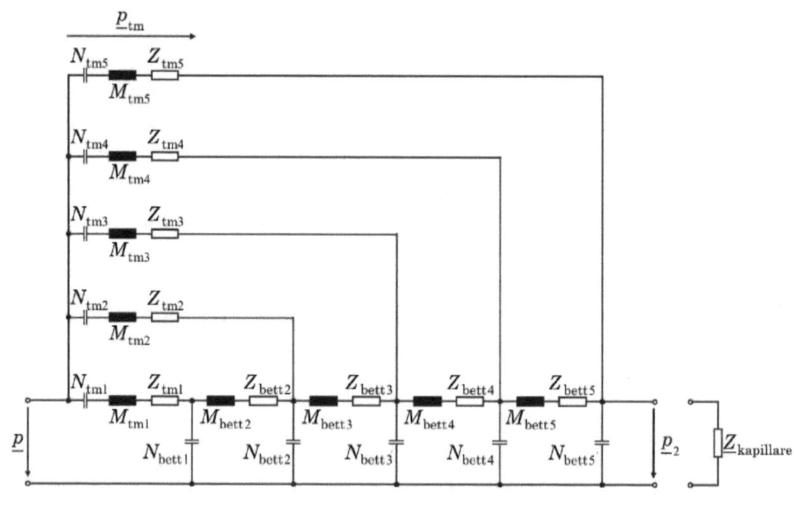

Fig. 6.35. Network model of separating diaphragm and bed volume consisting of five finite network elements

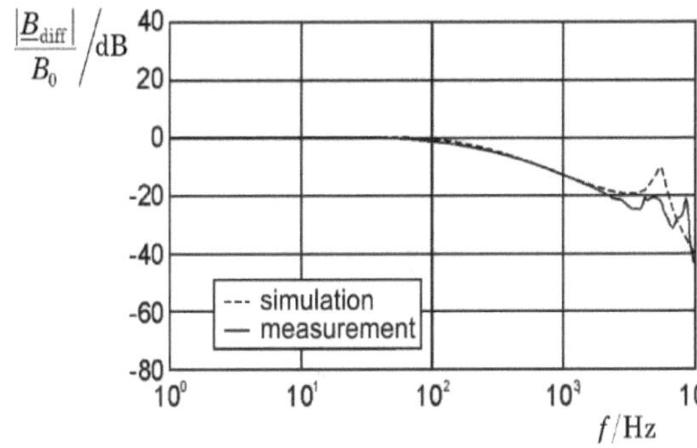

Fig. 6.36. Progression of theoretically and experimentally determined pressure transfer function of differential pressure sensor $|\underline{B}_{\text{diff}}|$

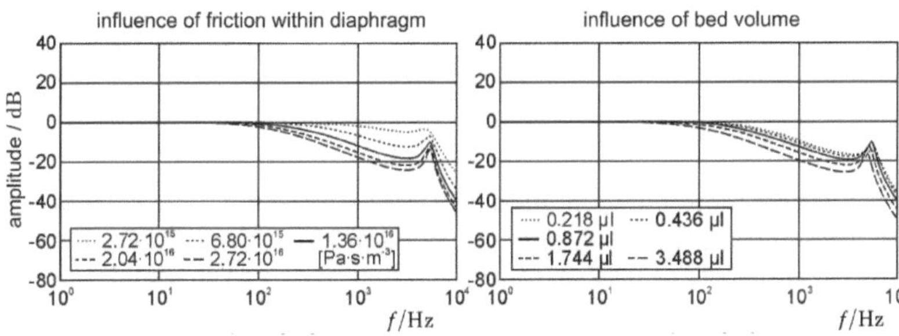

Fig. 6.37. Influence of the variation of capillary radius and bed volume of differential pressure sensor on frequency-response characteristic $|\underline{B}_{\text{diff}}|$

optimization of electromechanical systems. Here, in particular the requirement for an efficient precalculation concerning the necessary approximation ratio, the rapidity of model generation, the rapidity of design and optimization process and the handling by design engineers is in the foreground. For complex dynamic systems, this is only possible in a limited way with network methods in many cases. The combination of network and finite element methods (FE methods) at user level represents an advantageous possibility for an efficient precalculation of electromechanical system. With this so-called „combined simulation" – within this book this term is used as a proper name – the combination is realized by the user on the own user interfaces of common network programs (OrCAD CAPTURE, MICRO-CAP, etc.) and FE programs (ANSYS®, NASTRAN, ABAQUS, COMSOL, etc.). Compared to the *coupled simulation* [51], the *combined simulation* is independent of special simulation programs due to the restriction to user level in case of using these two methods. It is a teachable method which is suitable to apply both network models and FE models in a problem-specific manner in order to enable an efficient

precalculation.
In this section, at first the procedure concerning the combination of network methods and finite element networks is presented and illustrated by means of two examples subsequently.

6.4.1 Applied Combination of Network Methods and Finite Element Methods

A problem-oriented combination of network methods and finite element methods is realized with the *combined simulation*. For this purpose, the simulation task is separated into subtasks which are handled with the appropriate method. This may mean that the network model design of a complex structure is piecewisely realized by means of FE methods [52]. Similarly, an FE model can be simplified by means of network methods. For this purpose, the subsystems being not able or difficult to be modeled in the FE model are represented by equivalent structures being based on network methods [53, 54].

Finite Element Methods

The FE method represents a numerical method for solving physical problems in structures and continua spatially extended in at least one dimension. These spatially extended structures or continua are divided up into small parts. It is referred to these small parts as „finite elements". Their behavior is characterized by a mathematical model of the physical problem. In the simplest case the mathematical model uses a linear functional approach. The supporting points of the solution are the so-called nodes which are located – depending on the functional approach – at the corners and edges of the finite elements. Further considerations on theory and application of FE methods can be found e.g. in [55].
Primarily, FE methods were only used for solving structural problems. However, they are now suited for the solution of many other physical problems and interactions. The solvability of a specific physical problem or an interaction between physical domains depends on the implementability in the used FE program. Therefore, simulation possibilities are not provided for some simulation problems.
In practice, the solvability of FE models is furthermore limited by numerical stability and size of the mathematical model. Thus, in many cases the precalculation of the dynamic behavior of electromechanical systems is not possible in closed form or very time-consuming, respectively.

Generating Network Representations by means of FE Methods

In many cases the FE method represents a very efficient method to generate network representations. In particular, this enables a faster modeling and better approximation level of the network model for spatially extended structures and complex geometries. If the network structure of a configuration is already known a priori, simple FE analyses will suffice for the determination of component parameters of the network. However, also the network structure for a configuration can be determined. This is achieved e.g. by means of a harmonic FE analysis.

The procedural method for generating network representations using FE methods corresponds to the general method of generating network models for linear or approximately linearizable electromechanical systems (see Chap. 3 up to 5). At first, the simulation task has to be restricted to the quantity which is necessary for solving the particular problem. Subsequently, the system is divided up into subsystems and network coordinates are introduced at intersection points. For spatially extended continua, such as fluidic systems, the dividing up into subcontinua is realized by means of network coordinates spatially averaged at the sectional areas (see Sect. 3.3). Subsystems with known network representation are represented in the network domain. Unknown parts of the system are specified as general N-ports that are able to comprise several physical domains. For a convenient network modeling the subsystems have to be chosen to be solenoidal, in order that passive linear or linearizable N-ports are achieved. The behavior of these N-ports is characterized by suitable system equations [56]. The coefficients of these system equations are determined by means of an FE model of the subsystem. Since the FE model is limited to the subsystem, the calculating time is short in general.

Figure 6.38 shows how to generate a network representation using FE methods. For the acoustic subsystem of the presented piezoelectric signal generator, both the network structure and the component parameters are known. For the piezoelectric unimorph bending plate also a network structure can be specified. The appropriate determination of the component parameters of the circuit of the piezoelectric bending plate is achieved by simple FE analyses [54].

In the following, important procedural methods concerning the network preparation are presented for the practically important cases of the one- and two-port for harmonic processes in steady state.

For a one-port, the coefficient of the equation system results in a frequency-dependent complex impedance or admittance. In this form, it can already be realized in network analysis programs e.g. by storing in tabular form. For an efficient network model which enables also an understanding of functional principles, it has to be aimed for a complete or approximate representation of the impedance or admittance by basic components. But this is not always possible and it often necessitates further restriction or linearization of the model. In many cases, a priori knowledge about the network structure is existent, thus only the parameters of basic components must be determined from the FE solution.

In cases where the network structure is not known, it can be determined from the tabularly available results of the FE analysis using the methods of network analysis and network synthesis. For simple networks, usually a heuristic

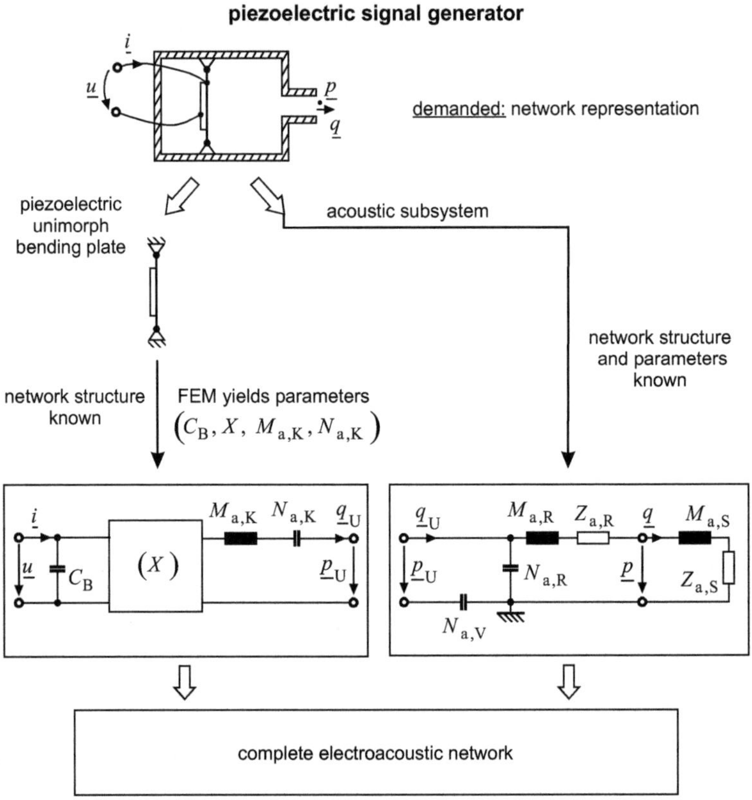

Fig. 6.38. Network representation by means of FE methods using the example of a piezoelectric signal generator
(*schematic representation of piezoelectric transducer is specified in Sect. 9.2*)

approach with piecewise interpretation of the frequency-dependent impedance or admittance by basic components is target-aimed. An approximate representation of the complex impedance by basic components can be carried out equivalently to the procedural method described in Sect. 6.1.

For a two-port network, there is the possibility of characterization by means of the impedance, admittance, hybrid or chain matrix. The choice of the used two-port matrix defines the necessary FE analyses for the calculation of the elements of the matrix. The two-port matrix of two-ports within one physical domain can be interpreted as a T-circuit or Π-circuit comprising complex impedances. As described in Sect. 6.1.4, two-port networks with different physical domains result in one of three possible network structures consisting of an ideal transformer and two complex impedances. The representation of complex impedances by means of interconnection of basic components is done by analogy with the method applied to one-port elements. At this point it should be noted that the minor deviations of the FE solution result in inaccurate or false network representations in combination with the choice of an inappropriate two-port matrix . For this reason, a second way for the approximate determination of a two-port network representation is specified in Sect. 6.4.2.

Expanding the Possibilities of FE Methods by means of Network Methods

Expanding the possibilities of FE methods by means of network methods represents a second field of application of the *combined simulation.* Often a closed simulation of the dynamic behavior of electromechanical systems in an FE model is not possible or extremely time-consuming. The reason for that is that the modeling of some physical effects and interactions is not implemented into the used FE program or their representation requires a large modeling effort e.g. by a very fine mesh. In many cases, a more effective modeling of the system in an FE model can be effected by means of network methods. The network methods are used either as an intermediate step during the generating process of the FE model or as part of the FE model.
The application of the network method as an intermediate step during modeling process is exemplified in Sect. 6.4.3. On the one hand, it is based on the application of knowledge known in network methods a priori. On the other hand, a network representation of a subsystem in the network domain can be transformed to another representation. The external effective terminal behavior of the network shows the same effect in the considered operating range. At first, the subsystem in the FE model is separated by introducing intersection points, sectional lines or sectional areas to the remaining model. Network coordinates are defined at intersection points, sectional lines or sectional areas, so that the subsystem can be described by a network. This represents already a model reduction in general. In the network domain, the network representation of the subsystem is transformed by means of network methods. Subsequently, the modified network representation of the subsystem is transformed into an „equivalent structure" which is represented in the FE model with finite elements.
The transformation of the network representation is realized in such a way that the resulting equivalent structure can be modeled efficiently with FE methods. The FE model of this equivalent structure with same external effects can be a model for other physical relations in the inside. The modeling of an acoustic friction by means of an equivalent structure comprising a mechanical friction represents an example. It is illustrated in Fig. 6.39. According to the assumptions made for derivation of the equivalent structure, the real structure is represented approximately only within a limited operating range.

The integration of network representations into the FE model represents a second way of expanding the possibilities of FE models by network methods. This is effected by finite elements which model basic network components. It

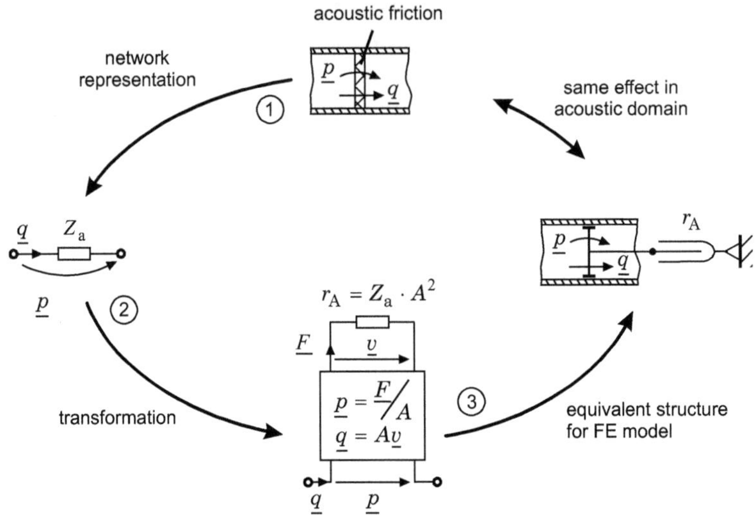

Fig. 6.39. Derivation of an equivalent structure with same effect by means of network methods

is an aim to represent approximately a substructure or subcontinuum by a network representation in order to achieve a rapidly predictable FE model of the general configuration. Most of the available FE programs include two-node finite elements for mechanical and also partly for electrical basic components. With their help, substructures can be modeled in an FE model by simple networks. By analogy with the approach mentioned above, the subsystem must be separated. The calculation of the network with basic components in FE representation is effected within the FE model, so that in accordance with the definition of finite elements the degrees of freedom and loads are existent at the basic components. The difference between these degrees of freedom and loads of the FE model and the difference between the network coordinates of the network model must be considered for the generation and evaluation of the network.
By integrating a transformed network representation into an FE model, also the combination of both presented ways is possible. Thus, e.g. an electromechanical transducer with known network representation in the network domain can be completely transformed into the mechanical domain. On the basis of available finite elements of the basic components, the mechanical network is integrated into the FE model of a mechanical structure. This enables the complete calculation of the electromechanical system in a mechanical FE model.

6.4.2 Combined Simulation using the Example of a Dipole Bass Loudspeaker

Loudspeaker systems based on the dipole principle have an enclosure in which both the front and the back side of the acoustic transducer are open in direction of the sound field. This configuration results in a directional effect in the form of a figure-of-eight pattern. Compared to a bass reflex loudspeaker, in the wanted signal frequency range there are no resonances in the enclosure. Such dipole systems are also used as dipole bass loudspeaker for bass transmission.

Structure and Modeling Approach of the Dipole Bass Loudspeaker

Figure 6.40 illustrates the structure of a dipole bass loudspeaker. For a frequency range of 20 Hz up to 200 Hz, a model is generated which enables both a fast simulation and an understanding of the operating mode of this system. The interaction of the electrodynamic loudspeaker with the acoustic subsystems „enclosure" and „sound field" determines the operating mode.

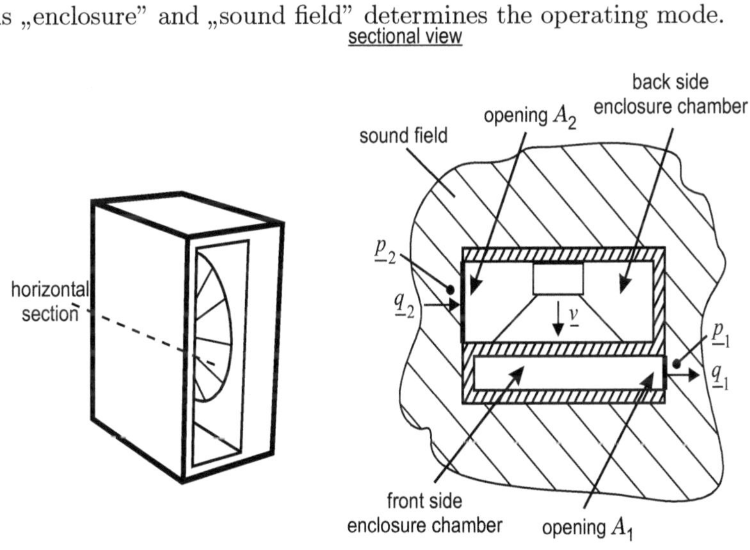

Fig. 6.40. Structure of a dipole bass loudspeaker (width 230 mm, height 380 mm, length 360 mm)

The approach for a network model is shown in Fig. 6.41. It consists of the known network representation of the electrodynamic loudspeaker (see left part of Fig. 6.41) that will be presented in Sect. 8.1. In order to model the acoustic area, this area is divided up into the sub-areas of the cuboidal enclosure chambers and the sound field at first. For that purpose, spatially averaged network coordinates $\underline{p}_1$ and $\underline{p}_2$ or $\underline{q}_1$ and $\underline{q}_2$ respectively are introduced at the chamber openings A_1 and A_2. Compared to the wavelength, the dimensions of

chamber openings of maximum 38 cm are small, so this model simplification represents a good approximation. Since network representations are known neither for the chambers nor for the sound field, acoustic two-ports are used in the network model. Due to the two openings of the dipole enclosure, the modeling of the sound field as two-port network (see also [16]) is required. Both openings emit into the sound field and interact simultaneously over the sound field. For the present geometry, the elements of the associated two-port matrix can not be achieved a priori by usual acoustic approximations.
By means of the FE method, the unknown parts of the network model can be determined very fast and accurately. For that purpose, comparatively simple FE models are sufficient which enable nevertheless a certain transformation of functional effects into the network domain.

Derivation of Network Representation of the Sound Field

Due to the out-of-phase acoustic volume velocity through both openings of the dipole enclosure and the system and due to the form of both openings, usual sound field representations can not be used here. But the derivation of a network model with very good approximation degree succeeds by means of FE methods.
The sound field of a loudspeaker enclosure which has two openings in the direction of the sound field, can be represented by an acoustic two-port network (Fig. 6.42). For processes in the field of linear acoustics, this is a passive linear reversible two-port matrix with impedance matrix

$$\begin{pmatrix} \underline{p}_1 \\ \underline{p}_2 \end{pmatrix} = \begin{pmatrix} \underline{Z}_{\mathrm{a,S1}} & -\underline{Z}_{\mathrm{a,M}} \\ \underline{Z}_{\mathrm{a,M}} & -\underline{Z}_{\mathrm{a,S2}} \end{pmatrix} \cdot \begin{pmatrix} \underline{q}_1 \\ \underline{q}_2 \end{pmatrix} .$$

The elements $\underline{Z}_{\mathrm{a,S}i}$ of the main diagonal represent the radiation impedances of the individual chamber openings. The secondary diagonal contains the element $\underline{Z}_{\mathrm{a,M}}$ with different sign. It can be perceived as a mutual radiation

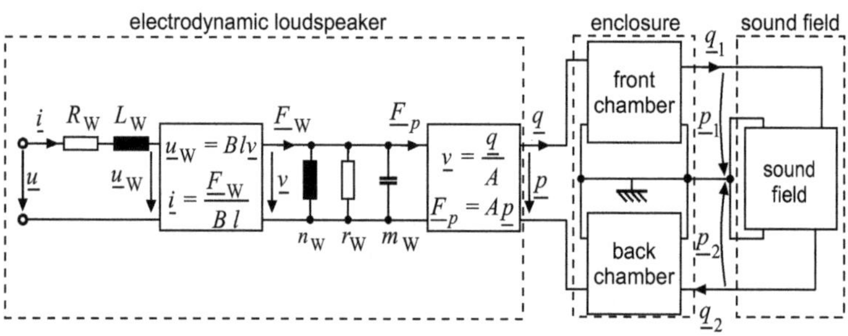

Fig. 6.41. Approach of a network model of dipole bass loudspeaker

impedance. It takes the effect on point 2 of a cause at point 1 and vice versa into account. For the determination of the impedance matrix elements no analytical solutions are available for the present geometry. Only for simple systems analytical solutions are available both for radiation impedances and for mutual radiation impedances.

In order to determine the radiation impedances $\underline{Z}_{a,Si}$ and the mutual radiation impedance $\underline{Z}_{a,M}$, the air surrounding the dipole enclosure is modeled in a hemispherical FE model area. Acoustic far field elements are arranged at the surface of the hemisphere. Figure 6.43 shows the symmetry plane of this model.

With the FE model, the radiation impedances of the rectangular enclosure openings can be determined at first. The resulting radiation impedance of an enclosure opening is exemplified in Fig. 6.44. Here, only one opening is considered in each case while the other opening is closed. A rigid and massless plate with mechanical elements is modeled at the considered opening and excited by a known force. A harmonic analysis yields the displacement of plate enabling at first the calculation of the mechanical impedance and subsequently the acoustic radiation impedance. The structure of the network representation of the sound field results from the analytical solution of the electrodynamic loud-

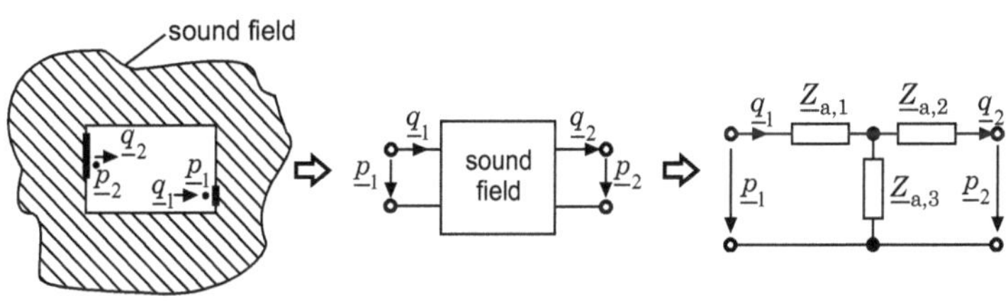

Fig. 6.42. Network representation for the sound field surrounding the dipole enclosure

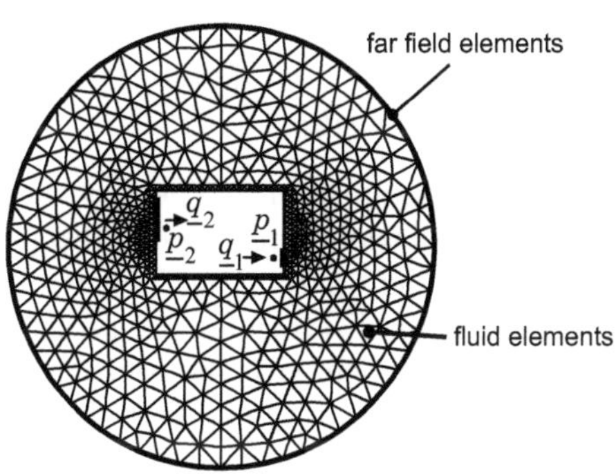

Fig. 6.43. FE model for the determination of network representation of sound field (only symmetry plane is shown)

speaker specified in Sect. 8.1.2. For low frequencies, the radiation impedance results in

$$\underline{Z}_{\mathrm{a,S}i} = Z_{\mathrm{a,S}i} + \mathrm{j}\omega \cdot M_{\mathrm{a,S}i} \qquad \text{with} \qquad Z_{\mathrm{a,S}i} = f(\omega) \sim \omega^2$$

corresponding to a series connection of an acoustic mass $M_{\mathrm{a,S}i}$ and a frequency-dependent acoustic friction $Z_{\mathrm{a,S}i}$ (Fig. 6.44 right). The parameters $M_{\mathrm{a,S}i}$ and $Z_{\mathrm{a,S}i}$ of the radiation impedance for rectangular enclosure openings can be derived from the frequency response of the complex radiation impedance (Fig. 6.44 left) calculated by means of FE methods.

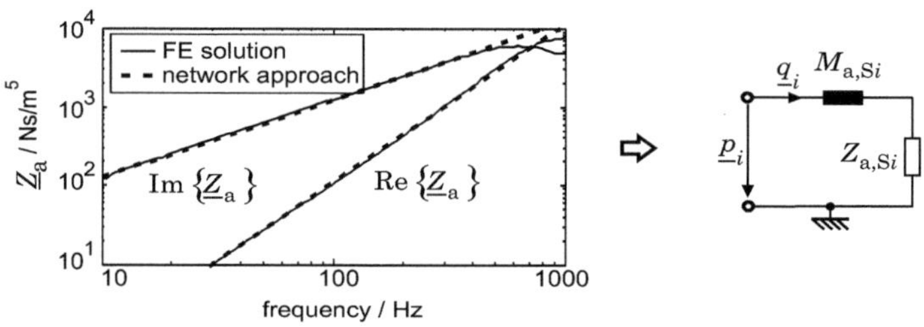

Fig. 6.44. Radiation impedance of an enclosure opening (left) and resulting network representation (right)

In order to determine the mutual radiation impedance, the area of an opening is provided with a rigid plate and excited with a velocity. The resulting mean pressure is determined at the closed area of the other opening. Afterwards the mutual radiation impedance can be calculated. In order to derive an approximate network representation for a large frequency range, a complex acoustic impedance according to following approach

$$\underline{Z}_{\mathrm{a,M}} = \mathrm{j}\rho_0 \frac{\omega}{4\pi \cdot d_{\mathrm{eff}}} \cdot \mathrm{e}^{\left(-\mathrm{j}\frac{\omega}{c_0} d_{\mathrm{eff}}\right)}$$

is chosen for the mutual radiation impedance. The approach is based on analytical solutions of the mutual radiation impedance of two point sound sources and has only one geometry-dependent parameter. This parameter is called the effective distance d_{eff}. The effective distance for the present geometry can be determined by comparing the approach with the solution of the FE model.
In order to achieve a manageable network model for the sound field from the two-port network, this two-port network can be represented in the form of a T-circuit (Fig. 6.42 right). The component parameters of the T-circuit

$$\underline{Z}_{\mathrm{a},1} = \underline{Z}_{\mathrm{a,S1}} - \underline{Z}_{\mathrm{a,M}} = Z_{\mathrm{a,S1}} + \mathrm{j}\omega \cdot M_{\mathrm{a,S1}} - \mathrm{j}\rho_0 \frac{\omega}{4\pi \cdot d_{\mathrm{eff}}} \cdot \mathrm{e}^{\left(-\mathrm{j}\frac{\omega}{c_0} d_{\mathrm{eff}}\right)}$$

$$\underline{Z}_{\mathrm{a},2} = \underline{Z}_{\mathrm{a,S2}} - \underline{Z}_{\mathrm{a,M}} = Z_{\mathrm{a,S2}} + \mathrm{j}\omega \cdot M_{\mathrm{a,S2}} - \mathrm{j}\rho_0 \frac{\omega}{4\pi \cdot d_{\mathrm{eff}}} \cdot \mathrm{e}^{\left(-\mathrm{j}\frac{\omega}{c_0} d_{\mathrm{eff}}\right)}$$

$$\underline{Z}_{\mathrm{a},2} = \underline{Z}_{\mathrm{a,M}} = \mathrm{j}\rho_0 \frac{\omega}{4\pi \cdot d_{\mathrm{eff}}} \cdot \mathrm{e}^{\left(-\mathrm{j}\frac{\omega}{c_0} d_{\mathrm{eff}}\right)}$$

result from the two-port matrix. These component parameters comprise complex functions which can be implemented into network analysis programs without any difficulty. In case of further restrictions with respect to the considered frequency range, these functions could also be represented by means of basic components.

Derivation of Network Representation of Enclosure Chambers

The unusual shape of both cuboidal chambers of the dipole enclosure necessitates the application of an FE model in order to derive the network representation. As already described before, at first the network approach is realized in the form of a two-port network. In the same manner, a systematic derivation of the two-port network parameters is made possible by numerical calculation of the terminal behavior of the two-port network with respect to different boundary conditions. However, it can be shown that depending on the choice of boundary condition numerical errors bias the results and make the derivation of correct two-port network parameters difficult. Here, a second, more heuristic approach is presented. For that purpose, an enclosure chamber is simulated with the realistic boundary condition of free field radiation. For this model of an enclosure chamber with sound field illustrated in Fig. 6.45, the acoustic input impedance $\underline{Z}_{\mathrm{a}}$ is calculated at the loudspeaker position.

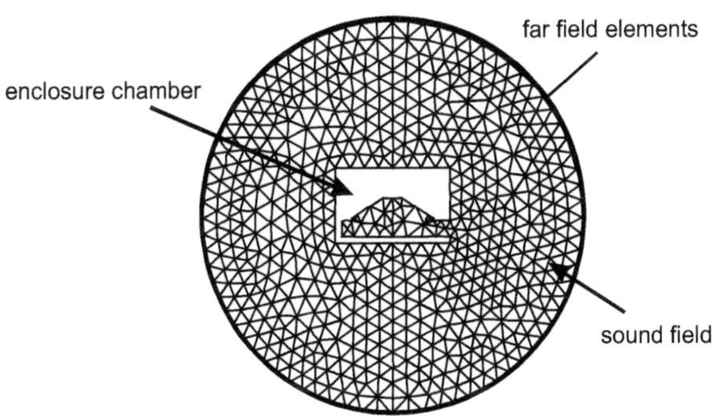

Fig. 6.45. FE model for determination of network representation of an enclosure chamber (only symmetry plane is shown)

The resulting input impedance is shown in the left part of Fig. 6.46. For low frequencies, the chamber only operates as acoustic mass. For fequencies higher than the resonant frequency of approximately 200 Hz, the enclosure chamber with sound field operates as an acoustic compliance (see phase-frequency response in Fig. 6.46 left). In the frequency range being of interest, a parallel connection of an acoustic mass and an acoustic compliance can be derived from this result. In combination with the known radiation impedance of the rectangular chamber opening, the result is a network structure of the chamber illustrated in the right part of Fig. 6.46. The parameters of both components can also be determined from the FE solution. Thus, a good approximation for the frequency range being of interest is achieved by means of the network model (Fig. 6.46 left).

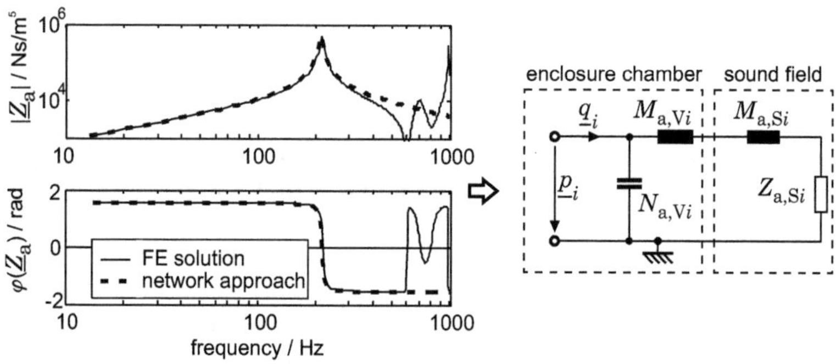

Fig. 6.46. Acoustic input impedance of a chamber of dipole enclosure (left) and resulting network structure (right)

Network Model of Dipole Bass Loudspeaker

Now a network model can be set up for the dipole bass loudspeaker by means of FE simulations. Figure 6.47 shows this network model. The resulting acoustic pressure at a certain distance can be determined analytically or from the acoustic volume velocities $\underline{q}_1$ and $\underline{q}_2$ resulting from an appropriate network representation. However, for this an additional parameter is required which denotes the effective distance between the openings of the dipole enclosure acting as acoustic source. This so-called effective source distance does not correspond to the geometrical distance between both openings, since refractive and diffractive effects at the enclosure must be taken into account. It can be shown that this effective source distance can be well approximated by the effective distance from the mutual radiation impedances. In the present case, the calculated effective source distance is 1.75 times larger than the geometrical distance.

The left part of Fig. 6.48 shows the resulting sound pressure level 1 m apart in front of the dipole bass loudspeaker at an input voltage of $\tilde{u} = 2.83\,\mathrm{V}$. The electrical input impedance of the dipole bass loudspeaker is represented in the right part of Fig. 6.48. The conformity with measurement results is very good both with respect to frequency response and absolute value. The frequency range being simulatable by the model reaches from 10 Hz up to 500 Hz and thus it is much larger than aimed.
The calculating time for the presented network model of dipole bass loudspeaker is less than one minute. Thus, it is possible to simulate different variants efficiently. For example, the influence of another electrodynamic loudspeaker can be quickly precalculated. In case of changes concerning the dipole enclosure, it must be estimated to what extent the derived network structures

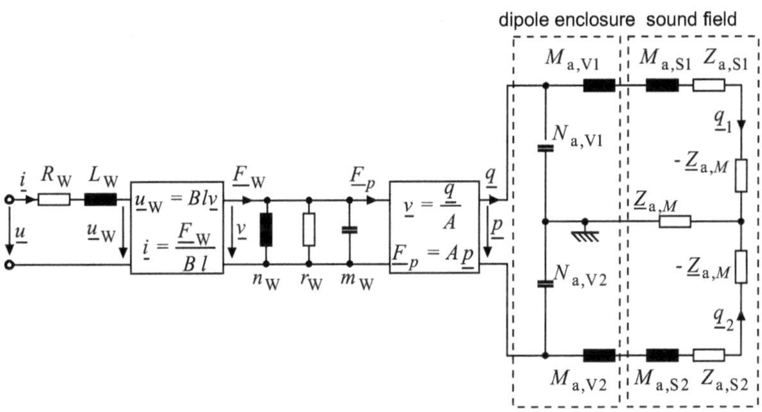

Fig. 6.47. Network model of dipole bass loudspeaker

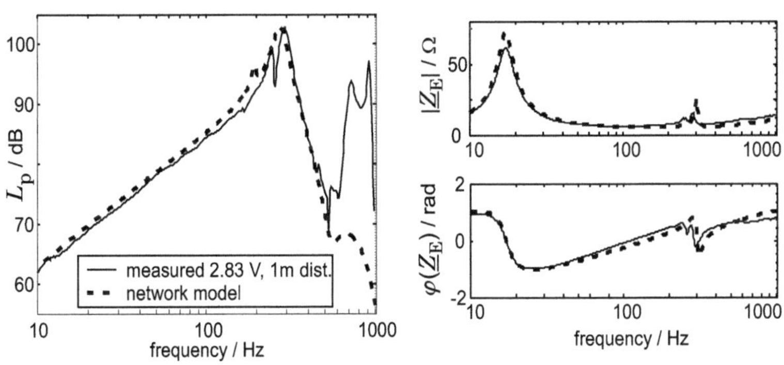

Fig. 6.48. Comparison of results of network model (dashed) with measurement results (continous)
L_p *sound pressure level,* $\underline{Z}_\mathrm{E}$ *input impedance*

and parameters still provide a good approximation or repeated FE simulations are necessary.

6.4.3 Combined Simulation using the Example of a Microphone with Thin Acoustic Damping Fabric

In this section, the *combined simulation* will be presented using the example of a microphone capsule with a thin acoustic damping fabric. Acoustic damping fabrics in the form of fleeces and fabrics are used in microphones for various function, such as for adjustment of the amplitude-frequency response. Acoustic damping fabrics are also used in simple microphone capsules with directional sensitivity.

Structure of Simplified Microphone Capsule

For a better understanding, the precalculation of the acoustic properties of a microphone capsule with a directional effect is discussed using the example of a simplified microphone capsule. For that purpose, a microphone with a rigid diaphragm, such as an electret codenser microphone, is assumed. A detailed representation of the diaphragm and transducer is omitted. The inside of the microphone capsule is only modeled as cavity. Figure 6.49 shows the structure and dimensions of the simplified microphone capsule. The diaphragm is arranged at the front side of the axisymmetric microphone capsule. It is simply modeled as a rigid structure. At the back side there is a further opening toward the sound field. It is covered by a thin acoustic damping fabric. The acoustic damping fabric has a thickness of 100 μm and a flow resistance of $\Xi l = 100\,\mathrm{Nsm}^{-3}$.

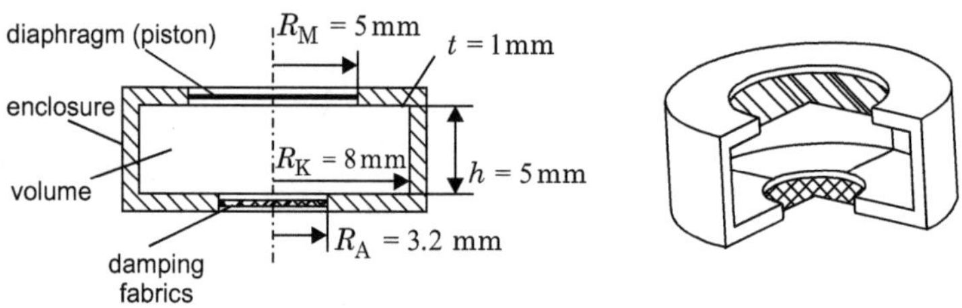

Fig. 6.49. Structure and dimensions of modeled microphone capsule

This represents a typical configuration for a microphone with directional effect. Depending on the geometry of the microphone capsule, the desired directional effect will be achieved by dimensioning the volume and damping fabric.

It can be shown that even for low frequencies the refraction and diffraction of sound at the microphone housing have a strong effect on the resulting directional effect of the microphone capsule. A modeling with network methods based on microphone geometry and simplified acoustic assumptions concerning the sound field yields incorrect results. In order to achieve sufficiently accurate results, it necessitates a modeling of the microphone capsule or the complete microphone in the sound field respectively by means of numerical methods, e.g. the FE method. However, the modeling of the acoustic damping fabric is currently not available in all standard FE programs. This does apply also for the considered FE program ANSYS®. Nevertheless, a calculable FE model can be created by means of the *combined simulation* of FE methods with network methods. That will be presented in the following.

Modeling of Porous Acoustic Damping Elements

Acoustic damping elements in the form of woven and nonwoven fabrics, as they are used in a variety of electroacoustic transducers, belong to the group of open-pored porous absorbers. Within a porous absorber, energy is dissipatively removed from the sound wave. Firstly, this is caused by frictional loss between the vibrating air particles and the absorber structure. Secondly, the air inside the absorber is alternately compressed and dilated. Thermal losses occur in the transition region of these two changes of state. A third loss mechanism may occur due to dissipative vibrations of the absorber skeleton.
Due to the complex microstructure of porous absorbers, it is not practicable to aim for a full description of the absorber skeleton in order to calculate the sound propagation within the absorber. Rather a number of structurally simple absorber models exists (see e.g. [57]) which enable an analytical description of sound propagation within the absorber on the basis of measured characteristics of the absorber. In spite of structural simplification, with these models a good approximation can be achieved for real absorbers. These analytical descriptions have in common that they assume a propagation of a plane wave within the absorber (Fig. 6.50) and that they describe this propagation by means of the complex propagation constant $\underline{\gamma}$ with

$$\underline{p}\,(x,t) = \hat{p}\,\mathrm{e}^{-\underline{\gamma} x}\,\mathrm{e}^{\mathrm{j}\omega t}$$

and the complex wave impedance

$$\underline{Z}_\mathrm{W} = \frac{\overline{\underline{p}\,(x,t)}}{\overline{\underline{v}\,(x,t)}}. \tag{6.27}$$

The specification of a wave impedance $\underline{Z}_\mathrm{W}$ for the absorber will be only sufficient, if it is spatially averaged over a region perpendicularly to the direction of sound propagation. Compared to structural dimensions of the absorber this region must be large, but it must be small compared to wavelength.

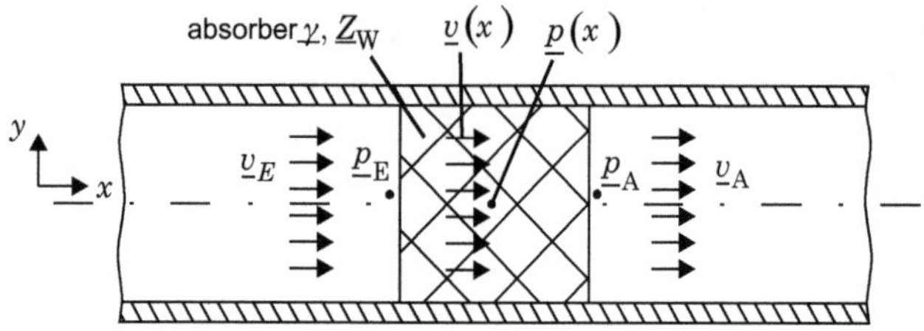

Fig. 6.50. Propagation of a plane wave in a tube with porous absorber

In the following, the acoustic quantities acoustic pressure $\underline{p}$ and acoustic volume velocity $\underline{q} = \underline{v} \cdot A$ ($A \ldots$ cross-sectional area) are used for all descriptions. Thus, the acoustic wave impedance

$$\underline{Z}_{a,W} = \frac{\overline{\underline{p}(x,t)}}{\overline{\underline{q}(x,t)}} = \frac{\underline{Z}_W}{A}$$

results from (6.27). In order to solve a multiplicity of acoustic problems, it is not necessary to describe completely the sound propagation within the absorber. The knowledge about the behavior of acoustic quantities at the boundary surfaces of the free fluid directed to the absorber will suffice. For this purpose, the absorber can be considered as one-dimensional dissipative acoustic waveguide with complex propagation constant $\underline{\gamma} = \alpha + j\beta$. Here, α denotes the damping constant and β denotes the propagation constant. In addition to the loss-free one-dimensional acoustic waveguide (see Sect. 6.2), the chain matrix of the dissipative waveguide yields

$$\begin{pmatrix} \underline{p}_E \\ \underline{q}_E \end{pmatrix} = \begin{pmatrix} \cosh(\underline{\gamma} l) & \underline{Z}_{a,W} \sinh(\underline{\gamma} l) \\ \frac{1}{\underline{Z}_{a,W}} \sinh(\underline{\gamma} l) & \cosh(\underline{\gamma} l) \end{pmatrix} \begin{pmatrix} \underline{p}_A \\ \underline{q}_A \end{pmatrix} .$$

For the modeling of the simplified microphone capsule it is sufficient to focus only on the consideration of thin acoustic damping fabrics. That means that the thickness l of the absorber is expected to be much smaller than the wavelength λ_A within the absorber. For real absorbers with thicknesses in the range of $\leq 100\,\mu$m this requirement is fulfilled in the complete audio frequency range ($f \leq 20\,$kHz). Furthermore, it is assumed that the absorber structure does not resonate completely. As a further requirement, the acoustic volume velocity is expected to be able to pass through the absorber, so that compression effects within the absorber can be neglected. Therefore, the acoustic volume velocity in front of the absorber $\underline{q}_E$ and the acoustic volume velocity behind the absorber $\underline{q}_A$ are almost equal ($\underline{q}_E \approx \underline{q}_A$).
In combination with the mentioned boundary conditions, the acoustic behavior of the absorber can be described in good approximation by the network representation shown in Fig. 6.51.

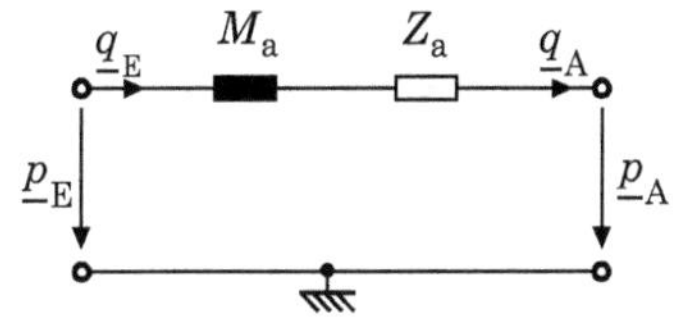

Fig. 6.51. Network representation of thin passed-through absorbers

In this network representation, the first term in the acoustic mass

$$M_a = \left(\rho_0 \frac{\sigma}{\chi} + \rho'_A\right) \frac{l}{A} = M_{a,L} + M_{a,A}$$

represents the effect of air mass

$$M_{a,L} = \rho_0 \frac{\sigma}{\chi} \cdot \frac{l}{A}$$

resonating within the absorber. Here, ρ_0 denotes the density of air, σ denotes the structure factor, χ denotes the porosity, l denotes the thickness of fabric and A denotes the area of fabric. For practical fabrics the factor σ/χ can mostly be set equal to 1 with sufficient accuracy. By means of the second term

$$M_{a,A} = \rho'_A \frac{l}{A},$$

mass fractions of parts of the absorber skeleton resonating within the absorber can be additionally taken into account. The value of the equivalent density ρ'_A has to be measured for the absorber in the particular configuration. The acoustic resistance

$$Z_a = \Xi \frac{l}{A}$$

represents the frictional losses within the absorber. Here, the quantity Ξ denotes the flow resistivity.

Derivation of an Acoustically Equivalent Structure for Modeling an Acoustic Damping Fabric

The illustrated network model for thin acoustic damping elements represents a well-known model reduced to simulation task. For an acoustic modeling of thin acoustic absorbers in the FE program ANSYS®, this network model must be represented in an FE representation. However, as already described before, a representation of acoustic damping elements is not possible in this FE program.

As already shown, under certain boundary conditions a thin acoustic damping fabric within a tube with plane waves (Fig. 6.52 a)) can be represented

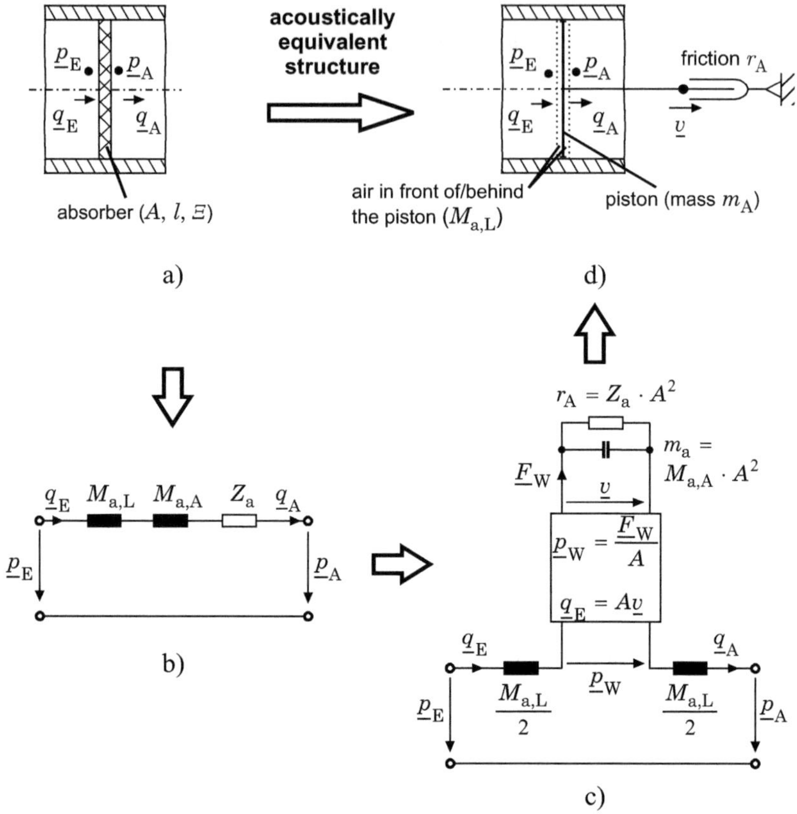

Fig. 6.52. Step-by-step derivation of an acoustically equivalent structure for modeling a thin acoustic damping fabric

as a simple acoustic network (Fig 6.52 b)). This circuit can arbitrarily be transformed within the network domain. Thus, also a representation can be chosen that can be realized with the used FE program. The transformation of the acoustic friction and a part of the acoustic mass to the mechanical domain by means of a mechanical-acoustic transducer represents a possibility (Fig. 6.52 c)). Afterwards, the transformed network is transfered into a mechanical-acoustic equivalent structure which can be represented in the FE program (Fig. 6.52 d)). The equivalent structure consists of a mechanical mass and a mechanical friction which can be realized in the FE-model with standard elements. The mechanical-acoustic transducer can be realized as rigid model structure (piston) with interconnection to fluid elements. The resonating air within the absorber $M_{a,L}$ is represented one half each by fluid elements in front and behind the piston. Thus, a travel time of the sound wave through the absorber of at least the travel time of the sound wave through the air is taken into account.

In consideration of the above mentioned boundary conditions, thus a thin acoustic damping fabric for acoustic considerations can be represented by an oscillating rigid piston with a mechanical friction. The effect of this representation is the same for the acoustic quantities outside the damping fabric. Thus, Fig. 6.52 d) shows an acoustically equivalent structure for the modeling of thin acoustic damping elements. It should be noted that this equality of effect applies only to acoustic processes. It does not apply to mechanical properties of the absorber or to the fluid flow.

Calculation of Directional Effect of Simplified Microphone Capsule

In combination with the presented equivalent structure, the simplified microphone capsule with acoustic damping fabric can now be modeled in the FE program ANSYS®.
In a good approximation, the electrical output voltage of the microphone is assumed to be proportional to the acoustic pressure difference between front side and back side of diaphragm being usually very stiff in reality. The directional effect can be calculated approximately by the acoustic pressure difference between front side and back side of an ideal rigid structure at diaphragm position.
In order to precalculate the directional effect of the microphone, the microphone is modeled in the field of a plane wave or a spherical wave. Since only a limited modeling region can be realized with FE methods, the field of the incident wave is defined by appropriate boundary conditions on the boundary of modeling region. The sound field in the inside results from constraints at the boundary. The limited region must be chosen so large that in spite of interference by the particular structure in the inside of the model region (here the configuration of the microphone) an undisturbed sound field (e.g. of the plane wave) can be assumed on the boundary in a good approximation. A transient FE analysis allows for the compliance of this boundary condition.
The rotation symmetry for a two-dimensional axisymmetric model must also apply to the wave, so that plane waves are possible with only one propagation direction parallel to the symmetry axis. It also applies to spherical waves that the position of source must coincide with a position on symmetry axis. Thus, the calculation of the directional effect of the microphone configuration is possible for sound incidence angles of $\sphericalangle = 0°$ and $\sphericalangle = 180°$. The FE model of the microphone capsule with the acoustic equivalent structure is illustrated in the left part of Fig. 6.53.
The transient calculation of the directional effect of the microphone for angles not equal to $\sphericalangle = 0°$ and $\sphericalangle = 180°$ requires a three-dimensional modeling. For that purpose, the mirror symmetry can be utilized for the present configuration of microphone, so that modeling of one half of the microphone and of a hemispherical region of air will suffice. The procedural method for providing an FE model corresponds to that of two-dimensional modeling. Therefore, a detailed description is omitted at this point. Based

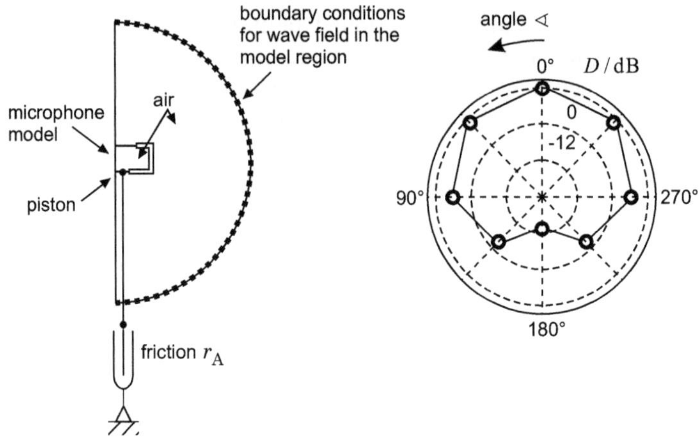

Fig. 6.53. Schematic representation of axisymmetric FE model and polar diagram of directional effect (for a better visibility the results are connected by lines)

on the results of three-dimensional transient analysis, the directional effect can be calculated by means of the acoustic pressure difference across the diaphragm of the microphone configuration. In the right part of Fig. 6.53 the resulting directional characteristic is illustrated as polar diagram for $\sphericalangle = [0°; 45°; 90°; 135°; 180°; 225°; 270°; 315°]$. The results in the polar diagram show a cardioid polar pattern of this microphone configuration. Now it is possible to understand the functioning of a microphone capsule with a cardioid polar pattern by means of the FE analysis. For that purpose, the acoustic pressure-time functions must be evaluated at positions in front of, behind and within the microphone capsule. The defined positions are shown in Fig. 6.54.

Considering the time function of acoustic pressures in case of a vertically incident plane wave from top ($\sphericalangle = 0°$) shown in the left part of Fig. 6.55, it can be realized that the acoustic pressure is present in front of the capsule at first. The acoustic pressure behind the capsule results from the time delay around

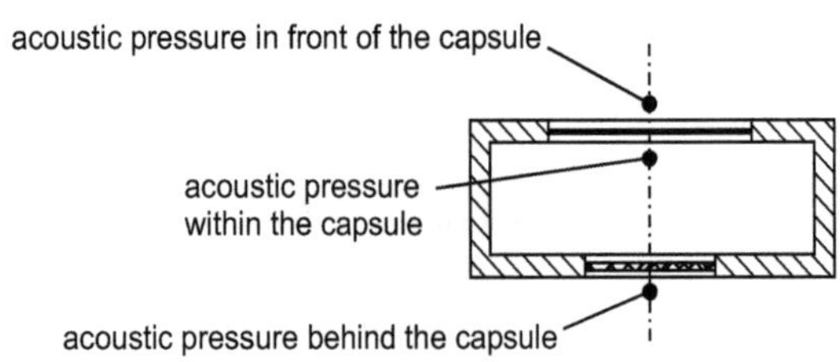

Fig. 6.54. Measuring point positions for acoustic pressure-time functions

the capsule enclosure. The acoustic pressure within the capsule is additionally delayed by the arrangement of absorber and the volume within capsule. Thus, the result is an acoustic pressure difference with a comparatively high amplitude.
Figure 6.55 on the right analogously shows the acoustic pressure-time functions for a vertically incident plane wave from bottom ($\sphericalangle = 180°$). This arrives at first the at the diaphragm's back side. The time delays around the capsule or through the absorber-volume arrangement are approximately equal, so that the acoustic pressure-time functions are nearly equal before and within the capsule. Both acoustic pressure-time functions represented in the diagram are visually not distinguishable. Therefore, the value for the acoustic pressure difference across the diaphragm is only very small.

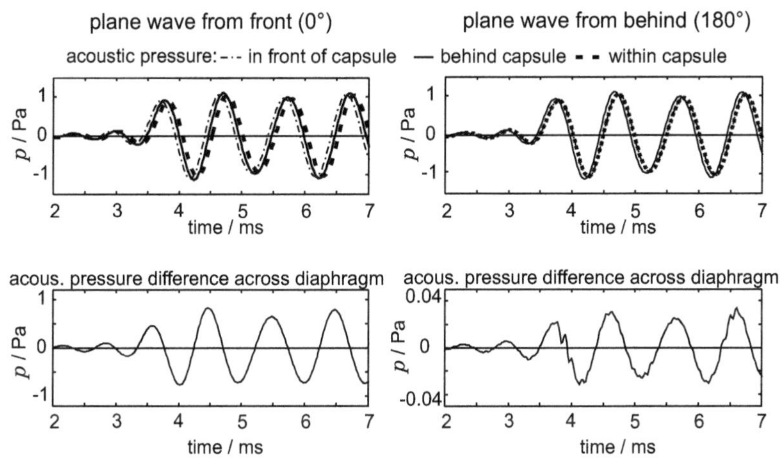

Fig. 6.55. Acoustic pressure-time functions of microphone capsule

Additionally, the time delay between front side (diaphragm) and back side (damping fabric) of the capsule can be estimated from the acoustic pressure-time functions. It amounts to approximately 60 μs which corresponds to an apparent distance of about 20 mm. Thus, the effective distance is almost 3 times as large as the height of microphone capsule of $h_K = 7$ mm. Even though the dimensions of the microphone configuration (diameter 18 mm) are significantly smaller than the wavelength ($\lambda = 344$ mm), the time delays for acoustic pressures at the diaphragm and the absorber are significantly larger than those that would expected from the geometrical distance. A simple analytical approach based on geometrical distances and with respect to neglected pressure stasis and diffraction effects would not provide suitable results.

The long computation time represents a disadvantage of the presented transient analysis. Individual calculations have to be made for each angle of inci-

dent sound, each sound source distance and each frequency.
The application of acoustic reciprocity (see Sect. 10.1) with a harmonic FE analysis represents an advantageous possibility for calculating the directional characteristics of a microphone. In linear acoustic systems, the ratio of an acoustic volume velocity at point 1 and the acoustic pressure at point 2 equals the ratio of an acoustic volume velocity at point 2 and the acoustic pressure at point 1. In order to calculate the acoustic pressure $\underline{p}_2$ at the microphone's diaphragm which is generated by a distant source of sound with an acoustic volume velocity $\underline{q}_1$, the acoustic pressure $\underline{p}_1$ at the primary position of the source of sound can also be calculated for an acoustic volume velocity $\underline{q}_2$ being available at the position of diaphragm. This provides the advantage for microphones that the directional effect is achieved for all angles and distances in model region with one calculation. For that purpose, in the FE model an acoustic volume velocity is applied to the diaphragm's position by means of an in-phase oscillating piston. This represents a good approximation for the calculation of the directional effect, as long as the influence of the vibrational mode of the diaphragm on the directional effect can be neglected.
A modeling as two-dimensional FE model is sufficient for the axisymmetric sample configuration. The harmonic analysis results in the distribution of sound pressure level within the model region. In order to represent the originating directional characteristics, the polar diagram represented in Fig. 6.56 can be derived from this solution. There is a very good correlation of the results of harmonic and transient solutions, thus the application of reciprocity for harmonic analysis is verified.

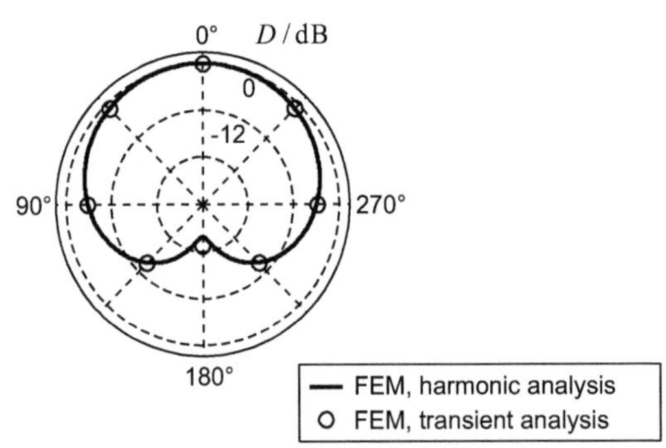

Fig. 6.56. Polar diagram of microphone configuration at $f = 1\,\text{kHz}$, 1 m apart

Comparison of Simulation and Measurement Results

In order to validate the modeling approach, an FE model of a real electret microphone capsule was generated and compared with measurement results.

The modeling was realized as two-dimensional axisymmetric model. For that purpose, only the geometry as well as the known flow resistance of the damping fabric were used as input data for the model. As an example, a polar diagram with curve shapes for 1 kHz and 4 kHz is shown in Fig. 6.57.

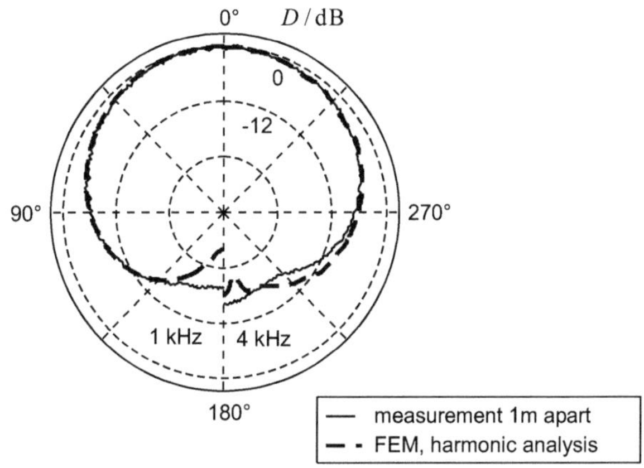

Fig. 6.57. Directional characteristic of an electret microphone capsule 1 m apart (measurements and combined simulation are compared)

It shows a very good correlation between precalculated and measured directional characteristic. Deviations become primarily evident in case of backward attenuation ($\sphericalangle \approx 180°$). They are significantly based on the used measurement setup. For higher frequencies, larger deviations occur which are caused by the simplistic representation of diaphragm. Calculations with a three-dimensional transient model verify a good correlation between model and measurement.

Part III

Electromechanical Transducers

7

Electromechanical Interactions

In the Chaps. 2 and 3 four types of linear networks were described, as there are electrical, mechanical-translational, mechanical-rotational and mechanical-acoustic networks. Based on modification options of these networks, the coupling elements between translational and rotational as well as translational and acoustic networks have been defined in Chap. 5. The question concerning the coupling of mechanical or acoustic subsystems with the electrical ones remained open.

Thus, based on the physical principles of electromechanical interactions presented in Sect. 2.4.2, the various structures of coupling between the electrical and mechanical subsystems are considered in detail in Chap. 7. The discovery and the circuit design interpretation of basic coupling elements in the form of loss-free two-port networks which correspond to the structure of mechanical transducers derived in Chap. 5 represents the aim of these considerations.

Due to restrictions made in Sect. 2.4.2, the considered coupling systems are also linear and passive, i.e. the transducers do not comprise internal energy sources.

7.1 Classification of Electromechanical Interactions

Concerning energy conversion, two groups of transducers (Fig. 7.1) can be distinguished in principle. With *transducers without auxiliary power* (*passive transducers*), the electrical energy is solely extracted from the mechanical subsystem and vice versa. With *transducers with auxiliary power* (*active transducers*), the control of an electric circuit is effected by means of a mechanical quantity. Here, the control of the electric circuit is based on a parameter variation of passive components especially of resistive, capacitive and inductive components. The parameter variation is enforced by the mechanical quantity. Therefore, the signal flow is possible only in one direction (from the mechanical to the electrical side).

A. Lenk et al., *Electromechanical Systems in Microtechnology and Mechatronics*,
Microtechnology and MEMS, DOI 10.1007/978-3-642-10806-8_7,

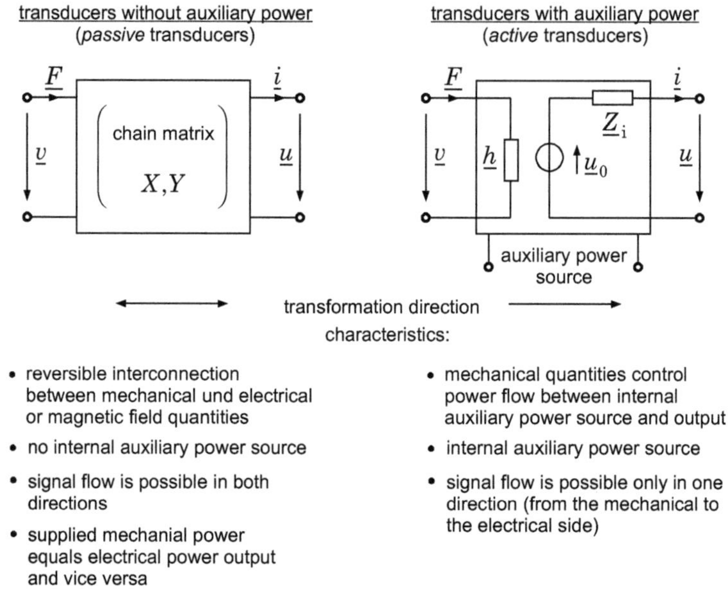

Fig. 7.1. Electromechanical transducers with and without auxiliary power
X, Y *transducer constants*

The main physical principles of loss-free electromechanical transducers without auxiliary power are shown in Table 7.1. Depending on whether electrical or magnetic quantities are associated with mechanical quantities, the interactions can be split into two groups, thus into *electrical* and *magnetic transducers*. The group of electrical transducers includes the electrostatic and piezoelectric transducer. The group of magnetic transducers includes the electromagnetic, electrodynamic and piezomagnetic transducer. Depending on the direction of transformation, these transducers are useable for sensor or actuator applications. However, sensors without auxiliary power are unsuitable for static measurements, since they do not have any electrical power output in case of a time-invariant input quantity and therefore they are not able to compensate energy losses caused by the evaluation circuit, e.g. finite input resistance.
In this case, transducers with auxiliary power are used. Since they possess a signal flow only from the mechanical to the electrical side, they are solely used as sensors, especially for *quasi-static* applications. Sensors with auxiliary power preferably use the *deflection* or *compensation method* as measuring method.
In case of the deflection method, the mechanical quantity to be measured deforms a deformation element (Fig. 7.2).

Table 7.1. Physical operating principles of electromechanical transducers without auxiliary power
Linearization is achieved by expansion around operating point

electrical transducers

electrostatic principle	piezoelectric principle
force change ΔF between two oppositely charged electrodes caused by charge change ΔQ or voltage change Δu caused by deflection ξ	charge separation ΔQ caused by applying a force ΔF or deformation ξ by applying a voltage Δu
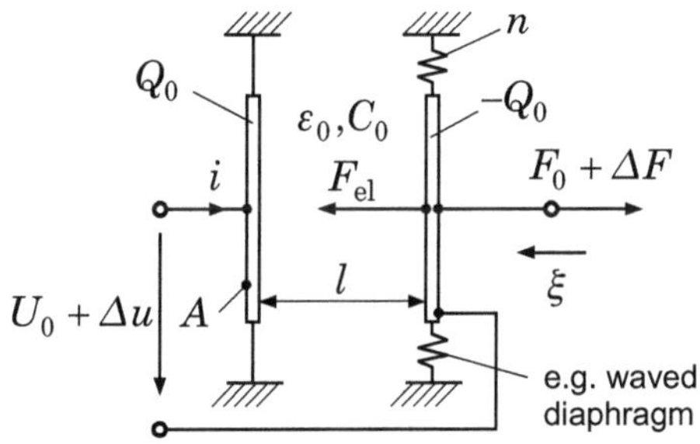	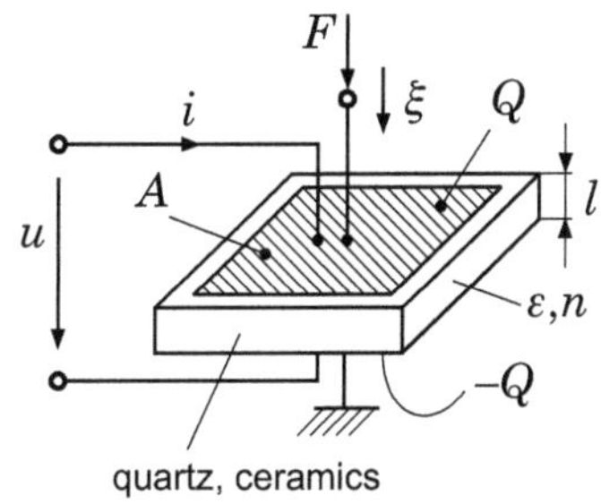
$\Delta F = \dfrac{Q_0}{\varepsilon_0 \cdot A} \Delta Q - \dfrac{1}{n} \cdot \xi$ $\Delta u = \dfrac{1}{C_0} \Delta Q - \dfrac{Q_0}{\varepsilon_0 \cdot A} \cdot \xi$	$Q = \varepsilon \dfrac{A}{l} u + d \cdot F$ $\xi = d \cdot u + n \cdot F$ d... charge constant

magnetic transducers

electromagnetic principle	electrodynamic principle	piezomagnetic principle
force change ΔF between two pole faces A (magnetic field B_0 inbetween) caused by a magnetic flux change $\Delta \Phi$ or current change Δi caused by deflection ξ	force change ΔF on wire with electric current passing through it in magnetic field B_0 caused by current change Δi or voltage change Δu caused by a motion at velocity Δv	magnetic flux change $\Delta \Phi$ by applying a force ΔF or deformation ξ by applying a current Δi
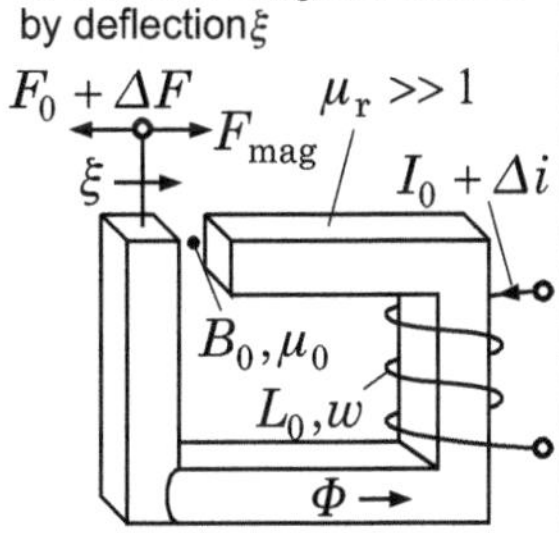	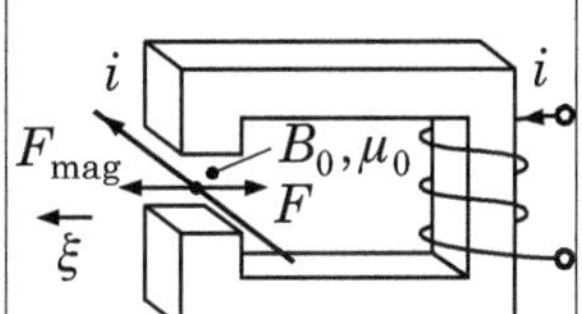	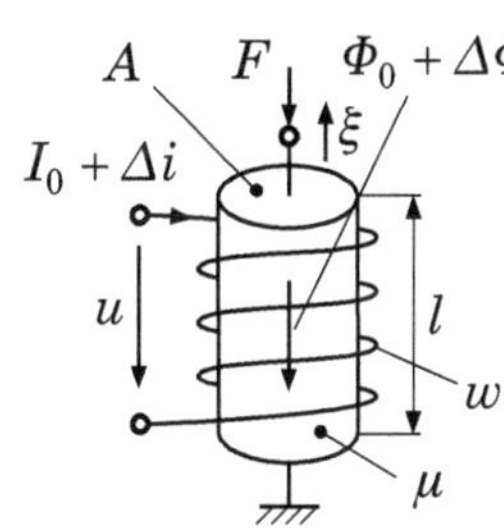
$\Delta F = \dfrac{B_0}{\mu_0} \cdot \Delta \Phi - \dfrac{1}{n} \xi$ $\Delta i = \dfrac{w}{L_0} \cdot \Delta \Phi + \dfrac{B_0}{w \cdot \mu_0} \xi$	$F = B_0 \cdot l \cdot i - \dfrac{1}{n} \xi$ $u = R \cdot i + B_0 \cdot l \cdot v$	$\Delta \Phi = \mu \cdot w \dfrac{A}{l} \Delta i - d \cdot F$ $\xi = w \cdot d \cdot \Delta i - n \cdot F$ d...piezomagn. constant

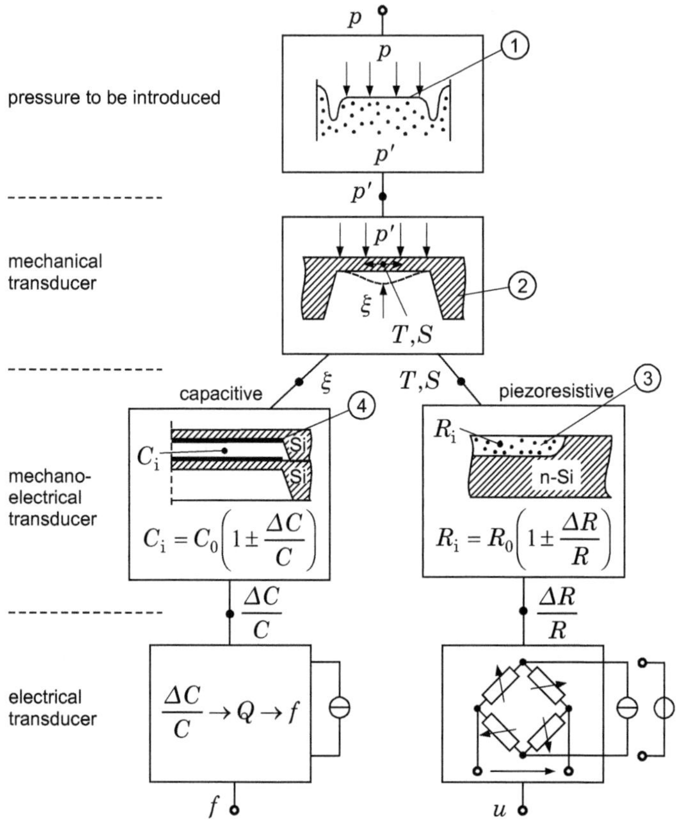

Fig. 7.2. Signal processing structure for piezoresistive and capacitive silicon pressure sensors using the deflection method
1 *separating diaphragm,* 2 *Silicon deformation element,* 3 *doped resistor,* 4 *electrodes*

The deformation causes a parameter variation of a resistive, capacitive or inductive transducer element which controls an electrical circuit. For resistive transducer elements, this parameter variation occurs almost without any reaction. The interactions in the electrostatic or electromagnetic field of capacitive or inductive transducer elements cause reactions that can be neglected for the majority of practical applications. The signal processing structures for silicon piezoresistive and capacitive pressure sensors shown in Fig. 7.2 exemplify the deflection method.
In case of the compensation method, a counteracting force is generated in the sensor. The very small deflection of the measuring element and thus the very small linearity errors represent an advantage of this method. Figure 7.3 illustrates the principle of a compensating pressure sensor with electrodynamic generation of a counteracting force.

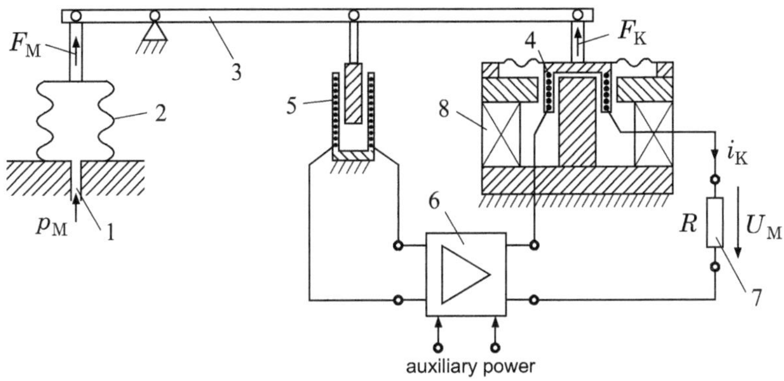

Fig. 7.3. Pressure sensor for measuring the gauge pressure using the compensation method
1 *gauge pressure,* 2 *metal bellows,* 3 *rigid lever,* 4 *moving coil,* 5 *inductive displacement sensor,* 6 *amplifier,* 7 *load resistor,* 8 *permanent magnet*

Even though electromechanical transducers with auxiliary power have great practical importance as sensors, they are not further discussed within the context of this book. In [3, 58–60] the design of such sensors is described in detail.
The situation is different with the initially mentioned transducers without auxiliary power. Here, real physical interconnections between different physical structures are utilized. They result in real coupling two-port networks. Their classification in a general linear dynamic electromechanical system is shown in Fig. 7.4.

7.2 Network Representation of Electromechanical Interactions

Based on Sect. 2.4.2 and the associated conditions, in this section the interaction between a mechanical pair of coordinates F, x (*port*) and an electrical pair of coordinates Q, u or i, μ respectively is considered at two selected reference points and is described by a schematic diagram. The charges or currents as well as the position coordinates of all other reference points will remain constant.

Electromechanical Coupling

According to the character of the acting force generation, the electromechanical coupling can be represented by one of the models specified in Fig. 7.5. They are based on the assumption defined in detail in Sect. 2.4.2 that with the

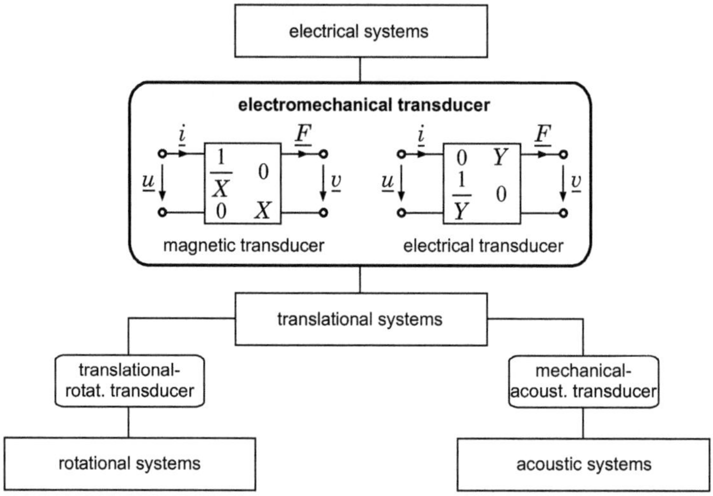

Fig. 7.4. General linear dynamic electromechanical system

coupling by means of *electric fields* the *equilibrium force* F at the mechanical reference point is additively composed of the *mechanical system force* $F_{\text{mech}} = \varphi(x)$ and the COULOMB *force* $F_{\text{el}} = \Psi(Q, x)$. The voltage u at the electric reference point depends not only on the charge Q but also on the position coordinate x. This results in the content of the model box shown in Fig. 7.5 bottom left.

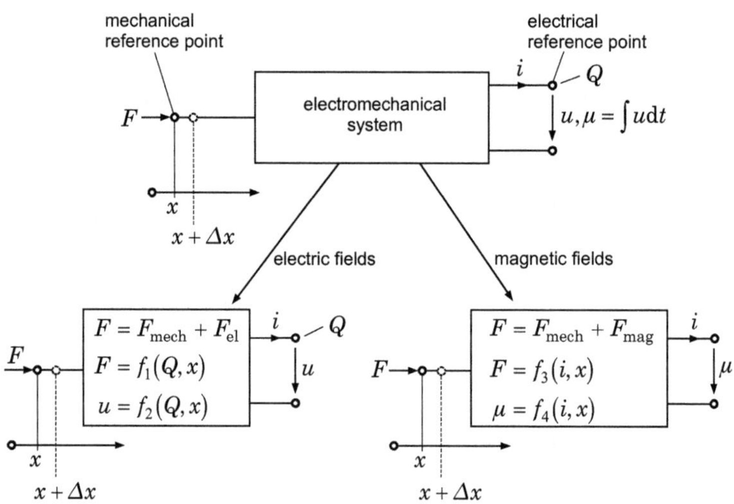

Fig. 7.5. Models of electromechanical coupling

In case of coupling by means of *magnetic fields*, the *mechanical system force* $F_{\mathrm{mech}} = \varphi(x)$ and the *magnetic field force* $F_{\mathrm{mag}} = \Psi(i,x)$ add up to the equilibrium force F at the mechanical reference point. In case of an available local model of the current loop i and its connection with the mechanical reference points, F_{mag} can be determined by means of the BIOT-SAVART *law*. In addition to the current, the voltage integral μ at the electrical reference point also depends on the position coordinate x. This results in the content of the model box shown in Fig. 7.5 bottom right.
In case of the linear system equation in the neighborhood of a reference point, the relations contained in Fig. 7.5 result in linear approximations (7.1) up to (7.4) in the neighborhood of x_0, F_0, u_0, μ_0, Q_0, i_0 with $\Delta x = x - x_0$, $\Delta F = F - F_0$, $\Delta Q = Q - Q_0$, $\Delta i = i - i_0$, $\Delta u = u - u_0$, $\Delta\mu = \mu - \mu_0$.
For the *electrical* interaction, it can be written for $F = f_1(Q,x)$ and $u = f_2(Q,x)$:

$$\Delta F = K_{\mathrm{el}}\Delta Q + \frac{1}{n}\Delta x \tag{7.1}$$

$$\Delta u = \frac{1}{C_{\mathrm{b}}}\Delta Q + K'_{\mathrm{el}}\Delta x \tag{7.2}$$

Here, n denotes the mechanical compliance of the clamping of the movable electrode plate. The quantity C_{b} denotes the capacitance between the electrode plates in the mechanically locked case ($v = 0$).
For the *magnetic* interaction, it can be written for $F = f_3(i,x)$ and $\mu = f_4(i,x)$:

$$\Delta F = K_{\mathrm{mag}}\Delta i + \frac{1}{n_{\mathrm{L}}}\Delta x \tag{7.3}$$

$$\Delta\mu = L_{\mathrm{b}}\Delta i + K'_{\mathrm{mag}}\Delta x \tag{7.4}$$

The quantity n_{L} denotes the mechanical compliance of the particular transducer element in case of open-circuit operation ($i = 0$), the quantity L_{b} denotes the transducer inductance in the mechanically locked case ($v = 0$).
In case of available simple structural models of the systems to be considered, the coefficients contained in (7.1) up to (7.4) can be determined by means of COULOMB forces and BIOT-SAVART forces respectively and by means of the network components in combination with their dependencies on geometry. Here, it will be found out that $K_{\mathrm{el}} = K'_{\mathrm{el}}$ and $K_{\mathrm{mag}} = K'_{\mathrm{mag}}$ is valid for the coefficients. The reason for this issue is described in Sect. 2.4.2. It may be appropriate in this case to use (7.5) and (7.6) with the field sizes as interaction between charges or current elements instead of the COULOMB and BIOT-SAVART equations in explicit form:

$$\mathbf{F}_{\mathrm{el}} = Q \cdot \mathbf{E} \tag{7.5}$$

$$\mathrm{d}\mathbf{F}_{\mathrm{mag}} = i \cdot \mathrm{d}\mathbf{l} \times \mathbf{B} \tag{7.6}$$

Derivation of Transducer Forces from Energy Balances

In those cases, where such a model is not available, for example, due to the occurrence of dielectric, piezoelectric, magnetic or piezomagnetic solids, the general requirement of *reversibility* of all running processes within the system helps for simplification. Formally, this assumption implies the requirement for the existence of a state function „internal energy" with total differential $\mathrm{d}W$:

electric field: $$\mathrm{d}W = F(x, Q)\,\mathrm{d}x + u(x, Q)\,\mathrm{d}Q \quad (7.7)$$

magnetic field: $$\mathrm{d}W = F(x, \mu)\,\mathrm{d}x + i(x, \mu)\,\mathrm{d}\mu \quad (7.8)$$

In case of coupling by means of magnetic field quantities, for the application of this axiomatic requirement it is advantageous to replace the structurally and physically founded form of the system equations (7.3) and (7.4) by the equivalent form

$$F_{\mathrm{mag}} = f_5(\mu, x) \quad (7.9)$$

$$i = f_6(\mu, x)\,. \quad (7.10)$$

The total energy supplied to the system through both ports from initial state $x_0, Q_0 = 0$ to state x_1, Q_1 results in:

electric field: $$W = \underbrace{\int_{x_0}^{x_1} F(x, Q(x))\,\mathrm{d}x}_{\mathrm{d}W_{\mathrm{mech}}} + \underbrace{\int_{Q=0}^{Q_1} u(Q, x(Q))\,\mathrm{d}Q}_{\mathrm{d}W_{\mathrm{el}}} \quad (7.11)$$

magnetic field: $$W = \underbrace{\int_{x_0}^{x_1} F(x, \mu(x))\,\mathrm{d}x}_{\mathrm{d}W_{\mathrm{mech}}} + \underbrace{\int_{\mu=0}^{\mu_1} i(\mu, x(\mu))\,\mathrm{d}\mu}_{\mathrm{d}W_{\mathrm{mag}}} \quad (7.12)$$

The functions $Q(x)$ or $\mu(x)$ follow the path, the state x_1, Q_1 is achieved. Reversibility or state function energy implie a path dependence of W_{mech} and W_{el} or W_{mag}, respectively. However, the sum $W_{\mathrm{mech}} + W_{\mathrm{el/mag}} = W$ is path-independent. Figure 7.6 shows the necessary sequence for two different paths in case of an electric field.

The total energy is supplied to the system in differential (virtual) steps through its two ports. In case of an available Q, in the neighborhood of the current value of x a partial energy $\Delta W_{\mathrm{el}} = F(x, Q)\,\Delta x$ is supplied through the mechanical port by the displacement Δx. Thus, from state point A, state point B is reached. Subsequently, a partial energy $\Delta W_{\mathrm{el}} = u(Q, x)\,\Delta Q$ is supplied through the electric port by the charge ΔQ. Thus, state point C is reached. The sum of both partial energies results in the change of the internal

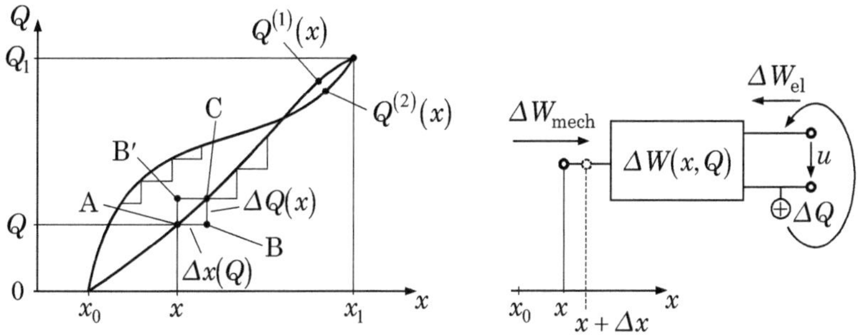

Fig. 7.6. Differential structure of internal energy consisting of mechanically and electrically supplied partial energies

energy ΔW. In a differential domain, the reversibility or the property of W as state function is characterized by the fact that the transition from point A to point C can also be realized along point B'. The same change ΔW results in both cases. The total differentials (7.7) and (7.8) formally result in (7.13) up to (7.16)

$$F(x, Q) = \frac{\partial W(x, Q)}{\partial x} \tag{7.13}$$

$$u(x, Q) = \frac{\partial W(x, Q)}{\partial Q} \tag{7.14}$$

$$F(x, \mu) = \frac{\partial W(x, \mu)}{\partial x} \tag{7.15}$$

$$i(x, \mu) = \frac{\partial W(x, \mu)}{\partial \mu}. \tag{7.16}$$

Due to equality of the mixed second derivative of W, (7.17) and (7.18)

$$\left.\frac{\partial F}{\partial Q}\right|_{x,Q} = \left.\frac{\partial u}{\partial x}\right|_{x,Q} \tag{7.17}$$

$$\left.\frac{\partial F}{\partial \mu}\right|_{x,\mu} = \left.\frac{\partial i}{\partial x}\right|_{x,\mu} \tag{7.18}$$

result from (7.13) up to (7.16).

Equation (7.17) also yields the identity of K_{el} and K'_{el} in the linearized system equations for electric fields. In order to be able to prove this also for the magnetic case, it is appropriate to rearrange and solve (7.3) and (7.4) for the variables $\Delta\mu$ and Δx.

This yields:

$$\Delta i = \frac{1}{L_b}\Delta\mu + \frac{K'_{mag}}{L_b}\Delta x \tag{7.19}$$

$$\Delta F = \frac{K_{mag}}{L_b}\Delta\mu + \left(\frac{1}{n_L} - \frac{K_{mag}K'_{mag}}{L_b}\right)\Delta x \tag{7.20}$$

In combination with (7.18) finally follows $K_{mag} = K'_{mag}$. The relations specified in (7.13) and (7.16) will also enable the derivatives of nonlinear system equations, if either the function $W(x, Q)$ and $W(x, \mu)$ respectively or in special cases the virtual changes $\mathrm{d}W$ are known as functions of virtual displacements $\mathrm{d}x$ or $\mathrm{d}Q$ and $\mathrm{d}x$ or $\mathrm{d}\mu$, respectively. The Chaps. 8 and 9 comprise concrete applications.
Figure 7.7 shows an example that will result in the determination of the nonlinear field force, if following system equations with known dependence $C(x)$ are available:

$$F = F_{el}(Q, x) + \frac{1}{n_L}(x - x_0) \tag{7.21}$$

$$u = Q/C(x) \tag{7.22}$$

For that purpose, the internal energy W is generated along two different paths and equality is used. Along path 1, at first a displacement $x - x_0$ supplied through the mechanical port results in an energy $W^{(1)}_{mech}$ and then for fixed x, an electrical energy $W^{(1)}_{el}$ is supplied through the electric port.

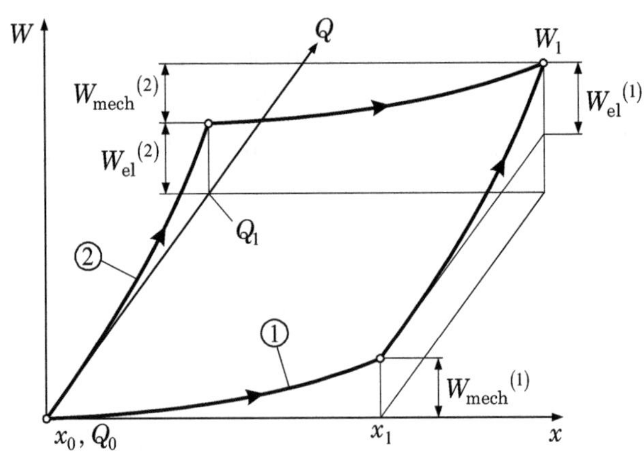

Fig. 7.7. Change of state of a system according to (7.11) along two different paths ① and ②

Thus, the overall supplied energy W results in:

$$W^{(1)} = \frac{1}{2}\frac{1}{n_{\mathrm{L}}}\left(x - x_0\right)^2 + \frac{1}{2}\frac{1}{C\left(x\right)}Q^2 \tag{7.23}$$

Along the second path, at first an electrical energy $W_{\mathrm{el}}^{(2)}$ is supplied at position $x = x_0$ and then for fixed Q, the mechanical energy $W_{\mathrm{mech}}^{(2)}$ is supplied. It should be noted that after the first part of operation an electric field force $F_{\mathrm{el}} = F_{\mathrm{el}}\left(Q, x_0\right)$ is existent. Thus, $W^{(2)}$ results in:

$$W^{(2)} = \frac{1}{2}\frac{1}{C\left(x_0\right)}Q^2 + \int\limits_{x_0}^{x} F_{\mathrm{el}}\left(Q, x_0\right) \mathrm{d}x + \frac{1}{2}\frac{1}{n_{\mathrm{L}}}\left(x - x_0\right)^2 \tag{7.24}$$

If it is now assumed that $x - x_0$ is so small that $F_{\mathrm{el}}\left(x, Q\right) = F_{\mathrm{el}}\left(x_0, Q\right)$ applies and thus $x - x_0$ represents a virtual displacement, the integral in (7.11) can approximately be replaced by

$$F_{\mathrm{el}}\left(x_0, Q\right)\left(x - x_0\right).$$

In combination with the condition $W_1 = W^{(1)} = W^{(2)}$, the unknown force $F_{\mathrm{el}}\left(x_0, Q\right)$ results in:

$$\begin{aligned}
\left(x - x_0\right) F_{\mathrm{el}}\left(x_0, Q\right) &= \frac{Q^2}{2}\left(\frac{1}{C\left(x_0 + \Delta x\right)} - \frac{1}{C\left(x_0\right)}\right) \\
&\approx \frac{Q^2}{2}\frac{\mathrm{d}}{\mathrm{d}x}\left(\frac{1}{C\left(x\right)}\right)\bigg|_{x=x_0} \cdot \left(x - x_0\right) \\
\Rightarrow F_{\mathrm{el}}\left(x_0, Q\right) &= \frac{Q^2}{2}\frac{\mathrm{d}}{\mathrm{d}x}\left(\frac{1}{C\left(x\right)}\right)\bigg|_{x=x_0}
\end{aligned} \tag{7.25}$$

With the knowledge of partial energies resulting from capacitances or field distributions it is possible in a similar manner to construct cyclic processes for concrete configurations that allow for the determination of reciprocal relations. These procedural methods are also analogously applicable to magnetic network or magnetic field problems. The Chaps. 8 and 9 comprise further examples.

Derivation of the Transducer Two-Port Networks

Taking the analytical methods of network theory into consideration, it is convenient to formulate (7.1) up to (7.4) for sinusoidal time dependences of the differential variations around the center of expansion and to introduce *complex amplitudes* according to Chap. 2:

$$\Delta F = \hat{F}\cos(\omega t + \varphi_F) \;\rightarrow\; \underline{F} = \hat{F}\,\mathrm{e}^{\mathrm{j}\varphi_F} \tag{7.26}$$

$$\Delta u = \hat{u}\cos(\omega t + \varphi_u) \;\rightarrow\; \underline{u} = \hat{u}\,\mathrm{e}^{\mathrm{j}\varphi_u} \tag{7.27}$$

$$\Delta Q = \hat{Q}\cos\left(\omega t + \varphi_Q\right) \;\rightarrow\; \underline{Q} = \hat{Q}\,\mathrm{e}^{\mathrm{j}\varphi_Q} \;\rightarrow\; \underline{i} = \mathrm{j}\omega\underline{Q} \tag{7.28}$$

$$\Delta \mu = \hat{\mu}\cos\left(\omega t + \varphi_\mu\right) \;\rightarrow\; \underline{\mu} = \hat{\mu}\,\mathrm{e}^{\mathrm{j}\varphi_\mu} \;\rightarrow\; \underline{u} = \mathrm{j}\omega\underline{\mu} \tag{7.29}$$

$$\Delta \xi = \hat{\xi}\cos\left(\omega t + \varphi_\xi\right) \;\rightarrow\; \underline{\xi} = \hat{\xi}\,\mathrm{e}^{\mathrm{j}\varphi_\xi} \;\rightarrow\; \underline{v} = \mathrm{j}\omega\underline{\xi} \tag{7.30}$$

In combination with these predefinitions, (7.1) and (7.2) pass into (7.31) and (7.32)

$$\underline{F} = \frac{1}{\mathrm{j}\omega}K_{\mathrm{el}}\underline{i} - \frac{1}{\mathrm{j}\omega n}\underline{v} \tag{7.31}$$

$$\underline{u} = \frac{1}{\mathrm{j}\omega C}\underline{i} - \frac{K_{\mathrm{el}}}{\mathrm{j}\omega}\underline{v} \tag{7.32}$$

and by transforming to $\underline{i}$, $\underline{F} = f(\underline{u}, \underline{v})$, (7.31) and (7.32) pass into (7.33) and (7.34)

$$\underline{i} = \mathrm{j}\omega C\underline{u} + K_{\mathrm{el}}C\underline{v} \tag{7.33}$$

$$\underline{F} = K_{\mathrm{el}}C\underline{u} - \frac{1}{\mathrm{j}\omega}\left(\frac{1}{n} - K_{\mathrm{el}}^2 C\right)\underline{v}. \tag{7.34}$$

In the same way, (7.3) and (7.4) are transfered into (7.35) and (7.36) in combination with the relations specified in (7.26) up to (7.30):

$$\underline{F} = K_{\mathrm{mag}}\underline{i} - \frac{1}{\mathrm{j}\omega n}\underline{v} \tag{7.35}$$

$$\underline{u} = \mathrm{j}\omega L\underline{i} + K_{\mathrm{mag}}\underline{v} \tag{7.36}$$

Using the transducer constants X and Y according to

$$X = \frac{1}{K_{\mathrm{mag}}} \quad \text{and} \quad Y = \frac{1}{K_{\mathrm{el}}C},$$

(7.33) and (7.36) result in (7.37) and (7.38) specifying the *electrical transducer*

$$\underline{i} - \mathrm{j}\omega C_{\mathrm{b}}\underline{u} = \underline{i}_{\mathrm{W}} = \frac{1}{Y}\underline{v} \tag{7.37}$$

$$\underline{F} + \frac{1}{\mathrm{j}\omega n_{\mathrm{K}}}\underline{v} = \underline{F}_{\mathrm{W}} = \frac{1}{Y}\underline{u} \tag{7.38}$$

and in (7.39) and (7.40) specifying *the magnetic transducer*

$$\underline{u} - \mathrm{j}\omega L_\mathrm{b}\underline{i} = \underline{u}_\mathrm{W} = \frac{1}{X}\underline{v} \tag{7.39}$$

$$\underline{F} + \frac{1}{\mathrm{j}\omega n_\mathrm{L}}\underline{v} = \underline{F}_\mathrm{W} = \frac{1}{X}\underline{i}. \tag{7.40}$$

By means of these equations, the circuits shown in Fig. 7.8 can be identified. It is remarkable that the transfer characteristics between an electrical and a mechanical pair of coordinates (port) of reversible (loss-free) electromechanical systems can be represented by one of both circuit structures shown in Fig. 7.8.

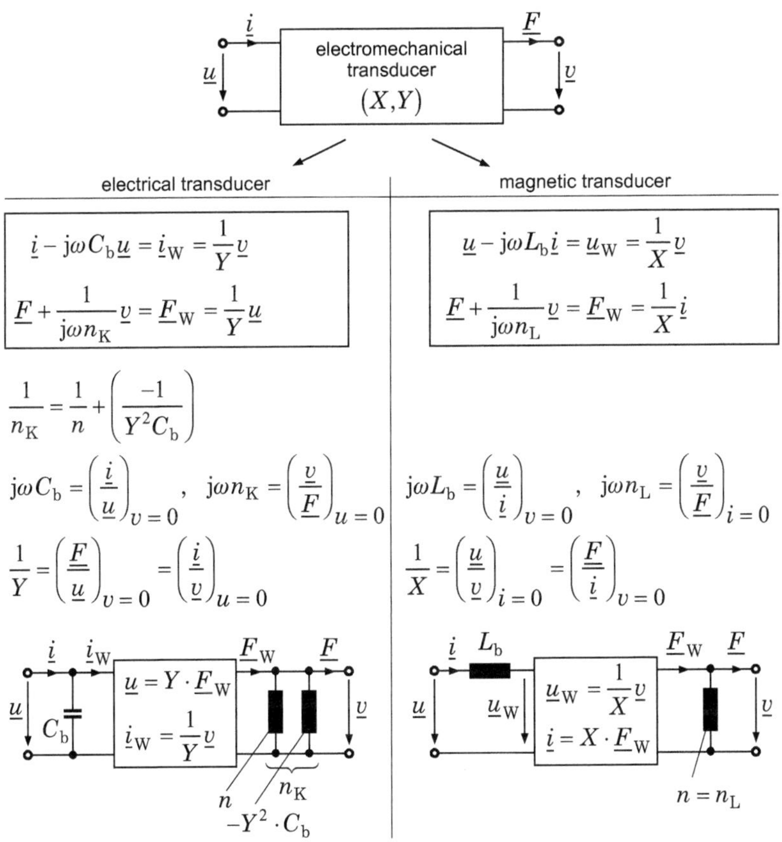

Fig. 7.8. Derivation of two-port circuits for the electrical and magnetic transducer *Indices:* b *mechanically locked system,* L *system during electrical open-circuit operation,* K *system during electrical short-circuit operation*

The left circuit is valid for coupling by means of electric fields. The right circuit is valid for systems, in which the coupling is realized by means of magnetic fields. In case of nonlinear system equations (Fig. 7.5), the circuits shown in Fig. 7.8 are valid for linear approximation of the system equations in the neighborhood of a chosen center of expansion (*small-signal behavior*). Due to generality of the assumptions made previously, these conclusions apply both to configurations with a known local structure, such as concentrated charges on reference points or current loops, and to configurations whose local microscopic structure is unknown, e.g. dielectric and magnetic solids with or without internal interactions.

The *coupling two-port networks* (X), (Y) in the circuits illustrated in Fig. 7.8 prove to be the two general coupling elements – electromechanical transducers – between mechanical and electrical networks and already specified in Chap. 2. The components L, C, n being available in the circuits illustrated in Fig. 7.8 must be assigned to the particular networks. Concerning the preconditions mentioned above – linear and loss-free couplings – the five electromechanical conversion principles represented in Table 7.1 can now be described by means of the basic circuits specified in Fig. 7.8. In Chap. 8, the calculation of the components L, C, n and *transducer constants* X, Y on the basis of the transducers' dimensions and material parameters is exemplified for magnetic transducers. The same is done for electrical transducers in Chap. 9.

As already mentioned in Chap. 1, the possibility of deriving a closed-form solution for the total system represents a significant advantage of description of electromechanical interactions by means of circuit networks. This closed-form solution is enabled by transformation of the mechanical components to the electrical side or vice versa by means of the *transducer matrix* (X) or (Y) of the coupling two-port networks.

Table 7.2 summarizes the transformation characteristics of the coupling two-port network of *magnetic transducers*. Here, the coupling two-port network has the character of a *transformer*. An impedance at the one side is represented as an impedance at the other side and vice versa. The relations shown in Table 7.4 are achieved from Table 7.2.

Table 7.3 summarizes the transformation characteristics of the coupling two-port network of *electrical transducers*. An impedance at the one side is represented as an admittance at the other side and vice versa. Thus, the coupling two-port network has the two-port theoretical character of a *gyrator*. The relations shown in Table 7.5 can be derived from Table 7.3.

Table 7.2. Transformation characteristics of magnetic coupling two-port network
X transducer constant of a transformer

$$P_{\mathrm{el}} = \widetilde{u}\,\widetilde{i} = \widetilde{F}\,\widetilde{v} = P_{\mathrm{mech}}$$

$$\underline{Z} = \frac{\underline{u}}{\underline{i}} = \frac{1}{X^2}\frac{\underline{v}}{\underline{F}} = \frac{1}{X^2}\underline{h}$$

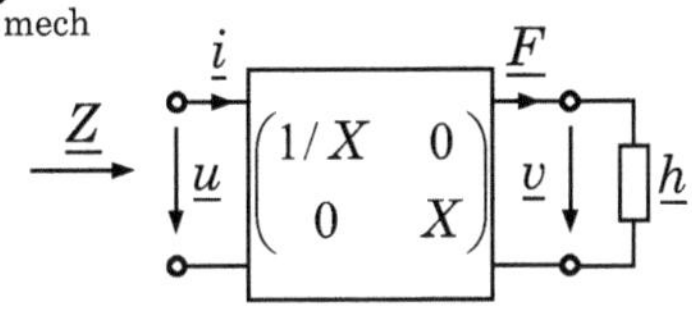

$$\underline{h} = \mathrm{j}\omega n \quad \rightarrow \quad \underline{Z} = \frac{1}{X^2}\mathrm{j}\omega n = \mathrm{j}\omega L$$

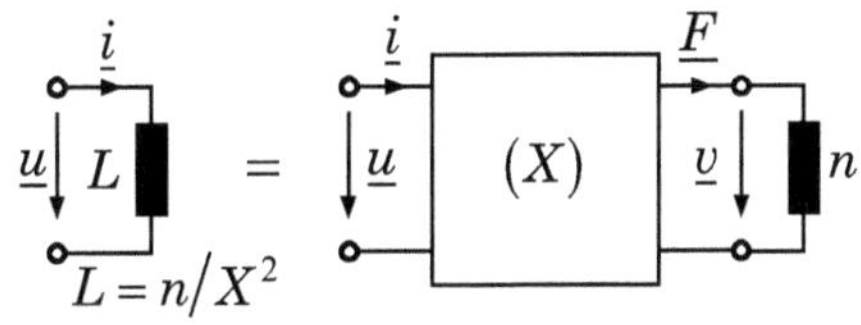

$$\underline{h} = \frac{1}{\mathrm{j}\omega m} \quad \rightarrow \quad \underline{Z} = \frac{1}{X^2\mathrm{j}\omega m} = \frac{1}{\mathrm{j}\omega C}$$

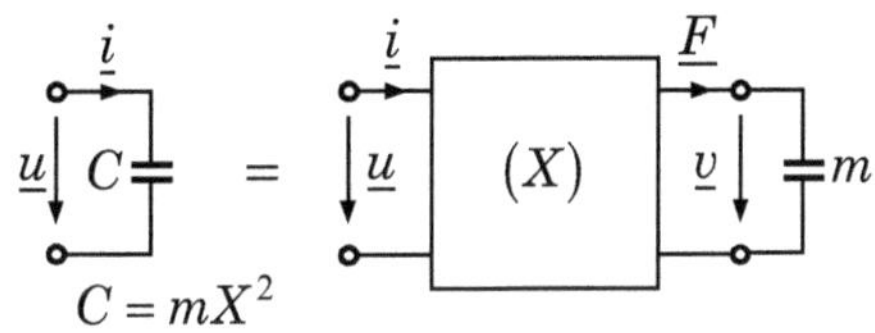

$$\underline{h} = \underline{h}_1 + \underline{h}_2 \quad \rightarrow \quad \underline{Z} = \frac{\underline{h}_1}{X^2} + \frac{\underline{h}_2}{X^2} = \underline{Z}_1 + \underline{Z}_1$$

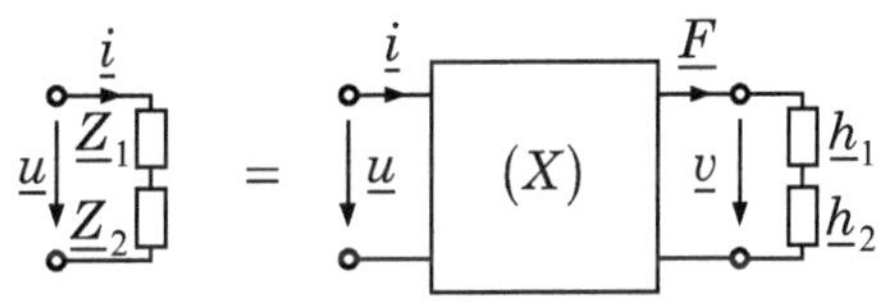

$$\underline{h} = \frac{1}{\underline{z}_1 + \underline{z}_2} \rightarrow \underline{Z} = \frac{1}{X^2\underline{z}_1 + X^2\underline{z}_2} = \frac{1}{\underline{Y}_1 + \underline{Y}_2}$$

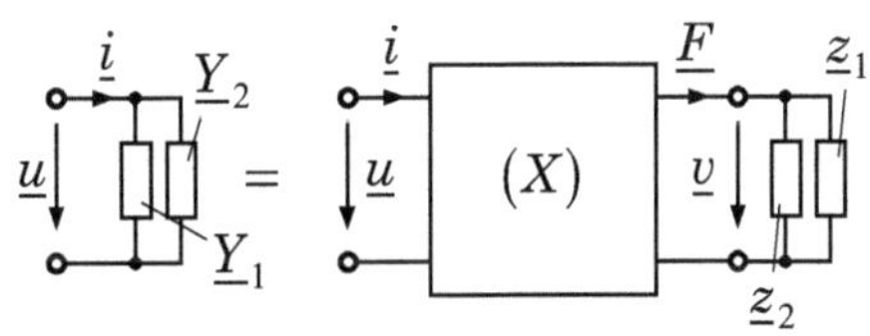

Table 7.3. Transformation characteristics of electrical coupling two-port network
Y transducer constant of a gyrator

$$P_{\mathrm{el}} = \tilde{u}\,\tilde{i} = \tilde{F}\,\tilde{v} = P_{\mathrm{mech}}$$

$$\underline{Z} = \frac{\underline{u}}{\underline{i}} = Y^2 \frac{\underline{F}}{\underline{v}} = Y^2 \frac{1}{\underline{h}}$$

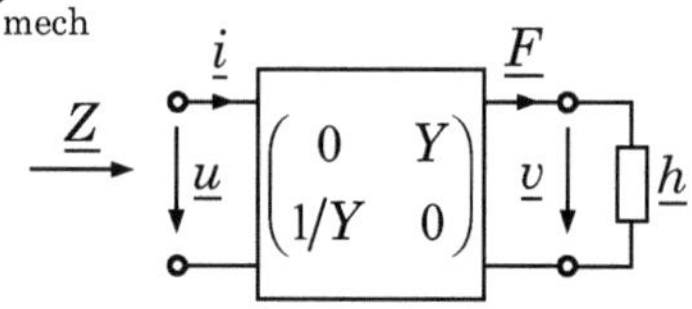

$$\underline{h} = \mathrm{j}\omega n \quad \rightarrow \quad \underline{Z} = \frac{Y^2}{\mathrm{j}\omega n} = \frac{1}{\mathrm{j}\omega C}$$

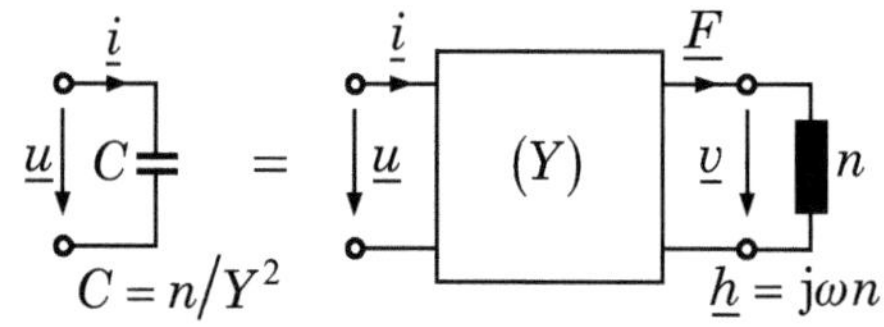

$$\underline{h} = \frac{1}{\mathrm{j}\omega m} \quad \rightarrow \quad \underline{Z} = Y^2 \mathrm{j}\omega m = \mathrm{j}\omega L$$

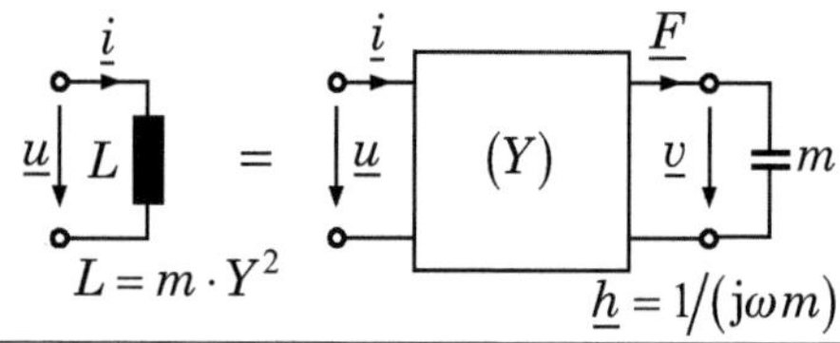

$$\underline{h} = \underline{h}_1 + \underline{h}_2 \rightarrow \underline{Z} = \frac{1}{\underline{h}_1/Y^2 + \underline{h}_2/Y^2} = \frac{1}{\underline{Y}_1 + \underline{Y}_2}$$

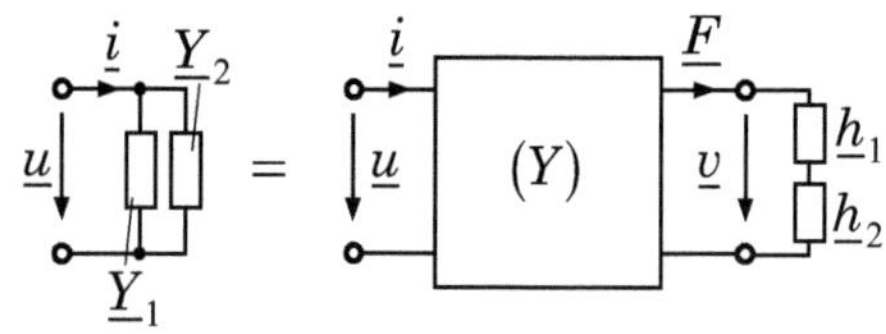

$$\underline{h} = \frac{1}{\underline{z}_1 + \underline{z}_2} \quad \rightarrow \quad \underline{Z} = Y^2(\underline{z}_1 + \underline{z}_2) = \underline{Z}_1 + \underline{Z}_2$$

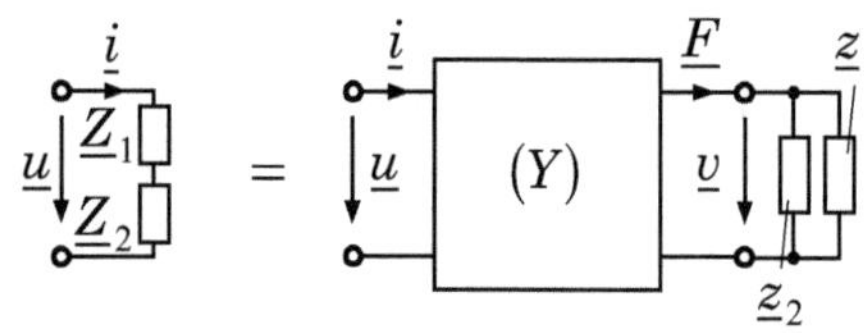

Table 7.4. Correlation between electrical and mechanical components concerning a transformer

electrics	transformer-like transducer			mechanics
inductance	L	o——o	n	compliance
capacitance	C	o——o	m	mass
resistance	R	o——o	h	frictional admittance
electrical impedance	$\underline{Z}$	o——o	$\underline{h}$	mechanical admittance
parallel connection		o——o		parallel connection
series connection		o——o		series connection
current source		o——o		force source

Table 7.5. Correlation between electrical and mechanical components concerning a gyrator

electrics	gyrator-like transducer			mechanics
inductance	L	o——o	m	mass
capacitane	C	o——o	n	compliance
resistance	R	o——o	r	frictional impedance
electrical impedance	$\underline{Z}$	o——o	$\underline{z}$	mechanical impedance
parallel connection		o——o		series connection
series connection		o——o		parallel connection
voltage source		o——o		force source

Within the scope of mechanical concepts of network theory, the terms „mechanical impedance" or „mechanical admittance" respectively are established terms with respect to the so-called *1st analogy.* However, they do not correspond to the predefinitions of this book that considers the impedance to be the quotient of the complex difference coordinate and the complex flow coordinate, but they characterize the reciprocal value. However, since these terms are very common with mechanical networks, they are here adopted regardless of the coordinate definition of velocity as difference coordinate and force as flow coordinate.

An important parameter of electromechanical transducers is defined by the degree of energy conversion. The *coupling factor* k characterizes the ratio of converted energy, i.e usable energy at the output and the total energy supplied to the input. The definition of the coupling factor is specified in Fig. 7.9 for both conversion directions. Since a reversible and thus a loss-free conversion mechanism is assumed, the coupling factor is higher than the *efficiency* η of transducer.

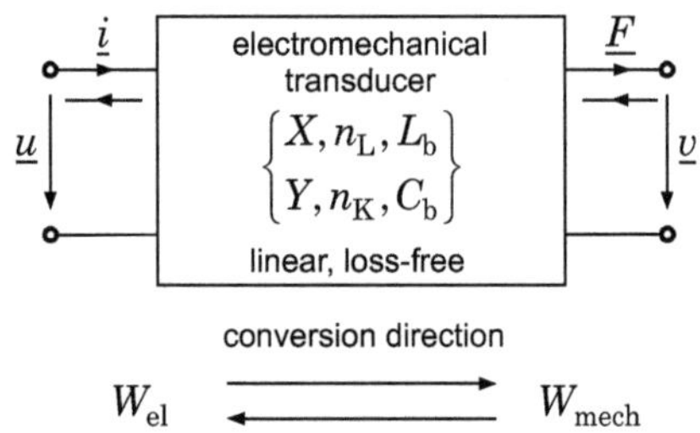

$$k^2 = \frac{\text{converted energy}}{\text{supplied energy}} \leq 1$$

$W_{el} \rightarrow W_{mech}$: $\quad k_{mech}^2 = \dfrac{W_{mech}}{W_{el}} \quad \rightarrow$ **actuators**

$W_{mech} \rightarrow W_{el}$: $\quad k_{el}^2 = \dfrac{W_{el}}{W_{mech}} \quad \rightarrow$ **sensors**

reversible conversion: $k_{el} = k_{mech}$

Fig. 7.9. Electromechanical coupling factor k

8

Magnetic Transducers

The magnetic transducers form the first large group of electromechanical transducers. *Electromagnetic*, *electrodynamic* and *piezomagnetic* transducers belong to the group of magnetic transducers [1, 2, 22]. In case of an electrodynamic transducer, a magnetic force – LORENTZ force – being proportional to coil current appears in the air gap between moving coil and fixed yoke. In contrast, in case of a magnetic transducer, a nonlinear magnetic force appears between armature and yoke. In order to achieve a linearization, an operating point adjustment by means of a constant magnetic field or a superimposed direct current is necessary. Also with the piezomagnetic transducer an operating point adjustment is achieved by a constant magnetic field or a direct current. Thus, an approximately linear small-signal behavior is achieved. Here, the connection between magnetic and mechanical field quantities is described by means of equations of state.
For all three magnetic transducers the mechano-electrical two-port networks are derived under practical marginal conditions. The application of these transducer networks during the design phase of electromechanical systems is illustrated by means of practical examples.

8.1 Electrodynamic Transducer

8.1.1 Derivation of the Two-Port Transducer Network

With the electrodynamic transducer, the force action – LORENTZ force – on a current-carrying conductor in a magnetic field is utilized as conversion principle. The magnetic field generated by a permanent magnet is constant and is not influenced by electrical and mechanical network coordinates.
Due to the linear relation of mechanical and electrical coordinates caused by the functional principle of the electrodynamic transducer, these transducers are used for *actuator* applications, such as small-power motors, shake tables,

A. Lenk et al., *Electromechanical Systems in Microtechnology and Mechatronics*,
Microtechnology and MEMS, DOI 10.1007/978-3-642-10806-8_8,

speakers, drives for positioning systems and magnetic or optical scanning systems as well as for *sensor* applications, such as microphones and velocity sensors.

The design principle of the moving coil forms the basis of the basic model of the electrodynamic transducer illustrated in Fig. 8.1. A cylindrical coil featuring mass m and wire length l and suspended by the spring n can move in an air gap in direction of its longitudinal axis. Since the circular air gap is permeated by the magnetic induction B_0 of a permanent magnet system, forces in axial direction are generated in case of a current flow in the coil. The *linear* relations for magnetic force F_{mag} and induced voltage u specified in Fig. 8.1 apply to the motion of coil in the homogeneous part of the magnetic field. Based on the balance of forces at the mechanical side and the mesh equation at the electrical side of the transducer, the linear coupling equations specified in Fig. 8.2 can be formulated.

With the derivation of the two-port transducer networks shown in Fig. 8.2 the magnetic resistance of the magnetic circuit with $\mu_e \gg 1$ and the magnetic leakage flux outside the circular gap are neglected. For the consideration of alternating quantities, the relations in the time domain are transformed into the frequency domain by introducing complex amplitudes.

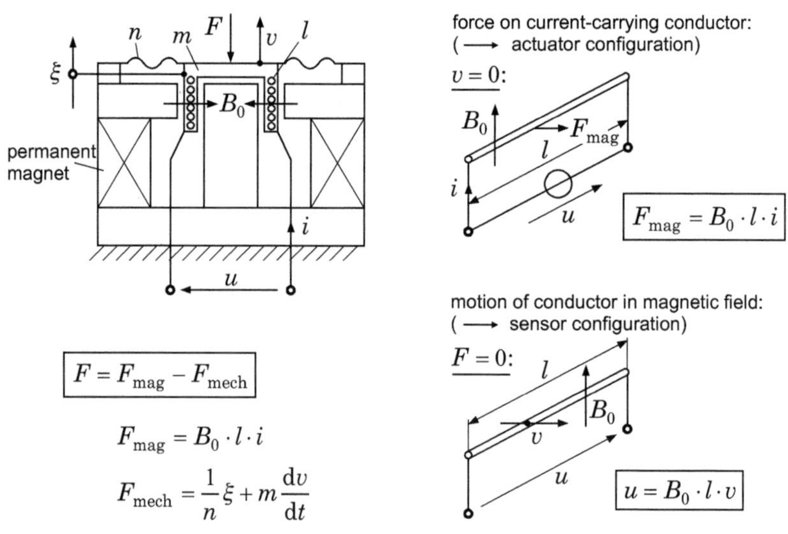

Fig. 8.1. Model of electrodynamic transducer

mechanical side of transducer: $i,\xi \neq 0$

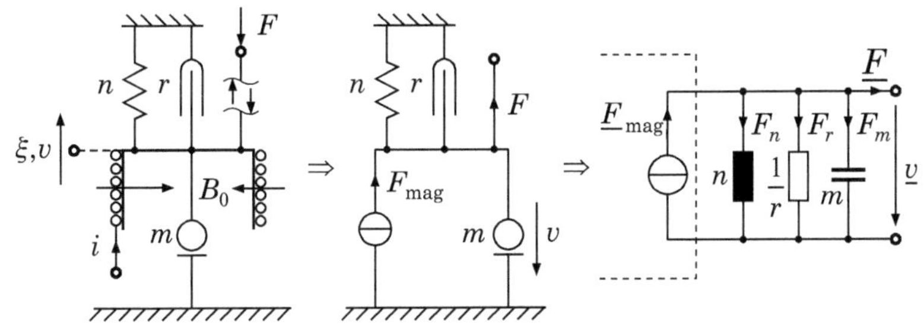

$$F = B_0\, l\, i - \frac{1}{n}\xi - m\frac{\mathrm{d}v}{\mathrm{d}t} - r\,v \;\Rightarrow\; \underline{F} = B_0\, l\, \underline{i} - \frac{1}{\mathrm{j}\omega n}\underline{v} - \mathrm{j}\omega m \underline{v} - r\,\underline{v}$$

electrical side of transducer: $u,\xi \neq 0$

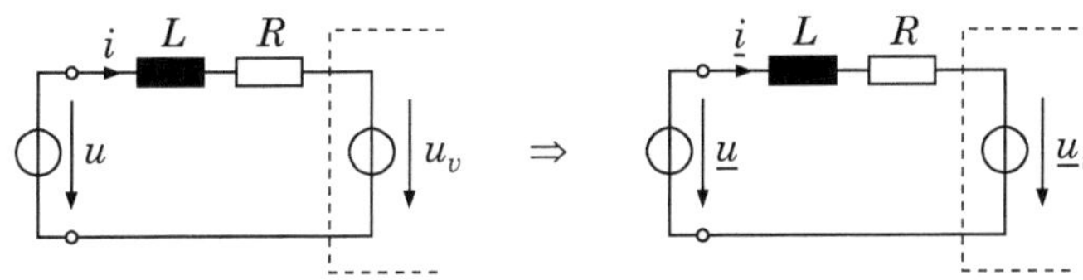

$$\left.\begin{aligned} u(t) &= R\cdot i(t) + L\frac{\mathrm{d}i}{\mathrm{d}t} + u_v(t) \\ u(t) &= R\cdot i(t) + L\frac{\mathrm{d}i}{\mathrm{d}t} + B_0\cdot l\cdot v(t) \end{aligned}\right\} \quad \underline{u} = R\cdot\underline{i} + \mathrm{j}\omega L\cdot\underline{i} + B_0\cdot l\cdot\underline{v}$$

Fig. 8.2. Coupling equations concerning the electrodynamic transducer

The circuit diagram of the transducer shown in Fig. 8.3 is achieved by rearranging the two equations specified in Fig. 8.2 with respect to the transducer voltage $\underline{u}_{\mathrm{W}}$ and transducer force $\underline{F}_{\mathrm{W}}$:

$$\underline{u} - (R + \mathrm{j}\omega L)\,\underline{i} = \underline{u}_{\mathrm{W}} = B_0 l \underline{v} \tag{8.1}$$

$$\underline{F} + \left(\frac{1}{\mathrm{j}\omega n} + \mathrm{j}\omega m + r\right)\underline{v} = \underline{F}_{\mathrm{W}} = B_0 l \underline{i} \tag{8.2}$$

The loss-free transducer features a *transformer-like* coupling $X = 1/B_0 l$ and corresponds to the basic circuit diagram of the magnetic transducer according to Fig. 7.8.

In the following a descriptive explanation of the transducer components is given.

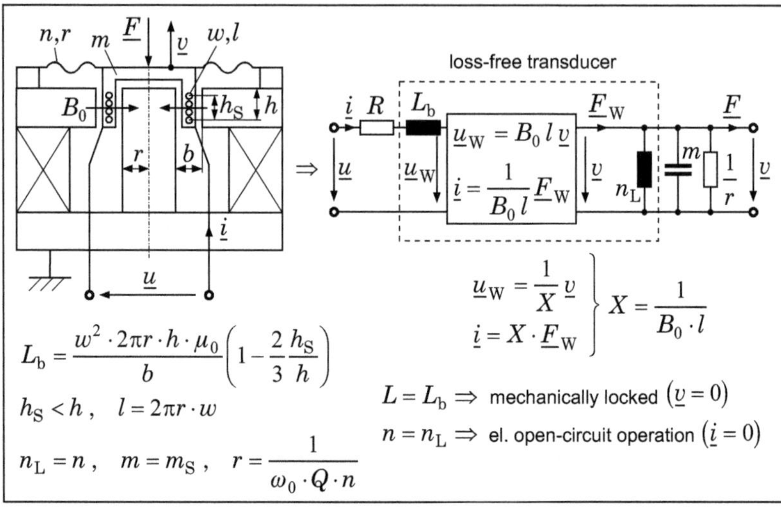

Fig. 8.3. Circuit diagram of electrodynamic transducer

In case of a *mechanically locked state*, $\underline{v} = 0$, the input impedance $\underline{Z} = R + \mathrm{j}\omega L_\mathrm{b}$ is measured at the electrical input terminals. Compared to the active component the reactive component can be neglected for very low-frequency applications. In order to prevent a coil displacement, the force

$$\underline{F}_\mathrm{W} = B_0 l \underline{i} = \frac{1}{X}\underline{i} \tag{8.3}$$

must be applied. In the opposite case of an *electrical open-circuit operation*, $\underline{i} = 0$, the transducer voltage

$$\underline{u}_\mathrm{W} = B_0 l \underline{v} = \frac{1}{X}\underline{v} \tag{8.4}$$

is achieved by supplying the velocity $\underline{v}$. In this case, the measurable compliance n_L at the mechanical side corresponds to the compliance n of the moving coil suspension. In the circuit diagram shown in Fig. 8.3, the moving coil mass $m = \rho V_\mathrm{S}$ and the moving coil friction $r = 1/\omega_0 Q n$ have already been taken into account.
The transformation relations of the *transformer* (X) apply to the transformation relations of the components from the mechanical to the electrical side or vice versa. For the different transformation directions it can be written:

mechanical → electrical

$$C_m = mX^2, \qquad L_n = \frac{n}{X^2}, \qquad R_r = \frac{1}{rX^2}$$

$$\underline{Z} = \frac{\underline{h}}{X^2} = \frac{1}{\underline{z}X^2}$$

electrical → mechanical

$$m_C = \frac{C}{X^2}, \qquad n_L = LX^2, \qquad r_R = \frac{1}{RX^2}$$

$$\underline{h} = \frac{1}{\underline{z}} = \underline{Z}X^2$$

Electrodynamic transducers are used for actuator and sensor applications for which a preferably linear transfer characteristic is desired. Depending on the design and the allowed supply current the achievable actuating forces are in the range of several mN up to 100 N. Considerably larger actuating forces can be achieved by means of the electromagnetic transducer (see Sect. 8.2).
The application of the electrodynamic transducer is illustrated in the following using the example of a drive system, a speaker, an active vibration damper, a vibration calibrator and a detection method of hip prosthesis loosening.

8.1.2 Sample Applications

Electrodynamic Drive Systems

Typical dimensions and characteristic values of an electrodynamic drive system, as it is used e.g. for *oscillating tables*, are specified in Fig. 8.4. The frictional coefficient r is calculated from the experimentally determined value of quality factor Q.

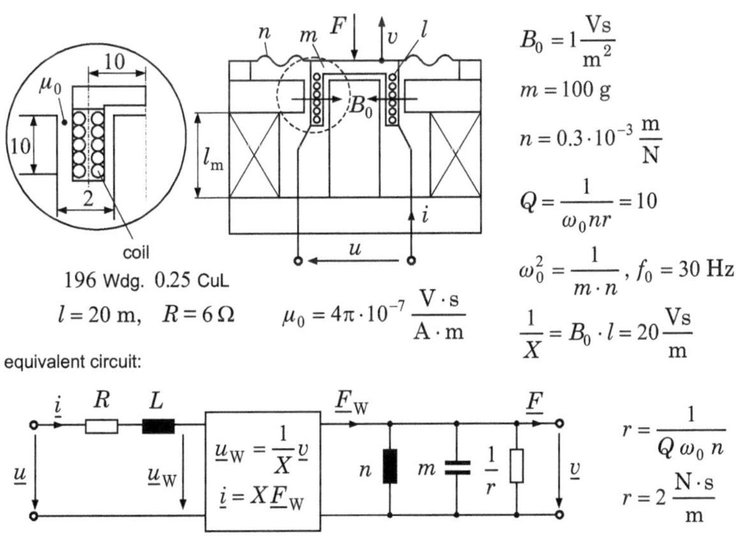

Fig. 8.4. Structure and circuit of an electrodynamic drive system

At first, the input impedance of the unloaded system, i.e. mechanical open-circuit operation ($\underline{F} = 0$), is determined against the frequency f. For that purpose, the circuit shown in Fig. 8.4 is assumed, thus as already mentioned before the inductance L is neglected and the mechanical components are transformed to the electrical side. The circuit simplified by the neglect of L is represented in Fig. 8.5 in combination with the frequency loci of the impedances. The left frequency locus shows the progression of the impedance $\underline{Z}_{\mathrm{m}}$ caused by mechanical components. The right frequency locus corresponds to the progression of the electrical input impedance $\underline{Z} = \underline{u}/\underline{i}$. The relatively high frequency-dependent component caused by electromechanical interactions is remarkable. On the other hand, this undesirable effect of strong frequency dependence of the electrical input impedance is specially utilized with piezoelectric transducers for the design of electrical resonators with very high quality factors for filter and oscillator applications (see Sect. 9.3.4).
If in case of voltage excitation the circuit illustrated in Fig. 8.5 is transformed into a mechanical circuit (Fig. 8.6), the transfer characteristics of the drive system can be calculated.
Based on the associated circuit shown in Fig. 8.6, it follows

$$\frac{\underline{v}}{\underline{F}} = \omega_0 n \frac{\mathrm{j}\,(\omega/\omega_0)}{1 + \mathrm{j}\,(\omega/\omega_0)\,1/Q - (\omega/\omega_0)^2}, \qquad \omega_0^2 = \frac{1}{mn}, \qquad Q = \frac{1}{\omega_0 n r}.$$

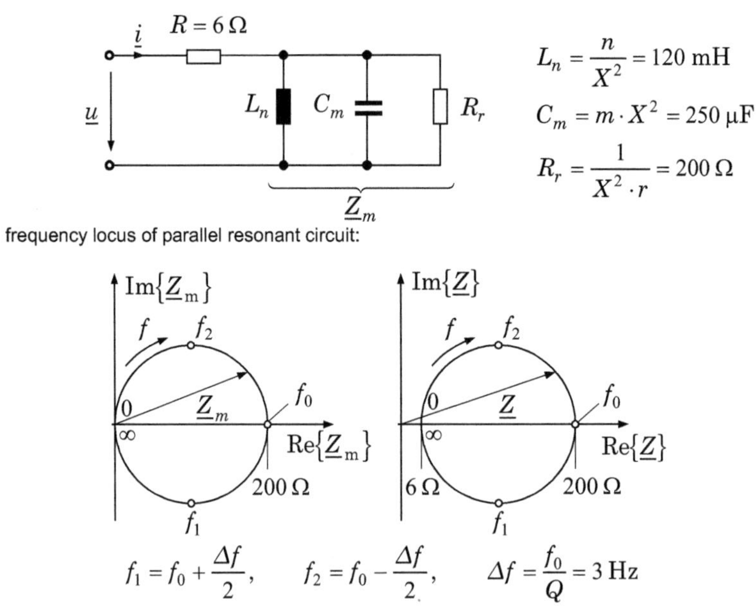

Fig. 8.5. Frequency loci of electrical impedances of electrodynamic drive system

If now the relation $\underline{F}_\mathrm{W} = \underline{u}/RX$ is inserted for the transducer force, the following relations for the transfer functions of displacement, velocity and acceleration can be determined:

$$\underline{B}_\xi = \frac{\underline{\xi}}{\underline{u}} = \frac{n}{RX}\frac{1}{1+\mathrm{j}\,(\omega/\omega_0)\,1/Q-(\omega/\omega_0)^2}, \qquad \underline{B}_{0\xi} = \frac{n}{RX} = 10^{-3}\,\frac{\mathrm{m}}{\mathrm{V}}$$

$$\underline{B}_v = \frac{\underline{v}}{\underline{u}} = \frac{\omega_0 n}{RX}\frac{\mathrm{j}\,(\omega/\omega_0)}{1+\mathrm{j}\,(\omega/\omega_0)\,1/Q-(\omega/\omega_0)^2}, \; \underline{B}_{0v} = \frac{\omega_0 n}{RX} = 5.6\cdot 10^{-3}\,\frac{\mathrm{m}}{\mathrm{s}}\cdot \mathrm{V}$$

$$\underline{B}_a = \frac{\underline{a}}{\underline{u}} = -\frac{\omega_0^2 n}{RX}\frac{(\omega/\omega_0)^2}{1+\mathrm{j}\,(\omega/\omega_0)\,1/Q-(\omega/\omega_0)^2}, \quad \underline{B}_{0a} = \frac{\omega_0^2 n}{RX} = 1.1\,\frac{\mathrm{m}}{\mathrm{s}^2}\cdot \mathrm{V}$$

The curve progression of the particular transfer functions is outlined as amplitude-frequency response in Fig. 8.6. For constant voltage excitation, below the resonant frequency the deflection of coil ξ and above the resonant frequency the acceleration of coil a is constant. Therefore, with the standardized IEC vibration test of components and devices, the mechanical sine sweep excitation is realized up to 50 Hz at constant displacement ξ and beyond that at constant acceleration a.

The transducer force $\underline{F}_\mathrm{W}$ represents another important characteristic value of the electrodynamic drive system. Since only the mass m appears at the mechanical side, above the resonant frequency f_0 of the oscillating table the acceleration is *frequency-independently* proportional to the transducer force and thus to the supplied current according to

$$\underline{a} = \mathrm{j}\omega\underline{v} = \mathrm{j}\omega\frac{\underline{F}_\mathrm{W}}{\mathrm{j}\omega m} = \frac{\underline{F}_\mathrm{W}}{m}$$

The maximum force is achieved in case of stationary locked state ($\underline{v} = 0$). In this case, the short-circuit force $\tilde{F}_\mathrm{W} = \tilde{F}_\mathrm{K}$ with

$$P_\mathrm{el} = \tilde{\imath}^2 R \qquad \text{and} \qquad \tilde{F}_\mathrm{W} = \tilde{\imath} B_0 l$$

can be estimated at

$$\frac{\tilde{F}_\mathrm{W}^2}{P_\mathrm{el}} = \frac{B_0^2 l^2}{R} = B_0^2\frac{V_\mathrm{S}}{\rho}, \qquad R = \rho\frac{A_\mathrm{D}}{l}, \qquad V_\mathrm{S} = lA_\mathrm{D},$$

where the quantity V_S denotes the volume of moving coil. The influence of the leakage field and the fill factor of copper have been neglected.
Here, P_el denotes the received electrical power. The parameter $\tilde{F}_\mathrm{W}^2/P_\mathrm{el}$ denotes a quality measure for electrodynamic drive systems.
Based on the conformity of the flux in the ferromagnetic circuit $\varPhi_\mathrm{e}$ and in the circular gap $\varPhi_0$

$$\varPhi_\mathrm{e} = B_\mathrm{m} A_\mathrm{m} = \varPhi_0 = B_0 2\pi r h$$

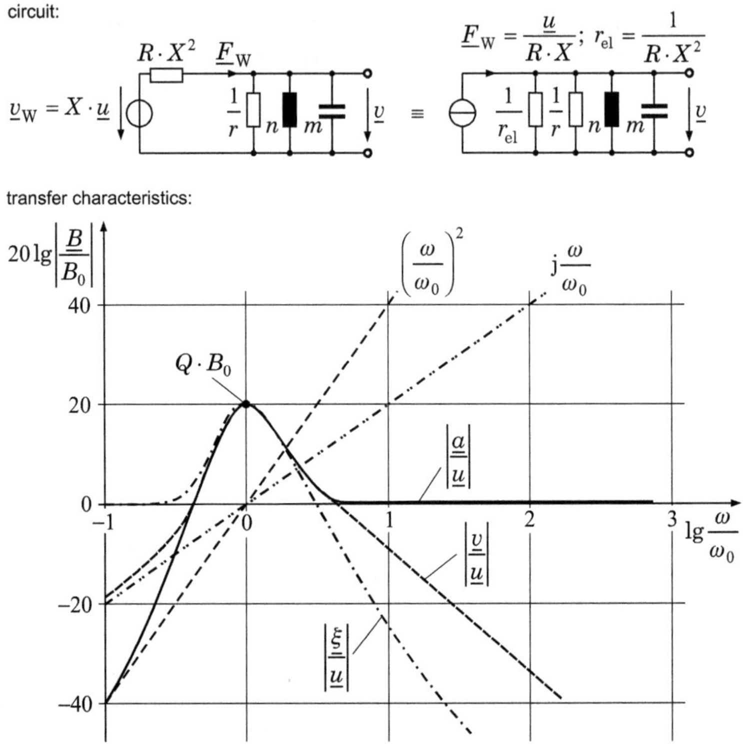

Fig. 8.6. Transfer characteristics of electrodynamic drive system

and based on the application of AMPÈRE's law

$$H_{\mathrm{m}} l_{\mathrm{m}} = \frac{B_0}{\mu_0} b$$

it follows

$$B_{\mathrm{m}} H_{\mathrm{m}} l_{\mathrm{m}} A_{\mathrm{m}} = \frac{B_0^2}{\mu_0} 2\pi r h b, \;\; V_{\mathrm{m}} = l_{\mathrm{m}} A_{\mathrm{m}}, \;\; V_{\mathrm{S}} = 2\pi r h b, \;\; B_0^2 V_{\mathrm{S}} = \mu_0 B_{\mathrm{m}} H_{\mathrm{m}} V_{\mathrm{m}}$$

and finally

$$\frac{\tilde{F}_{\mathrm{W}}^2}{P_{\mathrm{el}}} = \frac{\mu_0}{\rho} B_{\mathrm{m}} H_{\mathrm{m}} V_{\mathrm{m}}.$$

In the following example, the hard-magnetic material *neodymium* with a volume of $V_{\mathrm{m}} = 50\,\mathrm{cm}^{-3}$, $\{B_{\mathrm{m}} H_{\mathrm{m}}\}_{\mathrm{max}} = 250\,\mathrm{kJ\,m}^{-3}$ and and a conductivity of the coil wire of $1/\rho = 57 \cdot 10^6\,\mathrm{A\,V}^{-1}\,\mathrm{m}^{-1}$ were assumed. In combination with these values the characteristic value results in

$$\frac{\tilde{F}_{\mathrm{W}}^2}{P_{\mathrm{el}}} = 895\,\mathrm{N}^2\,\mathrm{W}^{-1}.$$

Assuming an electrical power consumption of $P_{\mathrm{el}} = 10\,\mathrm{W}$, a short-circuit force of

$$\tilde{F}_{\mathrm{W}} = \tilde{F}_{\mathrm{K}} = 95\,\mathrm{N}$$

is achieved in case of optimal dimensioning. Due to the allowed thermal stress and maximum flux density $B_{0\,\mathrm{max}}$ in the flux guiding parts, here the power limitation by i_{max} must be taken into account.

Compared to the oscillating table application, miniaturized electrodynamic drives for auto focus scanning systems have significantly lower transducer forces in the mN range. The arrangement of an electrodynamic transducer in an auto focus scanning system for CDs is illustrated in Fig. 8.7 as schematic diagram.

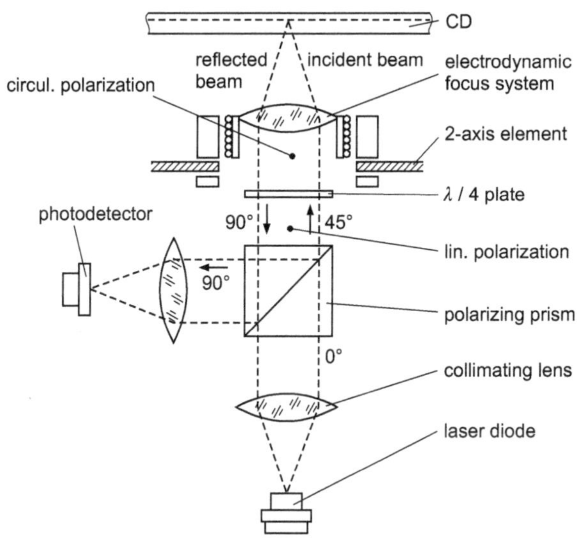

Fig. 8.7. Application of an electrodynamic drive system for CD scanning for auto focus alignment

Electrodynamic Loudspeaker

The design principle and the circuit of an electrodynamic loudspeaker including typical data are specified in Fig. 8.8. The circuit consists of an electrical, mechanical and acoustic subnetwork. In addition to the electrodynamic transducer, the connection between the mechanical and acoustic network is effected by means of the mechanical-acoustic transducer with transducer area $A = \pi a^2$.

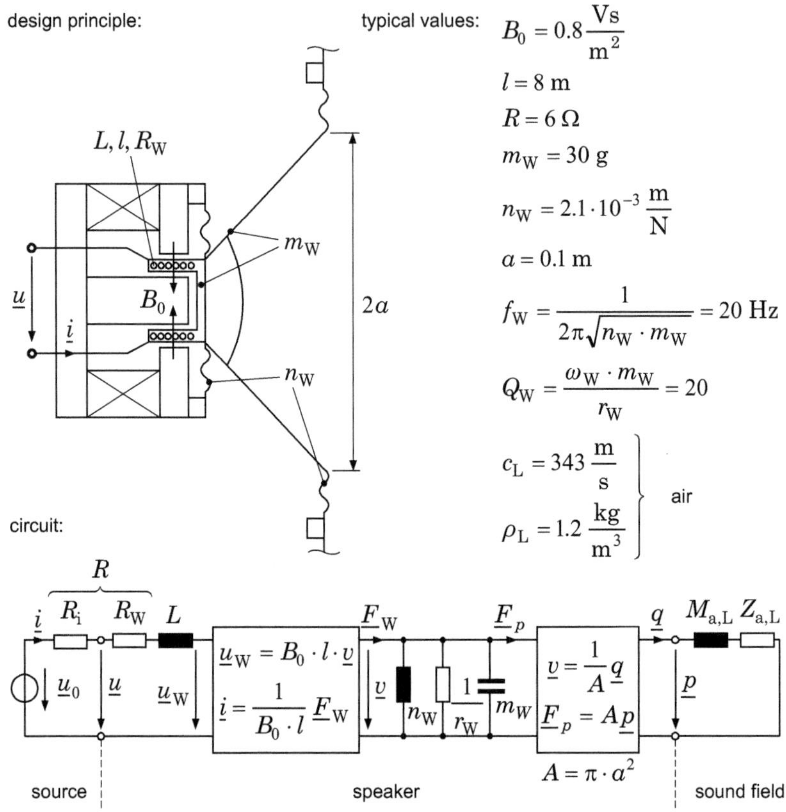

Fig. 8.8. Design principle and circuit diagram of an electrodynamic loudspeaker

The frequency-dependent acoustic friction $Z_{a,L}$ and the frequency-independent air mass $M_{a,L}$ are effective at the acoustic side [61]:

$$Z_{a,L} = \frac{1}{2}\frac{\rho_L c_L}{\pi a^2}\left(\frac{\omega}{c_L}a\right)^2, \quad M_{a,L} = \frac{8}{3}\frac{\rho_L}{\pi^2 a} \quad \text{for} \quad \omega < \omega_g = \sqrt{2}\frac{c_L}{a}.$$

At first, the acoustic components are transformed to the mechanical side. On the electrical side, the voltage source is replaced by a current source afterwards. Compared to the ohmic resistance, the coil inductance is neglected again. The transformation of the electrical components to the mechanical side results in the mechanical circuit diagram shown in Fig. 8.9. By combining the individual components in the total mass m and total friction r, the transfer function of speaker results in

$$\underline{B}_v = \frac{\underline{v}}{\underline{F}_0} = \frac{1}{\mathrm{j}\omega m + 1/\mathrm{j}\omega n_W + r} = B_0 \frac{\mathrm{j}\omega/\omega_0}{1 + \mathrm{j}\omega/\omega_0 Q - (\omega/\omega_0)^2}$$

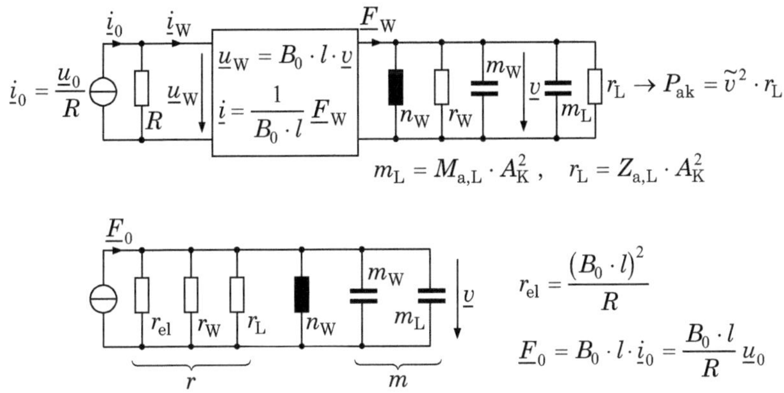

Fig. 8.9. Transformation of acoustic and electrical loudspeaker components to the mechanical side

with

$$B_0 = \sqrt{\frac{n_W}{m}}, \qquad m = m_W + m_L, \qquad r = r_W + r_{el} + r_L.$$

Taking the transformation relations and numerical values specified in Fig. 8.8 into consideration, the individual components can be calculated:

$$m_L = \left(\pi a^2\right)^2 M_{a,L} = \frac{8}{3}\rho a^3 = 3.2\,\mathrm{g}, \qquad m = m_L + m_W = 33.2\,\mathrm{g}$$

$$r_{el} = (B_0 l)^2 \frac{1}{R} = 6.82\,\mathrm{N\,s\,m^{-1}}, \qquad r_W = \frac{\omega_W m_W}{Q_W} = 0.19\,\mathrm{N\,s\,m^{-1}}$$

$$r_L(\omega) = \left(\pi a^2\right)^2 Z_{a,L} = \frac{1}{2}\pi a^2 \rho_L c_L \left(\frac{\omega}{c_L} a\right)^2 .$$

In combination with $r_L(\omega_0) = 8 \cdot 10^{-3}\,\mathrm{N\,s\,m^{-1}}$, f_0 and Q result in

$$f_0 = \frac{1}{2\pi\sqrt{n_W (m_L + m_W)}} = 19.1\,\mathrm{Hz} \approx f_W, \quad Q = \frac{\omega_0 (m_W + m_L)}{r_W + r_{el} + r_L} = 0.6.$$

Finally, the radiated acoustic power P_{ak} is calculated:
The radiated acoustic power $P_{ak} = \tilde{v}^2 r_L$ results in

$$P_{ak} = \frac{\tilde{v}^2}{\tilde{F}_0^2}\tilde{F}_0^2 \frac{1}{2}\pi a^2 \rho_L c_L \left(\frac{\omega}{c_L} a\right)^2 = \left(\left|\frac{\underline{v}}{\underline{F}_0}\right| \omega m\right)^2 \cdot P_0.$$

Figure 8.10 shows its curve characteristic. For a voltage $\tilde{u}_0 = 1\,\mathrm{V}$, P_0 amounts to 83 mW.

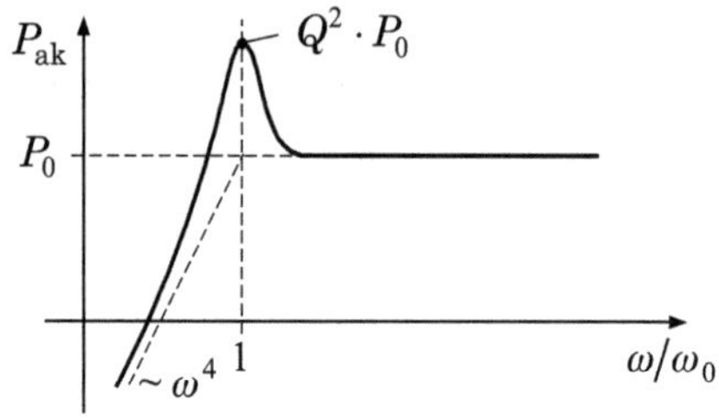

Fig. 8.10. Radiated acoustic power in the far sound field of the electrodynamic loudspeaker

The use of long-throw loudspeakers as acoustic anti-noise source for active noise control represents an interesting sample application for electromechanical transducers. The optimization of such acoustic sources is described in [62] in detail.

Electrodynamic Vibration Calibrator

In Fig. 8.11, the calibration of an acceleration sensor ② by comparison with a reference sensor ① is illustrated. An electrodynamic drive system is used for the vibrational excitation.
Assuming that the acceleration $\underline{a}_1$ and $\underline{a}_2$ match at the sensor to be calibrated and at the reference sensor, the transfer function of the measuring sensor is determined by measuring the output voltages $\underline{u}_{\mathrm{L}1}$ and $\underline{u}_{\mathrm{L}2}$ of both sensors and by knowing the transfer function $\underline{B}_{a1}$:

$$\underline{u}_{\mathrm{L}1} = \underline{B}_{a1}\underline{a}_1, \qquad \underline{u}_{\mathrm{L}2} = \underline{B}_{a2}\underline{a}_2, \qquad \underline{a}_1 = \underline{a}_2$$

$$\underline{B}_{a2} = \frac{\underline{u}_{\mathrm{L}1}}{\underline{u}_{\mathrm{L}2}}\underline{B}_{a1}$$

By means of the circuit diagram shown in Fig. 8.11, the frequency-dependent curve characteristic of accelerations resulting at both rod ends is calculated in combination with following assumptions:

$a_1 = a_2$: rigid rod, approximation a)
$a_1 \neq a_2$: rod as waveguide, approximation b)
$a_1 \neq a_2$: rod as waveguide, definite solution c)

In addition to the calculated characteristics, the cut-off frequency f_{g} is derived for which the assumption $a_1 = a_2$ is valid in combination with a permitted error F.

For *approximation a)* – rod is considered to be rigid – the swivel pin of the circuit diagram shown in Fig. 8.11 is represented as mass m_{S}. By transforming

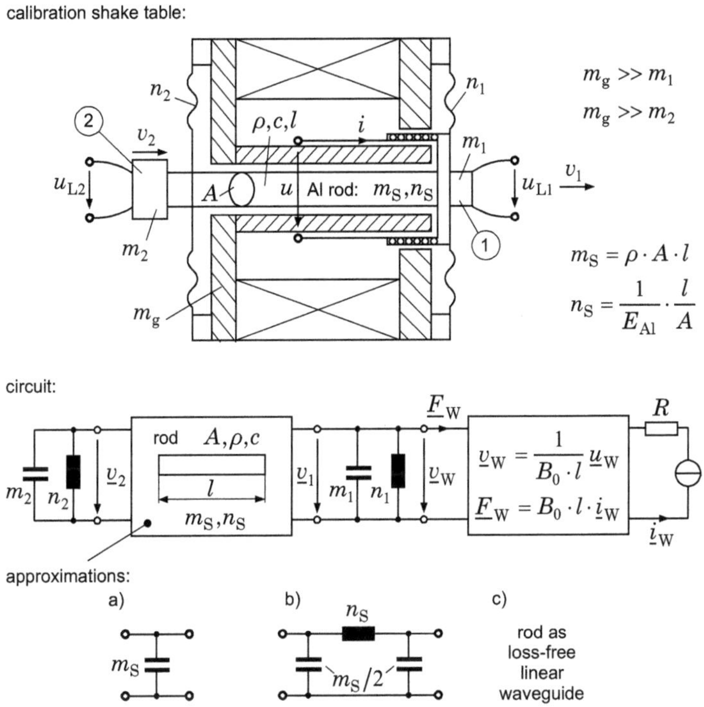

Fig. 8.11. Structure and circuit diagram of an electrodynamic vibration calibrator with aluminum rod

the electrical side to the mechanical side, the simplified circuit with concentrated mass m and compliance n shown in Fig. 8.12 is achieved. The rod velocity results from Fig. 8.12:

$$\underline{v} = \underline{F} \frac{1}{\mathrm{j}\omega m + 1/\mathrm{j}\omega n + r}$$

Finally, the acceleration $\underline{a} = \mathrm{j}\omega \underline{v}$ can be derived from velocity according to

$$\underline{a} = \frac{B_0 l}{m} \underline{i}_\mathrm{W} \frac{(\omega/\omega_0)^2}{1 + \mathrm{j}(\omega/\omega_0) 1/Q - (\omega/\omega_0)^2}$$

with

$$\omega_0^2 = \frac{1}{mn}, \qquad Q = \frac{1}{\omega_0 n r}.$$

The characteristic of $\underline{a} = \underline{a}_1 = \underline{a}_2$ related to the excitation $\underline{a}_0$ generated by the electrodynamic drive system

$$\frac{\underline{a}}{\underline{a}_0} = \frac{(\omega/\omega_0)^2}{1 + \mathrm{j}(\omega/\omega_0) 1/Q - (\omega/\omega_0)^2}, \qquad \underline{a}_0 = \frac{\underline{F}_\mathrm{W}}{m} = \frac{B_0 l}{m} \underline{i}_\mathrm{W}$$

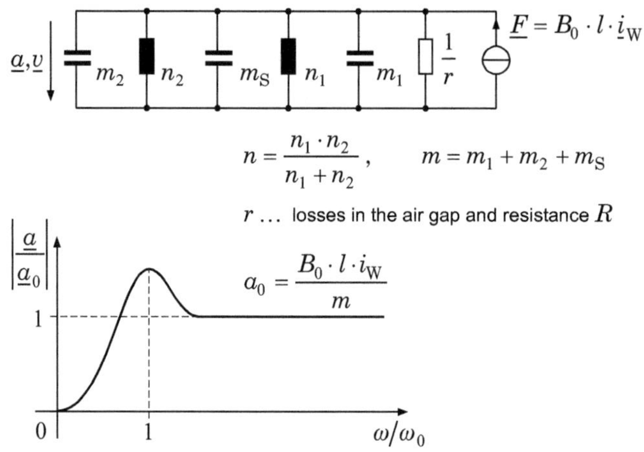

Fig. 8.12. Circuit diagram and acceleration characteristic at the swivel pin's ends for approximation a)

is also shown in Fig. 8.12. A characteristic high-pass response with resonant frequency f_0 can be identified.

If the swivel pin is considered to be a waveguide, there will be differences between the accelerations $\underline{a}_1$ and $\underline{a}_2$ at the rod ends. For *approximation b)* in Fig. 8.11 now the Π-circuit of the waveguide for low frequencies represented in Fig. 6.5 is used. Compared to the mass effect, the influence of friction r and compliances of the swivel pin suspension can be neglected above the resonant frequency f_0. With these assumptions the circuit diagram illustrated in Fig. 8.13 is achieved.
Based on the combination of the transducer masses m_1, m_2 and the rod mass m_S

$$m_1' = m_1 + m_S/2, \qquad m_2' = m_2 + m_S/2$$

the relations for the acceleration curve characteristics normalized with $\underline{a}_0$ can be specified. The acceleration at the end of rod ① is achieved with

$$\underline{a}_1 = \mathrm{j}\omega \underline{v}_1 \qquad \text{and} \qquad a_0 = \frac{\underline{F}_W}{m_1' + m_2'} = \frac{B_0 l}{m_1' + m_2'} \underline{i}_W$$

from the admittance derivable from Fig. 8.13

$$\frac{\underline{v}_1}{\underline{F}_W} = \frac{1}{\mathrm{j}\omega m_1' + \dfrac{1}{\mathrm{j}\omega n_S + 1/\mathrm{j}\omega m_1'}} = \frac{1}{\mathrm{j}\omega\,(m_1' + m_2')} \frac{1 - \omega^2 n_S m_2'}{1 - \omega^2 n_S \dfrac{m_1' m_2'}{m_1' + m_2'}}$$

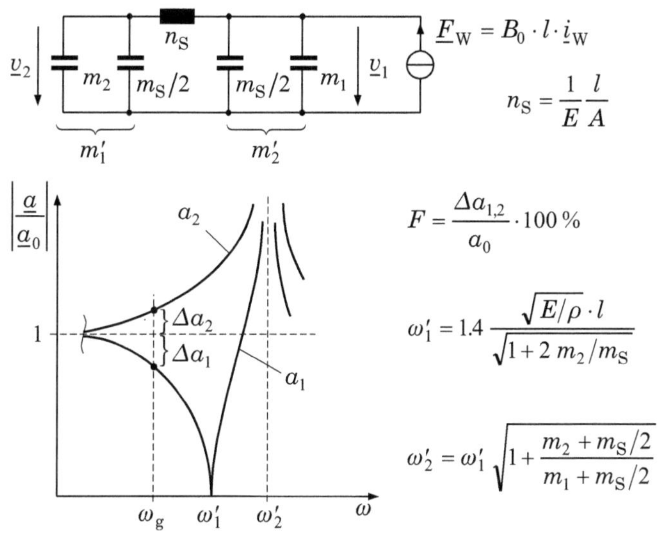

Fig. 8.13. Circuit diagram and acceleration characteristics at the ends of the swivel pin for approximation b)

and results in

$$\frac{\underline{a}_1}{\underline{a}_0} = \frac{1-(\omega/\omega_1')^2}{1-(\omega/\omega_2')^2}, \qquad \omega_1'^2 = \frac{1}{n_S m_2'}, \qquad \omega_2'^2 = \frac{1}{n_S m_2'}\left(1+\frac{m_2'}{m_1'}\right).$$

In the same way, the relation for the acceleration at the end of rod ② is achieved from

$$\frac{\underline{v}_2}{\underline{F}_W} = \frac{\underline{v}_2}{\underline{v}_1}\frac{\underline{v}_1}{\underline{F}_W} = \frac{j\omega m_1'}{j\omega n_S + 1/j\omega m_1'}\frac{\underline{v}_1}{\underline{F}_W} = \frac{1}{1-\omega^2 n_S m_1'}\frac{\underline{v}_1}{\underline{F}_W}$$

and results in

$$\frac{\underline{a}_2}{\underline{a}_0} = \frac{1}{1-(\omega/\omega_2')^2} \quad \text{or} \quad \frac{\underline{a}_2}{\underline{a}_1} = \frac{1}{1-(\omega/\omega_1')^2}.$$

For $\omega > \omega_0$ the acceleration characteristics specified in Fig. 8.13 show significant differences between $\underline{a}_1$ and $\underline{a}_2$. Compared to $a_1 = a_2$, the deviations Δa_1 and Δa_2 related to acceleration excitation $\underline{a}_0$ are characterized by the reduced error F.

Finally, the swivel pin is described by the definite solution of a loss-free, *linear waveguide*, thus by the chain matrix defined in (6.5). If friction r and compliances of the swivel pin suspension n_1, n_2 are neglected, the circuit shown in Fig. 8.14 will be achieved. The swivel pin is represented as two-port network in combination with the appropriate transducer matrix of the loss-free waveguide.

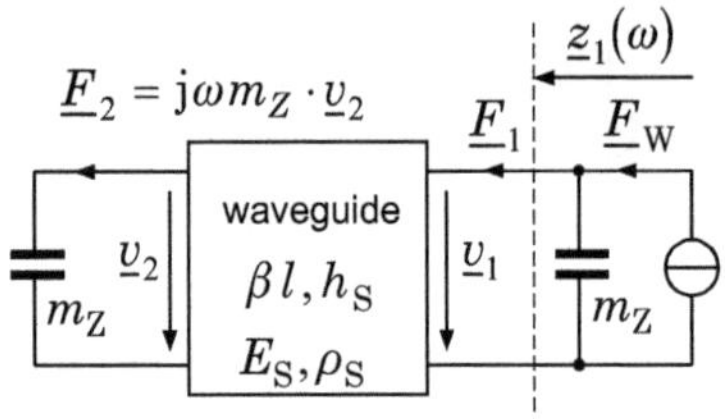

$$\begin{pmatrix} \underline{v}_1 \\ \underline{F}_1 \end{pmatrix} = \begin{pmatrix} \cos\beta l & jh_S \sin\beta l \\ j\frac{1}{h_S}\sin\beta l & \cos\beta l \end{pmatrix} \begin{pmatrix} \underline{v}_2 \\ \underline{F}_2 \end{pmatrix}$$

$$m_1 = m_2 = m_Z\,, \quad \beta = \omega/c_S = \omega\sqrt{\rho_S/E_S}\,, \quad h_S = \sqrt{n'_S/m'_S}\,,$$

$$n'_S = \frac{1}{A \cdot E_S}\,, \qquad m'_S = A \cdot \rho_S\,, \qquad c_S = \sqrt{E_S/\rho_S}$$

Fig. 8.14. Circuit diagram of vibration calibrator while representing the swivel pin as loss-free, linear waveguide

On the basis of this diagram the definite relations for frequency ω_1 of the zero are derived:

$$\underline{v}_1 = 0$$

$$\frac{\underline{v}_1}{\underline{F}_1} = \frac{\cos(\beta_1 l)\,\underline{v}_2 + jh_S \sin(\beta_1 l)\,\underline{F}_2}{j\frac{1}{h_S}\sin(\beta_1 l)\,\underline{v}_2 + \cos(\beta_1 l)\,\underline{F}_2} = 0 \qquad \text{with} \qquad \underline{F}_2 = j\omega m_Z \underline{v}_2$$

$$\cos(\beta_1 l) + jh_S j\omega m_Z \sin(\beta_1 l) = 0$$

$$h_S \omega m_Z = \frac{1}{A\rho_S c_S}\omega m_Z = \frac{1}{Al\rho_S}\frac{\omega l}{c_S} m_Z = \frac{m_Z}{m_S}\beta l$$

$$\cos(\beta_1 l) - \beta_1 l \frac{m_Z}{m_S}\sin(\beta_1 l) = 0 \qquad \text{with} \qquad \beta_1 = \frac{\omega_1}{c_S}$$

$$\beta_1 l \tan(\beta_1 l) = \frac{m_S}{m_Z} \to \omega_1 .$$

The pole of the acceleration characteristic at frequency ω_2 can be calculated in the following way:

$$v_1 \to \infty$$

$$\frac{\underline{v}_1}{\underline{F}_1} = \underline{z}_1(\omega_2) + j\omega_2 m_Z = 0$$

$$j\omega_2 m_Z + \frac{j\frac{1}{h_S}\sin(\beta_2 l)\,\underline{v}_2 + \cos(\beta_2 l)\,\underline{F}_2}{\cos(\beta_2 l)\,\underline{v}_2 + jh_S \sin(\beta_2 l)\,\underline{F}_2} = 0 \qquad \text{with} \qquad \underline{F}_2 = j\omega m_Z \underline{v}_2$$

$$2\mathrm{j}\omega_2 m_Z h_S \cos(\beta_2 l) + \mathrm{j}\sin(\beta_2 l)\left(1 - (\omega_2 m_Z h_S)^2\right) = 0$$

$$\omega_2 m_Z h_S = \frac{m_Z}{m_S}\beta_2 l$$

$$\frac{2\beta_2 l \frac{m_Z}{m_S}}{1 - \left(\beta_2 l \frac{m_Z}{m_S}\right)^2} + \tan(\beta_2 l) = 0 \rightarrow \omega_2 \qquad \text{with} \qquad \beta_2 = \frac{\omega_2}{c_S}.$$

In Table 8.1, the frequencies f_0, f_1 and f_2 are specified for the considered example. The small deviation between the frequencies f_1', f_2' calculated with the simple approximation b) and the definite frequencies f_1 and f_2 is remarkable.

Table 8.1. Numerical values for the frequencies f_0, f_1, f_2 and f_g of vibration calibrator

example:

$A = 4\ \mathrm{cm}^2$, $l = 100\ \mathrm{mm}$, $\rho = 2.7\cdot 10^3\ \mathrm{kg\cdot m^{-3}}\ (Al)$, $c_S = 5\cdot 10^3\ \mathrm{m\cdot s^{-1}}$

$E_S = 6.9\cdot 10^{10}\ \mathrm{N\cdot mm^{-2}}$, $m_1 = m_2 = m_Z = 50\ \mathrm{g}$, $m_S = 108\ \mathrm{g}$

$n_1 = n_2 = 10^{-4}\ \mathrm{m\cdot N^{-1}}$, $n_S = 3.6\cdot 10^{-9}\ \mathrm{m\cdot N^{-1}}$

	approximation b)	exact solution c)
$f_0 = 35$ Hz	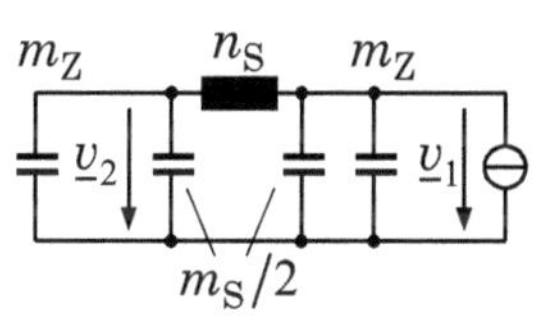	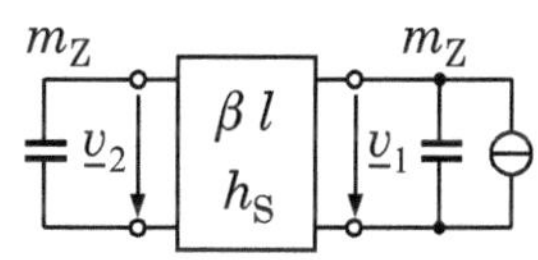
	f' / kHz	f / kHz
f_1 frequency of zero point of $\underline{a}_1$	8.11	8.75
f_2 frequency of zero point of $\underline{a}_1, \underline{a}_2$	11.47	16.55
f_g cut-off frequency at $F = 2\,\%$	1.12	1.14

Finally that cut-off frequency f_g is to be calculated, at which both acceleration curve shape characteristics $\underline{a}_1$ and $\underline{a}_2$ solely vary by a default error F.

approximation b):

$$\left|\frac{\underline{a}_2}{\underline{a}_1}\right|_{\omega=\omega_g} = \frac{1}{1 - (\omega_g/\omega_1')^2} = 1 + F \rightarrow \frac{\omega_g}{\omega_1'} = \sqrt{\frac{F}{1+F}}$$

definite solution:

$$\underline{v}_1 = \cos\left(\beta_{\mathrm{g}} l\right) \underline{v}_2 + \mathrm{j} h_{\mathrm{S}} \omega_{\mathrm{g}} m_{\mathrm{Z}} \sin\left(\beta_{\mathrm{g}} l\right) \underline{v}_2, \qquad h_{\mathrm{S}} \omega_{\mathrm{g}} m_{\mathrm{Z}} = \frac{m_{\mathrm{Z}}}{m_{\mathrm{S}}} \beta_{\mathrm{g}} l$$

$$\left|\frac{\underline{v}_1}{\underline{v}_2}\right| = \left|\frac{\underline{a}_1}{\underline{a}_2}\right| = 1 - F = \cos\left(\beta_{\mathrm{g}} l\right) - \frac{m_{\mathrm{Z}}}{m_{\mathrm{S}}} \beta_{\mathrm{g}} l \sin\left(\beta_{\mathrm{g}} l\right) \rightarrow \omega_{\mathrm{g}}$$

$$\text{with} \quad \beta_{\mathrm{g}} = \frac{\omega_{\mathrm{g}}}{c_{\mathrm{S}}}.$$

Detection of Hip Prosthesis Loosening by Vibration Analysis

The most serious common complication of a total hip replacement is a partial loosening of prosthesis in the thigh bone or femur. One possible method to detect hip prosthesis loosening is vibration analysis. In clinical practice mostly proximal, that means at body side, prosthesis loosening occurs. In this case there is no interface to the femur. That changes the natural frequencies of the femur-prosthesis compound which can be excited by an electrodynamic shaker fixed in the knee region at or near the inner medial femur condyle that is the knuckle of knee joint. The frequency spectrum of prosthesis acceleration measured by an acceleration sensor in the prosthesis shaft shows the resonance frequencies which change due to the loosening state.

Figure 8.15 shows the mechnical model of femur excitation, Fig. 8.15 a) shows the mechanical system and Fig. 8.15 b) illustrates the network. The shaker

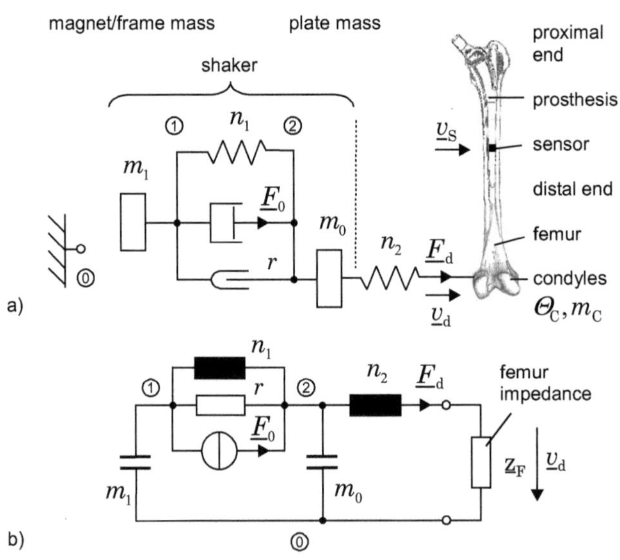

Fig. 8.15. Shaker and femur excitation model

model is basically that in Fig. 8.2 except that the compliance n_1 is connected to the shaker frame including the magnet m_1. Because of its large value compared to m_0, the mass m_1 acts almost as short cut between the system points ① and ⓪ for the considered frequencies. Femur and shaker are coupled by the dermis and subcutis which are modeled by the compliance n_2. It provides the distal excitation force $\underline{F}_d$ and velocity $\underline{v}_d$ to the inner femur condyle.

Figure 8.16 shows the model of the femur-prosthesis compound. Femur and prosthesis are modeled by dynamic bending beam elements represented in Fig. 6.21. The indices F and P label the femur and prosthesis network elements, respectively. Femur and prosthesis are coupled along the interface length $(h - k)\Delta x$ by the interface compliances Δn_I and rotational compliances Δn_{IR} which should be close to zero. In this case ($\Delta n_I = \Delta n_{IR} = 0$) the femur and the prosthesis velocities and rotational velocities are equal. Therefore, prosthesis and femur compliances are connected in parallel, as well as

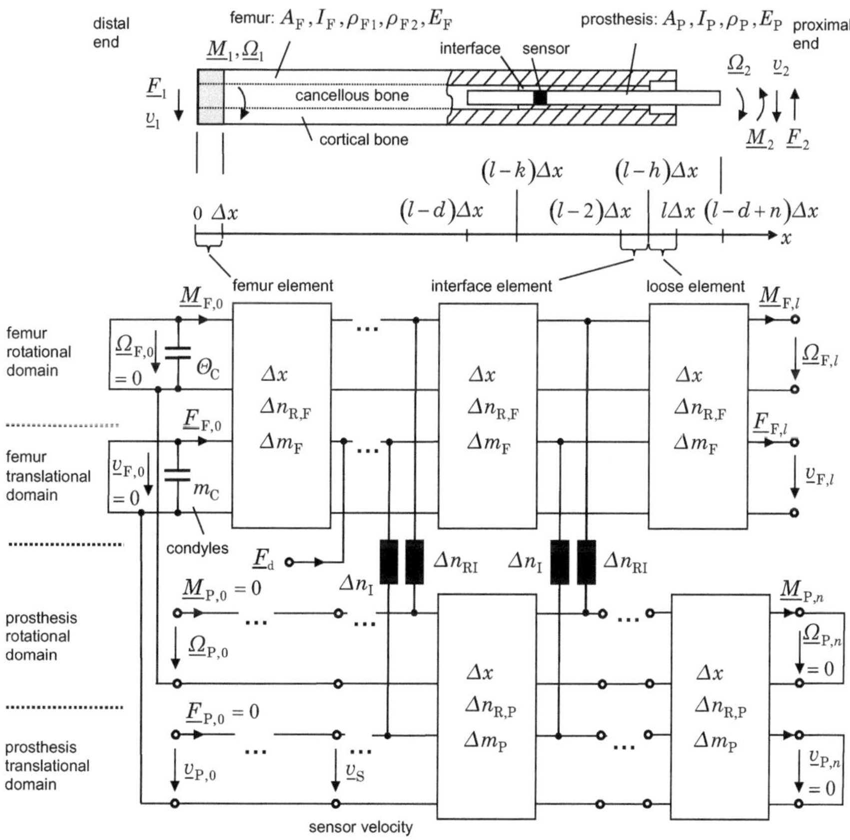

Fig. 8.16. Dynamic model of femur-prosthesis compound

both masses. The total compliance decreases while the mass increases. This influences the natural frequencies of the femur considering the prosthesis as part of the femur. The loose region and the uncovered part of prosthesis allow separat prosthesis vibrations with natural frequencies different from these of the femur. When the interface length varies, natural frequencies change which is the basis of loosening detection.

Figure 8.17 shows pSpice simulation results of prosthesis shaft acceleration and experimental results. For the network simulation, the femur of length $l = 380\,\mathrm{mm}$ without trochanter was discretized into 20 pieces leading to $\Delta x = 19\,\mathrm{mm}$. Due to its surface properties an interface between prosthesis shaft and femur can not be guaranteed. Therefore, in the model $(k-d) = 2$ prosthesis elements at the distal end are not connected to the femur and force $\underline{F}_{\mathrm{P},d}$ and moment $\underline{M}_{\mathrm{P},d}$ are zero. The sensor velocity $\underline{v}_{\mathrm{S}}$ was observed between the third and fourth element seen from the distal end and was differentiated to obtain the acceleration. The femur was modeled as thick-walled cylindrical tube with open ends and filled with water. The parameters of the network elements can be found in [63].

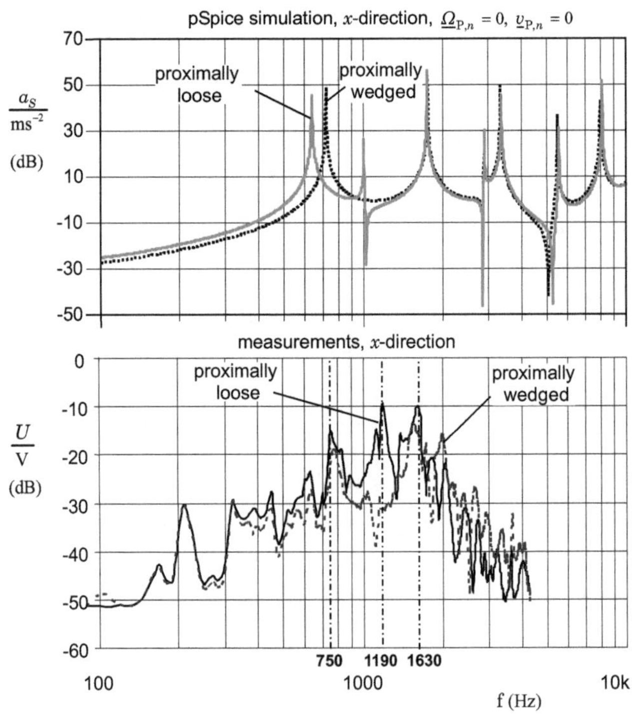

Fig. 8.17. Simulated and measured frequency response of acceleration in medical-lateral direction

The measurements were obtained from an experiment in which the middle section of a hip prosthesis was fixed in an artificial femur using cement such that a small part at the distal and proximal end has no interface and where damping is almost excluded [63]. The prosthesis was proximally wedged in the femur using a simple sheet to simulate the femur-prosthesis interface. The ceramics head of prosthesis was fixed both by a load head and the femur. Several resonances in the less sensitive band (500-1500 Hz) and sensitive band (1500-2500 Hz) as categorized by [64] occur. An additional resonance peak at 1190 Hz can be observed for the proximally loose prosthesis, while all other significant peaks in this bands exhibit only small changes. The amplitudes of higher vibrational modes are much lower in the highly sensitive band starting at 2500 Hz.
It can be concluded that the two simulated loosening states can be distinguished by vibrational analysis. The individual results in real patients will be very different, depending on implant size, properties and shape of the bone, static forces acting on the prosthesis during measurement and other influences. This means that no absolute criteria for loosing can be derived from frequency analysis. Therefore, each new frequency spectrum must be compared with a reference data set, e.g. obtained from a measurement right after hip replacement.

8.2 Electromagnetic Transducer

Due to a significantly higher energy density w_{mag} of the magnetic field compared to the energy density w_{el} of the electric field

$$\left.\begin{aligned} w_{\mathrm{mag}} &= \frac{1}{2}\frac{B_0^2}{\mu_0} = 0.4\cdot 10^6\,\mathrm{W\,s\,m^{-3}} \quad &&\text{with}\quad B_0 = 1\,\mathrm{V\,s\,m^{-2}} \\ w_{\mathrm{el}} &= \frac{1}{2}\varepsilon_0 E_0^2 = 4.4\,\mathrm{W\,s\,m^{-3}} \quad &&\text{with}\quad E_0 = 1\,\mathrm{kV\,m^{-1}} \end{aligned}\right\}\quad \frac{w_{\mathrm{mag}}}{w_{\mathrm{el}}} \approx 10^5$$

electromagnetic transducers are primarily used for *actuator applications*, e.g. as actuating drive in electromechanical relays and control devices, as pulling and lifting magnets as well as for small-power and stepping motors. The application in motors will not be discussed here, therefore it is referred to explanations given in [65] and [66]. Actuating devices, increasingly referred to as actuators, are used to affect technical processes. In addition to the sensors and information processing unit, they represent the third essential component concerning the control of technical processes. An overview of different actuator principles is specified in [22]. The further explanations focus on the linearization of the transducer characteristic, on the derivation of the two-port transducer network and on applications in micro and precision engineering as well as in mechatronics.

8.2.1 Derivation of the Two-Port Transducer Network

Neglecting force of inertia and frictional force, the balance of forces $F = F_{\mathrm{mag}} - F_{\mathrm{mech}}$ represented in Fig. 8.18 using the nonlinear magnetic force F_{mag} derived in Fig. 8.19 forms the basis for the *quasi-static* description of electromagnetic actuators.

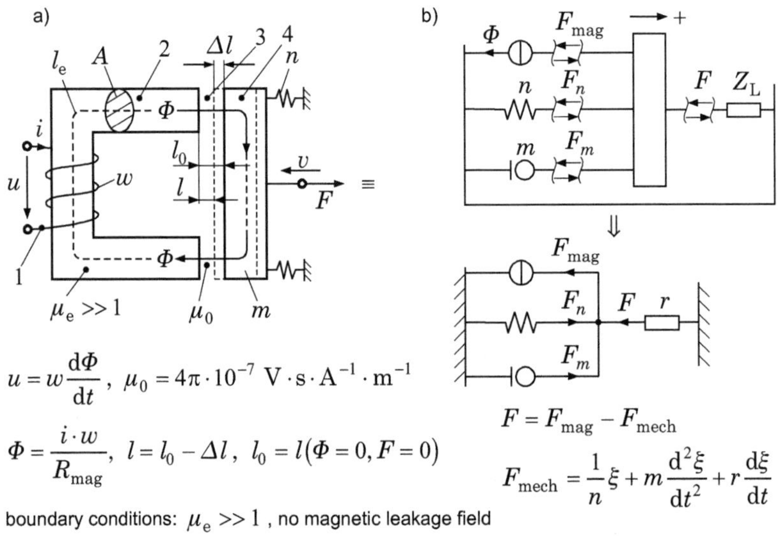

Fig. 8.18. Model of electromagnetic transducer
1 *coil*, 2 *soft-magnetic core - yoke*, 3 *air gap*, 4 *soft-magnetic actuating drive - armature*

However, the *dynamic* design is based on the linearization of the transducer characteristic. By considering the small-signal behavior around the adjusted operating point, also for the electromagnetic transducer a circuit representation can be derived. It corresponds to the basic circuit of magnetic transducers with transformer-like coupling already shown in Figure 7.8.
The descriptive derivation of the two-port network of the electromagnetic transducer from the transducer model shown in Fig. 8.18, the explanation of the assumed boundary conditions and the experimental determination of the components represent the topic of further discussion.

By supplying the coil having w turns with the current i, the magnetic flux Φ in the magnetic circuit consisting of the fixed yoke, the air gap l_0 and the armature with mass m and held by a spring n is generated. The magnetic force F_{mag} is effective between both pairs of pole faces limiting the air gap. The mechanical force F_{mech} caused by deflection Δl of the spring and the

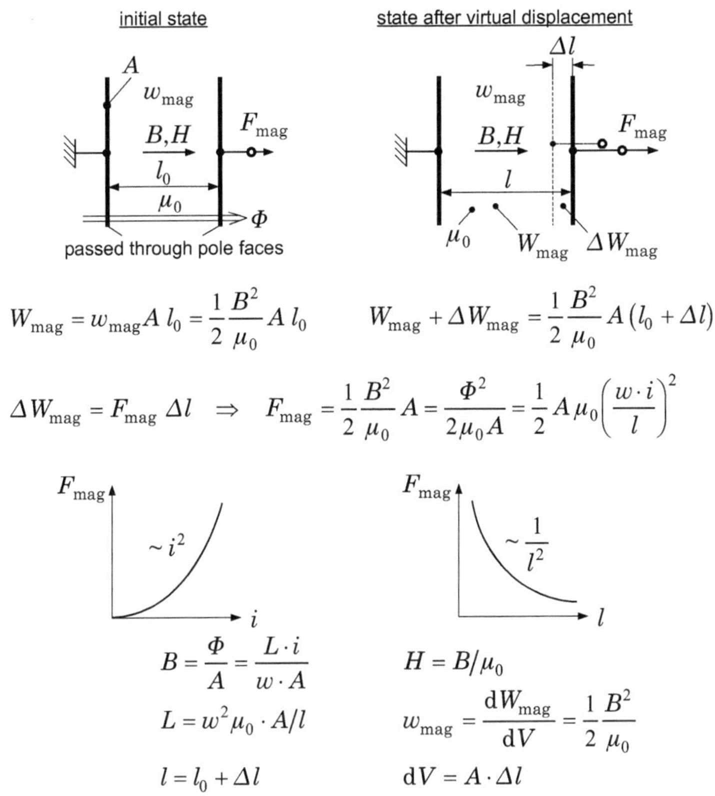

Fig. 8.19. Calculation of magnetic force using energy density and assuming a virtual displacement of armature

force of inertia counteracts the magnetic force. Compared to the air gap, the magnetic resistance of the iron parts is negligible.
The calculation of the magnetic force of attraction in the quasi-static case results from the application of energy balance after a *virtual displacement* Δl of armature as shown in Fig. 8.19. As a result, the magnetic force

$$F_{mag} = \frac{1}{2}\frac{i^2 L^2}{\mu_0 w^2 A} = \frac{1}{2} A \mu_0 \left(\frac{wi}{l}\right)^2 \tag{8.5}$$

depends on the square of the supply current i and reciprocally on the square of the air gap length l.
If now the coil is supplied with an alternating current, a distorted force characteristic curve with double frequency will be obtained. This is caused by the quadratic characteristic:

$$F_{mag} = \frac{1}{2}\frac{\hat{\imath}^2 L^2}{\mu_0 w^2 A}\left(1 - \cos\left(2\omega t\right)\right) \tag{8.6}$$

By means of the operating point adjustment caused by an

- additional DC component I_0 or by
- generating a direct flux component Φ_0 by means of a permanent magnet

and by means of an oscillation around the operating point with $\hat{\imath}\sin(\omega t)$ – *small-signal behavior* – an approximately linear relation can be assured. The magnetic force change ΔF_{mag} results in

$$\Delta F_{\text{mag}} = F_{\text{mag}} - F_0 = \frac{1}{2}\frac{L^2}{\mu_0 w^2 A}\left(2I_0\hat{\imath}\sin(\omega t) + \hat{\imath}^2\sin^2(\omega t)\right) \tag{8.7}$$

with

$$F_0 = \frac{1}{2}\frac{L^2 I_0^2}{\mu_0 w^2 A}$$

and the linearity error

$$F_{\text{lin}} = \frac{\hat{\imath}^2\sin^2(\omega t)}{2I_0\hat{\imath}\sin(\omega t)} < 10^{-2}.$$

In the large-signal behavior electromagnetic actuating devices are operated within the actuating range 5 up to 30 mm. Here, a limited linearization (linearity error in the working range approx. 5 up to 20%) results from the geometrical shaping of yoke and armature.

Concerning the derivation of the small-signal transducer circuit diagram for the dynamic case, the squared part of magnetic force is neglected due to the very small linearity error for operating point adjustment of the transducer. Based on the two basic transducer equations of the model represented in Fig. 8.20

$$F = F_{\text{mag}} - F_{\text{mech}} = \frac{\Phi^2}{2\mu_0 A} - \frac{1}{n}\xi - m\frac{\mathrm{d}^2\xi}{\mathrm{d}t^2} - r\frac{\mathrm{d}\xi}{\mathrm{d}t} \tag{8.8}$$

and with

$$i = \frac{\Phi w}{L} = \frac{l\Phi}{w\mu_0 A}, \tag{8.9}$$

the quantities ΔF and Δi yield in the neighborhood of the operating point

$$\Delta F = F - F_0 = \frac{\partial F}{\partial \Phi}(\Phi - \Phi_0) + \frac{\partial F}{\partial l}(l - l_0^*) \tag{8.10}$$

with

$$F_0 = F(\Phi_0, l_0^*) = \frac{\Phi_0^2}{2\mu_0 A} - \frac{1}{n}(l - l_0^*) = 0 \tag{8.11}$$

and

$$\Delta i = i - I_0 = \frac{\partial i}{\partial \Phi}(\Phi - \Phi_0) + \frac{\partial i}{\partial l}(l - l_0^*), \tag{8.12}$$

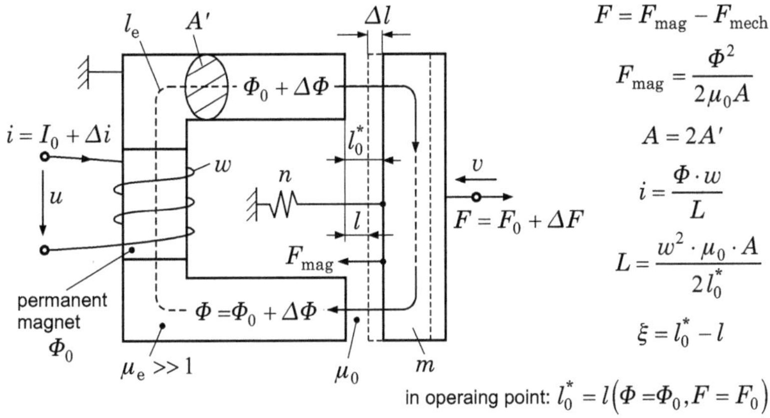

Fig. 8.20. Model of electromagnetic transducer with operating point adjustment by means of a direct current I_0 or permanent magnet Φ_0

where the quantity l_0^* denotes the air gap length after operating point adjustment.

For the quasi-static case, force of inertia and frictional force are neglected. All partial derivatives are calculated at point Φ_0 and point l_0^*. This yields following relations for the derivatives:

$$\left.\frac{\partial F}{\partial \Phi}\right|_{\Phi_0, l_0^*} = \frac{\Phi_0}{\mu_0 A}, \quad \left.\frac{\partial F}{\partial l}\right|_{\Phi_0, l_0^*} = \frac{1}{n}$$

$$\left.\frac{\partial i}{\partial \Phi}\right|_{\Phi_0, l_0^*} = \frac{w}{L}, \quad \left.\frac{\partial i}{\partial l}\right|_{\Phi_0, l_0^*} = \frac{\Phi_0}{w\mu_0 A}$$

Since a sinusoidal displacement around the operating point is assumed, in the next step complex amplitudes

$$l_0^* - l = \xi \rightarrow \underline{\xi}, \quad F - F_0 = \Delta F \rightarrow \underline{F}, \quad \Phi - \Phi_0 = \Delta\Phi \rightarrow \underline{\Phi}, \quad i - I_0 = \Delta i \rightarrow \underline{i}$$

are introduced and the transducer equations can be written in the following form:

$$\underline{F} = \frac{\Phi_0}{\mu_0 A}\underline{\Phi} - \frac{1}{n}\underline{\xi} \tag{8.13}$$

$$\underline{i} = \frac{w}{L}\underline{\Phi} + \frac{\Phi_0}{w\mu_0 A}\underline{\xi} \tag{8.14}$$

Finally, the transducer equations result in combination with $\underline{u} = \mathrm{j}\omega w \underline{\Phi}$ and $\underline{v} = \mathrm{j}\omega \underline{\xi}$ in:

$$\left.\begin{aligned} \underline{F} + \frac{1}{\mathrm{j}\omega n}\underline{v} = F_{\mathrm{W}} = \frac{1}{\mathrm{j}\omega}\frac{\Phi_0}{w\mu_0 A}\underline{u} \\ \underline{i} + \frac{1}{\mathrm{j}\omega L}\underline{u} = i_{\mathrm{W}} = \frac{1}{\mathrm{j}\omega}\frac{\Phi_0}{w\mu_0 A}\underline{v} \end{aligned}\right\} \quad Y^* = \mathrm{j}\omega\frac{w\mu_0 A}{\Phi_0}$$

The transducer is characterized by a gyrator-like coupling with imaginary transducer constant Y^*. Since the aim is an ensurance of a real-valued, frequency-independent transducer coupling, the transducer equations are rearranged for $\underline{u}, \underline{F} = f\,(\underline{i}, \underline{v})$:

$$\underline{u} - \mathrm{j}\omega L\underline{i} = \underline{u}_{\mathrm{W}} = \frac{\Phi_0 w}{l_0^*}\underline{v} \tag{8.15}$$

and

$$\underline{F} = \frac{1}{\mathrm{j}\omega}\frac{\Phi_0}{w\mu_0 A}\left(\mathrm{j}\omega L\underline{i} + \frac{\Phi_0 w}{l_0^*}\underline{v}\right) - \frac{1}{\mathrm{j}\omega n}\underline{v}$$

$$\underline{F} + \frac{1}{\mathrm{j}\omega}\left(\frac{1}{n} - \frac{\Phi_0^2}{\mu_0 A l_0^*}\right)\underline{v} = \underline{F}_{\mathrm{W}} = \frac{\Phi_0 w}{l_0^*}\underline{i} \tag{8.16}$$

Now there exists a *transformer-like coupling* between electrical and mechanical quantities defined by the real transducer constant

$$X = \frac{l_0^*}{\Phi_0 w}. \tag{8.17}$$

If now KIRCHHOFF's laws are applied to (8.15) and (8.16), the transducer circuit diagram specified in Fig. 8.21 will be the result.
What descriptive meaning do the transducer components have?
If the armature is fixed, mechanically locked case ($\underline{v} = 0$), at the electrical input terminals the impedance

$$\underline{Z} = R + \mathrm{j}\omega L_{\mathrm{b}} \quad \text{with} \quad L_{\mathrm{b}} = L\,(l_0^*) = \frac{w^2\mu_0 A'}{l_0^*}$$

can be measured. For this case, the force $\underline{F}_{\mathrm{W}} = (1/X)\,\underline{i}$ must be applied in order to avoid a displacement of the armature. Conversely, the transducer voltage $\underline{u}_{\mathrm{W}} = (1/X)\,\underline{v}$ will be generated, if the velocity $\underline{v}$ is applied.
For electrical open-circuit operation $\underline{i} = 0$ the compliance n_{L} can be measured at the mechanical side. The compliance n_{L} consists of the spring compliance n and a negative compliance n_I which is generated by the magnetic field. The negative field compliance $n_I = -X^2 L_{\mathrm{b}}$ results from the supportive effect of the magnetic field resulting from operating point adjustment.

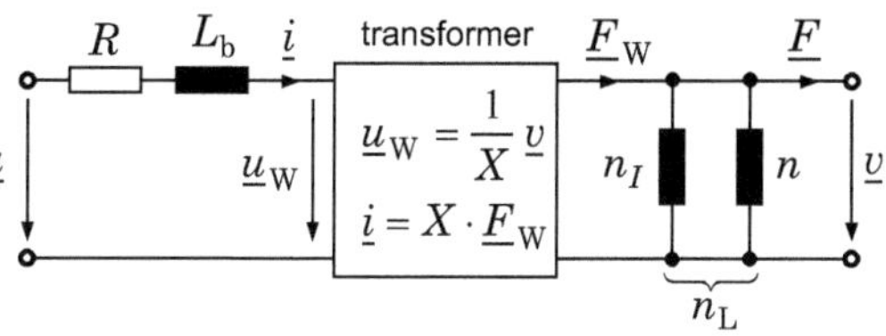

- resistance of excitation coil: R
- inductance in stationary locked state: $\underline{L}_b = \frac{w^2 \cdot \mu_0 \cdot A'}{l_0^*}, \quad A' = \frac{1}{2} A$
- transformer-like transducer constant: $X = \frac{l_0^*}{\Phi_0 \cdot w} = \frac{l_0^*}{B_0 \cdot A' \cdot w}$
- negative compliance caused by action of magnetic field: $n_I = -X^2 \cdot L_b$
- spring compliance: n
- resulting compliance for electrical open-circuit operation: $\frac{1}{n_L} = \frac{1}{n} + \frac{1}{n_I} = \frac{1}{n} - \frac{\Phi_0^2}{\mu_0 \cdot A' \cdot l_0^*}$

Fig. 8.21. Two-port network of loss-free electromagnetic transducer for small-signal behavior

In case of the electrical short-circuit $\underline{u} = 0$, only the mechanical compliance $n = n_K$ can be measured at the mechanical side, if R is neglected.
At first, in Fig. 8.21 the armature mass $m = \rho V_A$ and friction r in the air gap or losses of the spring respectively have not been taken into account. For the dynamic consideration of concrete sample applications, these components must be inserted parallel to compliance n_L at the mechanical side. Concerning the derivation of the transducer model, also the magnetic field leakage has been neglected due to $\mu_e \gg \mu_0$.
The relations already specified in Table 7.2 apply for the transformation relations between electrical and mechanical network side. Table 8.2 comprises a further summary of relations for components and transformations in comparison with the electrodynamic transducer.

When comparing the maximum possible forces F_{max} of both transducers, the different applications become understandable. Assuming the technical limiting values for the flux density and current density

$$B_0 = 2\,\mathrm{V\,s\,m^{-2}}, \qquad J = \frac{i}{A_D} = 10\,\mathrm{A\,mm^{-2}},$$

typical geometrical dimensions (A cross-sectional area of ferromagnetic circuit, d diameter of coil wire, A_D cross-sectional area of coil wire)

$$A = \pi r^2 = 100\,\mathrm{mm^2}, \qquad d = 0.3\,\mathrm{mm}, \qquad A_D = 0.07\,\mathrm{mm^2}$$

Table 8.2. Circuit diagram, component and transformation relations for loss-free magnetic transducer

electromagnetic transducer	electrodynamic transducer
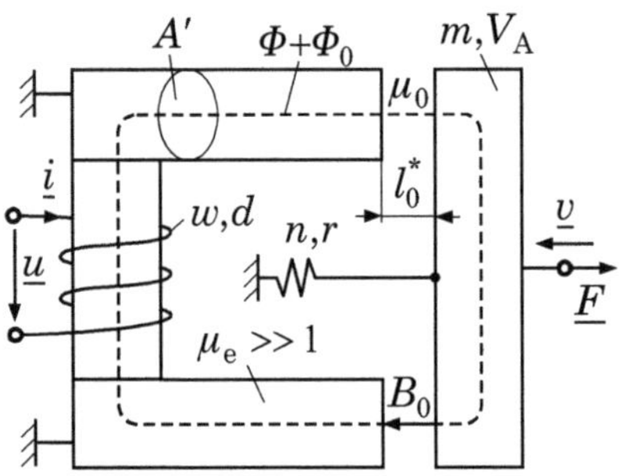	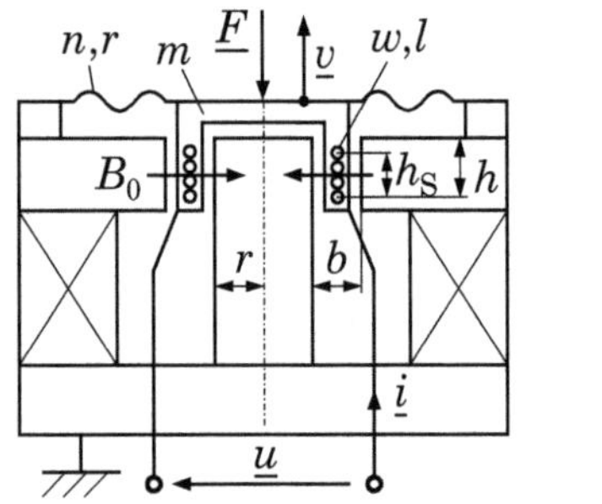

circuit diagram

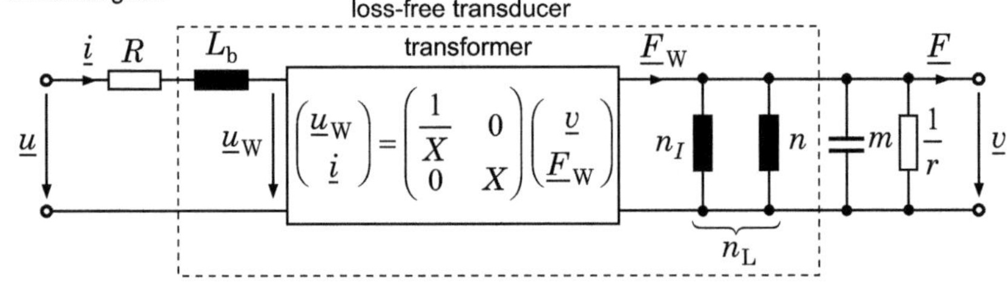

	electromagnetic transducer	electrodynamic transducer
X	$\dfrac{l_0^*}{B_0 \cdot A' \cdot w}$	$\dfrac{1}{B_0 \cdot l}$
L_b	$\mu_0 \cdot w^2 \cdot \dfrac{A'}{l_0^*}$	$\dfrac{\mu_0 \cdot w^2 \cdot 2\pi r \cdot h}{b}\left(1 - \dfrac{2}{3}\dfrac{h_S}{h}\right)$
n_L	n parallel to n_I, $\quad n_I = -X^2 \cdot L_b$	n
m	$\rho \cdot V_A$	$\rho \cdot V_S$
r	$\dfrac{1}{\omega_0 \cdot Q \cdot n_L}$, $\quad \omega_0^2 = \dfrac{1}{n_L \cdot m}$	
R	$\dfrac{\rho_D \cdot l}{A_D}$, $\quad l = 2\pi r \cdot w$	

transformation correlations

mechanical → electrical	electrical → mechanical
$C_m = mX^2, \quad L_n = n/X^2$	$m_C = C/X^2, \quad n_I = L \cdot X^2$
$R_r = \dfrac{1}{r \cdot X^2}, \quad \underline{Z} = \dfrac{\underline{h}}{X^2} = \dfrac{1}{\underline{z} \cdot X^2}$	$r_R = \dfrac{1}{R \cdot X^2}, \quad \underline{h} = \underline{Z} \cdot X^2, \quad \underline{z} = \dfrac{1}{\underline{Z} \cdot X^2}$

series connection → series connection
parallel connection → parallel connection

and a number of turns $w = 300$, then in combination with

$$F_{\text{dyn}} = B_0 li, \qquad i = JA_{\text{D}} = 0.7\,\text{A}, \qquad l = 2\pi rw = 10\,\text{m}$$

and

$$F_{\text{mag}} = B_0^2 A \frac{1}{\mu_0}$$

the maximum forces

$$F_{\text{dyn}} = 15\,\text{N}, \qquad F_{\text{mag}} = 300\,\text{N}$$

are achieved. Solely electromagnetic transducers are used for applications with large actuating forces. In contrast, electrodynamic transducers are used for actuator and sensor applications for which the main focus is put on a preferably linear transfer characteristic.

The armature motion can also happen parallel to the pole faces. Larger displacement and linear force-deflection characteristics without the need for any operating point adjustment by neglecting the magnetic field leakage represent the advantages of this configuration (*reluctance principle*). The dynamic behavior can also be calculated by means of the circuit diagram shown in Fig. 8.21. However, the component equations are modified.

8.2.2 Sample Applications

In the following, the dynamic design of an electromechanical relay and an electromagnetic vibrating conveyor is explained. They both represent sample applications of electromagnetic transducers. As a third sample application, the design of an electromagnetic actuator and classification of its performance characteristics compared to alternative actuator principles are considered.

Electromechanical Relay

Figure 8.22 shows the basic structure of an electromechanical relay and characteristic values. Compared to semiconductor switches, particularly the electrical isolation of the operating and switching circuit, low contact resistance, high puncture voltage and high insulation resistance are advantageous. The relay is driven by an electromagnetic transducer.
If the relay's armature mass m and friction r are neglected, the relay will show a low-pass characteristic. This approximate transfer characteristic is derived from the simplified relay circuit diagram shown in Fig. 8.23.
Figure 8.24 shows the switching on and switching off characteristics in the time domain. When closing the relay, the current flow having a characteristic which is defined by the time constant L/R is reduced by the induced counter voltage and a new time response with time constant L_1/R arises.

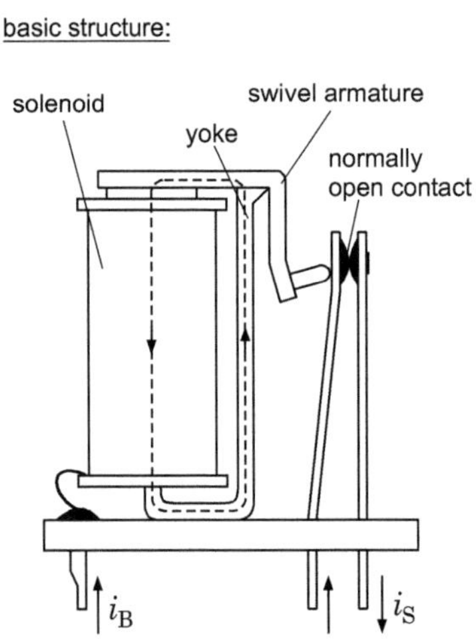

switching current range:	$(10^{-6} \ldots 10)$ A
switching voltage range:	$(10^{-7} \ldots 380)$ V
contact resistance:	$(1 \ldots 100)$ mΩ
driving power:	$(70 \ldots 4000)$ mW
max. switching frequency:	20 Hz
response time:	$(1 \ldots 30)$ ms
bounce time:	$(0.5 \ldots 10)$ ms
insulation resistance:	$(10^2 \ldots 10^6)\,\Omega$
puncture voltage:	$(0.5 \ldots 25)$ kV

Fig. 8.22. Basic structure and characteristic values of an electromechanical relay

Further developments of relays focus on miniaturized devices which also allow for array configurations. In the near future, the increasing application of *silicon microtechnology* will provide a high-potential alternative to conventional precision engineering realizations of relays. First prototypes comprising an electrostatic actuator have already been tested and have shown reproducible characteristic values [67].

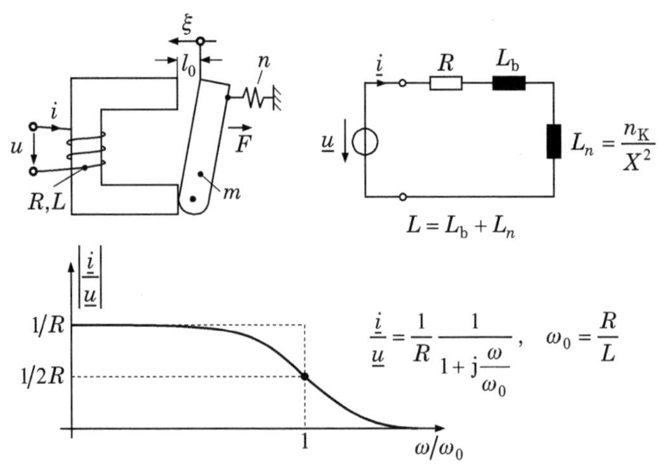

Fig. 8.23. Approximate transfer characteristic of the electromechanical relay (mass m and friction r are neglected)

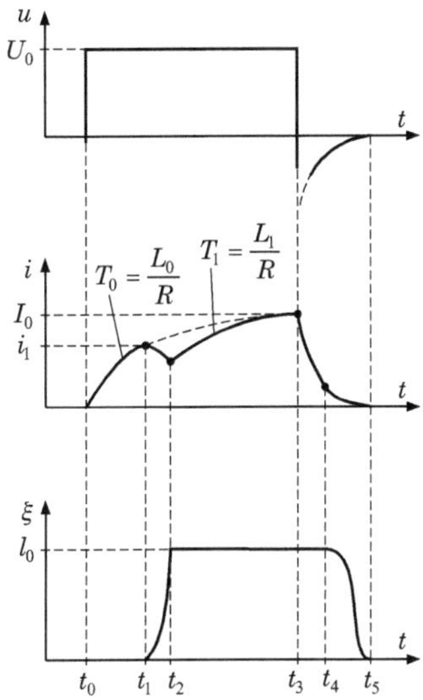

$t_2 - t_0$: total pull-in time

$t_1 - t_0$: pull-in time delay

$t_2 - t_1$: pull-in time

switching-on:

activation time t_0: $\xi = 0$	
$t > t_0$:	current increase $i = \frac{U_0}{R}\left(1 - e^{-R/L_0 \cdot t}\right)$ $L_0 = L(\xi = 0)$
$t = t_1$:	$F_{mag} = F_{mech} \rightarrow$ start of armature motion
$t_1 < t < t_2$:	increase of $L \rightarrow$ increase of Φ $\rightarrow$ induced counter voltage $\rightarrow$ decrease of i
$t = t_2$:	end of armature motion $\xi = l_0$, $L = L_1$
$t > t_2$:	current increase up to $I_0 = U_0/R$ with new time constant $T_1 = L_1/R$

switching-off:

switching-off time t_3 : $\xi = l_0$	
$t = t_4$:	$F_{mech} = F_{mag} \rightarrow$ start of armature motion
$t = t_5$:	$\rightarrow$ end of armature motion

Fig. 8.24. Time responses of operating current i and armature motion ξ for switching on and switching off a relay

The characteristic values of electromechanical relays are compared in Table 8.3 with those of solid state relays. With the electromechanical relay, the principal drawbacks like

- limited lifetime to 10^7 up to 10^9 switching cycles
- longer response time and
- contact aging

are offset however also by significant advantages such as

- low contact resistance
- high off-state resistance
- high load capacity and
- low temperature dependence of characteristic values.

In the future these advantages will ensure versatile application tasks for electromechanical relays.

Table 8.3. Comparison of selected characteristic values of electromechanical relays and solid state relays

characteristic values		electromechanical relay	solid state relay
max. switching frequency	Hz	$10 \ldots 1000$	10^8
driving power	mW	$10 \ldots 4000$	$5 \cdot 10^{-4} \ldots 300$
response time	ms	$0.5 \ldots 30$	$10^{-6} \ldots 1$
bounce time	ms	$0.5 \ldots 10$	0
contact resistance	mΩ	$1 \ldots 200$	$10 \ldots 10^5$
insulation resistance between open contacts between input and output	 MΩ MΩ	 $10^2 \ldots 10^6$ $10^2 \ldots 10^6$	 $10^{-2} \ldots 10^2$ $10^3 \ldots 10^6$
puncture voltage between open contacts between input and output	 kV kV	 $0.5 \ldots 25$ $0.5 \ldots 25$	 $10^{-2} \ldots 1$ $5 \cdot 10^{-3} \ldots 20$
lifetime	number of operations	$10^7 \ldots 10^9$	not limited

Electromagnetic Vibrating Conveyor

Figure 8.25 shows the design principle of the electromagnetic actuating drive of a vibrating conveyor. The linearization of the transducer is realized by operating point adjustment by means of a permanent magnet. In order to calculate the dynamic transfer characteristic, for deflections around this operating point the two-port transducer network shown in Fig. 8.21 can be approximately used. This circuit is also specified for current feed in Fig. 8.25 taking the oscillating mass m and friction r into account.

At first, the two-port parameters after operating point adjustment according to $\Phi = \Phi_0$, $l = l_0^*$ are calculated:

$$\Phi_0 = B_0 A = 2 \cdot 10^{-5}\,\mathrm{V\,s}, \qquad l_0^* = l_0 - n\frac{\Phi_0^2}{\mu_0 A} = 0.904\,\mathrm{mm}$$

$$\frac{1}{X} = \frac{w\Phi_0}{l_0^*} = 44\,\mathrm{V\,s\,m^{-1}}, \qquad L_\mathrm{b} = \frac{w^2 \mu_0 A}{l_0^*} = 140\,\mathrm{mH}, \qquad R = 28\,\Omega$$

$$n_\mathrm{L} = \frac{n_I n}{n + n_I} = 0.51 \cdot 10^{-4}\,\mathrm{m\,N^{-1}}$$

with

$$n_I = -X^2 L_\mathrm{b} = -7.2 \cdot 10^{-5}\,\mathrm{m\,N^{-1}} = -2.4n.$$

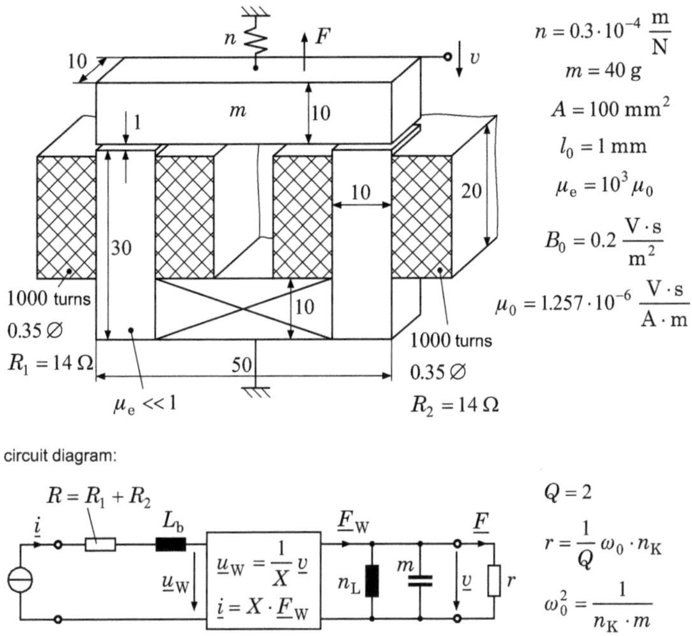

Fig. 8.25. Electromagnetic transducer for force generation at the vibrating conveyor

In the next step, as actuator characteristic value, the short-circuit force $\hat{F}_{\mathrm{K}}$ for sinusoidal excitation is calculated with

$$\left.\begin{aligned} \hat{F}_{\mathrm{K}} &= \frac{1}{X}\hat{\imath} \\ \hat{\Phi} &= \frac{\Phi_0}{3} \end{aligned}\right\} \hat{\Phi} = \frac{1}{w} i L_{\mathrm{b}} \rightarrow \hat{\imath}_{\max} = \frac{1}{3}\Phi_0 \frac{w}{L_{\mathrm{b}}}$$

and results in

$$\hat{F}_{\mathrm{K}} = 2.1\,\mathrm{N} \qquad \text{with} \qquad \hat{\imath}_{\max} = 48\,\mathrm{mA}.$$

The required electrical power results in

$$P = \tilde{\imath}^2_{\max} R = 33\,\mathrm{mW} \qquad \text{for} \qquad \tilde{u} = \frac{1}{\sqrt{2}}\hat{\imath}_{\max} R = 0.95\,\mathrm{V}.$$

The deflection for open-circuit operation $\left\{\hat{\xi}_{\mathrm{L}}\right\}_{\max}$ results in combination with

$$\underline{\xi}_{\mathrm{L}} = \frac{1}{\mathrm{j}\omega}\underline{v}_{\mathrm{L}}, \qquad \underline{v}_{\mathrm{L}} = \mathrm{j}\omega n_{\mathrm{L}} \underline{F}_{\mathrm{W}}, \qquad \underline{F}_{\mathrm{W}} = \frac{1}{X}\underline{i}$$

in

$$\left\{\hat{\xi}_{\mathrm{L}}\right\}_{\max} = \frac{n_{\mathrm{L}}}{X}\hat{\imath}_{\max} = 0.15\,\mathrm{mm}.$$

According to Fig. 8.26, it applies for the stability limit ξ_{g}

$$\xi_{\mathrm{g}} = l_0 - l_{\mathrm{g}} = l_0\left(1 - \sqrt[3]{\frac{n}{X^2 L_{\mathrm{b}}}}\right) = 0.57\,\mathrm{mm}.$$

The static deflection ξ_0 results in

$$\xi_0 = l_0 - l_0^* = 0.096\,\mathrm{mm}.$$

Thus, the maximum deflection for open-circuit operation satisfies the condition

$$\{\xi_{\mathrm{L}}\}_{\max} \leq \frac{1}{3}\left(\xi_{\mathrm{g}} - \xi_0\right).$$

The *stability limit* characterizes the distance between the pole faces, at which the increase of force characteristics shown in Fig. 8.26 reverses. The progression of magnetic and mechanical force characteristics including operating point adjustment $l = l_0^*$ is represented in the upper part of Fig. 8.26. The lower part of Fig. 8.26 shows the progression of total acting force and limit l_{g} for the stability limit.

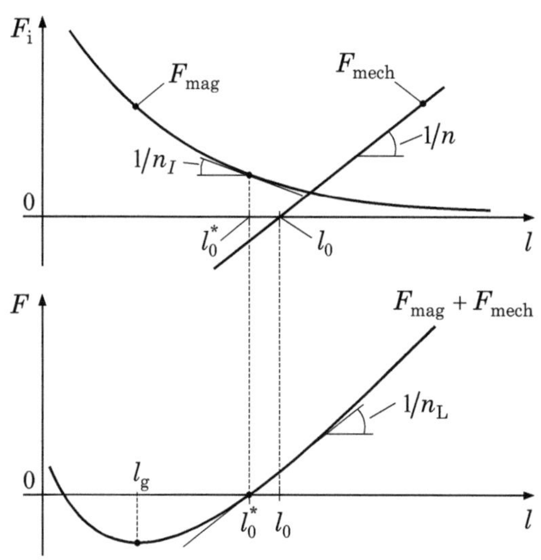

$$l\left(\Phi_0, F = 0\right) = l_0^*$$

$$l_0^* = l_0 - n\frac{\Phi_0}{\mu_0 \cdot A}$$

$$\xi = l_0^* - l$$

Fig. 8.26. Operating point adjustment and stability limit of electromagnetic transducer

The stability limit l_g is calculated by deriving the force characteristics at point $l = l_g$:

$$\left.\frac{\partial F}{\partial l}\right|_{l_g} = \frac{1}{n_L} = 0 \rightarrow \frac{1}{n} - \frac{1}{X^2 L_b}\left(\frac{l_0}{l_g}\right)^3 = 0, \quad \frac{l_g}{l_0} = \sqrt[3]{\frac{n}{X^2 L_b}}$$

In the third step, the dynamic transfer characteristics of the actuating drive are finally determined. For that purpose, the mechanical components of the circuit diagram illustrated in Fig. 8.25 are transformed to the electrical side. The transformed circuit diagram and the progression of the transfer characteristics $\underline{v}/\underline{i}$ and $\underline{\xi}/\underline{i}$ are specified in Fig. 8.27.

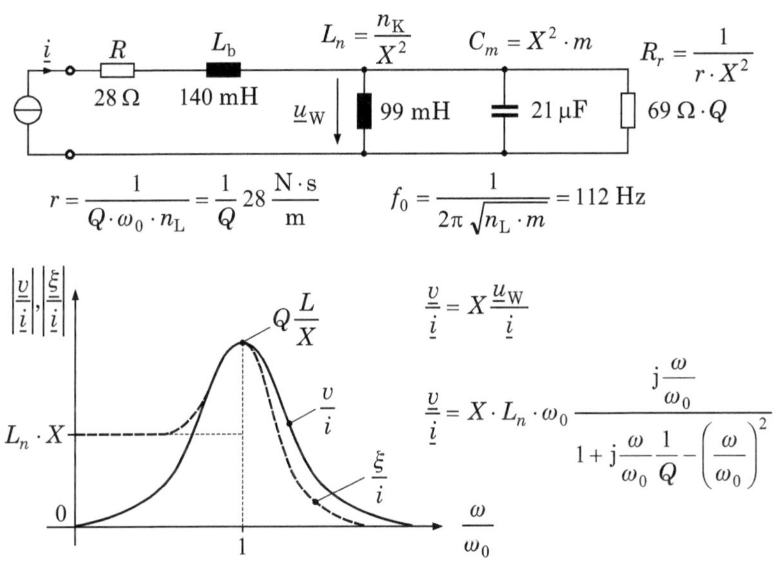

Fig. 8.27. Transfer characteristic of vibrating conveyor

In order to ensure maximum velocity or maximum deflection respectively, the vibrating conveyor is operated close to its resonant frequency $\omega \approx \omega_0$. In combination with

$$\hat{v}_{max} = \omega_0 \hat{\xi}_{max} = \{\hat{u}_W\} X = \hat{\imath}_{max} R_r X,$$

the current at resonance yields

$$\hat{\imath}_{max} = \frac{\omega_0 \hat{\xi}_{max}}{R_r X} = \frac{67}{Q}\,\text{mA} \quad \text{for} \quad \hat{\xi}_{max} = 0.15\,\text{mm}.$$

For $Q = 2$, the required power amounts to

$$P_{\text{el}} = \tilde{\imath}^2_{\max} (R + R_r) = 95\,\text{mW}$$

with an efficiency η of

$$\eta = \frac{P_{\text{mech}}}{P_{\text{el}}} = \frac{R_r}{R_r + R} = 0.83.$$

Electromagnetic Actuating Drive

The use as actuating drive in regulating units (valves) for influencing hydraulic or pneumatic fluid streams represents a particularly important application of the electromagnetic transducer. It is preferably used as short-stroke element in the form of a pot magnet (see Fig. 8.28). The current feed of the excitation coil is effected by means of a direct current (DC). Alternating current (AC) feeds are rather the exception.

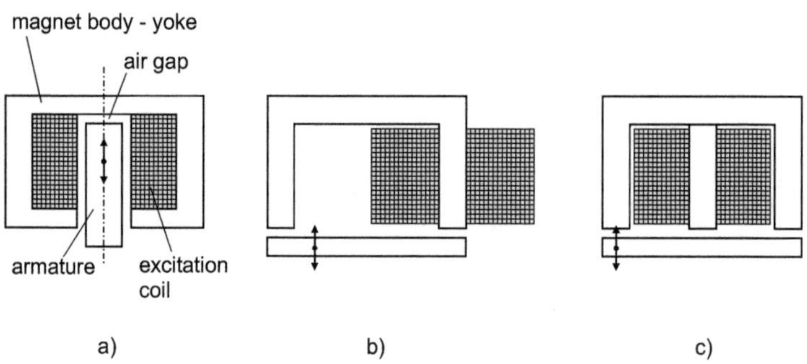

Fig. 8.28. Basic types of electromagnets for translational armature motions [22, 68] a) *pot magnet (plunger)*; b) *U-shape magnet (flat armature)*; c) *solenoid magnet (flat armature)*

In order to ensure an actuation being approximately proportional to the excitation current, magnetic force characteristics must be linearized. This is realized by a special shaping of the armature-yoke systems. While keeping constant the main dimensions of the DC magnet shown in Fig. 8.29, the magnetic conductance of the air gap and thus; force characteristics are changed by constructional variations.

In Table 8.4, electromechanical transducers are compared on the basis of selected properties with respect to their suitability as actuators. Here, the electromagnetic transducer represents the most advantageous actuating drive technology with respect to preferably high forces and large displacements. The maximum force is limited by maximum heat dissipation and saturation of the magnetic material.

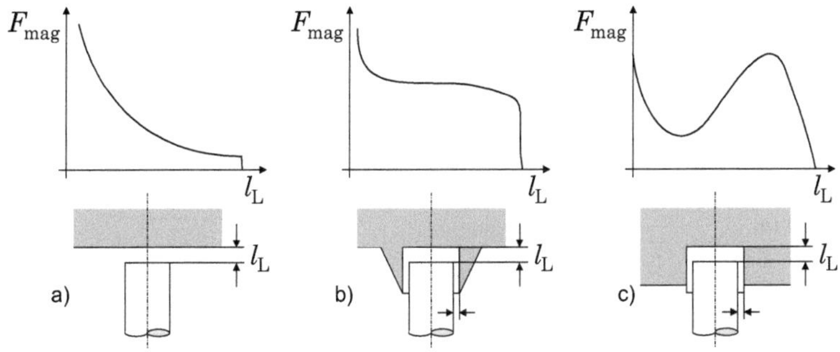

Fig. 8.29. Qualitative progression of force characteristics for different armature counterparts [65]

Table 8.4. Comparison of electromechanical transducers for actuator applications

transducer type	specific design	actuating forces [N]	displacements [mm]	characteristics
electromagnetic actuators	solenoid, pot form	10...300	1...30	- nonlinear force characteristics - force is dependent on deflection - being compensated by design to a certain extent
electrodynamic actuators	plunger coil, moving coil (voice coil actuator)	1...20	5...30	- linear force characteristics - force is independent of deflection - stroke limitation by leakage field
piezoelectric actuators (piezoceramics)	stack actuator, bending actuator	100...5000	0.01...0.20	- hysteresis behavior - aging
electrostatic actuators	comb and moving wedge configurations, stacked polymer actuators	0.02...2	0.1...5	- high miniaturization capability - manufacturing possibility in batch process - high operating voltages

The characteristic structure of an electromagnetic actuator as short-stroke element and its static characteristic curve in combination with its influencing and counter spring are specified in Fig. 8.30. The available actuation range is larger than the actuation range of that characteristic curve which is not influenced.

Figure 8.31 shows the quasi-static switching behavior of the electromagnetic actuator for the switching voltage and excitational current.

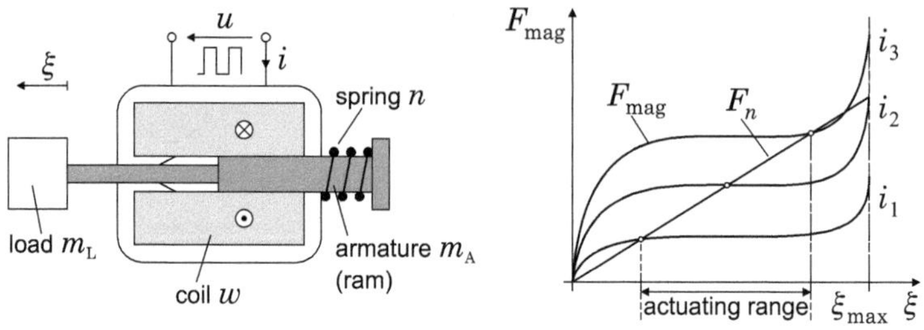

Fig. 8.30. Electromagnetic actuator as pot magnet and static characteristic curve for constructionally linearized operating condition

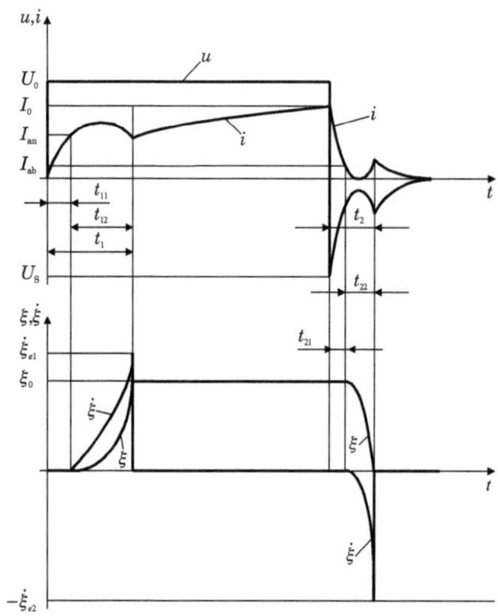

Fig. 8.31. Quasi-static switching behavior of electromagnetic DC actuator

Similar to the electromechanical relay, the armature motion is delayed due to spring force and induced countervoltage. In spite of a cut-off voltage peak, the switching off happens delayed analogously. The dynamic behavior of the electromagnetic actuator can be determined from the circuit represented in Fig. 8.32.

Compared to its resistance, the coil inductance was neglected for the low-frequency range. Velocity and acceleration transfer function result in:

$$\underline{B}_v = \frac{\underline{v}}{\underline{u}} = \frac{\omega_0 n_L}{RX} \frac{\mathrm{j}\,(\omega/\omega_0)}{1+\mathrm{j}\,(\omega/\omega_0)\,1/Q-(\omega/\omega_0)^2}, \quad \omega_0^2 = \frac{1}{n_L m}, \quad Q = \frac{1}{\omega_0 n_L r}$$

$$\underline{B}_\xi = \frac{\underline{\xi}}{\underline{u}} = \frac{n_L}{RX} \frac{1}{1+\mathrm{j}\,(\omega/\omega_0)\,1/Q-(\omega/\omega_0)^2}.$$

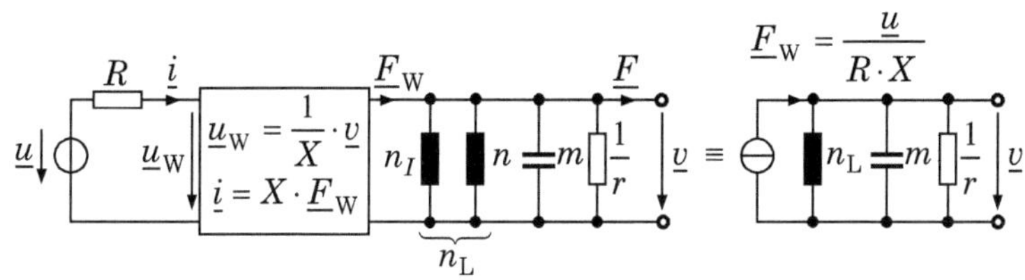

Fig. 8.32. Circuit of electrical solenoid used for calculation of the dynamic transfer characteristic

The corresponding amplitude-frequency responses are consistent with the curve progressions of the electrodynamic drive system specified in Fig. 8.6.

8.3 Piezomagnetic Transducer

The *piezomagnetic* transducers represent the third practically important class of magnetic transducers. Some ferromagnetic materials (metals and ceramics) show a distinct squared relation between mechanical and magnetic field quantities

$$S \sim H^2 \qquad \text{for} \qquad T = 0$$

and

$$B^2 \sim T \qquad \text{for} \qquad H = 0.$$

It is referred to the elongation generated in the magnetic field as *magnetostriction*. Conversely, in case of an elongation of a magnetostrictive material, an electrical voltage is induced in the surrounding coil. With the deformation illustrated in Fig. 8.33, a saturation occurs for high field strengths, to which it is referred as *saturation magnetostriction* S_S.

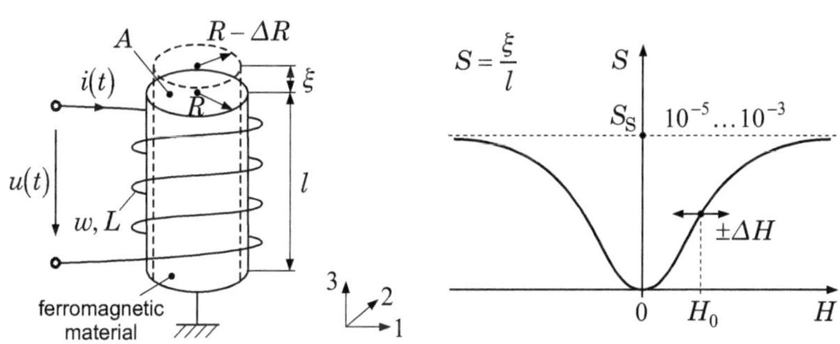

Fig. 8.33. Magnetostriction in ferromagnetic materials

The magnetostriction is primarily based on a directional change of magnetic domains within the ferromagnetic material depending on orientation of the external magnetic field – JOULE effect (1842). Even though this effect results in a deformation of the material, it is *volume invariant.*
In the following, the constitutive equations being valid after linearization are formulated. They form the basis for the derivation of the two-port network of the piezomagnetic transducer. Based on this circuit diagram and the characteristic values of technically important materials, applications in the field of micro and precision engineering are finally explained.

8.3.1 Derivation of the Two-Port Transducer Network

If the field quantities specified in Fig. 8.33 – for simplification here assumed to be position-independent – are replaced by corresponding integral quantities

$$\Phi = B_3 A, \quad \Phi = \frac{L}{w} i, \quad F = T_3 A,$$

where w and L denote the number of turns and the coil inductance respectively, the nonlinear magnetostrictive force will result in

$$F_{\text{mst}} = \frac{1}{2} K A B^2 = \frac{1}{2} K \frac{\Phi^2}{A} = \frac{1}{2} K \frac{1}{A} \left(\frac{L}{w} i \right)^2$$

for the mechanically locked state. Here, the quantity K denotes a material constant. For quasi-static actuator applications of these materials – *large-signal behavior* – the nonlinear magnetostrictive effect is often not disturbing. For technical applications as linear transducer (e.g. sensors), the squared dependence is linearized by the direct flux Φ_0. In the case that the remanent magnetization is too low, the direct flux Φ_0 is generated by an additive direct component I_0 in addition to the signal $i(t)$ or by an additional permanent magnet or solenoid. Similar to piezoelectric ceramics, in the material a preferred direction is generated by the linearizing direct quantity. In a macroscopic sense, this preferred direction gives the material the properties of an anisotropic body e.g. a crystal. By analogy with the piezoelectric case, a linear magnetomechanical coupling matrix between the field quantities

$$B_n = \mu^T_{nm} H_m + d_{nj} T_j, \qquad n, m = 1 \ldots 3^* \tag{8.18}$$

$$S_i = d_{mi} H_m + s^H_{ij} T_j, \qquad i, j = 1 \ldots 6 \tag{8.19}$$

can be defined. It is referred to the *linear* coupling between mechanical and magnetic field quantities as *piezomagnetic effect.* The field quantities are interrelated by the magnetic μ, elastic s and piezomagnetic d material constants.

---

*EINSTEIN summation convention: when an index variable appears twice in a single term, it implies a summation over all of its possible values

Depending on the combination of the variables

$$B, T = f(H, S), \tag{8.20}$$

$$H, S = f(B, T), \tag{8.21}$$

$$H, T = f(B, S) \tag{8.22}$$

further three systems of equations can be formulated each possessing a set of elastic, piezomagnetic and magnetic constants. However, the piezomagnetic coefficients of technically important materials are much less reported than in case of piezoelectric materials. Typically, only the *piezomagnetic constant* d_{33} as well as the roughly approximated constants $d_{31} \approx -\frac{1}{2}d_{33}$ and $d_{15} \approx (d_{33} - d_{31})$ assuming an invariant volume of the effect are specified.

As a special case of (8.18) and (8.19), in Table 8.5 both the direct and reciprocal piezomagnetic effect for corresponding mechanical and magnetic field directions are specified for different excitations.
Iron-nickel alloys, metal oxide compounds with ceramic properties and high magnetostrictive metallic specialty materials are specified as technically important materials in [66, 69, 70]. The main properties of these materials are summarized in the Tables 8.6 up to 8.9. For the super magnetostrictive material Terfenol-D, the characteristic curve progressions of magnetic polarization J ($J = B - \mu_0 H$) and strain S against the external magnetic field strength H are shown in Fig. 8.34 for the static loading case. The characteristic operating ranges are highlighted in gray.

Now it is possible to derive the circuit diagram of the piezomagnetic transducer by means of the piezomagnetic constitutive equations. For this purpose, the constitutive equations with piezomagnetic constant e_{33} are assumed in order to provide a real-valued,frequency-independent transducer constant X:

$$B_3 = \mu_{33}^S H_3 + e_{33} S_3$$

$$T_3 = -e_{33} H_3 + c_{33}^H S_3$$

The material constants μ_{33}^S and c_{33}^H are experimentally determined with respect to the conditions $S = 0$ (mechanically locked state) and $H = 0$ (electrical open-circuit operation).

Table 8.5. Direct and reciprocal piezomagnetic effect in x_3-direction (longitudinal direction)

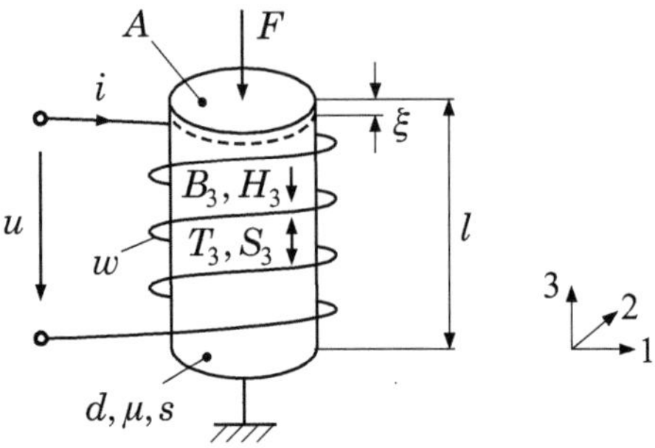

direct piezomagnetic effect:	→	sensor applications
excitation:	mechanical stress T_3 for $H_3 = 0$	strain S_3 for $H_3 = 0$
response:	$B_3 = d_{33} \cdot T_3$	$B_3 = \frac{d_{33}}{s_{33}} S_3\,,\quad e_{33} = \frac{d_{33}}{s_{33}}$
	with $B_3 = \Phi/A\,,\quad T_3 = F/A$ and $S_3 = \xi/l$ follows:	
	$\boxed{\Phi = d_{33} \cdot F}$	$\boxed{\Phi = e_{33} \frac{A}{l} \cdot \xi}$
	d, e piezomagnetic constants in x_3-direction	
reciprocal piezomagnetic effect:	→	actuator applications
excitation:	magnetic field H_3 for $T_3 = 0$	magnetic field H_3 for $S_3 = 0$
response:	$S_3 = d_{33} \cdot H_3$	$T_3 = -\frac{d_{33}}{s_{33}} H_3 = -e_{33} \cdot H_3$
	with $H_3 = w \cdot i/l\,,\quad S_3 = \xi/l$ and $T_3 = F/A$ follows:	
	$\boxed{\xi = d_{33} w i}$	$\boxed{F = e_{33} \frac{Aw}{l} i}$

Table 8.6. Properties of technically important piezomagnetic materials

iron-nickel alloy

<u>characteristics</u>: - saturation magnetostriction $S_S = f$ (alloy concentration and heat treatment)

- sign reversal of S

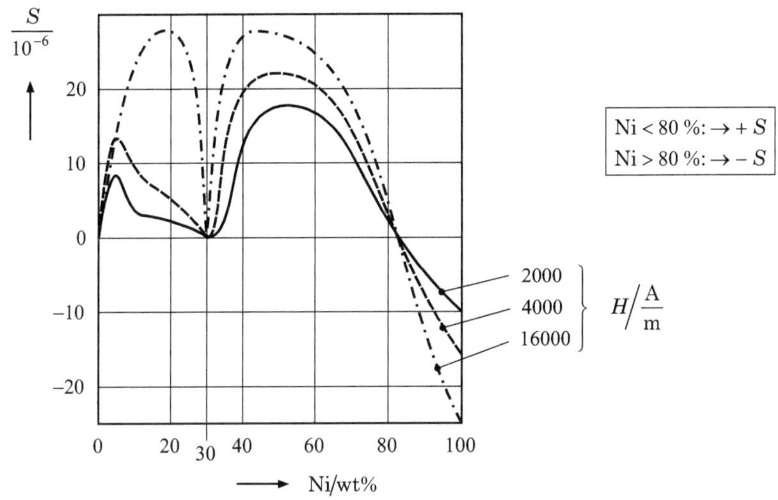

metal oxide compounds with ceramic properties: ferrites - magnetite, Ferroxcube

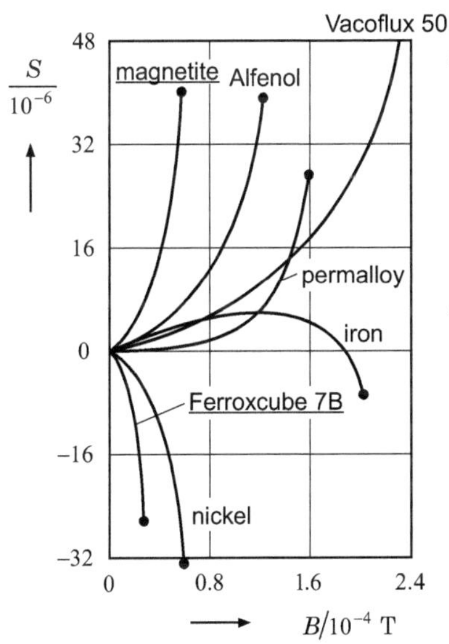

$B = \mu \cdot H$

<u>composition</u>:

$(Me)\,O \cdot Fe_2O_3$ (Me for bivalent metal ion)

e.g.: magnetite ≡ ferroferrite

$(FeO \cdot Fe_2O_3)$ semiconductor with positive magnetostriction

<u>advantages</u>:

- high resistivity up to 10^{12} Ωcm
- suitable for high frequencies due to the corresponding very low eddy-current losses
- relatively high coupling factors from 0.2 up to 0.5

Table 8.7. Properties of technically important piezomagnetic materials

	high magnetostrictive material: Terfenol-D
formula:	$Tb_{1-x}\,Dy_x\,Fe_2 \quad (x = 0.27)$ (Tb: Terbium, Fe: iron, nol: Naval Ordnance Laboratory, Clark 1975)
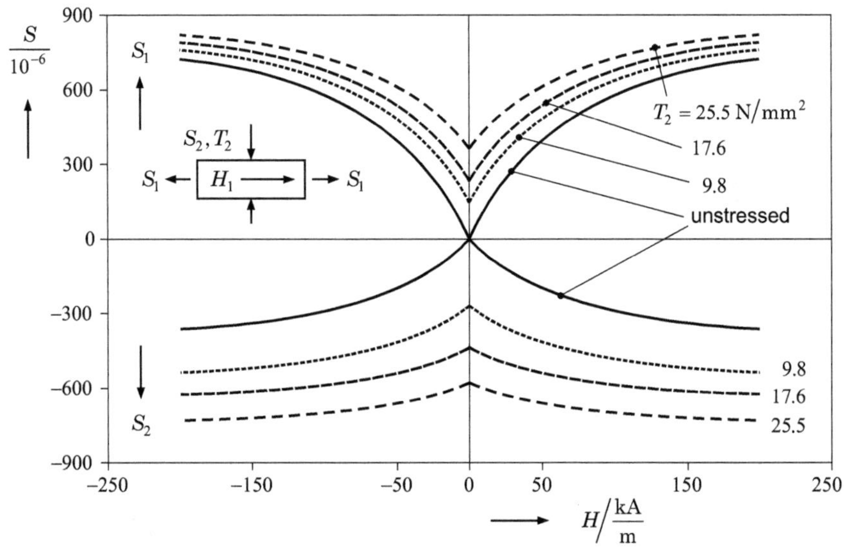	
manufacture:	- electric arc melting method, shielding gas: random crystal orientation, low magnetostriction - crystal growing process: Terfenol-D rod → nearly single-crystalline structure - or sintering process, e.g. MAGMEK 91 providing the phase equilibrium
properties:	- very high saturation magnetostriction of $(1.5\ldots 2)\cdot 10^{-3}$ at field strengths of 80 up to 200 kA/m - positive longitudinal and negative transverse magnetostriction - high operating temperature range up to 350°C - magnetostriction is influenced by mechanical longitudinal and transverse stress
attention:	- only applicable for low tensile stresses - increasing magnetostriction S with increasing compressive stress T_2 → compressive stress required optimum: $7 < T / \mathrm{N}\cdot\mathrm{mm}^{-2} < 12$ - avoiding contact with water due to risk of corrosion
application:	actuators

Table 8.8. Properties of technically important piezomagnetic materials

high magnetoelastic material: Metglas 2605 SC
source: Allied Signal Corporation, Parsippang (USA)

<u>formula</u>: $Fe_{81}B_{13.5}Si_{3.5}C_2$ (2605 SC) → (diagram)
$Fe_{78}B_{13}Si_9C_2$ (2605 S - 2)
basic material: amorphous Fe-B-Si alloy

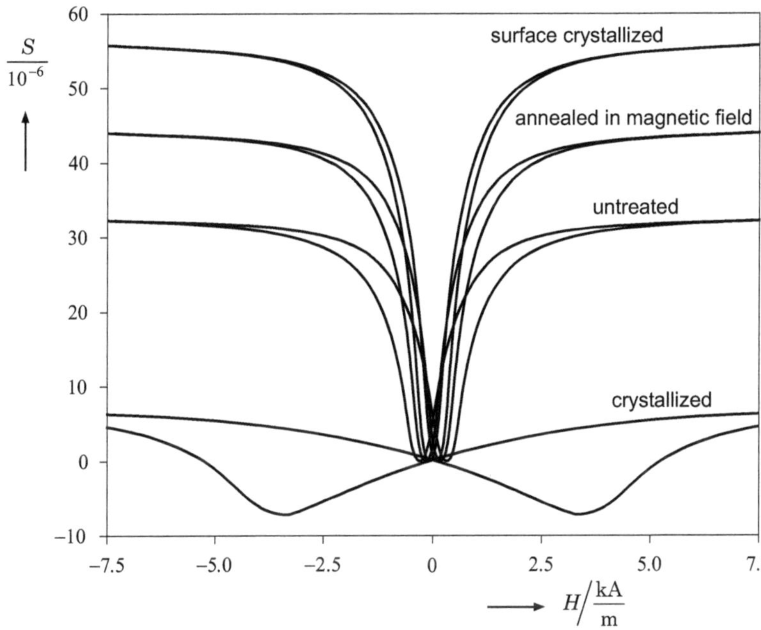

<u>manufacture</u>: - provoding the amorphous material state by extremely high cooling rate of more than 10^5 K/s

<u>properties</u>: - very low magnetostriction: $S_S \approx 3 \cdot 10^{-5}$ <u>but</u> extremely high magnetoelastic effect
- wire-shaped or ribbon-like semi-finished products

<u>application</u>: magnetoelastic strain gauge for sensors

Table 8.9. Properties of technically important piezomagnetic materials

materials characteristic values	Terfenol-D	MAGMEK 91 sintered Terfenol-D	Metglas 2605 SC amorphous Fe-B-Si alloy	Ni
saturation magnetostriction S_S	$1.5\cdot 10^{-3}$	$0.62\cdot 10^{-3}$	$30\cdot 10^{-6}$	up to $50\cdot 10^{-6}$
saturation induction B_S/T	1.0	0.71	1.61	0.61
coupling factor k	0.7…0.75	0.4	0.9…0.96	0,3
relative permeability μ_r $\mu = \mu_0 \cdot \mu_r = \mu_{33}$ $\mu_0 = 4\ \cdot 10^{-7} \mathrm{V}\cdot \mathrm{s}\cdot \mathrm{A}^{-1}\cdot \mathrm{m}^{-1}$	$\mu_r^T = 9.3$ $\mu_r^S = 4.5$	2.2…4.1	20 000 un-treated up to 300 000 annealed	
Curie temperature C	380	380	370	358
compressive strength $\mathrm{N/mm^2}$	350…700	250	1000	
tensile strength $\mathrm{N/mm^2}$	28	120 (yield strength)	1000	305
density $\mathrm{kg/m^3}$	$9.2\cdot 10^3$	$6.8\cdot 10^3$	$7.32\cdot 10^3$	$8.8\cdot 10^3$
Young's modulus $\mathrm{N/mm^2}$	$(2.5-3.5)\cdot 10^4$	$1.7-2.2\ \cdot 10^4$	$5.8-17.5\ \cdot 10^4$	$21.5\cdot 10^4$
resistivity $\Omega\cdot\mathrm{mm}$	$6.0\cdot 10^{-4}$	0.6	$13.5\cdot 10^{-4}$	$0.69\cdot 10^{-4}$
temperature coefficient of expansion 1/K	$12\cdot 10^{-6}$		$5.9\cdot 10^{-6}$	$13.3\cdot 10^{-6}$

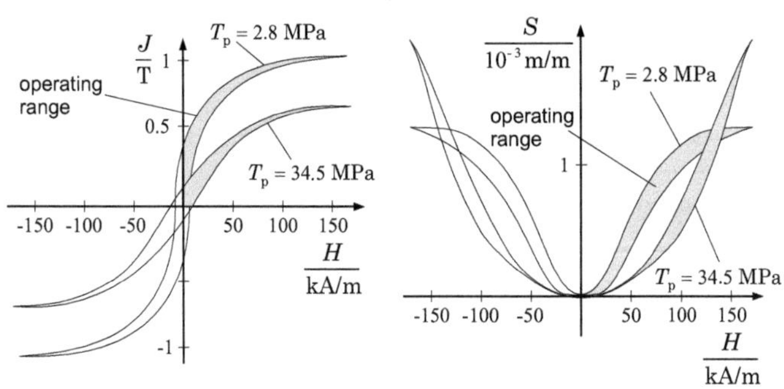

Fig. 8.34. Characteristic curve progressions of magnetic polarization J and strain S against field strength H for Terfenol-D [66]
T_p *mechanical compressive stress*

By introducing the integral quantities

$$\Phi = B_3 A, \qquad u = Aw\frac{\mathrm{d}B_3}{\mathrm{d}t}, \qquad i = \frac{l}{w}H_3,$$

$$\xi = S_3 l, \qquad v = \frac{\mathrm{d}\xi}{\mathrm{d}t}, \qquad F = -T_3 A$$

and by transforming into the frequency domain

$$\underline{u} = \mathrm{j}\omega Aw\underline{B}, \qquad \underline{v} = \mathrm{j}\omega\underline{\xi}, \qquad F \rightarrow \underline{F} \text{ and } i \rightarrow \underline{i},$$

the following two-port network equations are achieved:

$$\underline{u} = \mathrm{j}\omega Aw\underline{B} = \underline{u} = \mathrm{j}\omega\mu_{33}\frac{Aw^2}{l}\underline{i} + e_{33}\frac{Aw}{l}\underline{v}$$

$$\underline{u} = \mathrm{j}\omega L_\mathrm{b}\underline{i} + \frac{1}{X}\underline{v} = \underline{u}' + \underline{u}_\mathrm{W} \tag{8.23}$$

and

$$\underline{F} = -\underline{T}_3 A = e_{33}\frac{Aw}{l}\underline{i} - \frac{1}{\mathrm{j}\omega}\frac{c_{33}A}{l}\underline{v}$$

$$\underline{F} = \frac{1}{X}\underline{i} - \frac{1}{\mathrm{j}\omega n_\mathrm{L}}\underline{v} = \underline{F}_\mathrm{W} - \underline{F}'. \tag{8.24}$$

The application of KIRCHHOFF's laws results in the circuit diagram of the piezomagnetic transducer illustrated in Fig. 8.35.
The two-port transducer network corresponds to a transformer-like coupling and equals the network of the electromagnetic and electrodynamic transducer. The respective transformation relations have already been summarized in Table 7.4.
At the electrical side, the coil resistance R is added to the loss-free transducer. At the mechanical side of transducer, the friction r characterizing the internal losses within the piezomagnetic material is added. Its value is experimentally determined from the finite resonance rise according to $r = 1/\omega_0 nQ$.

In Table 8.10, the mechanical and piezomagnetic constants are specified for selected piezomagnetic materials.
It is also often referred to the *magnetoelastic effect* in connection with the magnetostrictive or piezomagnetic effect. This term does not characterize the interaction of magnetic and mechanical field quantities, but the dependence of the relative permeability μ_r in ferromagnetic materials on mechanical stress

$$B = \mu_\mathrm{r}(T)\,\mu_0 H \qquad \text{for} \qquad \mu_\mathrm{r}(T)\,\mu_0 H \gg d \cdot T.$$

This effect is particularly distinctive in ferromagnetic metals, such as CoSiB and the armorphous Fe-B-Si alloy *Metglas*. Compressive strain elements and

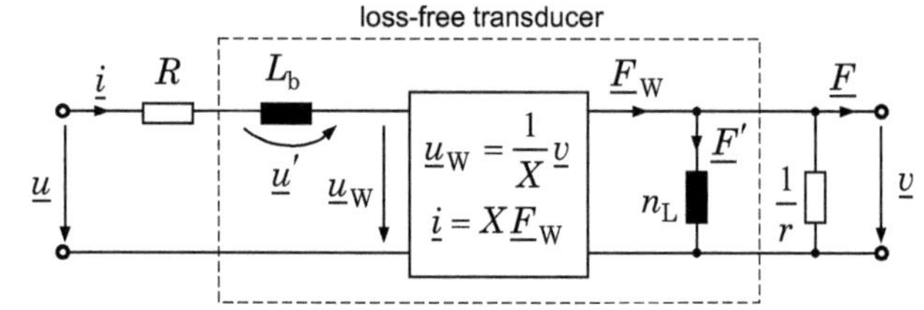

$$X = \frac{1}{e_{33}} \frac{l}{A \cdot w} = \frac{s_{33}}{d_{33}} \frac{l}{A \cdot w}, \quad s_{33} = \frac{1}{c_{33}}$$

coupling factor: $$k_{33} = e_{33} \frac{1}{\sqrt{c_{33} \cdot \mu_{33}}} = d_{33} \sqrt{\frac{c_{33}}{\mu_{33}}}, \quad \mu = \mu_0 \cdot \mu_r$$

$$L_b = \mu_{33}^S \frac{A \cdot w^2}{l}$$ (mechanically locked, v=0)

$$n_L = n = s_{33}^H \frac{l}{A} = \frac{l}{c_{33}^H \cdot A}$$ (electrical open-circuit operation)

Fig. 8.35. Circuit diagram of piezomagnetic transducer

Table 8.10. Piezomagnetic and magnetic constants of selected materials $\mu = \mu_r \mu_0$, $\mu_0 = 4\pi \cdot 10^{-7}\,\text{Vsm}^{-1}\text{A}^{-1}$, $\mu_r^T : T = 0$, $c_{33}^H : H = 0$

quantity ($B = B_{\text{rem}}$)	**Nickel** (100%)	**Alfenol** (13% Al, 87% Fe)	**Ferroxcube 7A2**	**Terfenol-D**
$d_{33}/10^{-9}\,\text{m A}^{-1}$	−1.5	−7.1	−2.5	15
μ_r^T	20	58	30...45	9.3
$c_{33}^H/10^{10}\,\text{N m}^{-2}$	ca. 20	ca. 14	15.1	2.5...3.5
k_{33}	0.14	0.25...0.31	0.15...0.20	0.65...0.75
$\rho/10^3\,\text{kg m}^{-3}$	8.8	6.5	5.35	9.25
$\vartheta_{\text{Curie}}/\,°\text{C}$	358	ca. 500	530	380

strain gauges consisting of magnetoelastic materials are industrially used in force and torque sensors [59].

For technical applications, the comparison of the properties of piezoelectric and piezomagnetic transducers is interesting. For that purpose, real limiting values

$$E = 1\,\text{kV mm}^{-1} \quad \text{and} \quad B = 0.1\,\text{T}$$

are considered with respect to beginning depolarization and possible breakdowns. The ceramic Ferroxcube 7A2 and high magnetostrictive material Terfenol-D represented in Table 8.10 are considered as an example of piezomagnetic material, the ceramic C-82 represented in Table 9.9 is considered as an example of piezoelectric material. The maximum stresses and strains result in

$$S_{\max} = s_{33} T_{\max}, \qquad T_{\max} = \frac{d_{33}}{s_{33}} \left(E_{\max}, H_{\max}\right), \qquad H_{\max} = \frac{B_{\max}}{\mu_{33}},$$

which are listed in Table 8.11.

Table 8.11. Comparison of maximum stress and strain values of selected piezomagnetic and piezoelectric materials

effect	piezomagnetic		piezoelectric
material	Ferroxcube 7A2	Terfenol-D	C-82
$T_{\max}/\,\mathrm{MPa}$	1.5	4.5	20
$S_{\max}/10^{-3}$	0.01	0.2	0.3
$w/\,\mathrm{kJ\,m^{-3}}$	2...4	14...25	1
$\rho/10^3\,\mathrm{kg\,m^{-3}}$	5.35	9.25	7.8
$\vartheta_{\mathrm{Curie}}/\,^{\circ}\mathrm{C}$	530	380	160...300

The calculated values of $S_{\max}$ correspond well in order of magnitude to the values of the saturation magnetostriction for Ferroxcube 7A2 $(0,03 \cdot 10^{-3})$ and Terfenol-D $(1,5 \cdot 10^{-4})$ as well as to the value of the saturation electrostriction for C-82 $(2 \cdot 10^{-4})$ specified in data sheets.
By applying the high magnetostrictive material Terfenol-D, similar maximum stresses and strains can be achieved as with piezoelectric ceramics. Also coupling factors of about 0.65 up to 0.75 as well as response times in the microsecond up to the millisecond range are comparable.
With the material selection for power transducers – actuators and ultrasonic transducers – the following distinctive features compared to piezoelectric materials should be considered:

- Considerably greater variety of piezoelectric ceramics. In addition to the longitudinal effect, the transversal and shear effect are used as well.
- Concerning piezoelectric transducers, the electric field is generated by electrodes directly placed on the ceramic. In contrast, the required space of the excitation coil or permanent magnet in order to generate the magnetic field is significantly larger.

- Piezoelectric transducers require very high electrical driving voltages in the range of 100 V up to 1 kV. However, they are able to maintain their static deflection almost without any electrical energy supply. Piezomagnetic transducers are operated with current drive in the range of 2 - 20 A. In order to maintain a constant deflection, a persistent magnetic bias by a direct current or permanent magnet is required.
- Piezoelectric oscillators show high resonant frequencies in the range of 100 kHz up to several MHz. Due to increasing eddy-current losses, the resonant frequency of metallic piezomagnetic oscillators with 10 - 100 kHz are significantly lower.
- The CURIE temperature of 380 °C for Terfenol-D is higher than the one of 160 - 300 °C for piezoelectric ceramics. This results in a higher operating temperature range for Terfenol-D. In addition, in contrast to piezoelectric ceramics, after exceeding the CURIE temperature a reconstitution of the magnetostrictive properties will occur, when the temperature falls again below the CURIE point.
- Due to the low tensile strength of Terfenol-D, for actuator applications a mechanical prestress is required. This also results in an increase of the saturation magnetostriction at the same time.

Due to the specified characteristics, the great variety and completely assembled types, piezoelectric transducers dominate as actuators and sensors at present. In contrast, high magnetostrictive and high magnetoelastic materials are offered only by a few suppliers at much higher prices.

8.3.2 Sample Applications

In the following, the static design of a magnetostrictive actuator and a magnetostrictive travelling wave motor is explained. The dynamic design is described using the example of an ultrasonic sonar working according to the sonar method (<u>So</u>und <u>Na</u>vigation and <u>R</u>anging). Here, the linearization of characteristics is realized by means of operating point adjustment using permanent magnets. Thus, the circuit of the piezomagnetic transducer shown in Fig. 8.35 is valid.

Magnetostrictive Actuator

The actuator represented in Fig. 8.36 is operated by a Terfenol rod which is biased by permanent magnets and mechanically prestressed by screws. The flux guide is provided by two soft magnetic pole shoes.
For the approximate calculation of mechanical quantities, the equation of state

$$S_3 = s_{33}T_3 + d_{33}H_3$$

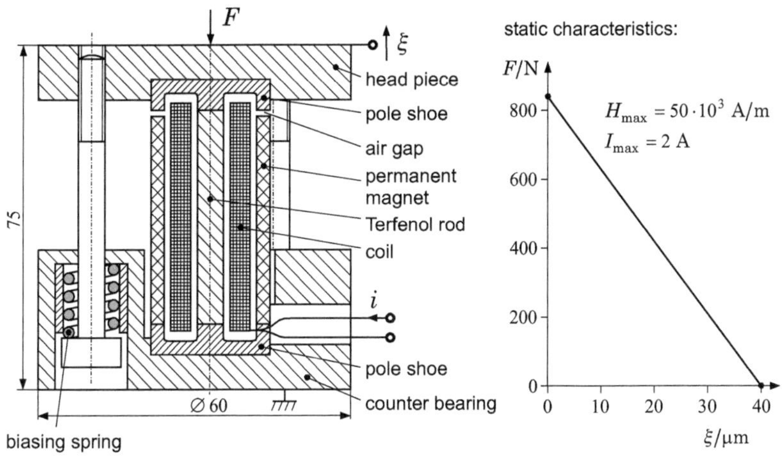

Fig. 8.36. Magnetostrictive transducer with static characteristics
Edge Technologies, Ames USA

is used for the mechanically locked case ($S_3 = 0 \rightarrow T_{\max}$) and mechanical open-circuit operation ($T_3 = 0 \rightarrow S_{\max}$). In combination with the constants specified in Table 8.10 and $H_{\max} = 5 \cdot 10^4 \,\mathrm{A\,m^{-1}}$, following calculation results are achieved:

$$T_{\max} = -\frac{d_{33}}{s_{33}} H_{\max} = 26 \cdot 10^6 \,\mathrm{N\,m^{-2}}$$

and

$$S_{\max} = s_{33} T_{\max} = 7.4 \cdot 10^{-4}.$$

Taking the dimensions of the Terfenol rod ($\varnothing$ 6.4 × 50) mm into consideration, it finally follows:

$$F_{\max} = 830\,\mathrm{N} \qquad \text{and} \qquad \xi_{\max} = 40\,\mu\mathrm{m}$$

Magnetostrictive Traveling Wave Motor (Inchworm Motor)

The functional principle of the magnetostrictive traveling wave motor is shown in Fig. 8.37. The discontinuous linear actuator uses the magnetostrictive effect caused by longitudinal and transverse extension of the Terfenol rod. The Terfenol rod expands in a magnetic field and narrows at the same time because of the approximately volume invariant magnetostrictive effect.
Based on this invariance of volume and in combination with the rod's volume $V = 1/4\left(\pi d^2 l\right)$, the magnetostrictive transverse strain $S_1 = \Delta l/l$ can be estimated in combination with the longitudinal strain $S_3 = \Delta l/l$ according to

$$V + \Delta V = \frac{1}{4}\pi \left(d + \Delta d\right)^2 \left(l + \Delta l\right)$$

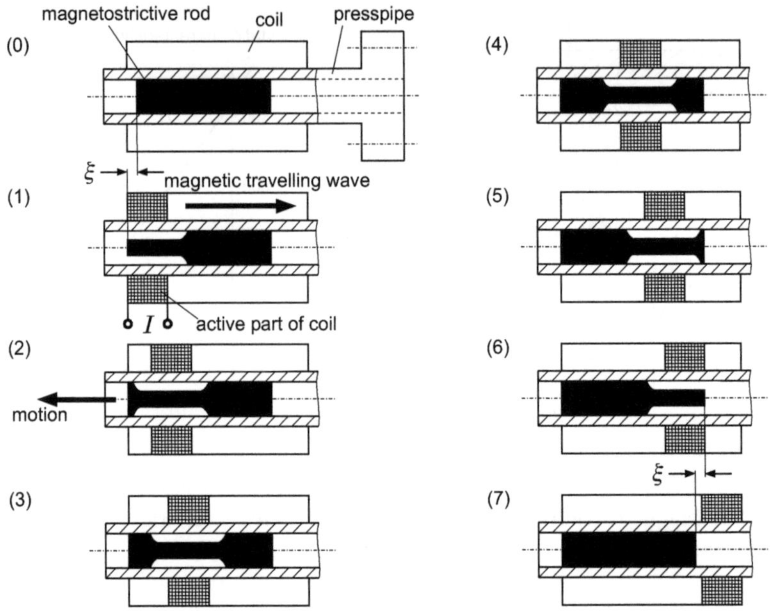

Fig. 8.37. Functional principle of magnetostrictive traveling wave motor „*Kiesewetter Motor*" during one iteration cycle [70]

and thus

$$\frac{\Delta V}{V} \approx 2\frac{\Delta d}{d} + \frac{\Delta l}{l} = 2S_1 + S_3 = 0 \quad \text{for} \quad \left(\frac{\Delta d}{d}\right)^2, \frac{\Delta d \Delta l}{dl} \ll 1.$$

The Terfenol rod is arranged within a press pipe with external coils. The press fit is compensated by cross-section diminution of the Terfenol rod in the area of the magnetic field. At the same time, the rod is able to expand in this area oppositely to the travelling direction of the magnetic field. This procedure is repeated for the particular control of the coil packages, so that an extensional wave propagates in the Terfenol rod oppositely to the control direction of the coil packages. After controling of all coil packages, the Terfenol rod moves with step length

$$\xi = d_{33} H_3 \Delta l,$$

where Δl denotes the length of a single magnetic field within the rod. Then the coil control happens again and the rod steps oppositely to coil control direction. At room temperature a step size of 50 μm is achieved with a functional model consisting of 6 coils and a Terfenol rod with a diameter of 20 mm and a length of 100 mm. With a step frequency of 30 Hz, the linear velocity amounts to 1.5 mm s^{-1} [71].

However, maintaining the required surface quality of press pipe and rod better than $2\,\mu$m is problematic. In addition, a thermally induced difference in diameter is found caused by different temperature coefficients of expansion of rod and press pipe. With higher room temperature this difference results in a decrease of step size. Thus, at 20 °C the step size amounts to $\xi = 50\,\mu$m, however, at 80 °C the step size amounts only to $\xi = 20\,\mu$m.

Piezomagnetic Ultrasonic Transmitter

Figure 8.38 shows the design principle of a piezomagnetic ultrasonic transmitter. The ultrasonic transmitter is used for detection of stationary or moving objects under the surface of sea according to the acoustic sonar method. Depending on the desired resolution and range, ultrasonic pulses at carrier frequencies from 20 kHz up to several 100 kHz are emitted. The travelling time of reflected pulses specifies the distance, the change of carrier frequency specifies the velocity of the object to be localized.

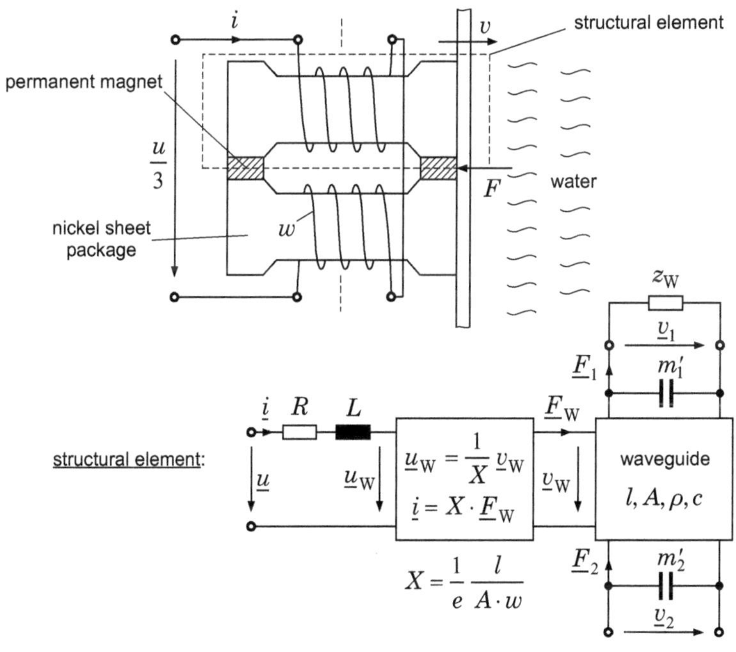

Fig. 8.38. Piezomagnetic ultrasonic transmitters for sonar devices and circuit diagram of a structural element

The ultrasonic transmitter is operated as $\lambda/2$-oscillator, i.e. the vibration node is located in the center of the element, the antinodes are located at the

oscillator's ends. Thus, due to its inertia amplified by resonance the left half of oscillator provides the fixed clamping point for the right emitting half of oscillator.
The complete ultrasonic oscillator consists of six structural elements which are electrically operated in series connection and mechanically operated in parallel connection. The structural elements are represented with dashed lines in the upper part of Fig. 8.38. The circuit diagram of the structure element is shown in the lower part of Fig. 8.38. It is based on the circuit diagram shown in Fig. 8.35, but with the difference that the bar surrounded by windings is considered as a waveguide. Since the flux in the end plates but also the piezomagnetic deformation due to $d_{31} \approx -\frac{1}{2}d_{33}$ are considerably smaller than in the bar, the end plates are only taken into account as discrete masses m_1' and m_2'. The radiation is provided by the plate with mass m_1'. At the mechanical side, the characteristic impedance of water $Z_{\mathrm{a,W}} = \rho_{\mathrm{W}} c_{\mathrm{W}} 1/A_{\mathrm{S}}$ is considered as friction $z_{\mathrm{a,W}} = \rho_{\mathrm{W}} c_{\mathrm{W}} A_{\mathrm{S}}$. The permanent magnets which are integrated in the structural elements help for the polarization of the transducer.

For further considerations, the oscillator units surrounded by the coils are considered to be waveguides. The Π-circuit shown in Fig. 6.5 is used for their specification.
As a rough approximation, the components $\underline{h}_1$ and $\underline{h}_2$ are considered for low frequencies, thus $\underline{h}_1 = m/2$ and $\underline{h}_2 = n$ apply. Taking the end pieces m_1' and m_2' into account, the circuit diagram illustrated in Fig. 8.39 is achieved.

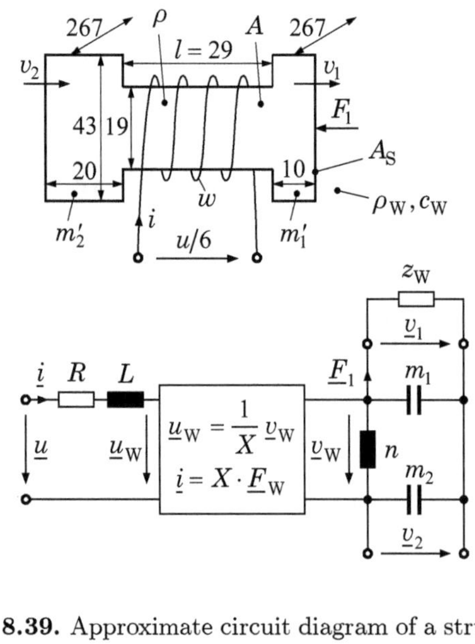

$\rho = 8.8 \cdot 10^3\ \mathrm{kg \cdot m^{-3}}$

$\rho_{\mathrm{W}} = 10^3\ \mathrm{kg \cdot m^{-3}}$

$c_{\mathrm{W}} = 1440\ \mathrm{m \cdot s^{-1}}$

$X = \frac{s}{d} \frac{l}{A \cdot w}, \quad n = s\frac{l}{A}$

$z_{\mathrm{W}} = \rho_{\mathrm{W}} \cdot c_{\mathrm{W}} \cdot A_{\mathrm{S}}$

$L_{\mathrm{b}} = \mu \frac{A \cdot w^2}{l}$

$m_1 = \rho/2 \cdot A \cdot l + m_1'$

$m_2 = \rho/2 \cdot A \cdot l + m_2'$

Fig. 8.39. Approximate circuit diagram of a structural element of ultrasonic transmitter

In combination with the dimensions of the structural elements shown in Fig. 8.39 and the constants for nickel listed in Table 8.10, the following component values are achieved:

$$\left.\begin{aligned} m_1' &= \rho \cdot 1\,\mathrm{cm} \cdot 4.3\,\mathrm{cm} \cdot 26.7\,\mathrm{cm} = 1.01\,\mathrm{kg} \\ m_2' &= \rho \cdot 2\,\mathrm{cm} \cdot 4.3\,\mathrm{cm} \cdot 26.7\,\mathrm{cm} = 2.02\,\mathrm{kg} \\ m &= \rho \cdot 2.9\,\mathrm{cm} \cdot 1.9\,\mathrm{cm} \cdot 26.7\,\mathrm{cm} = 1.30\,\mathrm{kg} \end{aligned}\right\} \begin{aligned} m_1 &= 1.66\,\mathrm{kg} \\ m_2 &= 2.67\,\mathrm{kg} \end{aligned}$$

$$n = \frac{sl}{A} = 2.8 \cdot 10^{-11}\,\mathrm{m\,N^{-1}}, \qquad z_\mathrm{W} = \rho_\mathrm{W} c_\mathrm{W} A_\mathrm{S} = 1.65 \cdot 10^4\,\mathrm{N\,s\,m^{-1}}$$

$$L_\mathrm{b} = \mu_0 \mu_\mathrm{r} A \frac{w^2}{l} = 0.16\,\mathrm{mH} \quad \text{for} \quad w = 6$$

The total inductance L results from the parallel connection of the inductance L_0 of the magnetic circuit parts being not involved in the conversion process and L_b. With

$$L_0 = \frac{2\mu_0 w^2 A_0}{l_0}$$

as well as with $l_0 = 6\,\mathrm{mm}$ and $A = 5.34 \cdot 10^3\,\mathrm{mm}^2$, the total inductance results in

$$L = \frac{L_\mathrm{b} L_0}{L_\mathrm{b} + L_0} = 0.04\,\mathrm{mH}, \qquad L_0 = 0.052\,\mathrm{mH}.$$

The resonant frequency of the approximate circuit diagram represented in Fig. 8.39 amounts to

$$\omega_0' = \frac{1}{\sqrt{\dfrac{m_1 m_2}{m_1 + m_2} n}} = 1.88 \cdot 10^5\,\mathrm{s}^{-1}, \qquad f_0' = 30\,\mathrm{kHz}.$$

The exact value of the resonant frequency which results from the two-port network equations of the loss-free extensional waveguide shown in Table 6.1, amounts to $f_0 = 33\,\mathrm{kHz}$.

Up till now, the oscillating bar was considered as a loss-free spring. In fact, internal losses occur in the nickel sheet package. Therefore, the friction

$$z_m = \frac{1}{\omega_0 n Q_\mathrm{L}}$$

must be connected in parallel with the compliance n. In combination with the measured value of $Q_\mathrm{L} = 89$ denoting the quality factor in air, the friction amounts to $r_\mathrm{m} = 1.94 \cdot 10^3\,\mathrm{N\,s\,m^{-1}}$. The experimentally determined value of the total quality factor of the ultrasonic transmitter in case of water coupling amounts to $Q = 17.5$. For a total voltage of $\tilde{u} = 716\,\mathrm{V}$ and a current $\tilde{\imath} = 9.5\,\mathrm{A}$ the radiated acoustic power amounts to 5 kW. This results in an efficiency of the ultrasonic transmitter of

$$\eta = \frac{P_\mathrm{a}}{P_\mathrm{el}} = 0.74.$$

8.3.3 Piezomagnetic Unimorph Bending Elements

Rotational Piezomagnetic Bending Model

A two-layer beam consisting of a magnetostrictive layer and a non-magnetic carrier or support beam as shown in Fig. 8.40 reacts with bending to an appropriate directed magnetic field with flux density B. It is called a *magnetic unimorph* since only one layer is active. In terms of temperature sensitivity of the two-layer beam it is a bimorph but this will be neglected further on. In Sect. 9.2.5, a piezoelectric bimorph is considered which consists of two bodies of same geometry and materials, but opposite polarization directions. Therefore, the neutral plane is located in the lamination area and the internal stress is nearly linear symmetric as shown in Fig. 5.6. This is not the case for the magnetostrictive unimorph. Here, the location of the neutral layer c depends on YOUNG's modulus and thickness of the beam layers. With known c which is given in Fig. 8.40, the bending moment M_{bx} can be calculated. It is a function of the average bending stiffness $\overline{EI}$ of the unimorph, the deflection angle variation $\mathrm{d}\varphi/\mathrm{d}x$ and the magnetostrictive generated moment M_0:

$$M_{bx} = -\overline{EI}\frac{\mathrm{d}\varphi}{\mathrm{d}x} + M_0$$

with

$$\overline{EI} = \frac{w}{12}\frac{E_1^2h_1^4 + E_2^2h_2^4 + E_1E_2h_1h_2\left(4h_1^2 + 6h_1h_2 + 4h_2^2\right)}{E_1h_1 + E_2h_2} \tag{8.25}$$

and

$$M_0 = \frac{w}{2}\frac{E_1E_2h_1h_2\left(h_1 + h_2\right)}{E_1h_1 + E_2h_2}\left(\alpha_1 - \alpha_2\right)B. \tag{8.26}$$

In (8.25) and (8.26), the quantity w denotes the beam width, the quantity E_2 denotes the YOUNG's modulus of carrier, the quantity $E_1 = E_{33}^B$ denotes the YOUNG's modulus of the magnetostrictive material, the quantities $\alpha_1 = d_{33}/\mu_{33}^T$ and $\alpha_2 = 0$ denote the translational transduction coefficient in the magnetostrictive layer and in the carrier, respectively. The rotational transduction coefficient $K_{\mathrm{mag,r}}$ can be derived from the actuation relation $M_0 = K_{\mathrm{mag,r}} \cdot \Phi = K_{\mathrm{mag,r}} \cdot B \cdot A$ between moment and magnetic flux. Due to a high permeability, the magnetic flux is concentrated in the magnetostrictive layer with cross-sectional area $A = w \cdot h_1$. This cancels the two geometrical quantities in the numerator. The transduction coefficient must be the same for the sensing relation $V_m = K_{\mathrm{mag,r}} \cdot \varphi$ between magnetic voltage $V_m = H \cdot x$ and deflection angle φ. It is obtained when the first complex equation of state specified in Fig. 8.40 is transposed to

$$H_3 = \frac{1}{\mu_{33}^T} \cdot B_3 - \alpha_1 \cdot T_3 = \frac{1}{\mu_{33}^S} \cdot B_3 - \alpha_1 E_{33}^B \cdot S_3 \tag{8.27}$$

and the distance $y = c + h_1/2$ in the middle of the magnetostrictive layer is the center of the effecting elongation, such that $S = y \cdot \mathrm{d}\varphi/\mathrm{d}x$ can be substituted:

$$H_3 = \frac{1}{\mu_{33}^S} \cdot B_3 - \underbrace{\alpha_1 E_{33}^B \left(c + \frac{h_1}{2}\right)}_{K_{\mathrm{mag,r}}} \cdot \frac{\mathrm{d}\varphi}{\mathrm{d}x}. \tag{8.28}$$

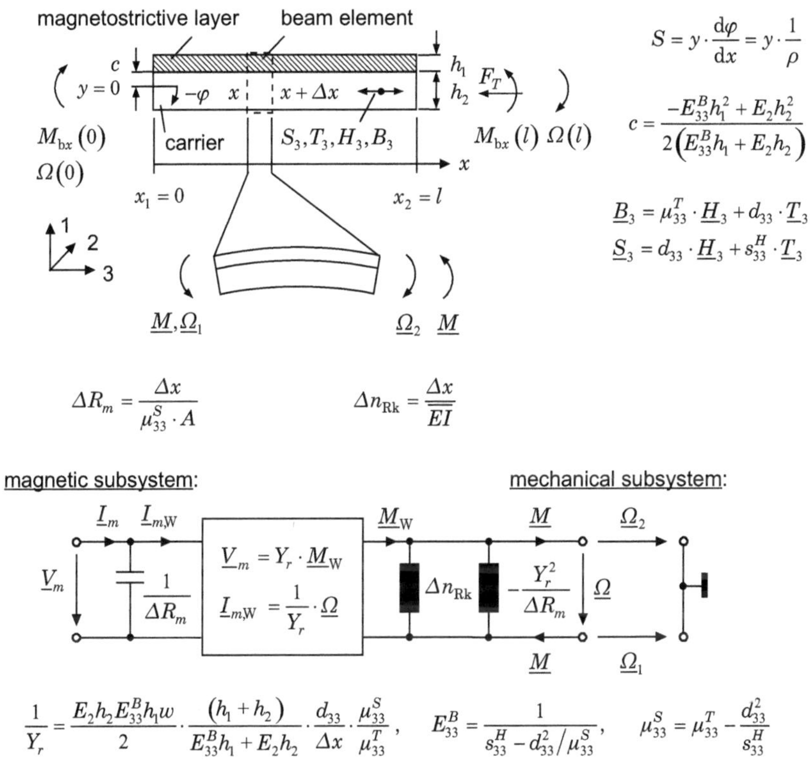

Fig. 8.40. Circuit of a piezomagnetic bending element

This assumption can be made when the bimorph is long and thin – to be conform with the BERNOULLI hypothesis – and the magnetostrictive layer is thin compared to the carrier thickness.

As second magnetic coordinate besides the magnetic voltage the magnetic flux rate $I_m = \mathrm{d}\Phi/\mathrm{d}t$ as time derivative of the magnetic flux is introduced. This is necessary since the product of network coordinates has to constitute a power which is not the case with the flux. It can be considered as analogon to the electrical current $i = \mathrm{d}Q/\mathrm{d}t$ which is the time derivative of charge Q. In combination with flux rate I_m, rotational compliance n_{Rk}, magnetic resistance

$R_m = \Delta x/(\mu_{33}^S \cdot A)$, and angular velocity $\underline{\Omega} = \underline{\varphi} \cdot \mathrm{j}\omega$, the following expression is achieved in the frequency domain:

$$\begin{pmatrix} \underline{M} \\ \underline{V}_m \end{pmatrix} = \begin{pmatrix} -1/\mathrm{j}\omega n_{\mathrm{Rk}} & K_{\mathrm{mag,r}}/\mathrm{j}\omega \\ -K_{\mathrm{mag,r}}/\mathrm{j}\omega & R_m/\mathrm{j}\omega \end{pmatrix} \cdot \begin{pmatrix} \underline{\Omega} \\ \underline{I}_m \end{pmatrix}. \tag{8.29}$$

In this model the magnetic conductance relates the magnetic voltage to the integrated flux rate. Due to isomorphism to the capacitance, the magnetic conductance is represented by a capacitance symbol. In (8.29) the transduction coefficient is imaginary and the transducer is a transformer. A real transduction coefficient can be obtained when (8.29) is rearranged to $(\underline{I}_m, \underline{M}) = f(\underline{\Omega}, \underline{V}_m)$. In combination with $1/Y_r = -K_{\mathrm{mag,r}}/R_m$, following relations are achieved:

$$\underline{I}_m = \frac{1}{Y_r}\underline{\Omega} + \frac{\mathrm{j}\omega}{R_m}\underline{V}_m \tag{8.30}$$

$$\underline{M} = -\left(-\frac{R_m}{\mathrm{j}\omega Y_r^2} + \frac{1}{\mathrm{j}\omega n_{\mathrm{Rk}}}\right)\underline{\Omega} + \frac{1}{Y_r}\underline{V}_m. \tag{8.31}$$

The Eqs. (8.30) and (8.31) form the basis for the derivation of the electromechanical network illustrated in Fig. 8.40. Here, the transducer is now a gyrator.

Cylindric Coil as Electromagnetic Transducer

A long and thin cylindric coil or solenoid (radius $r \ll l$) is introduced in Sect. 8.3.1 as ideal electromagnetic transducer where the magnetic field is concentrated in the coil and outside almost zero. FARADAY's law $\underline{u} = \mathrm{j}\omega \cdot A \cdot N \cdot \underline{B}$ and AMPÈRE's circuital law $\underline{H} = N/l \cdot \underline{i}$ yield its electromagnetic transduction coefficient $Y_{mS} = N$. The material equation $B = \mu \cdot H$ or $\underline{\Phi} = \mu \cdot A/l \cdot \underline{V}_m$ can be included as magnetic resistance $R_m = l/(\mu A)$ in the magnetic domain or as inductance $L = \mu \cdot A/l \cdot N^2$ in the electric domain and as depicted in Fig. 8.41 a) and b), respectively. It is transformed by N^2, the quadratic transduction coefficient.

These circuits show the ideal solenoid transducer model. It does not include the demagnetization field by which a magnetic material reacts to an applied magnetic field H_{app}. The demagnetization field is directed in opposite to the magnetization produced by the free poles which appear on its ends. Furthermore, it is proportional to the pole strength of the magnetized material and is a function of geometry. For a uniformly magnetized material it is defined by (8.32) where the quantities N_d and χ denote the demagnetization factor and the magnetic susceptibility, respectively [72]. The demagnetization phenomenon causes a lower magnetic field strength in the material H_{in} and can be described by a magnetic voltage divider which is shown in Fig. 8.41 c) with $R_m = R'_m(N_d\chi + 1)$.

$$\frac{H_{\mathrm{in}}}{H_{\mathrm{app}}} \cdot \frac{l}{l} = \frac{V_{m,\mathrm{in}}}{V_{m,\mathrm{app}}} = \frac{1}{1 + N_d\chi} = \frac{R'_m}{R'_m + R'_m N_d \chi} \tag{8.32}$$

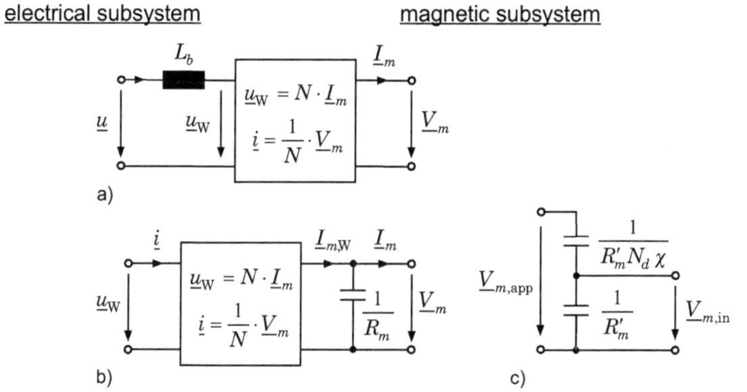

Fig. 8.41. Solenoid electromagnetic transducer models

Planar Coil as Electromagnetic Transducer

In miniaturized actuators and sensors magnetic fields are often generated or sensed by planar coils. The placement of a magnetic material of infinite thickness beside the planar turns has much less influence on the coil inductance than a magnetic core inside a solenoid coil. The solenoid inductance depends linearly on the permeability of the core material while the inductance L_∞ of a planar coil can be at the maximum twice the air coil inductance L_0 by a permeable layer underneath the turns [73]:

$$L_\infty = \frac{2\mu_r}{\mu_r + 1} L_0 \tag{8.33}$$

A simple electromagnetic model can be found when the influence of the magnetic layer on the H-field is analyzed. One possibility is to replace this layer by image turns in which the current has the same direction as the current in the real turns but the strength $(\mu_r - 1)i/(\mu_r + 1)$ [73]. The magnetic field strength in the air above the coil is the sum of the contributions of all turns. Therefore, the H-field is nearly doubled at the maximum for a small distance between both coils as illustrated in Fig. 8.42.
AMPÈRE's law relates the magnetic voltages in air and magnetic layer to its electrical current source:

$$\oint H \, \mathrm{d}s = N \cdot i = V_{m,\mathrm{Air}} + V_{m,\mathrm{m}} \tag{8.34}$$

Equation (8.34) explains the much lower field strength in the magnetic layer compared to that in air. It can be interpreted as magnetic voltage divider as shown in Fig. 8.43 a). Figure 8.43 b) and c) demonstrate the transformation of the related magnetic resistances into the electrical network domain where they result in two inductances.

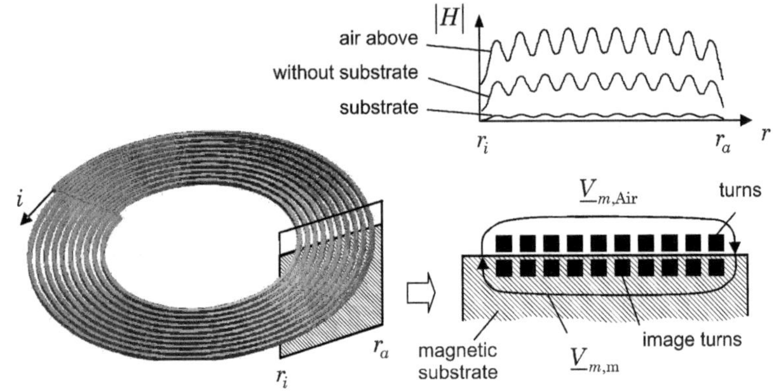

Fig. 8.42. Magnetic field strength and magnetic voltages around a planar coil on a thick magnetic substrate

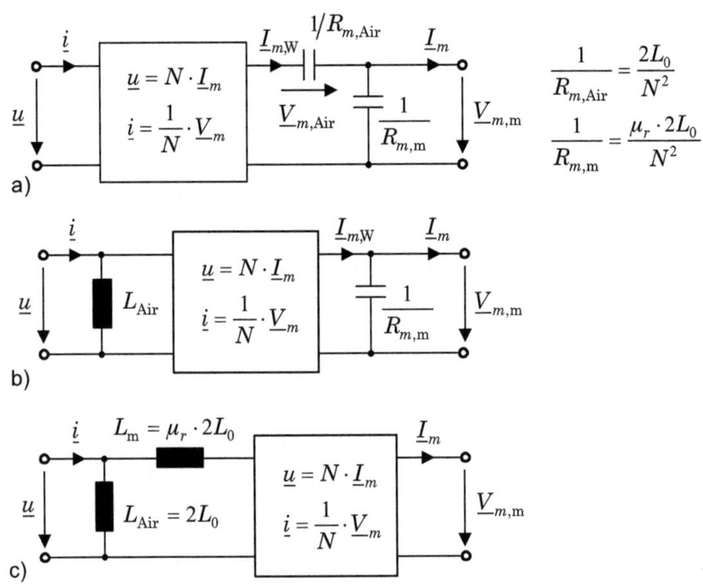

Fig. 8.43. Electromagnetic model of a planar coil on a thick magnetic substrate

The gyrator transforms an open magnetic circuit to an electrical short-circuit which connects L_{m} in parallel to L_{Air}. For $\mu_r = 1$ the magnetic layer vanishes and the planar coil inductance equals its minimum L_0. That means that both inductances must have the value $2L_0$.

Comparing this parallel connection with (8.33) yields

$$L = \frac{L_\mathrm{m} \cdot 2L_0}{L_\mathrm{m} + 2L_0} = \frac{\frac{L_\mathrm{m}}{2L_0}}{\left(\frac{L_\mathrm{m}}{2L_0} + 1\right)} 2L_0 = \frac{\mu_r}{\mu_r + 1} 2L_0 \tag{8.35}$$

and thus

$$L_\mathrm{m} = \mu_r \cdot 2L_0. \tag{8.36}$$

By using this model, both the magnetic field strength in the permeable layer and the inductance can be related to the relative permeability. Already a low relative permeability of 20 results in a 20 times lower magnetic resistance and thus voltage $V_{m,\mathrm{m}}$ and with $1.9 \cdot L_0$ almost doubled inductance.

In reality the magnetic layer has a finite thickness and the inductance L reaches only a value between L_0 and $2L_0$. That can be viewed as correction by means of L_C [74]:

$$L = L_\infty - L_\mathrm{C} \tag{8.37}$$

A thin magnetic layer of finite thickness does not concentrate the flux completely: a part is leaving the layer and enters it again. The magnetic resistances of layer $R_{m,\mathrm{m}}$ and the air below layer $R_{m,\mathrm{Ab}}$ are connected in parallel and thus experience the same magnetic voltage as depicted in Fig. 8.44. Due to the alomst identical geometrical dimensions, the magnetic resistances above the turns $R_{m,\mathrm{Aa}}$ and below are nearly equal. This circuit defines the ratio between the H-field in air and in layer. The magnetic resistances can be determined by a transformation of the resistances into the electrical domain. For a vanishing magnetic layer its magnetic resistance would be infinite large

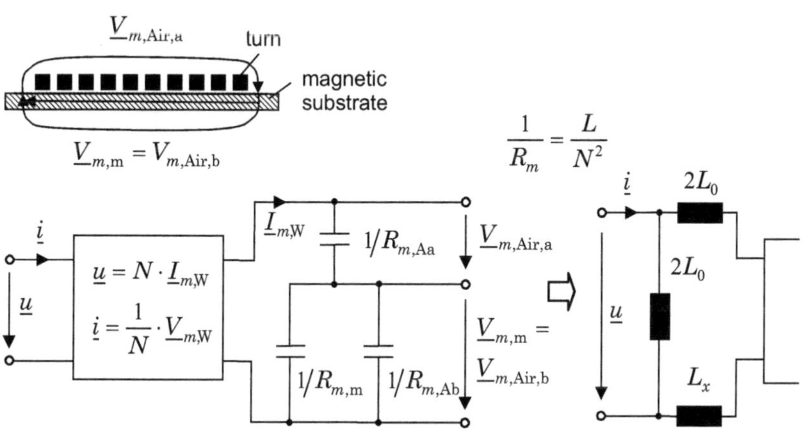

Fig. 8.44. Electromagnetic model of a planar coil on a thin magnetic substrate

and thus the related inductance would be zero. An open magnetic circuit is a short-circuit in the electrical domain. This results in two coils connected in parallel to each other with a total inductance equaling the air coil inductance. An existing magnetic layer increases the inductance. Based on this measured inductance and measured air coil inductance, the unknown inductance L_x can be calculated and thus $R_{m,\mathrm{m}}$ can be determined by means of reverse transformation. The magnetic circuit defines the ratio between the magnetic voltages and thus the magnetic field strength. The total magnetic voltage is given by the transducer relation. For the given example, the magnetic field strength in a layer with $\mu_r = 20$ is only 1.25 times smaller than the magnetic field strength in air directly above the turns.

Electromechanical Transducer Model

If the unimorph is surrounded by a solenoid and the small demagnetization factor is neglected then the following equations will be obtained from (8.29) and the solenoid transducer factor N:

$$\underline{u} = \frac{N}{Y_r}\underline{\Omega} + \mathrm{j}\omega \underbrace{\frac{N^2}{R_m}}_{L_b} \cdot \underline{i} \tag{8.38}$$

$$\underline{M} = -\left(-\frac{R_m}{\mathrm{j}\omega Y_r^2} + \frac{1}{\mathrm{j}\omega n_{\mathrm{Rk}}}\right)\underline{\Omega} + \frac{N}{Y_r}\underline{i} \tag{8.39}$$

The circuit is shown in Fig. 8.45. It should be noted that both the inductance and μ_{33}^S in this model are constant.

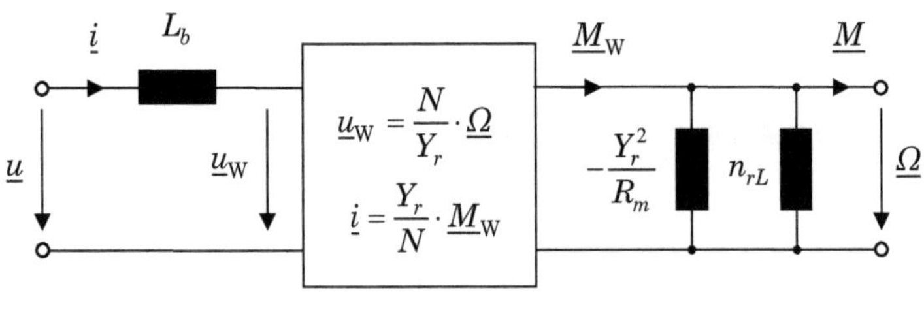

Fig. 8.45. Electromagnetic model of a magnetostrictive bimorph core in a solenoid coil

8.3.4 Example of a Parametric Magnetoelastic Bending Sensor

In Sect. 8.3.3, the permeability of the piezomagnetic material was assumed to be constant in the operating point. In reality the permeability changes

slightly with mechanical load. In certain cases this effect can be applied technically, e.g. when the mechanical quantities are quasi-static. Such transducers are called to be *parametric*. Since the mechanical feedback is not considered, one speaks of a magnetoelastic effect. A changing permeability of a magnetoelastic material can be detected as change of resonant frequency when the magnetoelastic material is part of a coil and this coil is connected to a capacitance. The wireless determination of resonant frequency is possible both with time and frequency domain measurement techniques [75]. Both techniques use inductive coupling between sensor and an additional measurement coil.
The magnetoelastic sensor using a Galfenol alloy can be fabricated in thin film technology. In [76], a technology route is described which allows a wide range of geometrical and material variations. A large number of small size sensors can be manufactured simultaneously. Figure 8.46 shows the top view of a resonant bending sensor and the cross-section including two turns. The sensitive Galfenol layer can be deposited onto an isolated silicon substrate and seed layer by DC magnetron sputtering. During cooling Galfenol contracts such that the silicon substrate acts as bending spring. For the manufactured sensors the average intrinsic tensile stress of Galfenol layers amounts to 400 MPa. Due to the planar coil geometry, the mechanical stress T and the magnetic field strength H are mainly oriented perpendicular to each other in the rectangular sensor plane. At both ends of the coil the mechanical and magnetic coordinates experience the same orientation.
The saturation magnetostriction is extremly sensitive to the Gallium content in the $Fe_{1-x}Ga_x$ ($10 \leq x \leq 30$) alloy. Distinct peaks of about 400 ppm are at 19% Ga and 29% Ga and a valley near 24% Ga for single crystal Galfenol [77]. The film composition for the above mentioned sensors is $Fe_{83}Ga_{17}$ and the thickness $3 \times 1\,\mu$m with an insolation layer in between to lower eddy currents. The copper coil on top of the insolated Galfenol is manufactured semiadditively by electroplating. This technology typically allows the same winding heigth and width but a lower heigth requires a lower effort. The manufactured sensor coil has 22 turns of $20\,\mu$m width and $10\,\mu$m height.

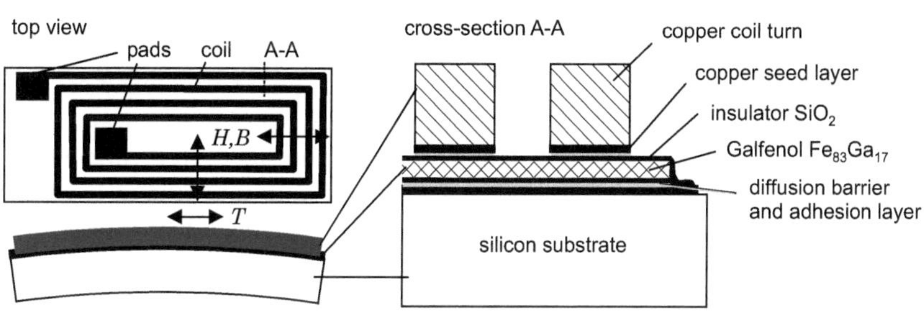

Fig. 8.46. Resonant bending sensor

Measurements at a frequency of 100 kHz showed that the $3 \times 1\,\mu$m Galfenol layers increased the inductance from $L_0 = 6.8\,\mu$H to $L = 7.5\,\mu$H. If these values are compared with the calculations of Rehfuß and co-workers [78], the average relative permeability of the Galfenol layer is determined to about 20. Using the transformation shown in Fig. 8.44 the inductances $L_x = 3.12\,\mu$H and $2L_0 = 13.6\,\mu$H are found and from reverse transformation $R_{m,\mathrm{Aa}} = R_{m,\mathrm{Ab}} =$ 35 MA/Wb and $R_{m,\mathrm{m}} = 156$ MA/Wb. This gives the ratio of the magnetic voltages $\underline{V}_{m,\mathrm{Air,a}} : \underline{V}_{m,\mathrm{Air,m}} = 1.25 : 1$. The small thickness of the magnetoelastic layer and its modarate permeability result in a magnetic field strength which is not much lower than the field strength in air close to the coil.

Figure 8.47 shows an experimental setup, where the sensor was glued on a $w \times t \times h = (10 \times 2 \times 100)\,\mathrm{mm}^3$ titanium beam and bent by an external force applied at one beam end. The other end of the titanium beam is mechanically fixed. An unloaded resonant frequency of 5.73 MHz and a frequency shift of 1 kHz/N up to 8 N were measured.

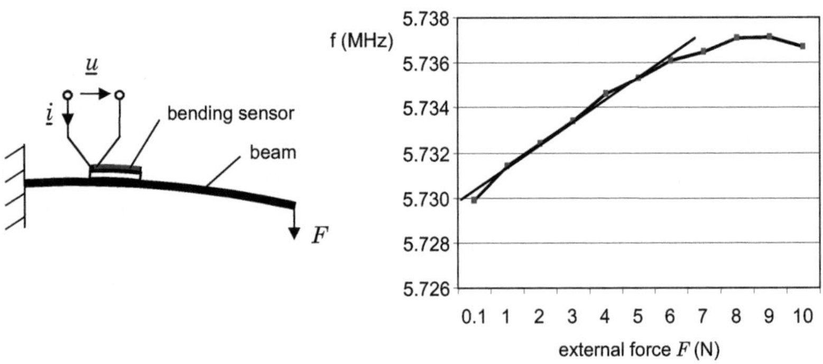

Fig. 8.47. Measured resonant frequency depending on an external load

In Fig. 8.48, the considered planar coil model is shown with the inductance of coil L, the inner resistance of coil turns R_L, the self capacitance of coil C_L and C_p, an additional capacitance to adjust the resonant frequency according to $C = C_L + C_p$. This model includes the induced voltage when an external magnetic field excites the sensor. Table 8.12 lists the calculated changes of inductance and relative permeabiliy caused by the sensor bending. The planar coil inductance shows only a small sensitivity to permeability change which is lowered by the quadratic relationship between resonant frequency and inductance. The change of resonant frequency of about 6 kHz is significant and can easily be evaluated by frequency demodulation.

As explained in the introduction to this section, the sensor reacts to quasi-static changes of the bending. When the piezomagnetic equations of state

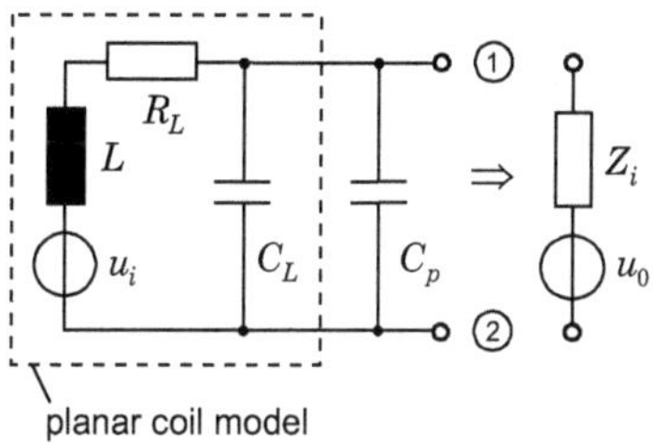

resonant frequency:

$$\omega_r = \omega_0 \cdot \sqrt{1 - R_L^2 / Z_0^2}$$

$$\omega_0 = \sqrt{1/(L \cdot C)}, \quad Z_0 = \sqrt{L/C}$$

resonant impedance: $Z_i = \left(1 + Q^2\right) \cdot R_L$

open circuit voltage:

$$|u_0| = |u_{i0} \cdot N \cdot (1 - \mathrm{j}Q)| \approx |u_i| \cdot Q$$

Fig. 8.48. Model of resonant bending sensor

Table 8.12. Measured changes of resonant frequency and calculated inductance and permeability change of a 50 nm Ti + 3x1000 nm $Fe_{83}Ga_{17}$-stack stressed with $\Delta S = 0.012\%$

f_r (MHz)	Δf_r (%)	L (μH)	ΔL (%)	$\Delta\mu_r$ (%)
5.73 . . . 5.736	0.105	7 . . . 7.0147	0.21	1

formulated in Fig. 8.40 are considered and are rearranged to

$$\underline{T}_3 = \frac{1}{s_{33}^H - d_{33}^2/\mu_{33}^T} \cdot \underline{S}_3 - \frac{d_{33}}{s_{33}^H \cdot \mu_{33}^T - d_{33}^2} \cdot \underline{B}_3 \tag{8.40}$$

$$\underline{H}_3 = \frac{d_{33}}{s_{33}^H \cdot \mu_{33}^T - d_{33}^2} \cdot \underline{S}_3 + \frac{1}{\mu_{33}^T - d_{33}^2/s_{33}^H} \cdot \underline{B}_3, \tag{8.41}$$

then a quasi-static change of strain which can be seen as a perodic change with a very large periode, causes a change of stress and field strength with same periode. A variation of the inductance is not explained by this model since μ is assumed to be constant. In the present case, the mechanical dynamics can be canceled by setting $\underline{S} = 0$ for a static load. Then only the relation between $\underline{B}$ and $\underline{H}$ remains where the permeability represents now the parameter.

9

Electrical Transducers

Depending on whether electrical or magnetic quantities are associated with mechanical quantities, in Chap. 7 it has been demonstrated that electromechanical interactions can be described by two basic types of transducers namely the electrical and magnetic transducer. The *electrostatic* and *piezoelectric transducer* belong to the group of electrical transducers.
With the electrostatic transducer, the electromechanical coupling is provided between moving electrodes – plates or diaphragms – of a capacitor comprising an isotropic dielectric. With the piezoelectric transducer, the coupling is provided within an anisotropic dielectric of a solid.
In the present book, the piezoelectric effect is described on the basis of a simplified phenomenological model representation by means of linear equations of state. For low frequencies – quasi-static case – the transition from field quantities inside the piezoelectric material to integral quantities and thus to the circuit consisting of concentrated components is very easy. With the *introduction of finite network elements*, also for piezoelectric oscillators considered to be waveguides approximate solutions in the form of circuits are achieved. The practical design of electrical transformers is explained by means of representative sample applications.

9.1 Electrostatic Transducer

9.1.1 Electrostatic Plate Transducer

The electrostatic transducer which is used for a variety of technical applications can be reduced to the basic circuit of *electrical transducers* represented in Fig. 7.8. In the following, it is dwelled on the derivation of this circuit from a transducer model, on assumed boundary conditions and on the descriptive meaning of circuit components.

A. Lenk et al., *Electromechanical Systems in Microtechnology and Mechatronics*,
Microtechnology and MEMS, DOI 10.1007/978-3-642-10806-8_9,

In the basic model shown in Fig. 9.1, two plates of a capacitor with air as dielectric $\varepsilon \approx \varepsilon_0$ are in opposition to each other at a distance of l. One of the two plates of mass m is movable and for $Q = 0$ it is kept stably in rest position (distance l_0) by the spring n. When applying a charge Q on the capacitor's plate, the COULOMB *force* F_{el} is generated, the spring is deflected about Δl and an equilibrium of forces arises for a plate distance of l.

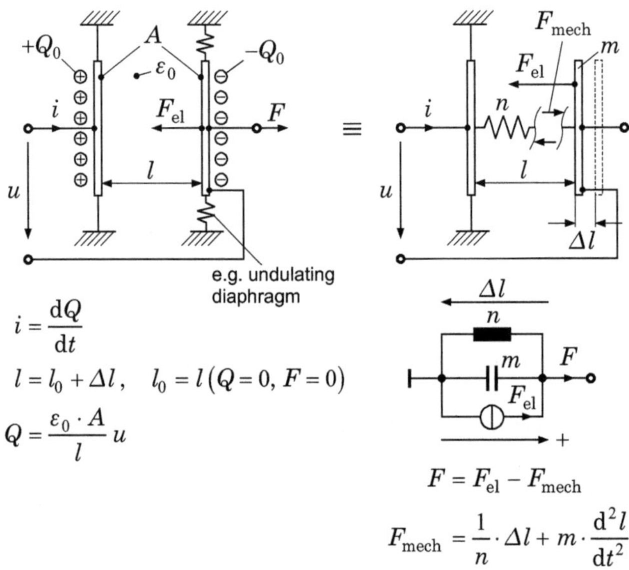

Fig. 9.1. Balance of forces at a movable capacitor plate

The calculation of electrostatic force of attraction F_{el} between the capacitor plates can be effected by means of two methods already discussed in Sect. 7.2,

- solving the electrostatic field equations or
- applying the energy balance for a virtual displacement Δl of the movable plate.

In case of knowing the charge density profile $\rho(x)$ at the plate bounding surfaces, the electrostatic field equations can be solved as it is shown in Fig. 9.2.

The application of energy balance yields the same result for the electrostatic force of attraction F_{el}. The calculation of force F_{el} is made on the basis of energy change of the forced virtual displacement Δl of the movable capacitor plate (Fig. 9.3).

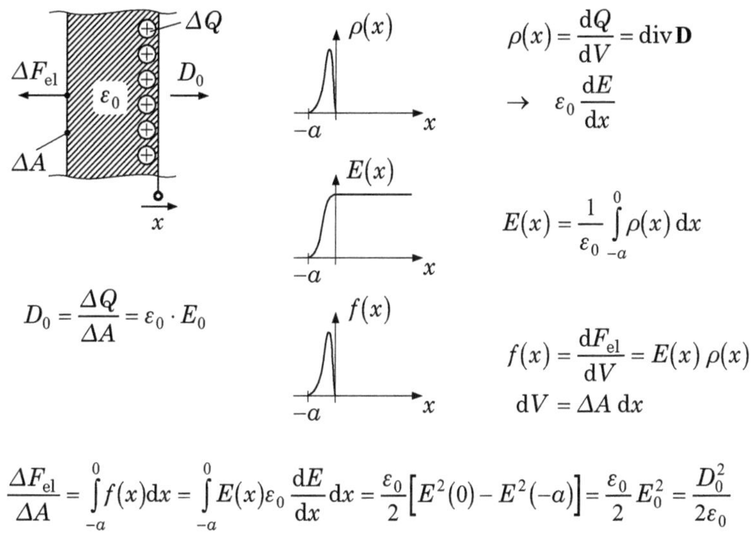

Fig. 9.2. Calculation of COULOMB force from electrostatic field equations in case of knowing the charge density distribution $\rho(x)$

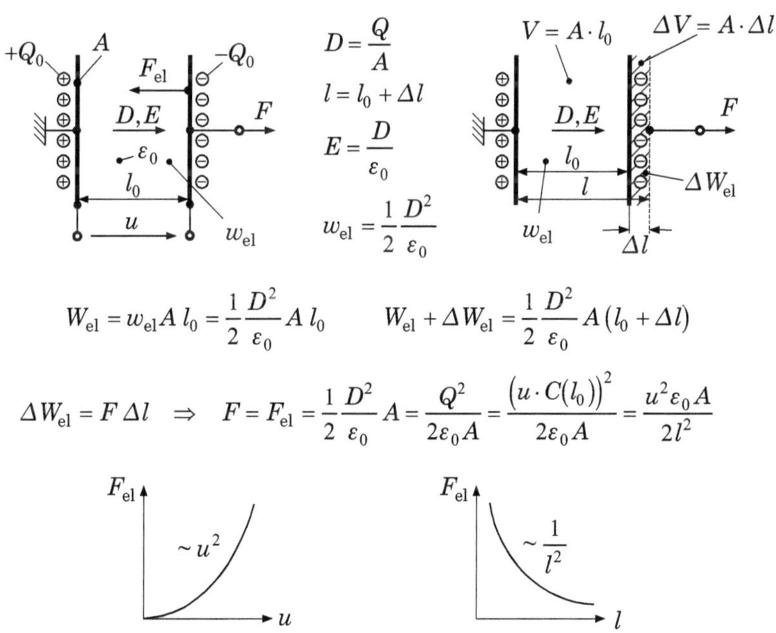

Fig. 9.3. Calculation of COULOMB force from energy balance for a virtual displacement Δl of movable capacitor plate

The force depends on the square of applied voltage u and the reciprocal square of plate distance l:

$$F_{\mathrm{el}} = \frac{Q^2}{2\varepsilon_0 A} = \frac{1}{2}\frac{u^2\varepsilon_0 A}{l^2} \tag{9.1}$$

If now an AC voltage

$$u(t) = \hat{u}\sin(\omega t)$$

is supplied to the capacitor plates, a distorted force progression at double frequency

$$F_{\mathrm{el}}(t) = \frac{\hat{u}^2 C^2}{4\varepsilon_0 A}\left(1-\cos(2\omega t)\right)$$

will be generated due to the squared characteristic. The linear relation being required for electromechanical transducers according to Sect. 2.1 is achieved by adding a DC voltage U_0 to the output signal $u(t)$. According to Fig. 9.4 this results in a force F_{el} which is superimposed by an alternating component $F_{\mathrm{el}}(t)$ being approximately linearly dependent on u:

$$F_{\mathrm{el}} = F_0 + F_{\mathrm{el}}(t),$$

$$F_{\mathrm{el}}(t) = \frac{C^2}{2\varepsilon_0 A}\left(2U_0\hat{u}\sin(\omega t) + \hat{u}^2\sin^2(\omega t)\right), \qquad F_0 = \frac{1}{2}\frac{C^2(l_0^*)\,U_0^2}{\varepsilon_0 A}$$

In order to derive the transducer circuit, the square part is neglected for $U_0 \gg \hat{u}$. Due to operating point adjustment with U_0, the plate distance decreases from l_0 to l_0^*. The constant charging of the capacitor plates in Fig. 9.5

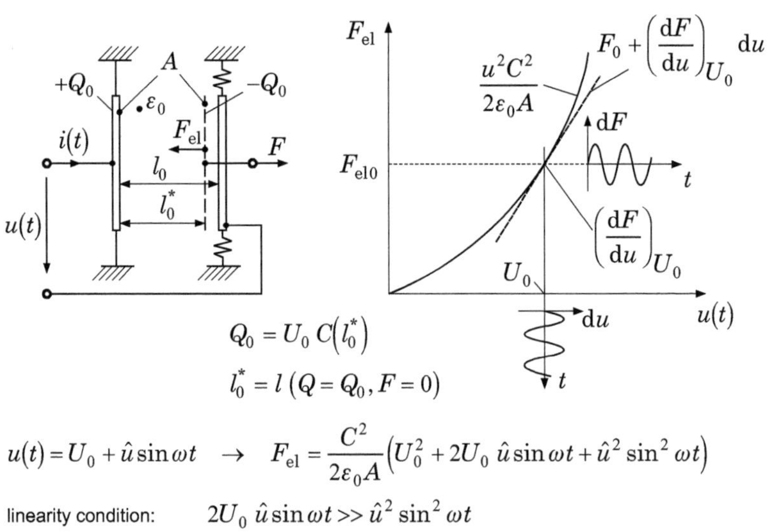

Fig. 9.4. Linearization of force characteristic

is generated by the *polarization voltage source* U_0 which comprises a very high internal resistance R. In combination with $R \gg 1/(\omega C)$, it will be assured that there is no current flow through the polarization voltage source in the operating frequency range of the electrostatic transducer. When considering the dynamic transfer characteristic of the transducer, this leg can be considered to be nonexistent.

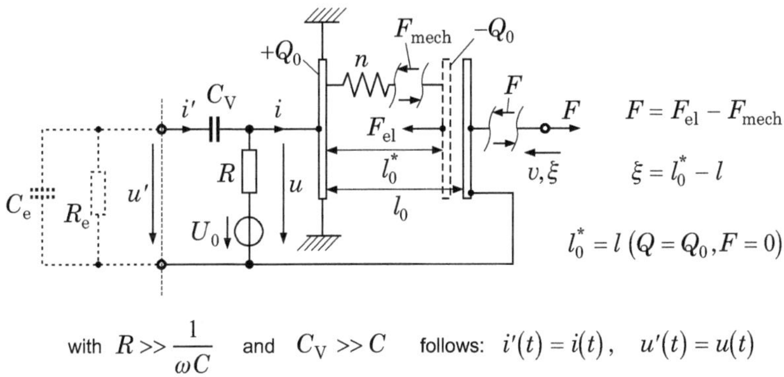

Fig. 9.5. Model of linearized decoupled electrostatic transducer

The capacitor C_V which shows a much higher capacitance than the capacitor C between the transducer plates provides for electrical potential separation in case of connecting real input resistances R_e and capacitances C_e of primary electronics as well as for charging state Q_0 of the plates in case of electrical short-circuit.

Based on the real transducer model shown in Fig. 9.5 the *quasi-static circuit* can now be derived. Based on the transducer equations

$$F = F_{el} - F_{mech} = \frac{Q^2}{2\varepsilon_0 A} - \frac{1}{n}\xi = \frac{Q^2}{2\varepsilon_0 A} + \frac{1}{n}\left(l - l_0^*\right) \quad \text{and} \quad u = \frac{Q}{C\,(l)} = \frac{Q}{\varepsilon_0 A}l$$

the quantities ΔF and Δu can be calculated in the neighborhood of the operating point U_0, Q_0, F_0, l_0^* for the quasi-static case ($F_0 = 0$) according to

$$\Delta F = F - F_0 = \frac{\partial F}{\partial Q}\left(Q - Q_0\right) + \frac{\partial F}{\partial l}\left(l - l_0^*\right)$$

$$F_0 = F\left(Q_0, l_0^*\right) = \frac{Q_0^2}{2\varepsilon_0 A} - \frac{1}{n}\left(l_0 - l_0^*\right) = 0$$

$$\Delta u = u - U_0 = \frac{\partial u}{\partial Q}\left(Q - Q_0\right) + \frac{\partial u}{\partial l}\left(l - l_0^*\right).$$

Due to the linearized transducer characteristic, only the first terms of Taylor series are considered. All derivatives are calculated at Q_0, l_0^*. If the basic transducer equations are derived, following relations will be achieved:

$$\left.\frac{\partial F}{\partial Q}\right|_{Q_0,l_0^*} = \frac{Q_0}{\varepsilon_0 A}, \qquad \left.\frac{\partial F}{\partial l}\right|_{Q_0,l_0^*} = \frac{1}{n}$$

$$\left.\frac{\partial u}{\partial Q}\right|_{Q_0,l_0^*} = \frac{1}{C\left(l_0^*\right)}, \qquad \left.\frac{\partial u}{\partial l}\right|_{Q_0,l_0^*} = \frac{Q_0}{\varepsilon_0 A}$$

Since, as already mentioned in Sect. 7.2, sinusoidal displacements around the operating point are assumed, complex amplitudes according to (7.26) - (7.30)

$$-\left(l - l_0^*\right) = \xi \rightarrow \underline{\xi}, \qquad F - F_0 \rightarrow \underline{F}, \qquad Q - Q_0 \rightarrow \underline{Q}, \qquad u - U_0 \rightarrow \underline{u}$$

are introduced for further considerations. In combination with $\underline{\xi} = (1/\mathrm{j}\omega) \cdot \underline{v}$ and $\underline{Q} = (1/\mathrm{j}\omega) \cdot \underline{i}$, the transducer equations result in a transformer-like and frequency-dependent coupling

$$\underline{F} = \frac{1}{\mathrm{j}\omega}\frac{Q_0}{\varepsilon_0 A}\underline{i} - \frac{1}{\mathrm{j}\omega n}\underline{v}$$

$$\underline{u} = \frac{1}{\mathrm{j}\omega C\left(l_0^*\right)}\underline{i} - \frac{1}{\mathrm{j}\omega}\frac{Q_0}{\varepsilon_0 A}\underline{v}$$

between network coordinates. The occurrence of an imaginary transducer constant $\underline{X} = \mathrm{j}\omega\left(\varepsilon_0 A/Q_0\right)$ can be avoided by transformation of $\underline{F}, \underline{u} = f\left(\underline{i}, \underline{v}\right)$ into $\underline{i}, \underline{F} = f'\left(\underline{u}, \underline{v}\right)$. Thus, it can be written:

$$\underline{i} - \mathrm{j}\omega C\left(l_0^*\right)\underline{u} = \underline{i}_{\mathrm{W}} = \frac{Q_0}{l_0^*}\underline{v} \tag{9.2}$$

$$\underline{F} = \frac{1}{\mathrm{j}\omega}\frac{Q_0}{\varepsilon_0 A}\left[\mathrm{j}\omega C\left(l_0^*\right)\underline{u} + \frac{Q_0}{l_0^*}\underline{v}\right] - \frac{1}{\mathrm{j}\omega n}\underline{v}$$

$$\underline{F} + \frac{1}{\mathrm{j}\omega}\left[\frac{1}{n} - \frac{Q_0^2}{l_0^*\varepsilon_0 A}\right]\underline{v} = \underline{F}_{\mathrm{W}} = \frac{Q_0}{l_0^*}\underline{v} \tag{9.3}$$

The equations (9.2) and (9.3) comprise now a *gyrator-like* coupling with real transducer constant $Y = l_0^*/Q_0$.

If now KIRCHHOFF's nodal rule is applied, the circuit diagram of the electrostatic transducer specified in Fig. 9.6 will result from (9.2) and (9.3). The relations specified in Table 7.3 are valid for transformation relations between the mechanical and electrical network side and vice versa.

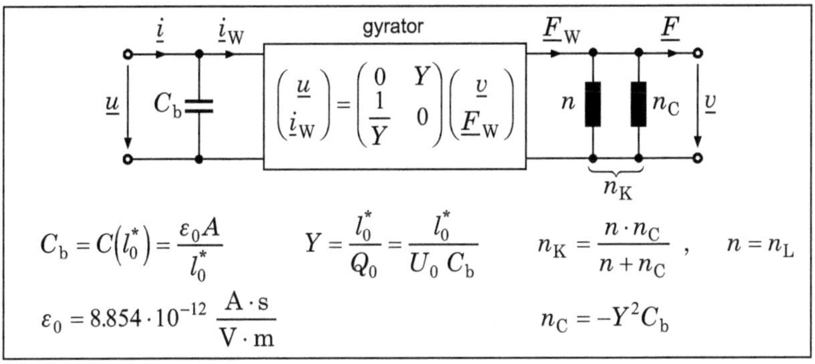

Fig. 9.6. Circuit diagram of loss-free electrostatic transducer
index K: *electrical short-circuit operation, index* L: *electrical open-circuit operation, index* b: *mechanically locked case*

What *descriptive meaning* do the components C_b, Y and n_K have?

If the movable plate is fixed ($\underline{v} = 0, l = l_0^*$), mechanically locked case, the capacitance

$$C_b = C\left(l_0^*\right) = \frac{\varepsilon_0 A}{l_0^*}$$

can be measured at the electrical input terminals. In this case, only the electrical short-circuit force F_K is in action as total force F. With respect to Fig. 9.4, it can be written for the assumed sinusoidal modulation around the operating point

$$\hat{F}_K = \frac{C_b U_0}{l_0^*}\hat{u} = \frac{1}{Y}\hat{u}.$$

The other way round, the transducer constant can be determined from the ratio of a velocity $\underline{v}$ forced at mechanical side and resulting short-circuit current $\underline{i}_K$. With $\underline{u} = 0$ it immediately follows from (9.2)

$$\underline{i} = \underline{i}_K = \frac{Q_0}{l_0^*}\underline{v} = \frac{1}{Y}\underline{v}.$$

In case of an electrical short-circuit $\underline{u} = 0$, the compliance n_K can be measured at the mechanical side. In addition to mechanical compliance n the electrostatic field reaction must be taken into consideration in the form of *field compliance* n_C. The action of field compliance can be explained by means of balances of forces which can be derived from Fig 9.7.
For the mechanically locked case $\xi = 0$ it can be written:

$$F = F_{\mathrm{el},0} - F_{\mathrm{mech},0} = 0$$

$$F_{\mathrm{el},0} = \frac{U_0^2 C^2}{2\varepsilon_0 A} = \frac{U_0^2 \varepsilon_0 A}{2 l_0^{*2}} = \frac{1}{n}\left(l_0 - l_0^*\right)$$

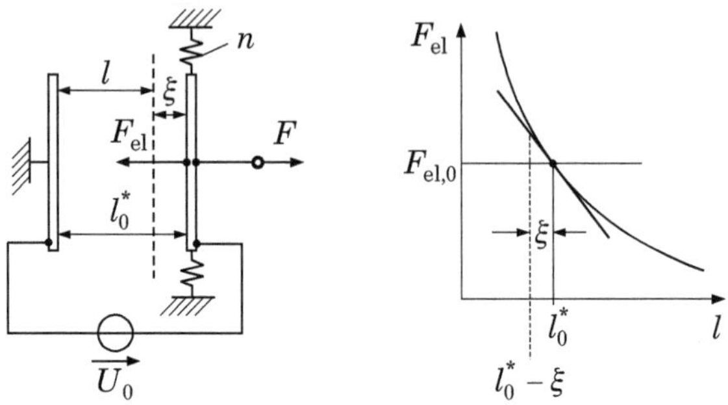

Fig. 9.7. Influence of a displacement ξ of capacitor plate on COULOMB force F_{el}

If now a displacement ξ is forced according to Fig. 9.7, F_{el} will increase:

$$F_{el} = \frac{U_0^2 \varepsilon_0 A}{2\left(l_0^* - \xi\right)^2}$$

The total force F results in

$$F = F_{el,0} - F_{mech,0} + \left(\frac{dF_{el}}{d\xi}\right)_{l_0^*} \xi - \frac{1}{n}\xi \qquad \text{with} \qquad F_{el,0} = F_{mech,0}$$

$$F = \frac{U_0^2 C_b}{l_0^{*2}}\xi - \frac{1}{n}\xi = -\left(\frac{1}{n_C} + \frac{1}{n}\right)\xi$$

with

$$\frac{U_0^2 C_b}{l_0^{*2}} = \frac{1}{Y^2}\frac{1}{C_b} = -\frac{1}{n_C}.$$

Thus, in case of short-circuit operation ($u = 0$), the force being necessary for generating a deflection ξ of the movable electrode consist of the partial force being required for spring deformation minus the increase of electrical force of attraction F_{el} generated by voltage U_0.
If the movable plate is moved toward counter electrode ($\xi > 0$), an increasingly less force $-F$ will be required for the generation of deflection ξ due to the electrostatic force of attraction F_{el} increasing with ξ. Thus, the differential stiffness $\Delta(-F)/\Delta\xi$ shown in Fig. 9.7 becomes increasingly less and equals zero at a *deflection limit* ξ_g.

If the applied source being necessary for mechanical excitation of system represents an ideal source of force for this experiment, at this point an unstable equilibrium is existent. Such a case could be realized e.g. by the configuration illustrated in Fig. 9.8 where the force is generated quasi-statically by a weight $F = mg$. For $\xi > \xi_g$, the mechanical counteracting force consisting of spring force and inertia force of mass is not able to compensate the electrical force of

attraction. In order to maintain equilibrium of forces, it necessitates an inertia force which causes an increasing acceleration of the movable plate toward the fixed counter electrode till both plates clash.
The deflection limits ξ_g/l_0 normalized with the rest plate distance l_0 are specified in Fig. 9.9 for different values of n/n_C. The deflection limit increases with increasing spring stiffness.

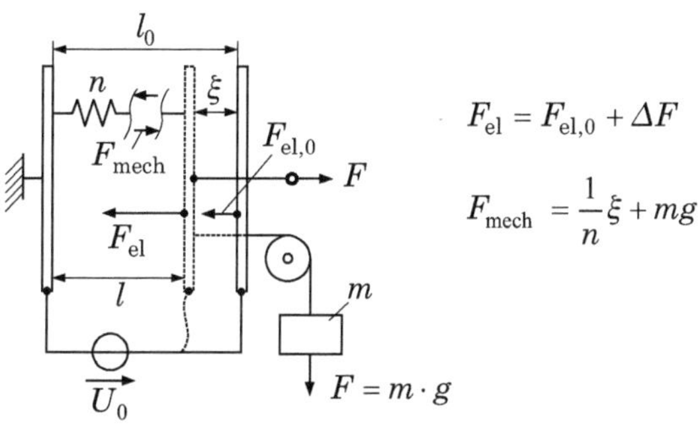

Fig. 9.8. Equilibrium of forces in combination with an ideal source of force

A further type of electrostatic transducer is based on a horizontal plate arrangement with $l_0 = const.$ as illustrated in Fig. 9.10 a). If the plates carry the charge $\pm Q$, it will necessitate a force F_{el} in order to maintain balance. As specified in Fig. 9.3, this force can be determined by means of a virtual displacement of the movable plate in rightward direction at constant charge.

Taking Fig. 9.10 b) into consideration, the energy balance

$$F_{\text{el}} \cdot (-\Delta b) = \Delta W_{\text{el}} \qquad \text{with} \qquad \Delta W_{\text{el}} = \Delta\left(\frac{Q^2}{2C}\right)$$

results in the equilibrant force

$$F_{\text{el}} = -\frac{\Delta W_{\text{el}}}{\Delta b} = -\frac{\mathrm{d}\left(\frac{Q^2}{2C}\right)}{\mathrm{d}b} = \frac{Q^2}{2C^2}\frac{\varepsilon_0 a}{l_0} = \frac{1}{2}u^2\varepsilon_0\frac{a}{l_0}.$$

For the configuration shown in Fig. 9.10 a) a reference point is defined which is determined by the condition $F_0, b_0^* = b\,(F = 0, U_0, Q_0)$.

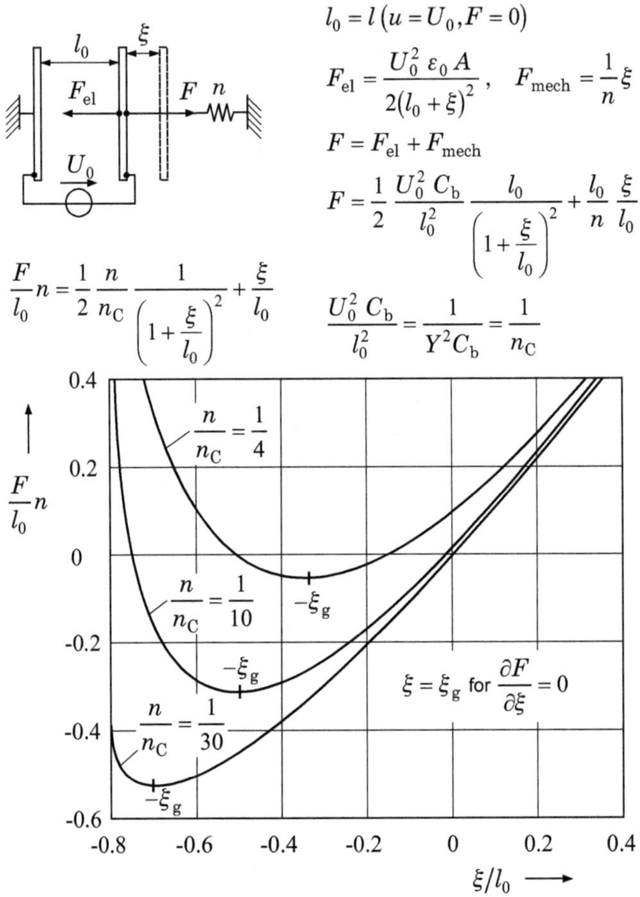

Fig. 9.9. Stability limits ξ_g of electrostatic transducer

In combination with balance of forces and component equations specified in Fig. 9.10 a), the deviations $\varDelta F$ and $\varDelta Q$ from reference state result in

$$\varDelta F = \left(\frac{\partial F}{\partial u}\right)_{U_0, b_0^*} \cdot \varDelta u + \left(\frac{\partial F}{\partial b}\right)_{U_0, b_0^*} \cdot \varDelta b = \frac{U_0 \varepsilon_0 a}{l_0}\varDelta u + \frac{1}{n}\xi \tag{9.4}$$

$$\varDelta Q = \left(\frac{\partial Q}{\partial u}\right)_{U_0, b_0^*} \cdot \varDelta u + \left(\frac{\partial Q}{\partial b}\right)_{U_0, b_0^*} \cdot \varDelta b = C_b \varDelta u + U_0 \frac{\mathrm{d}C_b}{\mathrm{d}b}\xi \tag{9.5}$$

for $\xi = b - b_0^*$.

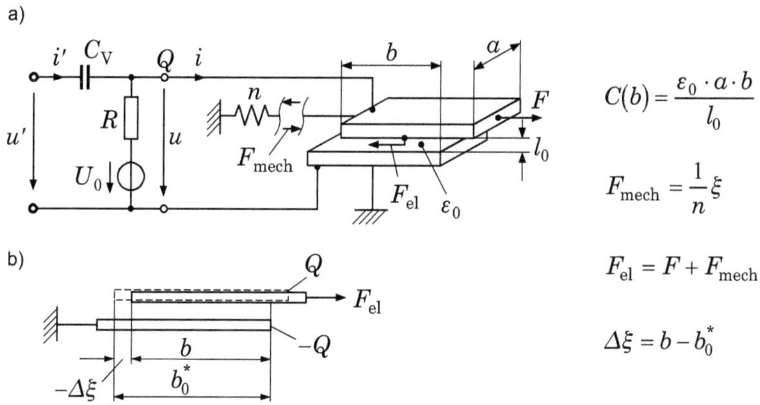

Fig. 9.10. Transformer model for a horizontal plate arrangement with $l_0 = const.$

The introduction of complex amplitudes results in relations between the coordinates $\underline{u}$, $\underline{i}$, $\underline{F}$, $\underline{v}$

$$\underline{F} = \frac{\varepsilon_0 U_0 a}{l_0}\underline{u} + \frac{1}{j\omega n}\underline{v}, \tag{9.6}$$

$$\underline{i} = j\omega C_b \underline{u} + \frac{\varepsilon_0 U_0 a}{l_0}\underline{v}. \tag{9.7}$$

Like the equations (9.2) and (9.3), the equations (9.6) and (9.7) can be interpreted by the circuit diagram illustrated in Table 9.1.

The two types of electrostatic transducer represented in Table 9.1 show a gyrator-like coupling for a real transducer constant Y. The characteristic values of the transducer and the transformation relations for the two types of electrostatic transducer are also summarized in Table 9.1.
In the following, the application of electrostatic plate transducers in real electromechanical systems and their dimensioning will be explained by means of selected examples.

9.1.2 Sample Applications

As sample applications, it is concerned with the dynamic design of an electrostatic sensing probe, with a flexible cantilever beam characterized by electrostatic excitation and sensing as well as with a micromechanical filter characterized by electrostatic excitation and sensing. The circuit of electrostatic plate transducer with vertical or parallel plate arrangement shown in Table 9.1 represents the basic part of all three sample applications.

Table 9.1. Circuit diagram and components of electrostatic plate transducer taking mass of plate and losses into account

plate arrangement with A = const.	plate arrangement with l_0 = const.
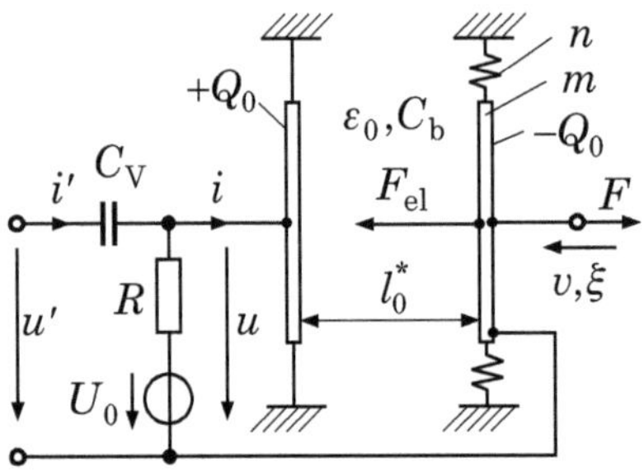	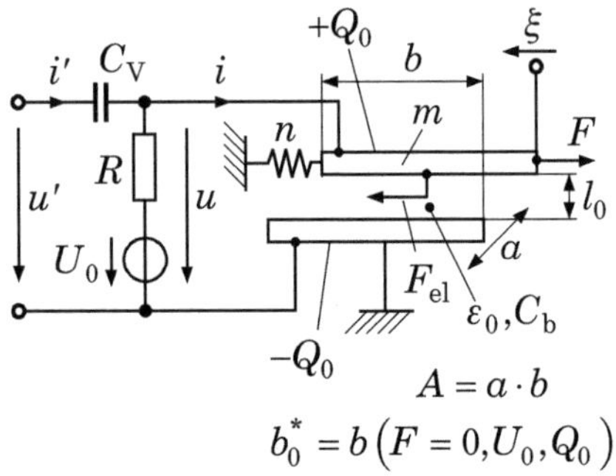

equivalent circuit diagram

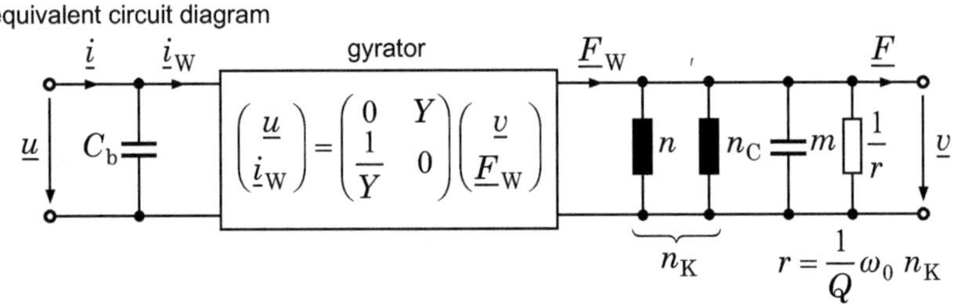

transducer constant und components

Y	$l_0^*/Q_0 = l_0^*/(U_0 \cdot C_b)$	$b_0^*/(U_0 \cdot C_b) = l_0/(\varepsilon_0 \cdot a \cdot U_0)$
C_b	$(\varepsilon_0 \cdot A)/l_0$	
n_C	$-Y^2 \cdot C_b$	—
n_K	n parallel to n_C	n

transformation correlations

mechanical → electrical	electrical → mechanical
$C_n = n/Y^2$	$n_C = Y^2 \cdot C$
$L_m = Y^2 \cdot m$	$m_L = L/Y^2$
$R_r = Y^2 \cdot r$	$r_R = R/Y^2$
$\underline{Z} = Y^2/\underline{h} = Y^2 \cdot \underline{z}$	$\underline{h} = Y^2/\underline{Z}, \quad \underline{z} = \underline{Z}/Y^2$

series connection → parallel connection
parallel connection → series connection

Electrostatic Sensing Probe

The tip deflection ξ of a flexible steel cantilever beam is measured by means of an electrostatic plate transducer (Fig. 9.11). The additionally connected components C_a and R_a define the input impedance of the electrical evaluation device.

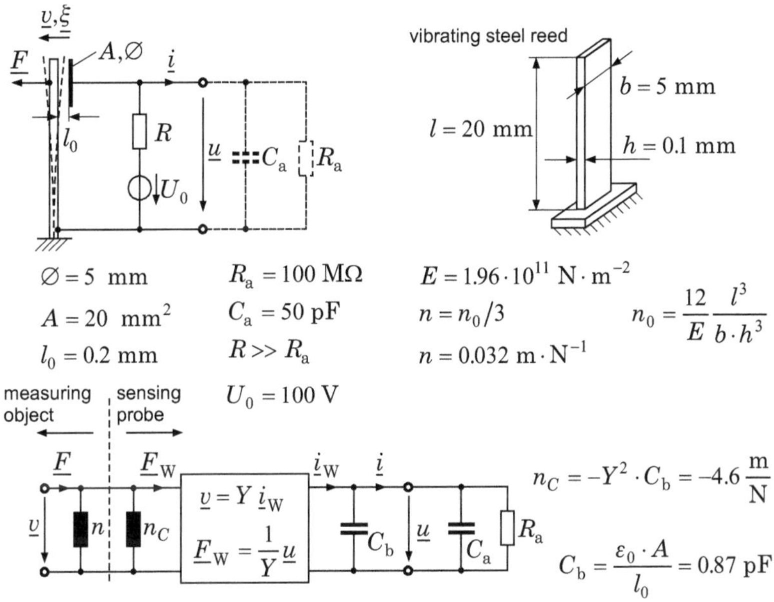

Fig. 9.11. Tip deflection measurement of a vibrating steel reed by means of an electrostatic sensing probe

First of all, the transfer function $\underline{B} = \underline{u}/\underline{\xi}$ of configuration is determined. For that purpose, the circuit diagram shown in Fig. 9.12 is assumed.
The source of motion $\underline{v}_0 = \mathrm{j}\omega\underline{\xi}_0$ is effective at the mechanical side. In the simplified circuit represented in Fig. 9.12 the mechanical components n and n_C are not taken into account, since the force that has to be generated by source of motion is initially not of interest. In combination with Fig. 9.12, the output voltage $\underline{u}$ yields

$$\underline{u} = \frac{\underline{v}_0}{Y} \frac{1}{\mathrm{j}\omega\left(C_\mathrm{a} + C_\mathrm{b}\right) + \dfrac{1}{R_\mathrm{a}}} = \frac{1}{Y} \frac{\mathrm{j}\omega\underline{\xi}_0}{\mathrm{j}\omega C_\mathrm{b}\left(1 + \dfrac{C_\mathrm{a}}{C_\mathrm{b}}\right)} \cdot \frac{1}{\left(1 + \dfrac{1}{R_\mathrm{a}\mathrm{j}\omega\left(C_\mathrm{a} + C_\mathrm{b}\right)}\right)}$$

$$\underline{u} = \underline{\xi}_0 \frac{U_0}{l_0} \cdot \frac{C_\mathrm{b}}{C_\mathrm{a} + C_\mathrm{b}} \cdot \frac{\mathrm{j}\left(\omega/\omega_0\right)}{1 + \mathrm{j}\left(\omega/\omega_0\right)}$$

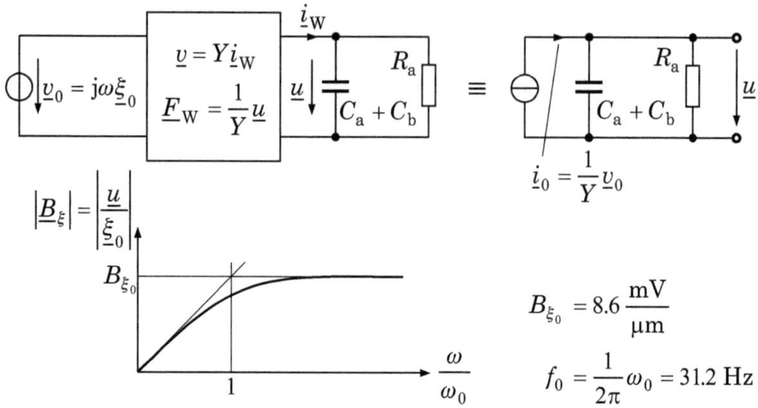

Fig. 9.12. Circuit diagram used for determination of the transfer function

with

$$Y = \frac{l_0}{U_0 C_b} \quad \text{and} \quad \omega_0 = \frac{1}{R_a (C_a + C_b)}$$

and thus the transfer function $\underline{B}_\xi$ results in

$$\underline{B}_\xi = \frac{\underline{u}}{\underline{\xi}_0} = \frac{U_0}{l_0} \cdot \frac{C_b}{C_a + C_b} \cdot \frac{j(\omega/\omega_0)}{1 + j(\omega/\omega_0)} \quad \text{with} \quad B_{\xi 0} = \frac{U_0}{l_0} \cdot \frac{C_b}{C_a + C_b}.$$

Its curve progression is shown in Fig. 9.12.
Afterwards, the additional mechanical load on the measuring object „flexible cantlever beam" will be determined by means of the sensing probe. The mechanical impedance $\underline{z}$ which can be measured into the circuit on the right of separation line illustrated in Fig. 9.13 corresponds to the additional load. Based on the circuit illustrated in Fig. 9.13, for $C_a \gg C_b$ an additional mechanical load on the flexible cantilever beam is generated by the quantity $n_C = -143n$. Thus, the compliance n_{mech} which is measured at the tip of the flexible cantilever beam in compliance with the conditions specified in Fig. 9.13

$$n_{mech} = \frac{\underline{v}}{j\omega\underline{F}} = \frac{n n_C}{n + n_C} = 1.007n$$

is only slightly larger due to $n_C < 0$. By the influence of the sensing probe, the frequency of free vibration of the flexible cantilever beam $f_0 \sim 1/\sqrt{n}$ would be measured too low approximately about 0.5%.

Flexible Cantilever Beam with Electrostatic Excitation and Sensing

In Fig. 9.14, the tip of a flexible aluminum cantilever beam is excited by an electrostatic transducer. On the other side the motion being generated

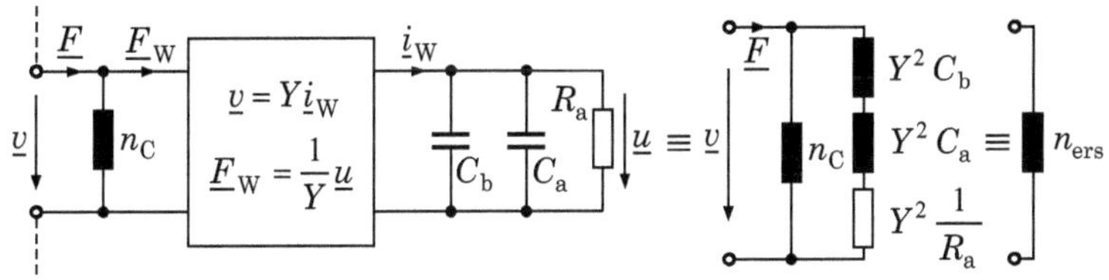

$$\underline{z} = -\frac{1}{\mathrm{j}\omega Y^2 C_\mathrm{b}} + \frac{1}{\mathrm{j}\omega Y^2 \left(C_\mathrm{a} + C_\mathrm{b} + \dfrac{1}{\mathrm{j}\omega R_\mathrm{a}} \right)}, \qquad \frac{1}{\omega R_\mathrm{a}} \ll 1$$

$$\underline{z} = -\frac{1}{\mathrm{j}\omega Y^2 C_\mathrm{b}} \cdot \frac{C_\mathrm{a}}{C_\mathrm{a} + C_\mathrm{b}} = \frac{1}{\mathrm{j}\omega n_\mathrm{ers}} = \frac{1}{\mathrm{j}\omega n_\mathrm{C}} \cdot \frac{C_\mathrm{a}}{C_\mathrm{a} + C_\mathrm{b}}$$

$$\underline{z} = \begin{cases} 0 & \text{for } C_\mathrm{a} = 0 \\ \dfrac{1}{\mathrm{j}\omega n_\mathrm{C}} & \text{for } C_\mathrm{a} \gg C_\mathrm{b}, \quad n_\mathrm{C} = -Y^2 C_\mathrm{b} = -4.59\ \dfrac{\mathrm{m}}{\mathrm{N}} \end{cases}$$

Fig. 9.13. Determination of mechanical input impedance of sensing probe

is sensed by a second electrostatic transducer. A parasitic partial capacitance $C_\mathrm{S} = 10^{-3} C_\mathrm{b}$ is existent between excitation and sensing electrode. For higher excitation frequencies the flexible cantilever beam will be represented by a flexural cantilever waveguide (see Sect. 9.3.5, Table 9.14) with $n_{ers} = 0.971 n_0/3 \approx n$, $m_{ers} = m/4$, $h = \omega_0 n Q_0$ and $Q_0 = 1/\eta$. In order to calculate the transfer function $\underline{B} = \underline{u}_2/\underline{u}_1$ of the electromechanical system, the mechanical components are transformed to one of both electrical sides (see Fig. 9.15). The cascade connection of the two electrostatic gyrators yields a transformer with transformation ratio 1. The resultant and simplified circuit shown in Fig. 9.15 yields the wanted transfer function:

$$\underline{B}_u = \frac{\underline{u}_2}{\underline{u}_1} = \frac{\mathrm{j}\omega C_\mathrm{S} + \dfrac{1}{\mathrm{j}\omega L + R + 1/(\mathrm{j}\omega C_n) - 2/(\mathrm{j}\omega C_\mathrm{b})}}{\mathrm{j}\omega C_\mathrm{b} + \mathrm{j}\omega C_\mathrm{S} + \dfrac{1}{\mathrm{j}\omega L + R + 1/(\mathrm{j}\omega C_n) - 2/(\mathrm{j}\omega C_\mathrm{b})}}$$

$$\underline{B}_u = B_0 \frac{1 - \left(\dfrac{\omega}{\omega_\mathrm{N}}\right)^2 + j\dfrac{\omega}{\omega_\mathrm{N}}\dfrac{1}{Q_\mathrm{N}}}{1 - \left(\dfrac{\omega}{\omega_\mathrm{P}}\right)^2 + j\dfrac{\omega}{\omega_\mathrm{p}}\dfrac{1}{Q_\mathrm{p}}} \quad \begin{cases} B_0 \approx \dfrac{C_\mathrm{S}}{C_\mathrm{S} + C_\mathrm{b}}, \qquad \omega_0 = 1/\sqrt{n\dfrac{m}{4}} \\ \omega_\mathrm{p} \approx \omega_0 \sqrt{1 - \dfrac{C_\mathrm{n}}{C_\mathrm{b}}}, \ Q_\mathrm{p} = (\omega_\mathrm{P}/\omega_0)\, Q_0 \\ \omega_\mathrm{N} \approx \omega_0 \sqrt{1 + \dfrac{C_\mathrm{n}}{C_\mathrm{b}}}, \ Q_\mathrm{N} = (\omega_\mathrm{N}/\omega_0)\, Q_0 \end{cases}$$

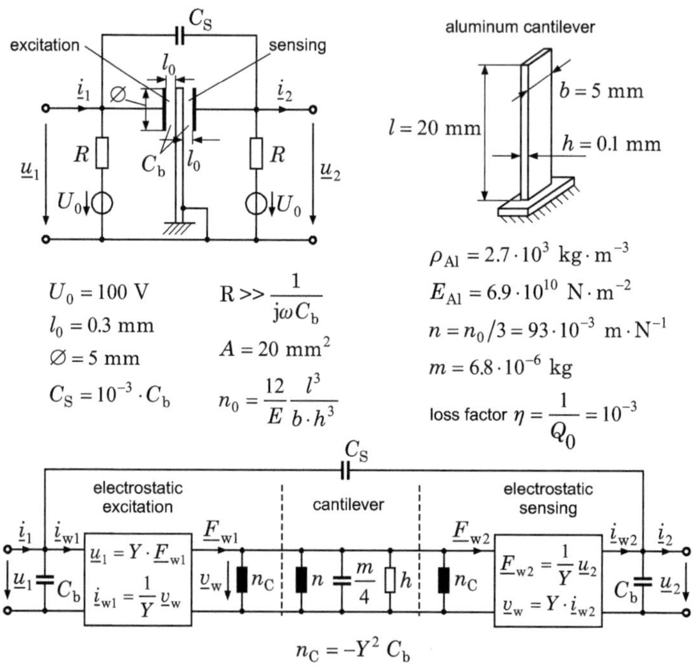

Fig. 9.14. Electrostatic excitation and sensing of a flexible aluminum cantilever beam and associated circuit diagram

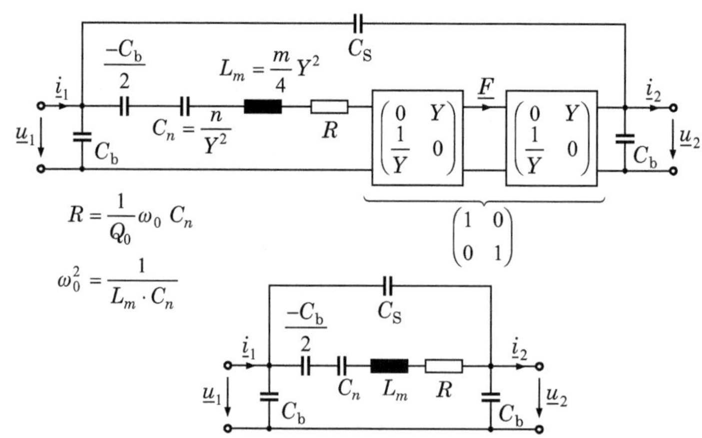

Fig. 9.15. Simplified circuit of electrostatically excited and sensed flexible cantilever beam

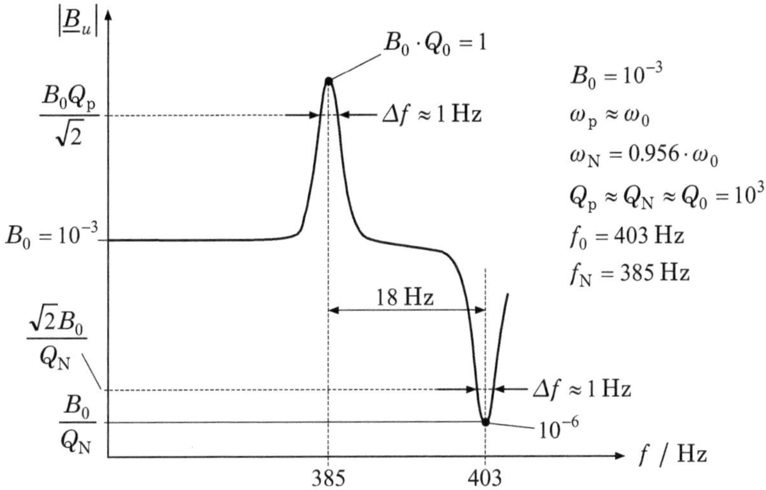

Fig. 9.16. Amplitude frequency response of the transfer function $\underline{B}_u$ of flexible cantilever beam

Taking the specific values of components into account, finally the frequency response of $|\underline{B}_u|$ represented in Fig. 9.16 is achieved.

Micromechanical Filter with Electrostatic Excitation and Sensing

Figure 9.17 illustrates the design principle of an electromechanical filter described in [79]. The filter was manufactured using the technology of *silicon surface micromechanics*. The dimensions of the mechanical components are also specified in Fig. 9.17. Due to their high quality factor and thus narrow bandwith and due to their low losses and very high signal-to-noise ratio, micromechanical filters will become more important particularly in the field of telecommunication technology.

The excitation and sensing of the spring-mass system is provided by an electrostatic transducer in each case. The movable electrodes, the bending elements of resiliences and the mass elements can move freely above the silicon substrate. They consist of polysilicon and their thickness amounts to $h = 2\,\mu$m. In Fig. 9.17, the junction points with the substrate are marked in black. The circuit diagram illustrated in Fig. 9.18 can be derived from the schematic diagram.

For the observed frequency range, the components can be considered in good approximation to be concentrated due to their small dimensions in the micrometer range. After transforming the mechanical components to the electrical side, multiplying the transducer matrices in such a way that the product equals 1 and neglecting the parasitic capacitance C_S, the simplified circuit also illustrated in Fig. 9.18 is achieved.

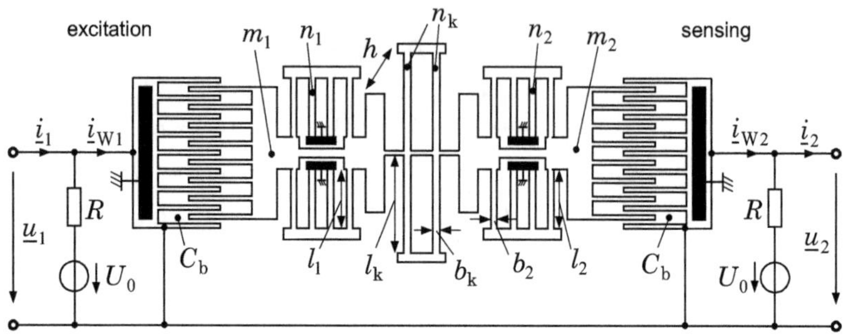

$h = 2\ \mu\text{m}\,,\ l_1 = l_2 = 150\ \mu\text{m}\,,\ b_1 = b_2 = 2\ \mu\text{m}\,,\ l_k = 150\ \mu\text{m}\,,\ b_k = 1\ \mu\text{m}$

Fig. 9.17. Design principle and dimensions of an electromechanical filter manufactured using the technology of silicon micromechanics (topview of microstructure)

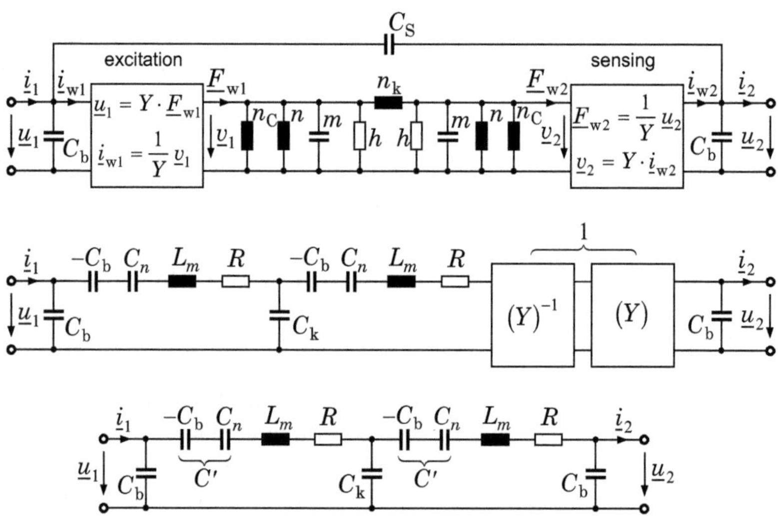

$$n_C = -Y^2\,C_b\,,\quad h = Q\cdot\omega_0\cdot n'\,,\quad n' = \frac{n\cdot n_C}{n+n_C}\,,\quad C_n = \frac{n}{Y^2}\,,\quad L_m = m\cdot Y^2$$

$$C_k = \frac{n_k}{Y^2}\,,\quad \omega_0^2 = \frac{1}{L_m\cdot C'}$$

Fig. 9.18. Derivation of simplified circuit of micromechanical filter

The calculation of the transfer function $\underline{B}_u = \underline{u}_2/\underline{u}_1$ was made by using a network simulation program. Figure 9.19 shows the measured and calculated transfer function of the filter. The center frequency amounts to 18.7 kHz for a quality factor of 16. The 3 dB bandwidth amounts to 1.2 kHz. If the filter is vacuum-encapsulated, the center frequency will increase to 24 kHz for a

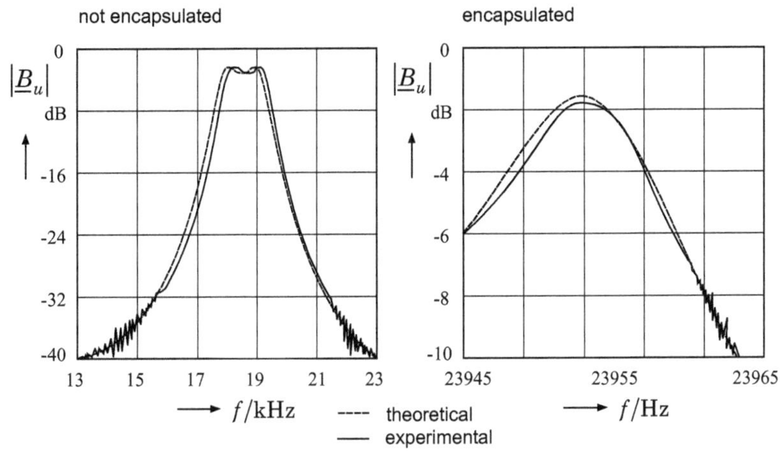

Fig. 9.19. Theoretical and experimental curve progression of $|\underline{B}| = \underline{u}_2/\underline{u}_1$ of a not encapsulated and a vacuum-encapsulated micromechanical filter [79]

quality factor of 2200 while the 3 dB bandwidth will decrease to 11 Hz. By varying mechanical components, i.e. further reduction of beam benders' length and their stiffening, bandpass filters up to a center frequency of about 1 MHz can be produced.

9.1.3 Electrostatic Diaphragm Transducer

So far, the movable electrode has been considered as a rigid plate which is kept in its rest position l_0 by a spring n. If the rigid resiliently clamped plate is replaced by a *diaphragm* which shows a compliance that is solely determined by *mechanical prestress* T_0 and not by its flexural rigidity $1/n$, the basic model of electrostatic transducer with variable plate distance shown in Fig. 9.5 will pass into the model with circular diaphragm as movable electrode shown in Fig. 9.20. The deflection of diaphragm is generated by the acoustic coordinate differential pressure p and the COULOMB force F_{el} related to diaphragm area A.

According to [2] the acoustic two-port network properties of a thin circular diaphragm are summarized in Fig. 9.21.

Based on basic equations for the deflected diaphragm volume V

$$V = N_{\text{a}} p_{\text{ges}} = N_{\text{a}} \left(p + p_{\text{el}}\right) = N_{\text{a}} p + \frac{Q^2}{2\varepsilon_0 A^2} N_{\text{a}}$$

and applied charge Q

$$Q = \int\limits_0^R D\left(r\right) 2\pi r \mathrm{d}r, \qquad D\left(r\right) = \varepsilon_0 \frac{u}{l_0 - \xi} \approx \varepsilon_0 \frac{u}{l_0} \left(1 + \frac{\xi\left(r\right)}{l_0}\right)$$

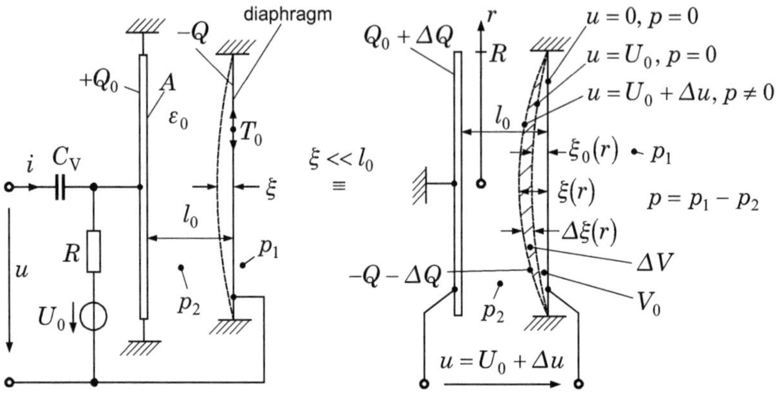

$$Q = \int_0^R D(r) \cdot 2\pi r \cdot \mathrm{d}r\,, \quad D(r) = \varepsilon_0 \cdot E(r)$$

$$p_{\mathrm{ges}} = p + p_{\mathrm{el}}\,, \quad p_{\mathrm{el}} = \frac{F_{\mathrm{el}}}{A} = \frac{u^2 \cdot \varepsilon_0}{2(l_0 - \xi)} \approx \frac{u^2 \cdot \varepsilon_0}{2 l_0} = \frac{Q^2}{2\varepsilon_0 \cdot A^2}$$

$$i = -\frac{\mathrm{d}Q}{\mathrm{d}t}\,, \qquad C_0 = \frac{\varepsilon_0 \cdot A}{l_0} \approx C_{\mathrm{b}}(u = U_0)$$

Fig. 9.20. Model of electrostatic transducer with circular diaphragm

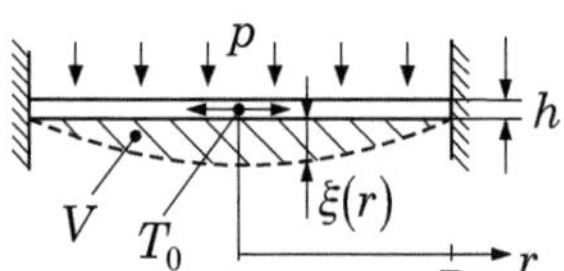

$$\xi(r) = \xi_0 \left(1 - \left(\frac{r}{R} \right)^2 \right)$$

$$\xi_0 = \frac{R^2}{4 T_0 \cdot h} p$$

$$V = \int_0^R 2\pi r \cdot \mathrm{d}r \cdot \xi(r) = \frac{1}{2} \pi \cdot R^2 \cdot \xi_0$$

$$N_{\mathrm{a}} = \frac{V}{p} = \frac{1}{8} \frac{\pi \cdot R^4}{T_0 \cdot h}$$

$$M_{\mathrm{a}} = 1.33\, \rho \frac{h}{\pi \cdot R^2}$$

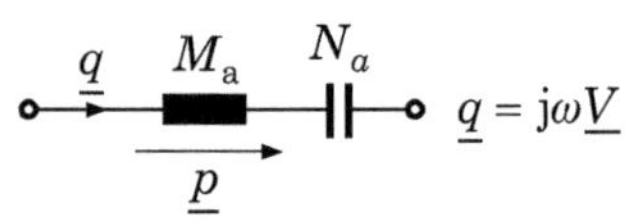

Fig. 9.21. Two-port network properties of a thin circular diaphragm prestressed with T_0

and thus

$$Q = \frac{\varepsilon_0 A}{l_0} u + \frac{\varepsilon_0 u}{l_0^2} \int_0^R \xi(r)\, 2\pi r \mathrm{d}r = C_\mathrm{b} u + \frac{\varepsilon_0 u}{l_0^2} V$$

the derivation of the transducer circuit diagram is realized in the same way as with the plate transformer. For small variations $\mathrm{d}p$, $\mathrm{d}Q$, $\mathrm{d}u$ and $\mathrm{d}V$, the expansion of ΔV and ΔQ is carried out in the neighborhood of the operating point U_0, Q_0, $p = 0$, $\xi_0(r)$ at first:

$$\Delta V = N_\mathrm{a} \mathrm{d}p + \frac{Q_0}{\varepsilon_0 A^2} N_\mathrm{a} \mathrm{d}Q$$

$$\Delta Q = C_\mathrm{b} \mathrm{d}u + \frac{\varepsilon_0 U_0}{l_0^2} \mathrm{d}V$$

Subsequently, complex amplitudes are introduced

$$\mathrm{d}u \to \underline{u}, \qquad \mathrm{d}Q \to -\frac{\underline{i}}{\mathrm{j}\omega}, \qquad \mathrm{d}V \to \frac{\underline{q}}{\mathrm{j}\omega}, \qquad p = p_1 - p_2 \to \underline{p}$$

and the basic equations are transformed with respect to electrical and acoustic network coordinates

$$\underline{q} = \mathrm{j}\omega N_\mathrm{a} \underline{p} - \frac{Q_0}{\varepsilon_0 A^2} N_\mathrm{a} \underline{i}$$

$$\underline{i} = -\mathrm{j}\omega C_\mathrm{b} \underline{u} - \frac{\varepsilon_0 U_0}{l_0^2} \underline{q}.$$

In order to ensure a real transducer constant Y, the equations are transformed into the form $\underline{i}, \underline{p} = f(\underline{u}, \underline{q})$

$$\underline{i} + \mathrm{j}\omega C_\mathrm{b} \underline{u} = \underline{i}_\mathrm{W} = -\frac{\varepsilon_0 U_0}{l_0^2} \underline{q} = -\frac{1}{YA} \underline{q} \tag{9.8}$$

and

$$\underline{p} = \mathrm{j}\omega M_{\mathrm{a,M}} \underline{q} + \frac{1}{\mathrm{j}\omega N_{\mathrm{a,M}}} \underline{q} - \frac{Q_0}{\mathrm{j}\omega \varepsilon_0 A^2} \left(\mathrm{j}\omega C_\mathrm{b} \underline{u} + \frac{\varepsilon_0 U_0}{l_0^2} \underline{q} \right)$$

$$\underline{p} = \mathrm{j}\omega M_{\mathrm{a,M}} \underline{q} + \frac{1}{\mathrm{j}\omega} \left(\frac{1}{N_{\mathrm{a,M}}} - \frac{1}{(YA)^2 C_\mathrm{b}} \right) \underline{q} - \frac{1}{YA} \underline{u}$$

$$\underline{p} - \mathrm{j}\omega M_{\mathrm{a,M}} \underline{q} - \frac{1}{\mathrm{j}\omega} \left(\frac{1}{N_{\mathrm{a,M}}} + \frac{1}{N_{\mathrm{a,C}}} \right) \underline{q} = \underline{p}_\mathrm{W} = -\frac{1}{YA} \underline{u}. \tag{9.9}$$

By applying KIRCHHOFF's laws, (9.8) and (9.9) provide the circuit diagram of the electrostatic transducer with prestressed diaphragm illustrated in Fig. 9.22.

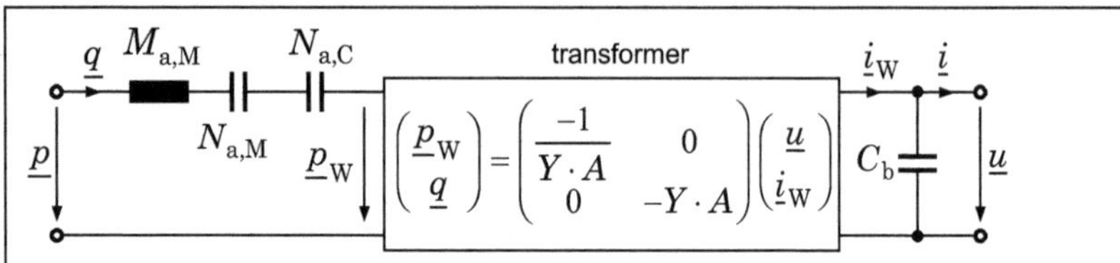

$$M_{\mathrm{a,M}} = 1.33\frac{\rho h}{\pi R^2},\ N_{\mathrm{a,M}} = \frac{\pi R^4}{8T_0 h},\ N_{\mathrm{a,C}} = -(YA)^2 C_{\mathrm{b}},\ Y = \frac{l_0^*}{U_0\, C_{\mathrm{b}}},\ C_{\mathrm{b}} = \frac{\varepsilon_0 A}{l_0^*}$$

Fig. 9.22. Circuit diagram of electrostatic transducer with prestressed circular diaphragm

Compared to the plate transformer, the electrostatic diaphragm transducer shows a *transformer-like* coupling. This is caused by excitation of the transducer with acoustic coordinates $\underline{p}$ and $\underline{q}$ in contrast to mechanical coordinates $\underline{v}$ and $\underline{F}$ with the plate transformer. The transformation relations from electrical to acoustic side and vice versa correspond to the relations of the transformer-like coupling two-port network represented in Table 7.2.
Transducers with circular diaphragms are used for electrostatic microphones and speakers. The *strip diaphragm* with acoustic compliance

$$N_{\mathrm{a,M}} = \frac{1}{12}\frac{l^3 b}{T_0 h}$$

represents another technically important application. If the width of diaphragm b reaches the dimension of diaphragm thickness h, it will be referred to the strip diaphragm as *tensioned string. Resonant sensors* with vibrating strings consisting of quartz or silicon become more important for high-resolution measurements of forces and pressures.

9.1.4 Sample Applications

As sample applications of diaphragm transducer the dynamic design of condenser microphones is explained. The operating frequency range of these acoustic pressure sensors manufactured in very large quantities is within the acoustic noise range in the first case and in the second case within the ultrasonic range.

Condenser Microphone for Acoustic Noise Range ($f < 20\,\mathrm{kHz}$)

The design principle of a condenser microphone with prestressed circular diaphragm and circular pressure compensation is specified in Fig. 9.23. In order to avoid electrical bias U_0, condenser microphones with permanently charged dielectric – *electrets* – are used (*electret microphones*). As a dielectric, here an electrically charged teflon foil (charge density of about $2 \cdot 10^{-4}\,\mathrm{C\,m^{-2}}$) with extremely high time constant of charge drain (about 200 years) is used. The calculation of the transfer function $\underline{B}_p = \underline{u}_1/\underline{p}$ is made separately for low and high frequencies.

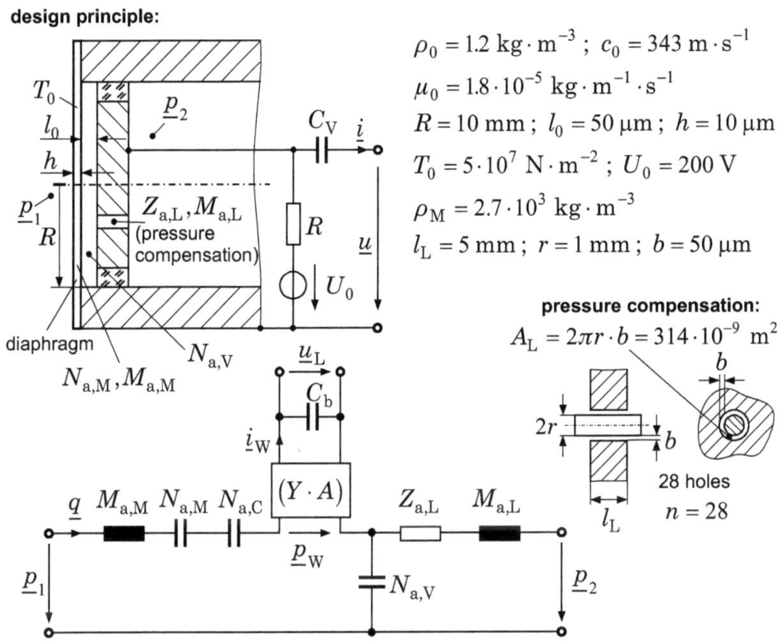

components:

$$M_{a,M} = 1.33\frac{\rho_M h}{\pi R^2} = 114.3\ \mathrm{kg\cdot m^{-4}} \qquad Z_{a,L} = \frac{1}{n}\frac{12\mu_0}{b^3}\frac{l_L}{2\pi r} = 49.1\cdot 10^{9}\ \mathrm{N\cdot s\cdot m^{-5}}$$

$$N_{a,M} = \frac{\pi}{8}\frac{R^4}{T_0\cdot h} = 7.9\cdot 10^{-12}\ \mathrm{m^5N^{-1}} \qquad M_{a,L} = \frac{1}{n}\frac{\rho_0\cdot l_L}{A_L} = 682.1\ \mathrm{kg\cdot m^{-4}}$$

$$N_{a,C} = -(Y\cdot A)^2\cdot C_b \qquad N_{a,V} = \frac{\pi R^2\cdot l_0}{\rho_0\cdot c_0^2} = 1.1\cdot 10^{-13}\ \mathrm{m^5N^{-1}}$$

Fig. 9.23. Schematic diagram and circuit diagram of a condenser microphone with metal diaphragm prestressed by T_0

For *low frequencies*, a pressure compensation $\underline{p}_1 = \underline{p}_2 = \underline{p}$ occurs. Furthermore, the acoustic mass of diaphragm $M_{a,M}$ and the acoustic mass of moving air in the pressure compensation bore $M_{a,L}$ can be neglected. By transforming the capacitance C_b to the acoustic side $(YA)^2\,C_b$, the negative compliance

$N_{a,C}$ will be compensated. The transfer function of resultant simplified circuit illustrated in Fig. 9.24 finally results in combination with

$$\underline{u}_L = \frac{1}{j\omega} \frac{1}{C_b Y A} \underline{q} = \frac{1}{j\omega} \frac{U_0}{l_0 A} \underline{q}$$

and

$$\underline{q}' = \frac{\underline{p}}{\dfrac{1}{j\omega N_{a,V}} + \dfrac{1}{j\omega N_{a,M} + \dfrac{1}{\underline{Z}_{a,L}}}}, \qquad \underline{q} = \frac{j\omega N_{a,M}}{j\omega N_{a,M} + \dfrac{1}{\underline{Z}_{a,L}}} \underline{q}'$$

$$\underline{q} = j\omega \frac{N_{a,M} N_{a,V}}{N_{a,M} + N_{a,V}} \cdot \frac{1}{1 + \dfrac{1}{j\omega \left(N_{a,M} + N_{a,V}\right) \underline{Z}_{a,L}}} \underline{p}$$

in

$$\underline{B}_p = \frac{\underline{u}_L}{\underline{p}} = B_0 \frac{j\left(\omega/\omega_1\right)}{1 + j\left(\omega/\omega_1\right)}$$

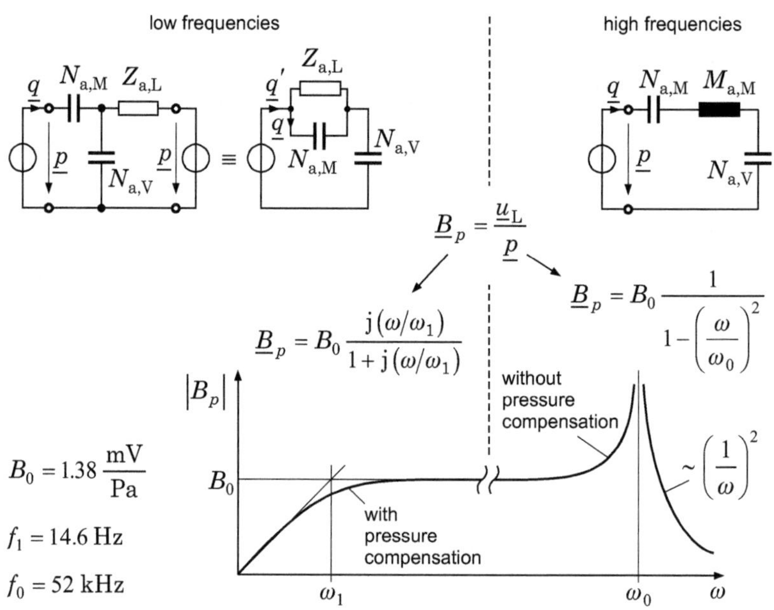

Fig. 9.24. Simplified circuit diagrams and curve progression of the transfer function of a condenser microphone for low and high frequencies within acoustic noise range

with

$$B_0 = \frac{N_{\mathrm{a,M}} N_{\mathrm{a,V}}}{N_{\mathrm{a,M}} + N_{\mathrm{a,V}}} \frac{U_0}{A l_0}$$

$$\omega_1 = \frac{1}{Z_{\mathrm{a,L}} \left(N_{\mathrm{a,M}} + N_{\mathrm{a,V}}\right)}.$$

For *high frequencies*, the pressure compensation does not work. The impedance of diaphragm must now be supplemented by the mass effect. The simplified circuit diagram for high frequencies is also shown in Fig. 9.24.
In combination with

$$\underline{q} = \frac{1}{\mathrm{j}\omega M_{\mathrm{a,M}} + \cfrac{1}{\cfrac{1}{\mathrm{j}\omega N_{\mathrm{a,M}}} + \cfrac{1}{\mathrm{j}\omega N_{\mathrm{a,V}}}}} \underline{p} = \mathrm{j}\omega \frac{N_{\mathrm{a,M}} N_{\mathrm{a,V}}}{N_{\mathrm{a,M}} + N_{\mathrm{a,V}}} \frac{1}{1 - \left(\cfrac{\omega}{\omega_0}\right)^2} \underline{p}$$

and

$$\underline{u}_{\mathrm{L}} = \frac{1}{\mathrm{j}\omega} \frac{U_0}{l_0 A} \underline{q},$$

it applies to the transfer function

$$\underline{B}_p = \frac{\underline{u}_{\mathrm{L}}}{\underline{p}} = B_0 \frac{1}{1 - \left(\omega/\omega_0\right)^2}$$

with

$$B_0 = \frac{N_{\mathrm{a,M}} N_{\mathrm{a,V}}}{N_{\mathrm{a,M}} + N_{\mathrm{a,V}}} \frac{U_0}{A l_0}$$

$$\omega_0^2 = \frac{N_{\mathrm{a,M}} + N_{\mathrm{a,V}}}{M_{\mathrm{a,M}} N_{\mathrm{a,M}} N_{\mathrm{a,V}}}.$$

Taking losses within the rear air volume and pressure within the pressure compensation bores into account, for practical realizations the quality factors are in the range of $0.5 \leq Q \leq 2$.

Micromechanical Silicon Microphone for Ultrasonic Range

The structure and principle of a condenser microphone comprising a silicon diaphragm chip and a chip with silicon counter electrodes presented in [80] are shown in Fig. 9.25. Both chips are realized with bulk micromachining technology using anisotropic etching processes. The bonding technology represents a suitable interconnection technology of both silicon elements. Both chips are adhesively bonded and the resulting device represents an electrostatic transducer with an edge dimension of $(2 \times 2)\,\mathrm{mm}^2$ and a thickness of about 0.5 mm.

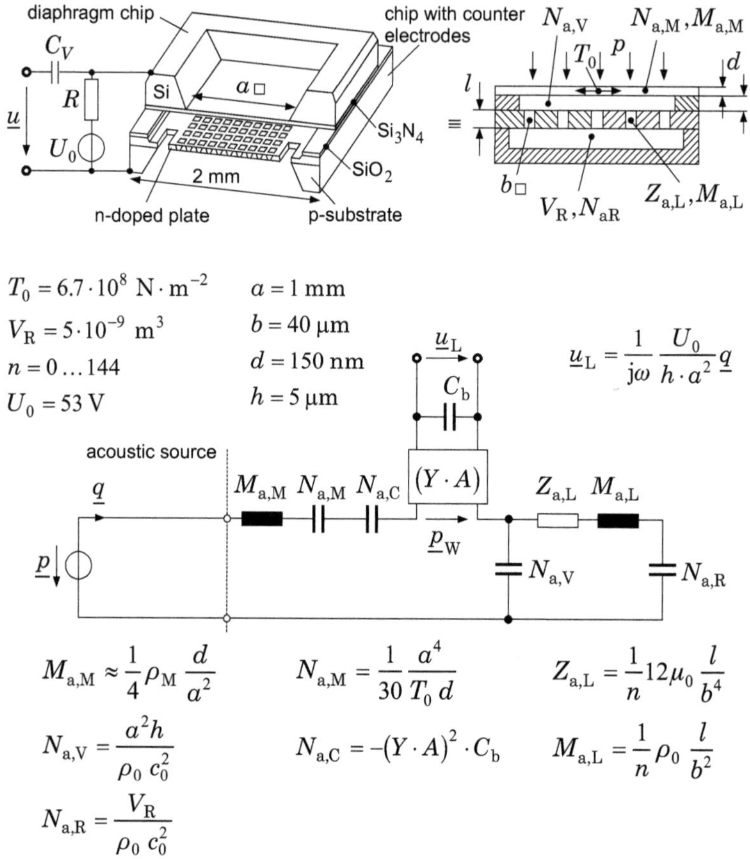

Fig. 9.25. Structure, schematic diagram and circuit of a condenser microphone realized with silicon micromachining technology

The diaphragm stress T_0 is adjusted by the polarization voltage U_0 for specified diaphragm dimensions a. The circuit of this microphone and the values of acoustic components are also specified in Fig. 9.25. Particular attention should be given to the minimum thickness of the silicon nitride diaphragm of only 150 nm and to the small air gap distance of (3 up to 5) μm.

The silicon microphone is expected to be affected by the acoustic pressure load $\underline{p}$. The silicon nitride diaphragm is represented by its mass $M_{a,M}$ and its compliance $N_{a,M}$. The compliances of air gap volume and rear volume are taken into account by the quantities $N_{a,V}$ and $N_{a,R}$. The influence of air flow within the n holes is described by $Z_{a,L}$ and $M_{a,L}$.
The transfer function $\underline{B}_p = \underline{u}_L/\underline{p}$ of the circuit was calculated by means of a network simulation program. In Fig. 9.26, the normalized transfer function

characteristics are shown for a variation of diaphragm dimension a, polarization voltage U_0 and number of holes n. With decreasing diaphragm area the transfer factor B_0 of microphone decreases, however the operating frequency range is extended. In order to provide a constant diaphragm stress T_0, the polarization voltage U_0 must be increased in case of a reduction in diaphragm dimension a. The extension of operating frequency range with increasing number of holes n and thus with the decrease of $Z_{a,L}$ as shown in the right part Fig. 9.26 without significant decrease of transfer factor B_0 is remarkable.

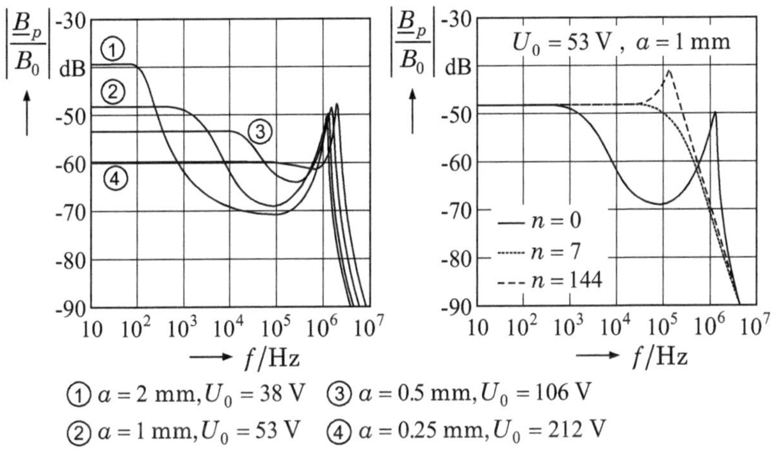

Fig. 9.26. Influence of diaphragm dimension a, polarization voltage U_0 and number of holes n on frequency response of the transfer function $\underline{B}_p/B_0$ of silicon microphone ($B_0 = 1\,\mathrm{V\,Pa^{-1}}$) [80]

This example shows that even for complex electromechanical systems, the network representation with concentrated components is appropriate. Instead of an analytical calculation of the transfer behavior, the application of network simulation programs should be preferred for more comprehensive network structures as they are shown in the present example.

9.1.5 Electrostatic Solid Body Transducers

For the transducer models discussed in Sect. 9.1.1 and 9.1.3 air or vacuum is used as dielectric. Now in Fig. 9.27 an *isotropic* dielectric with $\varepsilon_r > 1$ is inserted between the electrode plates.

When avoiding the strain of the dielectric, the MAXWELL stress T_M in the dielectric and thus the force F_M at the bounding surfaces is generated by the electric field E (Fig. 9.27 a)). If the dielectric shows a strain-dependent permittivity according to $d\varepsilon/dS \neq 0$, an additional mechanical stress T_E or

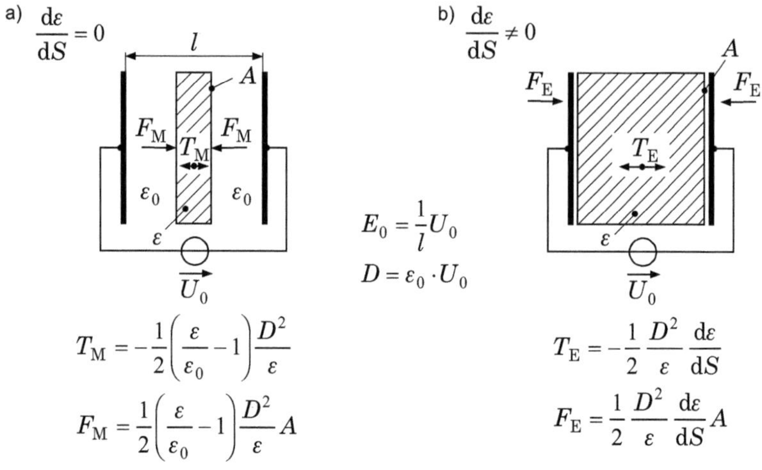

Fig. 9.27. Interactions in isotropic dielectrics

force F_E will be generated as shown in Fig. 9.27 b). Therefore, the total stress in the dielectric consists of MAXWELL stress and stress T_E that is referred to as *electrostriction*:

$$T = T_M + T_E = -\frac{1}{2}\left(\frac{\varepsilon}{\varepsilon_0} - 1\right)\frac{D^2}{\varepsilon} - \frac{1}{2}\frac{D^2}{\varepsilon}\frac{d\varepsilon}{dS} \tag{9.10}$$

The utilization of interactions in isotropic dielectrics is subject of research activities at present. As sample application an electrostatic polymer actuator is presented in the following section.

The situation is completely different with *anisotropic* piezoelectric materials, e.g. piezoelectric crystals and ferroelectric ceramics. In these materials an internal *polarization* P_i is already existent even there is no external electric field E_0. The mechanical stress T is generated by the already existing polarization P_i and polarization P_0 induced by the external field E_0:

$$T = -\frac{1}{2}E_0\left(P_0 + P_i\right), \qquad P_0 \sim E_0$$

Due to linear dependence of induced polarization on field strength E_0, the relation specified in (9.11) is achieved after introducing the constants K_1 and K_2

$$T = -\frac{1}{2}K_1E_0^2 + K_2E_0. \tag{9.11}$$

For piezoelectric materials $K_2 \gg K_1E_0$ is valid. Thus, piezoelectricity can also be considered as *linearized electrostriction.* In Sect. 9.2 it is dwelled on the description, properties and applications of piezoelectric materials.

9.1.6 Sample Application

Electrostatic Polymer Actuator

Electroactive polymers belong to the group of active polymers that are deformed by the influence of electrical quantities. An electric field being existent in the solid causes deformation. Electroactive polymers can be divided into two groups, into

- ionic electroactive polymers and
- electronic electroactive polymers.

In the first group, an ion transport causes a deformation even at low voltages of about 1 up to 5 V. In order to provide for the ion transport, an electrolyte is required, e.g. an aqueous solution. In addition to the required electrolytes the very slow reactivity with time constants in the range of 0.1 up to 1 s is disadvantageous [81].
The second group is represented, on the one hand, by ferroelectric polymers showing a piezoelectric effect, e.g. polyvinylidene fluoride (PVDF), and, on the other hand, by dielectric polymers which are excited electrostatically. In the following section 9.2, it is dwelled on anisotropic piezoelectric materials.

The considered polymer actuator is based on a *dielectric* polymer showing an *isotropic* deformation in case of an electrostatic excitation. The basic principle of a single-layer electrostatic polymer actuator is shown in Fig. 9.28. The elastic, isotropic dielectric (polymer or elastomer) is placed between two movable electrodes. By supplying electrical energy, a volume-invariant deformation results from the appearing force of attraction of the electrodes.

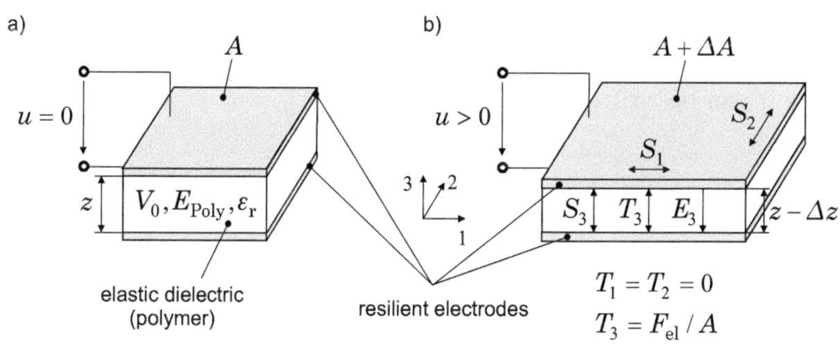

Fig. 9.28. Deformation of an elastic dielectric caused by an electrostatically generated mechanical stress
a) *without electric field* ($E_3 = 0$), b) *with electric field* E_3
E_{Poly} YOUNG*'s modulus of polymer*

Even though only the mechanical stress T_3 is generated, three strains S_1, S_2 and S_3 arise, i.e. the distance z between electrodes decreases about Δz and the electrode surface increases about ΔA. The strains can be calculated from the volume invariance of deformation and the isotropic material behavior of dielectric according to

$$\frac{\Delta V}{V_0} = S_1 + S_2 + S_3 = 0, \qquad S_1 = S_2 = -\nu_{\text{Poly}} \cdot S_3$$

$$S_1 = S_2 = -\frac{1}{2} S_3 = -\frac{1}{2} \frac{1}{E_{\text{Poly}}} T_3, \qquad T_3 = -p. \tag{9.12}$$

The mechanical compressive stress T_3 results from the force of attraction of the electrodes $T_3 = F_{\text{el}}/A$. The force of attraction can be calculated from the energy balance of the electrostatic transducer

$$W = u \cdot Q + \frac{1}{2} \frac{D^2}{\varepsilon} \cdot V_0 \tag{9.13}$$

(the term $u \cdot Q$ denotes the energy of electrical supply and term $\left(D^2/2\varepsilon\right) \cdot V_0$ denotes the field energy) and by applying the principle of virtual displacements

$$W + \Delta W = u \cdot Q + \frac{1}{2} \frac{D^2}{\varepsilon} (A + \Delta A)(z - \Delta z)$$

$$\Delta W = F_{\text{el}} \cdot \Delta z = \frac{1}{2} \frac{D^2}{\varepsilon} (\Delta A \cdot z - \Delta z \cdot A), \qquad \Delta A \cdot \Delta z \ll V_0$$

with

$$F_{\text{el}} = \frac{1}{2} \frac{D^2}{\varepsilon} z \cdot A \left(\frac{\Delta A}{A} - \frac{\Delta z}{z} \right) \cdot \frac{1}{\Delta z}. \tag{9.14}$$

The volume invariance of deformation results in

$$\Delta V = A \cdot \Delta z + z \cdot \Delta A = 0$$

and thus it can be written

$$\frac{\Delta z}{z} = -\frac{\Delta A}{A}. \tag{9.15}$$

Inserting (9.15) into (9.14) yields the force of attraction

$$F_{\text{el}} = -\frac{D^2}{\varepsilon} z \cdot A \cdot \frac{\Delta z}{z} \cdot \frac{1}{\Delta z} = -\frac{D^2}{\varepsilon} \cdot A = -\varepsilon_0 \varepsilon_{\text{r}} E^2 \cdot A. \tag{9.16}$$

Finally, the mechanical stress results in

$$T_3 = -p = \frac{F_{\text{el}}}{A} = -\varepsilon_0 \varepsilon_{\text{r}} E^2 \quad \text{with} \quad E = \frac{u}{z}. \tag{9.17}$$

Compared to (9.10) twice the value for the MAXWELL stress will be obtained, if electrostriction is neglected. The reason for this is the non-consideration of electrode surface change when deriving (9.10).

The silicone rubber *Elastosil P7670* (Wacker) is used as dielectric for the considered example of a polymer actuator. The relative permittivity ε_r amounts to approximately 3 according to [81]. For a layer thickness of $z = 30\,\mu\text{m}$ and an electrode voltage of $u = 1\,\text{kV}$ the pressure can be calculated in combination with (9.17) according to

$$p = 2.9 \cdot 10^4\,\text{Nm}^{-2} = 29\,\text{kPa}.$$

In combination with YOUNG's modulus of Elastosil of about $3 \cdot 10^5\,\text{Nm}^{-2}$, the strain results in

$$S_3 = -9.5 \cdot 10^{-2} = -9.5\%.$$

This value corresponds well with experimental results specified in Fig. 9.29.

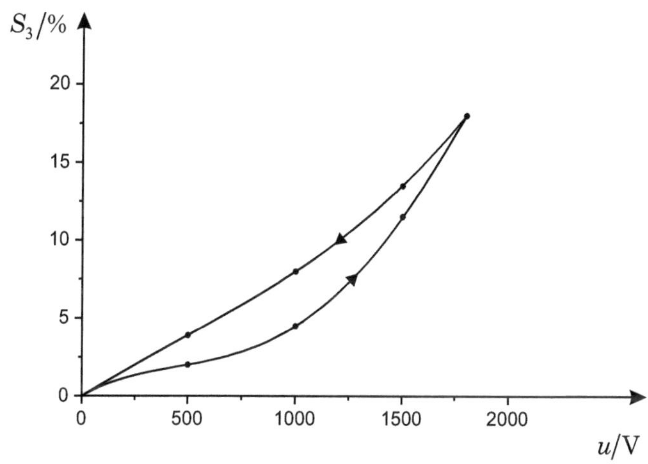

Fig. 9.29. Strain S_3 of dielectric as a function of electrode voltage [81]
Polymer: Elastosil P7679 (Wacker), thickness $30\,\mu\text{m}$

The very large strains up to approximately 20% and the strain characteristic with hysteresis represent a typical characteristic of electrostatic polymer actuators. Due to this large surface strain S_A

$$S_A = \frac{\Delta A}{A} = S_1 + S_2 = -S_3 = 9.5\%,$$

resilient electrodes must be utilized for polymer actuators. For that purpose graphite electrodes are recommended in [82].
For a strain of $S_3 = 7\%$ and a layer thickness of $30\,\mu\text{m}$ the decrease of electrode distance of the polymer actuator specified in Fig. 9.28 amounts to $\Delta z = 2.1\,\mu\text{m}$.

In order to increase the actuator stroke, in [81, 82] stack actuators are prepared and tested. The chosen materials and dimensions as well as the achievable stroke are specified in Fig. 9.30.

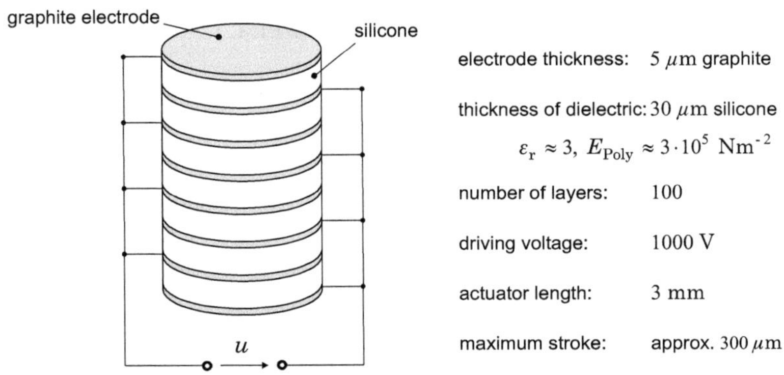

Fig. 9.30. Electrostatic polymer stack actuator [81]

The characteristics of dielectric polymer actuators are summarized in Table 9.2. Repeatable production of elastic electrodes and reduction of driving voltage represent the subject of current research studies.

Table 9.2. Characteristics of dielectric polymer actuators [81]

realizability of very large strains:	up to 20% (compared with piezoceramics approx. factor 100)
high electrostatically generated pressures:	up to 7 MPa
low density:	approx. 1 g/cm^3
very high breakdown field strength:	30 V/μm (compared with air approx. factor 10)
low risk of contamination:	solid
integration of functions possible:	dielectric, return spring, electrodes
batch processing:	possible

In the future stack actuators can be used for tactile displays, e.g. BRAILLE displays for blind persons [82]. The roll actuator represents another type of polymer actuators. It is a rolled-up polymer actuator in the form of a solid or hollow cylinder that can be used for artificial muscles in arm prostheses. The development of miniaturized peristaltic pumps for medical engineering by means of polymer actuators represents another field of research [83].

9.2 Piezoelectric Transducers with Lumped Parameters

9.2.1 Model Representation of the Piezoelectric Effect

Piezoelectric transducers represent the second practically meaningful group of electrical transducers. Also here, the electromechanical coupling is effected between mechanical and electrical field quantities. The transition to integral quantities which can be found spatially concentrated at a point or a surface, is not so easy as with the electrostatic transducers at first. However, considering low frequencies at first – *quasi-static* case – piezoelectric materials can be considered to be virtually massless. On this condition, the field quantities inside the piezoelectric material have the same value at each point. The field quantities are integrable and it can be passed on to the description with integral quantities and concentrated components.
In a second step, simple piezoelectric oscillators are considered whose mass and spatial dependence of component parameters can no longer be neglected. The transfer characteristics of oscillators are determined by applying solutions for *one-dimensional waveguides* and components with *distributed parameters* represented in the Sects. 6.1 and 6.2. For advanced studies it is referred to special scientific literature [2, 84, 85].
Within the scope of the present book, a detailed consideration of the theory of piezoelectric materials is abandoned. Here, it is referred to more detailed descriptions in special literature, too [86–88]. In Sect. 9.2.2, the derivation of the circuit structure of the quasi-static piezoelectric transducer is based on the phenomenological description of the piezoelectric effect (greek: *piezein* $\equiv$ *to press*).

In case of simultaneous action of mechanical and electrical quantities, no electromechanical couplings will be existent in *isotropic materials*, if the MAXWELL strain and electrostriction are neglected (9.10) as done in Sect. 9.1.5. When applying a mechanical stress T (Fig. 9.31), solely the elastic constant s allows for generating a strain S. When supplying a field strength E, the permittivity ε allows for generating a dielectric displacement D. However, selected *anisotropic materials* such as piezoelectric crystals and ferroelectric materials show a distinct coupling of mechanical and electrical quantities. Due to the molecular structure, in piezoelectric crystals, e.g. in *quartz*, a linear coupling is existent between electrical and mechanical quantities in principle. In

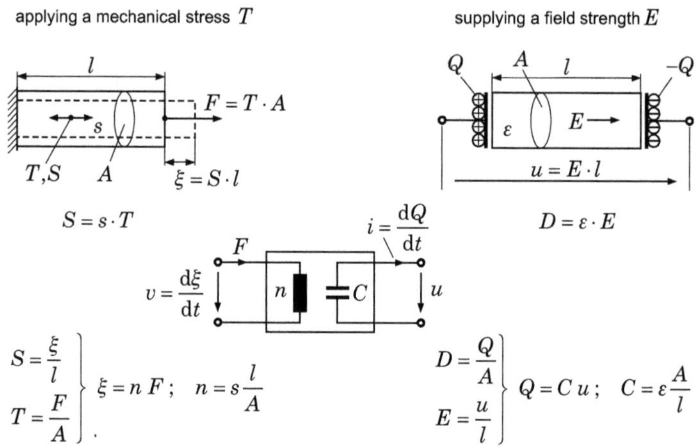

Fig. 9.31. Effect of mechanical and electrical quantities in isotropic materials

ferroelectric materials, e.g. *piezoelectric ceramics*, an internal polarization is generated by applying an electric constant field (see Sect. 9.2.6). This polarization linearizes the square effect of electrostriction according to (9.11) (see Sect. 9.1.5). Toward outside, the piezoelectric effect (Fig. 9.32) is character-

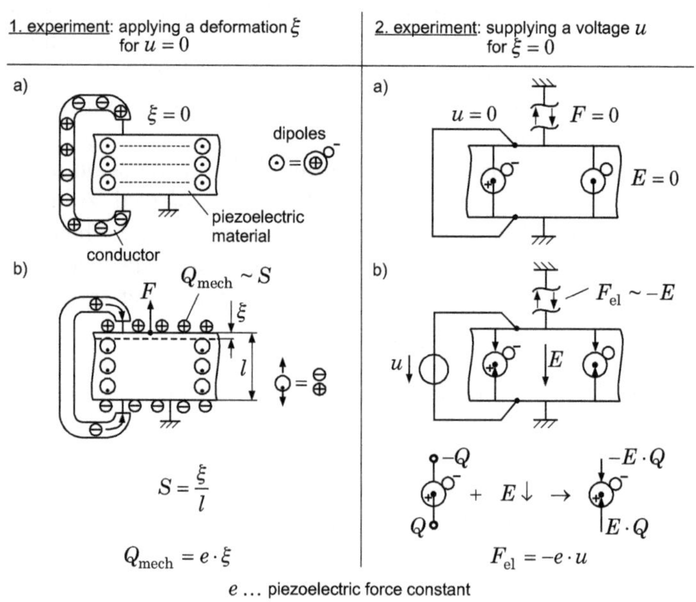

Fig. 9.32. Phenomenological model description of the piezoelectric effect

ized by the ability to displace the charge in case of a mechanical excitation by forces or deformations, or vice versa by a material deformation in case of an electrical excitation by a voltage or current. For piezoelectric materials the first experiment in Fig. 9.32 shows phenomenologically the charge displacement in case of applying a deformation. The second experiment illustrates the force generation in case of supplying an electrical voltage with respect to an avoided deformation. The piezoelectric effect is based on an elastic deformation of electric dipoles in a crystal lattice.
Depending on direction of action shown in Table 9.3, the piezoelectric effect results in a charge change or a deformation change. In case of a mechanical excitation, the generation of a charge is technically used for sensors which are applied for measuring mechanical quantities. Conversely, actuators for generating deformations or forces by electrical excitation can be devised.

Table 9.3. Basic equations of the piezoelectric effect for sensor and actuator applications

sensor applications

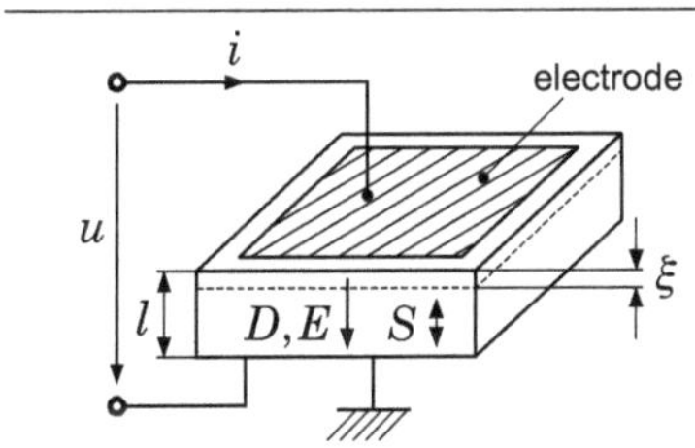	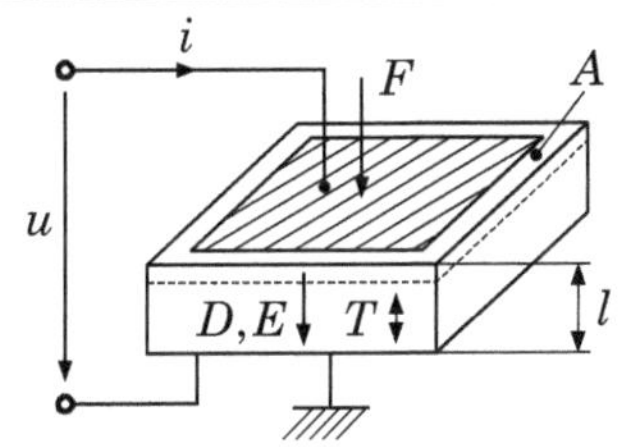
excitation: strain S for $E = 0$	excitation: mechanical stress T for $E = 0$
response: $D = e \cdot S$	response: $D = d \cdot T$

with $D = \frac{Q}{A}$, $T = \frac{F}{A}$ and $S = \frac{\xi}{l}$ follows:

$Q = e \cdot \frac{A}{l} \cdot \xi$	$Q = d \cdot F$
e … piezoelectric force constant	d … piezoelectric charge constant

actuator applications

excitation: electric field E for $S = 0$	excitation: electric field E for $T = 0$
$T = -e \cdot E$	$S = d \cdot E$

with $E = \frac{u}{l}$, $T = \frac{F}{A}$ and $S = \frac{\xi}{l}$ follows:

$F = -e \cdot \frac{A}{l} \cdot u$	$\xi = d \cdot u$

9.2.2 Piezoelectric Equations of State and Circuit Diagram for Longitudinal Coupling

The electromechanical couplings in piezoelectric materials can be described by means of equations of state. By means of two simple thought experiments, the *piezoelectric equations of state* are set up using the piezoelectric constant e (Fig. 9.33). In addition to this form, three other forms of equations of state with their special piezoelectric, elastic and dielectric constants are derivable. They are represented in Sect. 9.2.3.
In the first experiment, the voltage u is supplied and the charge Q_{el} is generated with respect to an avoided deformation $\xi = 0$. In case of short-circuit operation, a deformation is applied and the charge Q_{mech} can be measured. The total charge results from addition of both partial charges.

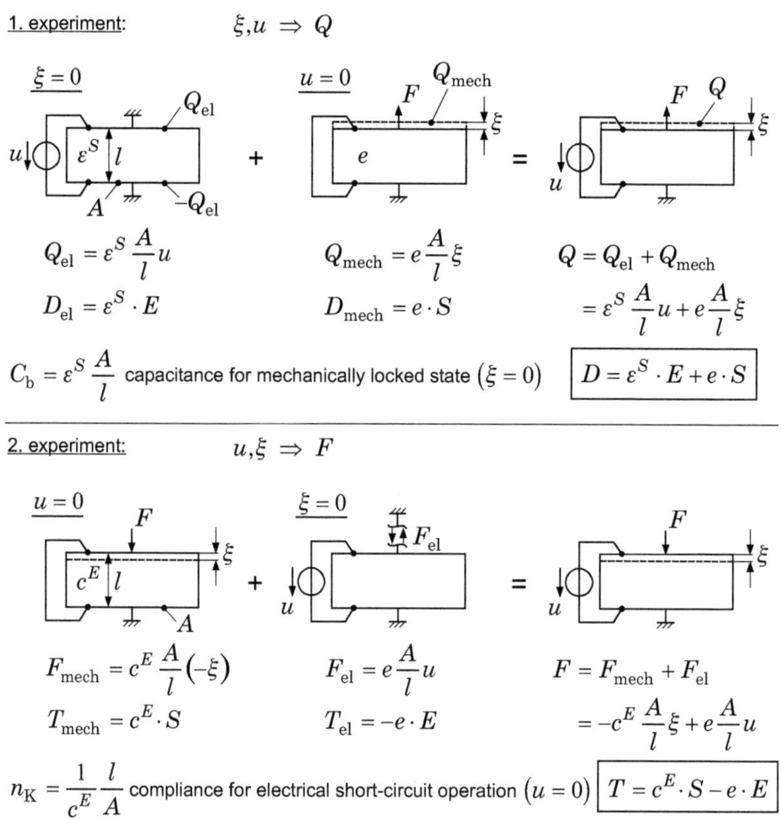

Fig. 9.33. Thought experiments for the definition of the piezoelectric constant e and derivation of piezoelectric equations of state

The first equation of state

$$D = \varepsilon^S E + eS \tag{9.18}$$

will be achieved after inserting the field quantities into this relation.

The second equation of state is achieved with the second thought experiment concerning the superposition of mechanically and electrically generated partial forces F_{mech} and F_{el}:

$$T = c^E S - eE \tag{9.19}$$

Using these equations of state, the derivation of the circuit diagram for the piezoelectric transducer shown in Fig. 9.34 is now possible.
The initial equations are formed by integral notation of the equations of state:

$$Q = \varepsilon^S \frac{A}{l} u + e \frac{A}{l} \xi \tag{9.20}$$

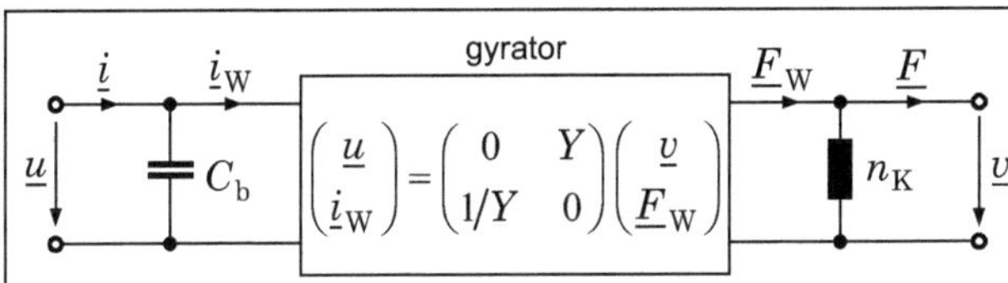

	with piezoelectric force constant e		with piezoelectric charge constant d
C_b	$\varepsilon^S \frac{A}{l}$	for $v = 0$	$\left(\varepsilon^T - d^2 \cdot c^E\right)\frac{A}{l}$
n_K	$\frac{1}{c^E} \frac{l}{A}$	for $u = 0$	$s^E \frac{l}{A}$
Y	$\frac{1}{e} \frac{l}{A}$	$e = d \cdot c^E = \frac{d}{s^E}$	$\frac{s^E}{d} \frac{l}{A}$
k^2	$\frac{e^2}{\varepsilon^S \cdot c^E}$		$\frac{d^2}{\varepsilon^S \cdot s^E}$

alternative circuit:

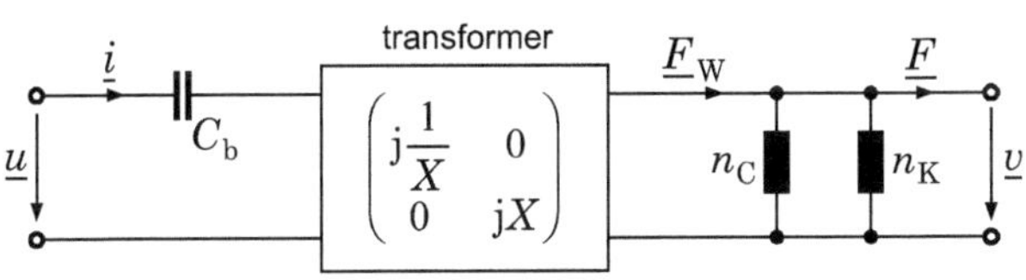

$$X = \frac{1}{\omega C_b \cdot Y}, \quad n_C = Y^2 \cdot C_b$$

Fig. 9.34. Circuit diagrams of piezoelectric transducer and component parameters for longitudinal effect

$$F = e\frac{A}{l}u - c^E\frac{A}{l}\xi \tag{9.21}$$

Here, it is also passed on to the notation with complex amplitudes

$$Q \to \underline{Q} = \frac{1}{\mathrm{j}\omega}\underline{i}, \qquad u \to \underline{u}, \qquad F \to \underline{F}, \qquad \xi \to \underline{\xi} = \frac{1}{\mathrm{j}\omega}\underline{v}$$

and (9.20) and (9.21) are rearranged with respect to $\underline{i}_{\mathrm{W}}$ and $\underline{F}_{\mathrm{W}}$. Thus, it can be written:

$$\underline{i} - \mathrm{j}\omega\varepsilon^S\frac{A}{l}\underline{u} = \underline{i}_{\mathrm{W}} = e\frac{A}{l}\underline{v} \tag{9.22}$$

$$\underline{F} + \frac{1}{\mathrm{j}\omega}c^E\frac{A}{l}\underline{v} = \underline{F}_{\mathrm{W}} = e\frac{A}{l}\underline{u} \tag{9.23}$$

The equations (9.22) and (9.23) correspond to the structure of the general electrical transducer. The numerical value of the material parameters s, c, ε is dependent on experimental boundary conditions. In each case, the „zero-forced" boundary value is written as a superscript symbol and put at the characteristic value. The gyrator-like transducer constant Y results in

$$Y = \frac{1}{e}\frac{l}{A}. \tag{9.24}$$

The circuit diagram and relations which are necessary for the calculation of component parameters from the material's characteristic values e, c, ε and the transducer's dimensions l and A are summarized in Fig. 9.34. Due to parallel direction of action of mechanical and electrical field quantities, it is referred to this coupling as *piezoelectric longitudinal effect*.

In addition to the circuit diagram with real-valued gyrator-like coupling, the alternative circuit diagram with imaginary transformer-like coupling is specified in Fig. 9.34. In Sect. 9.3.3 this alternative circuit will be used for the derivation of the circuit of the piezoelectric thickness oscillator as waveguide.

Depending on the direction of energy conversion – actuator or sensor operation – in Fig. 9.35 different sign specifications can be found in the circuit diagram illustrated in Fig. 9.34.

9.2.3 General Piezoelectric Equations of State

So far, in Sect. 9.2.2 solely a special case of the general piezoelectric coupling has been considered. Both the electrical and mechanical field have been one-dimensional and have had same direction of action. In the general case, all stress and strain components are interrelated to all field strength and dielectric displacement components.

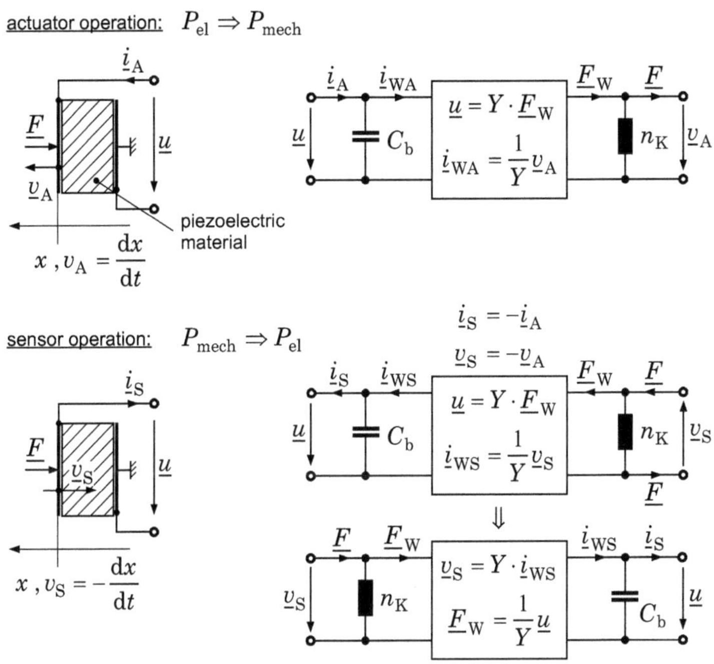

Fig. 9.35. Sign convention for different transfer directions

Using the piezoelectric force constant e for the linear relation of all electrical and mechanical field quantities it can be written [2, 84–86]:

$$\begin{array}{rl}
D_1 = & \varepsilon_{11}^S E_1 + \varepsilon_{12}^S E_2 + \varepsilon_{13}^S E_3 + e_{11} S_1 + e_{12} S_2 \cdots e_{16} S_6 \\
D_2 = & \varepsilon_{21}^S E_1 + \varepsilon_{22}^S E_2 + \varepsilon_{23}^S E_3 + e_{21} S_1 + e_{22} S_2 \cdots e_{26} S_6 \\
D_3 = & \varepsilon_{31}^S E_1 + \varepsilon_{32}^S E_2 + \varepsilon_{33}^S E_3 + e_{31} S_1 + e_{32} S_2 \cdots e_{36} S_6 \\
T_1 = & -e_{11} E_1 - e_{21} E_2 - e_{13} E_3 + c_{11}^E S_1 + c_{12}^E S_2 \cdots c_{16}^E S_6 \\
\vdots & \qquad\qquad\qquad\qquad\qquad\qquad \vdots \qquad \vdots \\
T_6 = & -e_{16} E_1 - e_{26} E_2 - e_{36} E_3 + c_{61}^E S_1 + c_{62}^E S_2 \cdots c_{66}^E S_6
\end{array}$$

This matrix notation can be shortened with introduction of summation signs according to:

$$D_n = \sum_{m=1}^{3} \varepsilon_{mn}^S E_m + \sum_{j=1}^{6} e_{nj} S_j \qquad n = 1 \ldots 3,\ m = 1 \ldots 3,\ j = 1 \ldots 6$$

$$T_i = \sum_{m=1}^{3} -e_{mi} E_m + \sum_{j=1}^{6} c_{ij}^E S_j \qquad i = 1 \ldots 6,\ m = 1 \ldots 3,\ j = 1 \ldots 6$$

By means of the EINSTEIN summation convention – when an index variable appears twice in a single term it implies a summation over all its possible values – finally the abbreviated form is achieved

$$D,T = f(E,S): \qquad D_n = \varepsilon^S_{mn} E_m + e_{nj} S_j \qquad n = 1\dots3,\ m = 1\dots3,\ j = 1\dots6 \tag{9.25}$$

$$T_i = -e_{mi} E_m + c^E_{ij} S_j \qquad i = 1\dots6,\ m = 1\dots3,\ j = 1\dots6. \tag{9.26}$$

The coefficients ε^S_{mn} denote dielectric constants on the condition that $S = 0$, i.e. they are determined experimentally for the mechanically locked state. The coefficients c^E_{ij} denote elastic constants on the condition that $E = 0$, i.e. for electrical short-circuit operation.
Using the piezoelectric charge constant d, the second form of piezoelectric equations of state is achieved similarly

$$D,S = f(E,T): \qquad D_n = \varepsilon^T_{mn} E_m + d_{nj} T_j \qquad n = 1\dots3,\ m = 1\dots3,\ j = 1\dots6 \tag{9.27}$$

$$S_i = d_{mi} E_m + s^E_{ij} T_j \qquad i = 1\dots6,\ m = 1\dots3,\ j = 1\dots6. \tag{9.28}$$

In order to determine the permittivities ε^T_{mn} and elastic constants s^E_{ij} experimentally, the mechanical stress T and field strength E were set equal to zero respectively, i.e. mechanical open-circuit and electrical short-circuit operation. In [2], [84] and [87] the third and fourth form of piezoelectric equations of state

$$E,S = f(D,T) \qquad \text{and} \qquad T,E = f(D,S)$$

are specified, but they will not be required within the scope of the present book. The following relations exist between the constants of the equations of state (9.25) up to (9.28):

$$e_{nj} = d_{nj} c^E_{ij}, \qquad d_{nj} = e_{ni} s^E_{ij}$$

$$c_{ij} = c_{ji}, \qquad s_{ij} = s_{ji}$$

$$\varepsilon^T_{nm} - \varepsilon^S_{nm} = e_{ni} e_{mj} s^E_{ij} = c^E_{ij} d_{mi} d_{nj}$$

9.2.4 Piezoelectric Transducers and Corresponding Equivalent Parameters

Based on the general equations of state (9.25) up to (9.28) presented in Sect. 9.2.3 now the task is to determine the component parameters of the circuit shown in Fig. 9.34 from the sets of constants ε, s, c, d, e and geometrical dimensions for technically important configurations by introducing electrical and mechanical boundary conditions. The compilation of required piezoelectric, elastic and dielectric constants of technically important materials is provided in Sect. 9.2.6.
The principle approach will be explained by means of two examples.

Free Thickness Oscillator (Longitudinal Effect)

Figure 9.36 illustrates the configuration of the free thickness oscillator and appropriate boundary conditions.

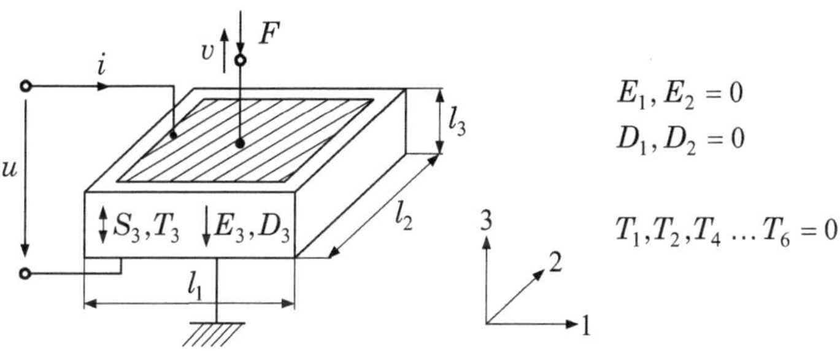

Fig. 9.36. Free thickness oscillator (longitudinal effect)

In the first step, the general equations of state $D, T = f(E, S)$ are rearranged taking the boundary conditions between the four field quantities in two equations into account:

$$D_3 = \varepsilon_{33}^T E_3 + d_{33} T_3, \qquad S_3 = d_{33} E_3 + s_{33}^E T_3$$

In the following, these equations of state are transformed to the form $D, T = f(E, S)$ with

$$T_3 = \frac{1}{s_{33}^E} S_3 - \frac{d_{33}}{s_{33}^E} E_3 = c_{33}^E S_3 - e_{33} E_3$$

and

$$D_3 = \varepsilon_{33}^T E_3 + d_{33} \left(\frac{1}{s_{33}^E} S_3 - \frac{d_{33}}{s_{33}^E} E_3 \right)$$

$$= \varepsilon_{33}^T \left(1 - \frac{d_{33}^2}{\varepsilon_{33}^T s_{33}^E} \right) E_3 + \frac{d_{33}}{s_{33}^E} S_3 = \varepsilon_{33}^T E_3 + e_{33} S_3$$

which form the starting point for deriving the circuit shown in Fig. 9.34 of Sect. 9.2.3.
As already performed in Sect. 9.2.2, in a second step it is passed on to the integral coordinates $\underline{v}$, $\underline{F}$, $\underline{u}$ and $\underline{i}$. The circuit diagram shown in Fig. 9.34 will be the result. Taking the electrically and mechanically effective dimensions

$$l_{\mathrm{el}} = l_{\mathrm{mech}} = l_3 \qquad \text{and} \qquad A_{\mathrm{el}} = A_{\mathrm{mech}} = l_1 l_2$$

into account, after insertion of ε, c, e the equivalent parameters finally result in:

$$C_{\mathrm{b}} = \varepsilon_{33}^T \left(1 - \frac{d_{33}^2}{\varepsilon_{33}^T s_{33}^E}\right) \frac{l_1 l_2}{l_3}, \qquad n_{\mathrm{K}} = s_{33}^E \frac{l_3}{l_1 l_2}$$

$$Y = \frac{s_{33}^E}{d_{33}} \frac{l_3}{l_1 l_2}, \qquad k_{33}^2 = \frac{d_{33}^2}{\varepsilon_{33}^T s_{33}^E}$$

In Sect. 9.2.7, a free thickness oscillator consisting of PZT ceramics is used in an accelerometer.

Free Longitudinal Oscillator (Transverse Effect)

The configuration of the free longitudinal oscillator is specified in Fig. 9.37 in combination with boundary conditions.

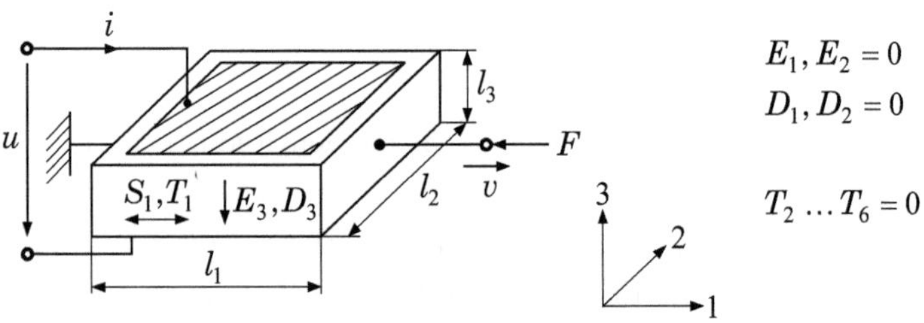

Fig. 9.37. Free longitudinal oscillator (transverse effect)

Also here, the general equations of state are rearranged in the first step in two equations with four coupled field quantities

$$D_3 = \varepsilon_{33}^T E_3 + d_{31} T_1, \qquad S_1 = d_{31} E_3 + s_{11}^E T_1 .$$

These equations of state are transformed again to the form $D, T = f(E, S)$ with the constants ε, c, e according to

$$T_1 = \frac{1}{s_{11}^E} S_1 - \frac{d_{31}}{s_{11}^E} E_3 = c_{11}^E S_1 - e_{31} E_3$$

and

$$D_3 = \varepsilon_{33}^T E_3 + d_{31}\left(\frac{1}{s_{11}^E} S_1 - \frac{d_{31}}{s_{11}^E} E_3\right)$$
$$= \varepsilon_{33}^T \left(1 - \frac{d_{31}^2}{\varepsilon_{33}^T s_{11}^E}\right) E_3 + \frac{d_{31}}{s_{11}^E} S_1 = \varepsilon_{33}^T E_3 + e_{31} S_1 .$$

With the transverse effect, the electrically and mechanically effective dimensions

$$l_{\text{el}} = l_3, \qquad l_{\text{mech}} = l_1 \qquad \text{and} \qquad A_{\text{el}} = l_1 l_2, \qquad A_{\text{mech}} = l_2 l_3$$

are not identical anymore. After insertion of constants ε, c, e and dimensions into the relations specified in Fig. 9.34 the equivalent parameters are achieved:

$$C_{\text{b}} = \varepsilon_{33}^T \left(1 - \frac{d_{31}^2}{\varepsilon_{33}^T s_{11}^E}\right) \frac{l_1 l_2}{l_3}, \qquad n_{\text{K}} = s_{11}^E \frac{l_1}{l_2 l_3}$$
$$Y = \frac{s_{11}^E}{d_{31}} \frac{1}{l_2}, \qquad k_{31}^2 = \frac{d_{31}^2}{\varepsilon_{33}^T s_{11}^E} .$$

In the Tables 9.4 and 9.5, the boundary conditions and characteristic values of free thickness and longitudinal oscillator and of clamped thickness and shear oscillator are specified. The approach for determining the parameters corresponds to the approach applied to the described examples.
Free thickness and flexural oscillators are especially used in accelerometers, clamped-free thickness oscillators are used for ultrasonic transmitter elements, face-shear oscillators are used as transmitter in the sonar technology and thickness-shear oscillators are used as filter elements and accelerometers without pyroelectric effect. The characteristic operating frequency ranges of these oscillators are represented in Fig. 9.38.

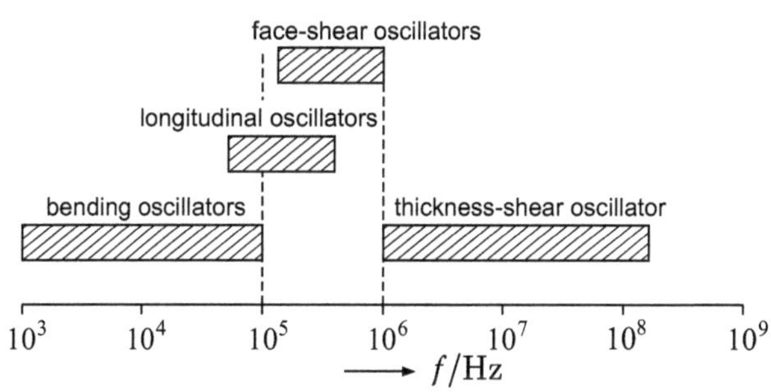

Fig. 9.38. Characteristic operating frequency ranges of piezoelectric oscillators

Table 9.4. Selected vibration modes of piezoelectric oscillators and appropriate characteristic values

	free thickness oscillator	clamped thickness oscillator	longitudinal oscillator
geometrical configuration	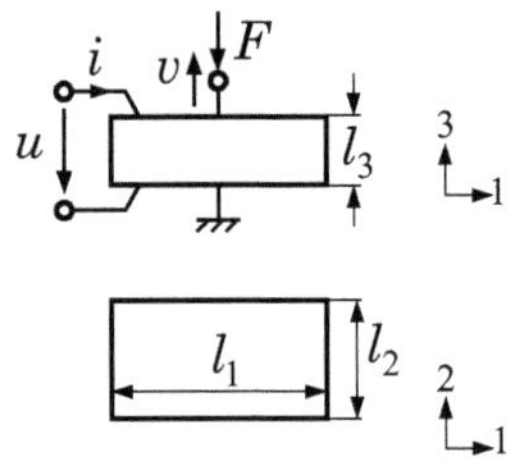	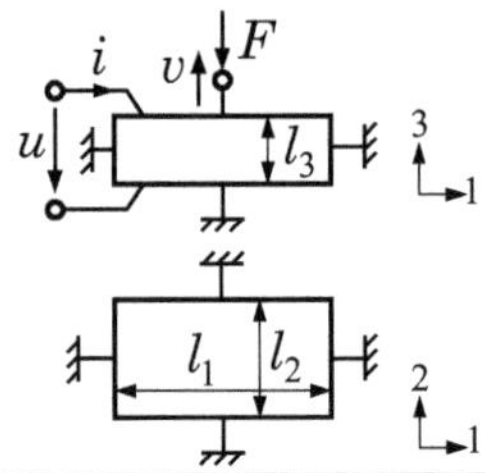	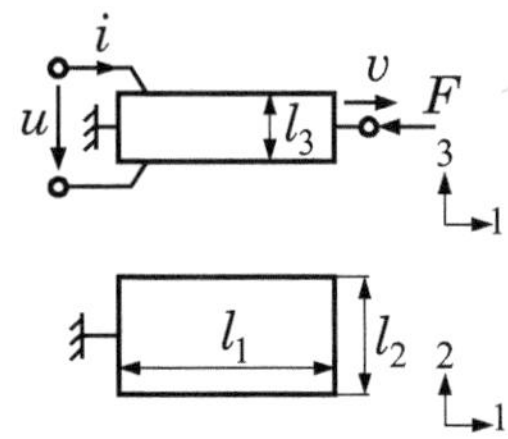
mechan. and electr. boundary conditions	$T_1, T_2, T_4 \ldots T_6 = 0$ $F = -T_3\, l_1 \cdot l_2$ $\underline{v} = \mathrm{j}\omega\, \underline{S}_3 \cdot l_3$	$S_1, S_2, S_4 \ldots S_6 = 0$ $F = -T_3\, l_1 \cdot l_2$ $\underline{v} = \mathrm{j}\omega\, \underline{S}_3 \cdot l_3$	$T_2 \ldots T_6 = 0$ $F = -T_1\, l_2 \cdot l_3$ $\underline{v} = \mathrm{j}\omega\, \underline{S}_1 \cdot l_1$
	$E_1, E_2 = 0$, $\quad \underline{u} = l_3 \cdot \underline{E}_3$, $\quad \underline{i} = \mathrm{j}\,\omega\, \underline{D}_3\, l_1 \cdot l_2$		
A	$A_{\mathrm{el}} = A_{\mathrm{mech}} = l_1 \cdot l_2$	$A_{\mathrm{el}} = A_{\mathrm{mech}} = l_1 \cdot l_2$	$A_{\mathrm{el}} = l_1\, l_2, A_{\mathrm{mech}} = l_2\, l_3$
l	$l_{\mathrm{el}} = l_{\mathrm{mech}} = l_3$	$l_{\mathrm{el}} = l_{\mathrm{mech}} = l_3$	$l_{\mathrm{el}} = l_3,\ l_{\mathrm{mech}} = l_1$
e	$\dfrac{d_{33}}{s_{33}^E}$	e_{33}	$\dfrac{d_{31}}{s_{11}^E}$
ε	$\varepsilon_{33}^T\left(1-\dfrac{d_{33}^2}{\varepsilon_{33}^T \cdot s_{33}^E}\right)$	ε_{33}^S	$\varepsilon_{33}^T\left(1-\dfrac{d_{31}^2}{\varepsilon_{33}^T \cdot s_{11}^E}\right)$
c	$\dfrac{1}{s_{33}^E}$	c_{33}^E	$\dfrac{1}{s_{11}^E}$
k^2	$\dfrac{d_{33}^2}{\varepsilon_{33}^T \cdot s_{33}^E}$	$\dfrac{e_{33}^2/\left(\varepsilon_{33}^S \cdot c_{33}^E\right)}{1+e_{33}^2/\left(\varepsilon_{33}^S \cdot e_{33}^E\right)}$	$\dfrac{d_{31}^2}{\varepsilon_{33}^T \cdot s_{11}^E}$
two-port network representation	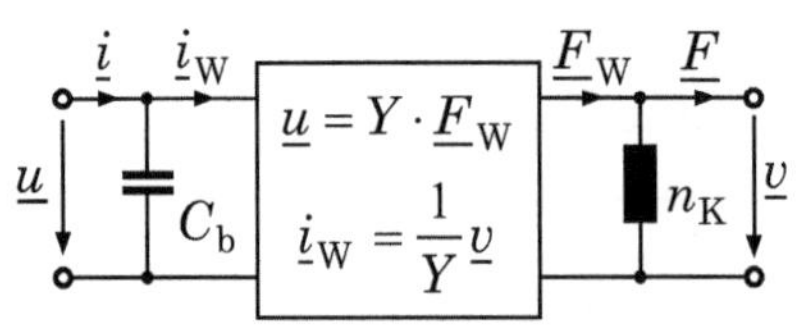 $C_{\mathrm{b}} = \varepsilon \dfrac{A_{\mathrm{el}}}{l_{\mathrm{el}}}$ $\quad n_{\mathrm{K}} = \dfrac{1}{c}\dfrac{l_{\mathrm{mech}}}{A_{\mathrm{mech}}}$		$\dfrac{1}{Y} = e\dfrac{A_{\mathrm{el}}}{l_{\mathrm{mech}}} = e\dfrac{A_{\mathrm{mech}}}{l_{\mathrm{el}}}$ $k^2 = \dfrac{e^2/(c\cdot\varepsilon)}{1+e^2/(c\cdot\varepsilon)} = \dfrac{1}{1+Y^2 C_{\mathrm{b}}/n_{\mathrm{K}}}$

Table 9.5. Selected vibration modes of piezoelectric oscillators and appropriate characteristic values

	face-strain oscillator	thickness-shear oscillator	face-shear oscillator
geometrical configuration	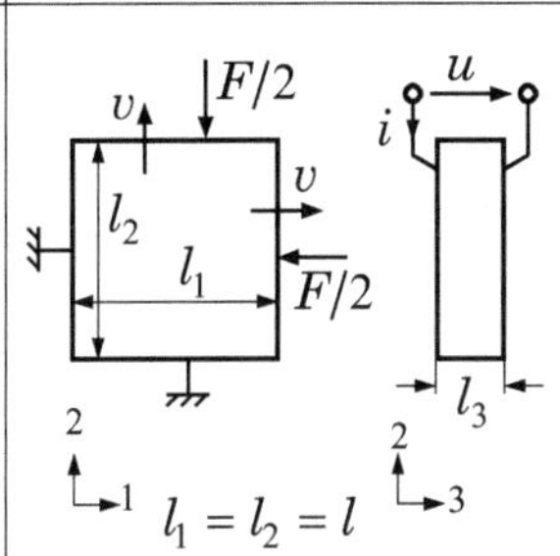	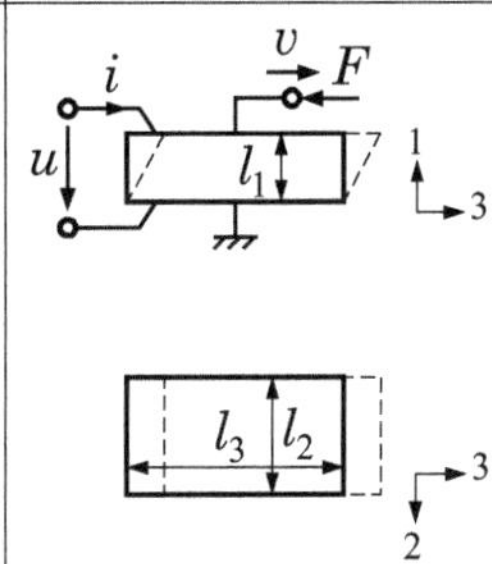	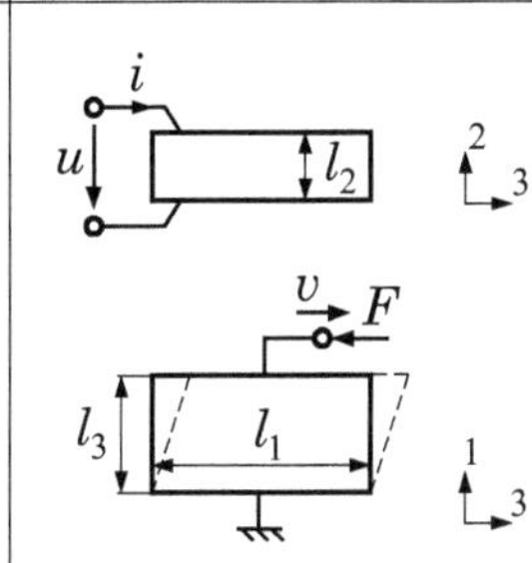
mechan. and electr. boundary conditions	$T_3 \ldots T_6 = 0,\ T_1 = T_2 = T/2$ $F = T \cdot l_1 \cdot l_3$ $\underline{v} = \mathrm{j}\omega\, \underline{S} \cdot l,\quad S_1 = S_2 = S$	$T_1 \ldots T_4, T_6 = 0$ $F = -T_5 \cdot l_2 \cdot l_3$ $\underline{v} = \mathrm{j}\omega\, \underline{S}_5 \cdot l_1$	$T_1 \ldots T_4 = 0$ $F = -T_5\ l_2 \cdot l_3$ $\underline{v} = \mathrm{j}\omega\ l_1 \cdot \underline{S}_5$
	$E_1 = E_2 = 0$ $\underline{u} = l_3 \underline{E}_3,\ \underline{i} = \mathrm{j}\omega \underline{D}_3 l_1 l_2$	$E_2 = E_3 = 0$ $\underline{u} = l_1 \underline{E}_1,\ \underline{i} = \mathrm{j}\omega \underline{D}_1 l_2 l_3$	$E_1 = E_3 = 0$ $\underline{u} = l_2 \underline{E}_2,\ \underline{i} = \mathrm{j}\omega \underline{D}_2 l_1 l_3$
A	$A_{\mathrm{el}} = l^2,\ A_{\mathrm{mech}} = l\ l_3$	$A_{\mathrm{el}} = A_{\mathrm{mech}} = l_2\ l_3$	$A_{\mathrm{el}} = l_1\ l_3,\ A_{\mathrm{mech}} = l_2\ l_3$
l	$l_{\mathrm{el}} = l_3,\ l_{\mathrm{mech}} = l$	$l_{\mathrm{el}} = l_{\mathrm{mech}} = l_3$	$l_{\mathrm{el}} = l_2,\ l_{\mathrm{mech}} = l_3$
e	$\dfrac{2d_{31}}{s_{11}^E + s_{12}^E}$	$\dfrac{d_{15}}{s_{55}^E}$	$\dfrac{d_{25}}{s_{55}^E}$
ε	$\varepsilon_{33}^T\left(1\ \ k_p^2\right)$	$\varepsilon_{11}^T\left(1 - k_s^2\right)$	$\varepsilon_{22}^T\left(1 - k_F^2\right)$
c	$\dfrac{2}{s_{11}^E + s_{12}^E}$	$\dfrac{1}{s_{55}^E}$	$\dfrac{1}{s_{55}^E}$
k^2	$\dfrac{2d_{31}^2}{\varepsilon_{33}^T\left(s_{11}^E + s_{12}^E\right)}$	$\dfrac{d_{15}^2}{\varepsilon_{11}^T \cdot s_{55}^E}$	$\dfrac{d_{25}^2}{\varepsilon_{22}^T \cdot s_{55}^E}$
two-port network representation	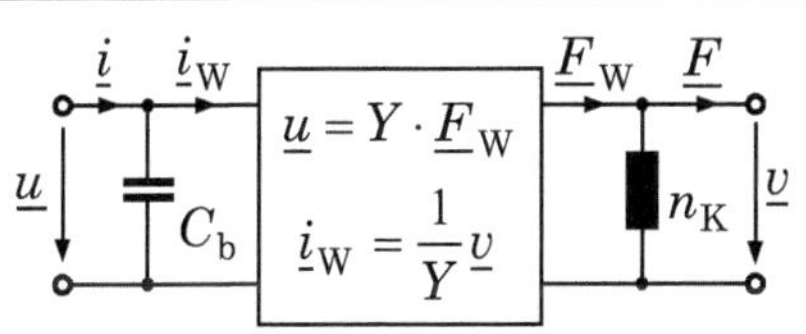 $C_{\mathrm{b}} = \varepsilon \dfrac{A_{\mathrm{el}}}{l_{\mathrm{el}}}$	$n_{\mathrm{K}} = \dfrac{1}{c} \dfrac{l_{\mathrm{mech}}}{A_{\mathrm{mech}}}$	$\dfrac{e^2}{\varepsilon\, c} = \dfrac{k^2}{1 - k^2} = \dfrac{C_{\mathrm{n}}}{C_{\mathrm{b}}}$ $\dfrac{1}{Y} = \dfrac{k}{\sqrt{1 - k^2}} \sqrt{\dfrac{C_{\mathrm{b}}}{n_{\mathrm{K}}}}$

9.2.5 Piezoelectric Bending Bimorph Elements

If two piezoceramic elements with same polarization direction are firmly connected (glueing or cementing) and if they are operated in electrical parallel connection or in electrical series connection, in case of opposite polarization direction, a *parallel* or *series bimorph* will be obtained. In addition to the bimorph, Fig. 9.39 represents the *monomorph* and *trimorph type* each with a carrier layer. Piezoelectric bending elements are used both in actuators generating large deflection amplitudes up to 2 mm and in highly sensitive accelerometers.

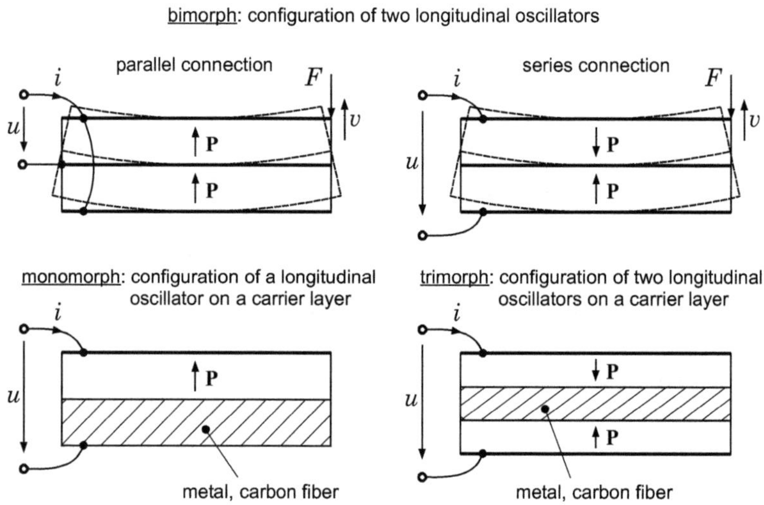

Fig. 9.39. Piezoelectric flexural vibrators
P *polarization direction*

The circuit diagram of the bending element has been derived in Sect. 5.1. Supplemented by the piezoelectric transducer, the circuit diagram of the piezoelectric bending element specified in Fig. 9.40 is achieved. At first, the loss-free piezoelectric transducer interconnects electrical and mechanical-rotational coordinates. By means of the transformer-like coupling between rotational and translational mechanical network of the bending rod represented in Sect. 5.1.2, finally the five-port network of the piezoelectric bending bimorph element represented in Fig. 9.41 is achieved. This circuit is valid for low frequencies. In Sect. 9.3.5 it will be made use of this bending element as finite network element taking one-dimensional bending waves into account.

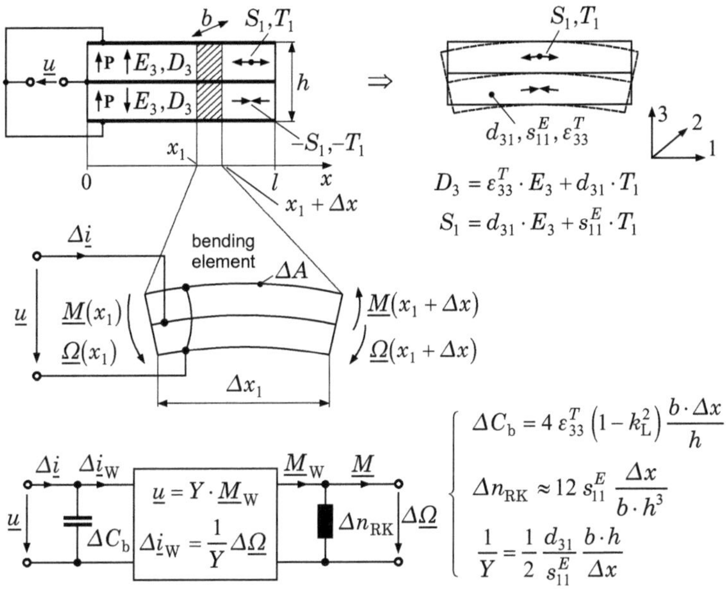

Fig. 9.40. Circuit diagram of piezoelectric bending element in case of interconnecting electrical and mechanical-rotational coordinates

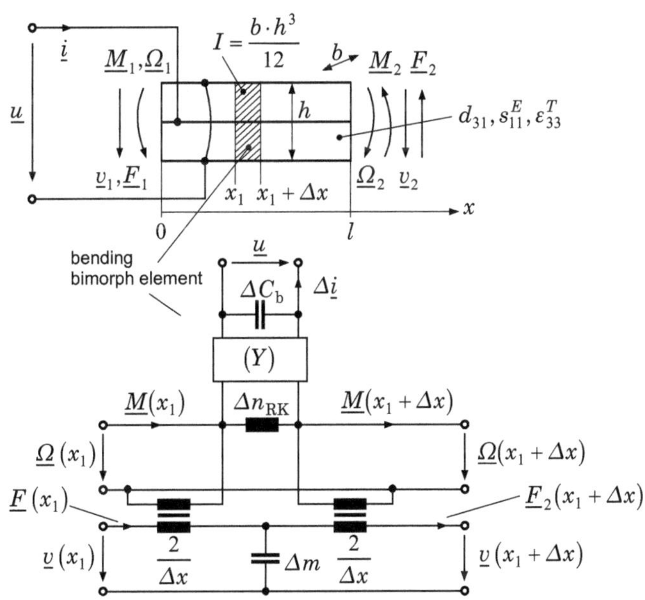

Fig. 9.41. Complete quasi-static two-port network representation of a piezoelectric bimorph element

As a special case, in Fig. 9.42 the circuit of the commonly used cantilevered bimorph is specified. The constants of characteristic values are conform to those of the piezoelectric longitudinal oscillator.

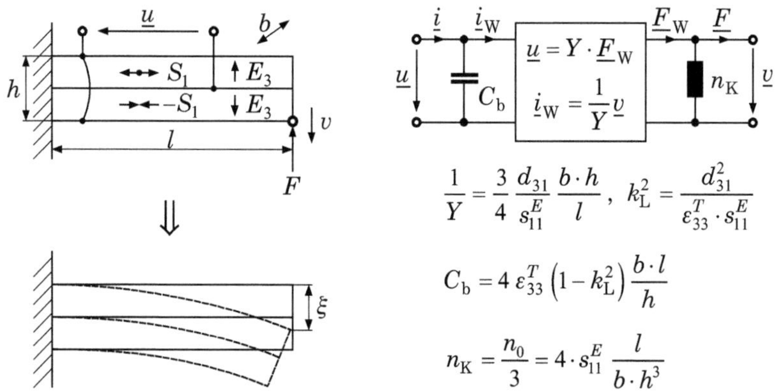

Fig. 9.42. Quasi-static circuit diagram and characteristic values of cantilevered piezoelectric bimorph
k_L *coupling factor of the longitudinal oscillator*

9.2.6 Piezoelectric Materials

The technologically most important piezoelectric materials are represented

- by *single crystals*:
  quartz, gallium orthophosphates, langasite, lithium niobate, lithium tantalate
- by *polycrystalline ferroelectrics*:
  piezoelectric ceramics (e.g. $BaTiO_3$, $Pb(Zr, Ti)O_3$)
- by *crystalline amorphous plastics*:
  polyvinylidene fluoride PVDF

The most important properties of these materials – except for the crystals lithium niobate and lithium tantalate being solely important for telecommunications – are specified in the Tables 9.6 up to 9.8. Further information can be found in suitable technical literature [66, 88].

Single Crystals

Quartz represents a piezoelectric crystal and belongs to the symmetry group 32 of the trigonal system. Quartz is characterized by its very high mechanical quality factor and the long-term stability of its coefficients. In addition, quartz (as well as gallium orthophosphate and langasite) does not show a *pyroelectric effect*. Therefore, it is used in precision sensors for dynamic measurement of acceleration, force and pressure. High stability resonators and filters represent further important fields of application for quartz. In particular the X-and AT-cuts in the quartz crystal are of practical importance (Table 9.6).
Langasite-LGS is a synthetic crystal ($La_3Ga_5SiO_{14}$) and along with quartz it also belongs to the symmetry group 32 of the trigonal crystal system. Compared with quartz (573 °C) Langasite shows no phase transition up to the melting point of 1470 °C. In addition, the coefficient of thermal expansion is lower (Langasite: $\alpha_{11} = 5.1 \cdot 10^{-6}\,\mathrm{m \cdot K^{-1}}$; quartz: $\alpha_{11} = 13.7 \cdot 10^{-6}\,\mathrm{m \cdot K^{-1}}$) and the piezoelectric coefficients are larger than those of quartz. Such as quartz also Langasite does not show any pyroelectricity. The industrial crystal growing is based on the CZOCHRALSKI process. Since the 1990s 4 inch wafers (100 mm) have been commercially available as longitudinal-, transverse- and shear-cut wafers. Langasite is used as basic material in resonant BAW (Bulk Acoustic Wave) and SAW (Surface Acoustic Wave) sensors which are applied for measurement of forces, pressures and torques.

Polycrystalline Ferroelectrics

Piezoelectric ceramics e.g. **lead zirconate titanate (PZT)** are polycrystalline materials produced by sintering ceramic powders (Table 9.7). Initially these materials are inhomogeneously polarized, i.e. the directions of permanent polarization are randomly distributed and there is no resultant linear piezoelectric effect. The materials show an electrostrictive behavior. When applying a sufficiently high field strength above the respective CURIE temperature, for example $2\,\mathrm{kV\,mm^{-1}}$, the polarization of the individual domains is mainly turned in one direction, i.e. the material becomes piezoelectric. After switching off the field, the polarization will remain, if the temperature is not raised above the CURIE temperature. Both the internal polarization P and mechanical strain S against field strength E of a polarized ceramic material including the operating ranges are illustrated in Fig. 9.43. By varying the titanium/zirconium ratio, a multiplicity of variants of ceramics can be generated. Due to their high coupling factors, piezoelectric ceramics are used for power transducers, e.g. ultrasonic transmitters, but also for sensors and filter elements. Compared to quartz, piezoelectric ceramics have a lower long-term stability and show *pyroelectricity*.

Table 9.6. Material properties of technically important piezoelectric materials

quartz (left-handed)

material:

- monocrystalline, hexagonal material; chemical: silicic acid (SiO_2);
- found in several crystal modifications;
- technically important: α-quartz (ϑ < 573 °C)
- α-quartz: anisotropic material properties;

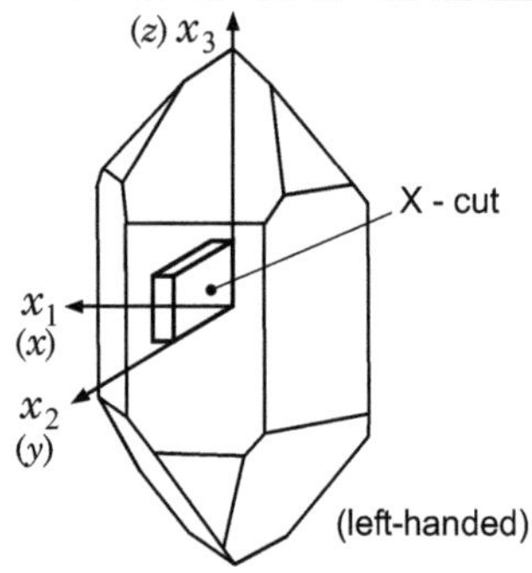

α-quartz cuts:

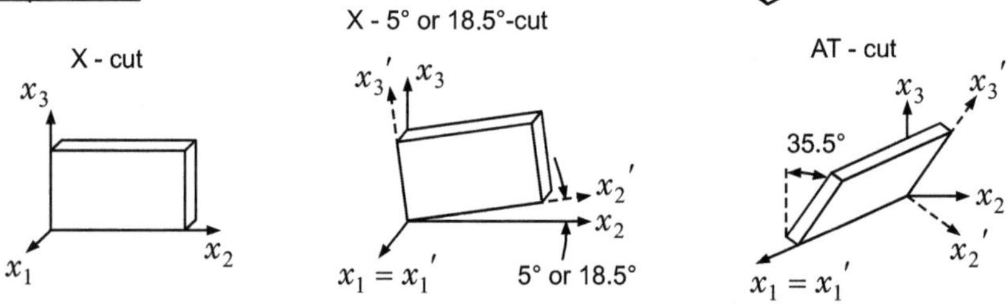

production: crystal growing in autoclaves at 450 °C and 1000 bar

properties:

- showing a polarization being proportional to the mechanical stress (no polarization in case of a stress-free state)
- very good long-term stability of the piezoelectric coefficients, since a depolarization is not possible
- very good linearity, no hysteresis
- high compressive strength ($4 \cdot 10^5$ N / cm^2)
- no pyroelectricity
- high insulation resistance up to approx. 400 °C ($10^{16}\ \Omega \cdot$ cm)
- very high quality factor $Q = 5 \cdot 10^3$ up to 10^6

but:

- only small coupling factor (approx. 0.1)
- tendency to twinning above 400 °C or in case of mechanical overloading → change of the transfer characteristics

Table 9.7. Material properties of technically important piezoelectric materials

	piezoelectric ceramics
material:	• ferroelectric material, e.g. barium titanate ($BaTiO_3$) or lead zirconate titanate ($Pb(ZrTi)O_3$)
production:	• sintering (1200 °C), sawing und grinding of polycrystalline backing material • applying the metal electrodes • polarization above the Curie temperature (200 up to 350)°C with a constant field strength of about 2 kV/mm, freezing of the aligned dipoles during cooling, polarization direction: x_3-direction
properties:	• compared to quartz piezoelectric ceramics show no spontaneous polarization but they must be polarized artifically • Strain-field strength curve characteristics with hysteresis in polarized state • high mechanical compressive strength, but only low tensile strength and shear strength, thus a mechanical preload is necessary for actuator applications • significantly higher coupling factor than quartz: up to 0.7 • high resistivity: up to $10^{12}\ \Omega \cdot \mathrm{cm}$ • types: - plates, disks: h = (0.1 … 2) mm - tubes: R_a = (1 … 10) mm, R_i = (0.5 … 9) mm - foils: (20 ... 100) μm (multilayer technology)
but:	• piezoelectric coefficients depend on the mechanical stress (nonlinearity), on the mechanical prehistory (hysteresis) and on the frequency (fatigue) • pyroelectric effect has to be taken into account • lower long-term stability than quartz due to tendency to depolarization • higher temperature dependency of the piezoelectric coefficients

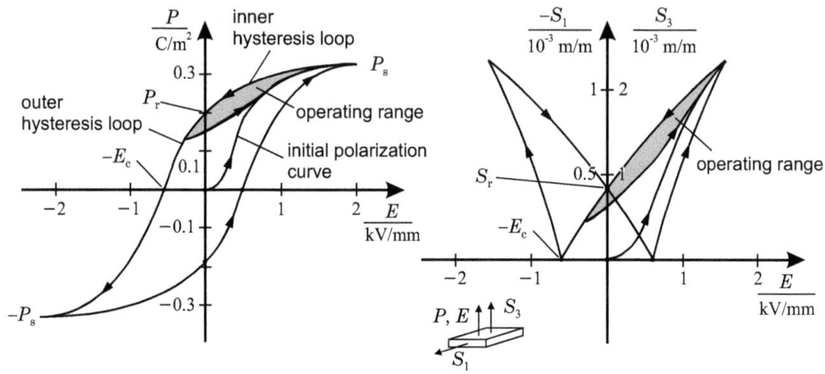

Fig. 9.43. Operating characteristics $(P = f(E), S = f(E))$ of polarized PZT ceramic for $T = 0$ [66, 85]

Crystalline Amorphous Plastics

Polyvinylidene fluoride (**PVDF**, Table 9.8), a crystalline amorphous plastic, consists of long molecular chains of carbon, hydrogen and fluorine. By polymerization the VDF monomers form large crystalline domains in an amorphous environment. Initially the crystalline domains are existent in a non-

Table 9.8. Material properties of technically important piezoelectric materials

	PVDF foils (Polyvinylidene Fluoride)	
material:		• ferroelectric material, crystalline amorphous plastic
production:		• extruding and tape casting • mechanical stretching by approx. five times of initial length • polarization above the Curie temperature of approx. 200°C with a constant field strength of about 100 kV/mm
properties:		• production of very thin foils up to a thickness of 5 μm
	but:	• possibility of hygroscopic deposition of water, risk of depolarization • smaller coupling factor k = 0.1 ... 0.2 than piezoelectric ceramics

polarizable α-phase. Only by a mechanical stretching by five times of the initial length the molecular chains are transfered into a polarizable β-phase. Also here, the polarization happens above the CURIE temperature of about 200 °C at very high field strengths of about 100 kV mm^{-1}. Due to small foil thickness at distinct piezoelectric properties, PVDF foils are applied in sensors, e.g. miniaturized hydrophones and as a foil bending oscillator in hi-fi headphones.

The material constants and coupling factors of quartz (X-cut), of selected piezoelectric ceramics for sensor and actuator applications as well as of PVDF foils are summarized in Table 9.9. The materials are supplemented by adding the piezoelectric semiconductor material zinc oxide (ZnO) which is existent in a hexagonal crystal structure. The semiconducting properties enable applications in silicon micromachining, e.g. for the generation and reception of mechanical bulk or surface waves in resonant sensors. The ZnO layers are deposited in thicknesses of a few micrometers.
The piezoelectric ceramic PIC155 shows a low temperature dependence of the permittivity at a high k-value. Therefore, it is particularly suitable for ultrasonic transmitters and receivers in pulsed mode, but also for accelerometers with broad operating temperature range. Due to its high k-value, the piezoelectric ceramic C82 is particularly suitable as material for actuators.

9.2.7 Sample Applications

The application of the circuit diagram for piezoelectric transducers shown in Fig. 9.34 and in the Tables 9.4 and 9.5 will be explained using the example of an accelerometer and a microphone and using the material constants listed in Table 9.9.

Piezoelectric Accelerometer

The piezoelectric accelerometer can be described by the design principle and circuit diagram shown in Fig. 9.44 a). The force $F_0 = ma_0$ generated by the seismic mass m acts on a piezoelectric thickness element made of PZT-4 ceramic and causes the voltage u being measurable across the input resistance R of sensor electronics. Only the compliance n_K of the ceramic is taken into account, i.e. contact compliances between mass m and ceramic or housing base respectively are neglected. Additionally, the mechanical losses in the form of frictional impedance r and cable capacitance C_K are taken into consideration. Based on the total circuit diagram illustrated in Fig. 9.44 b), in Fig. 9.44 c) the voltage source is replaced by a current source. By applying transformation relations of the gyrator, the mechanical subsystem can now be completely transformed to the electrical side as shown in Fig. 9.44 d).

Table 9.9. Material properties of technically important piezoelectric materials
PZT Brush Clevite Comp., PIC PI Ceramic, C Fuji Ceramics, PXE Morgan Electroceramics
$\varepsilon_0 = 8,854 \cdot 10^{-12}\,\mathrm{A\,s\,V^{-1}\,m^{-1}}$

constants	quartz	gallium orthophosphate	langasite	ZnO	PZT-4	PZT-5a	PIC 155	C 82	PXE52	PXE54	PVDF
d_{33} 10^{-12} m/V	2.3 (d_{11})	4.5 (d_{11})	6.2 (d_{11})	12.3	289	374	360	540	580	> 450	-27
d_{31} 10^{-12} m/V	-2.3 (d_{12})	—	5.3 (d_{14})	-5.1	-123	-171	-165	-260	-270		20
e_{33} / $\mathrm{As/m^2}$	0.181 (e_{11})	—	0.45 (e_{11})	1.7	15.1	15.8	18.3	28.1	—	—	108
e_{31} / $\mathrm{As/m^2}$	-0.181 (e_{12})	—	—	-2	-5.2	-5.4	-10.6	-15.4	—	—	—
s_{33}^E 10^{-12} m²/N	12.78 (s_{11})	17.93 (s_{11})	—	6.9	15.4	18.8	19.7	19.2	20	—	—
s_{11}^E 10^{-12} m²/N	9.74 (s_{33})	11.35 (s_{33})	—	7.9	12.3	16.4	15.6	16.9	16	—	—
c_{33}^E 10^{10} N/m²	8.68 (c_{11})	10.21 (c_{11})	—	1.4	6.5	5.3	5.1	5.2	6...9	—	—
c_{11}^E 10^{10} N/m²	10.58 (c_{33})	6.66 (c_{33})	—	-4.3	8.1	6.1	6.4	5.9	—	—	—
$\frac{\varepsilon_{33}^T}{\varepsilon_0}$; $\frac{\varepsilon_{33}^S}{\varepsilon_0}$	4.68 ; –	6.6 ; –	18.9 ; –	8.2 ; –	1300 ; 635	1730 ; 960	1700 ; –	3400 ; –	3500 ; –	3000 ; –	12 ; 12
$\frac{\varepsilon_{11}^T}{\varepsilon_0}$; $\frac{\varepsilon_{11}^S}{\varepsilon_0}$	4.52 ; –	6.1 ; –	50.7 ; –	8.1 ; –	1475 ; 730	1700 ; 830	1500 ; –	3100 ; –	3000 ; –	—	—
k_{33}	0.1 (k_{11})	0.14 (k_{11})	0.15 (k_{11})	0.23	0.7	0.71	0.69	0.72	0.74	> 0.6	0.20
k_{31}	0.12 (k_{12})	—	—	0.05	0.33	0.34	0.35	0.36	0.39	> 0.3	0.15
$\vartheta_{\mathrm{Curie}}$/°C	573	~ 700	—	—	328	365	345	190	165	220	80
$\rho/\mathrm{kg\,m^{-3}}$	2655	3570	5750	5680	7500	7500	7700	7400	7800	7900	1790

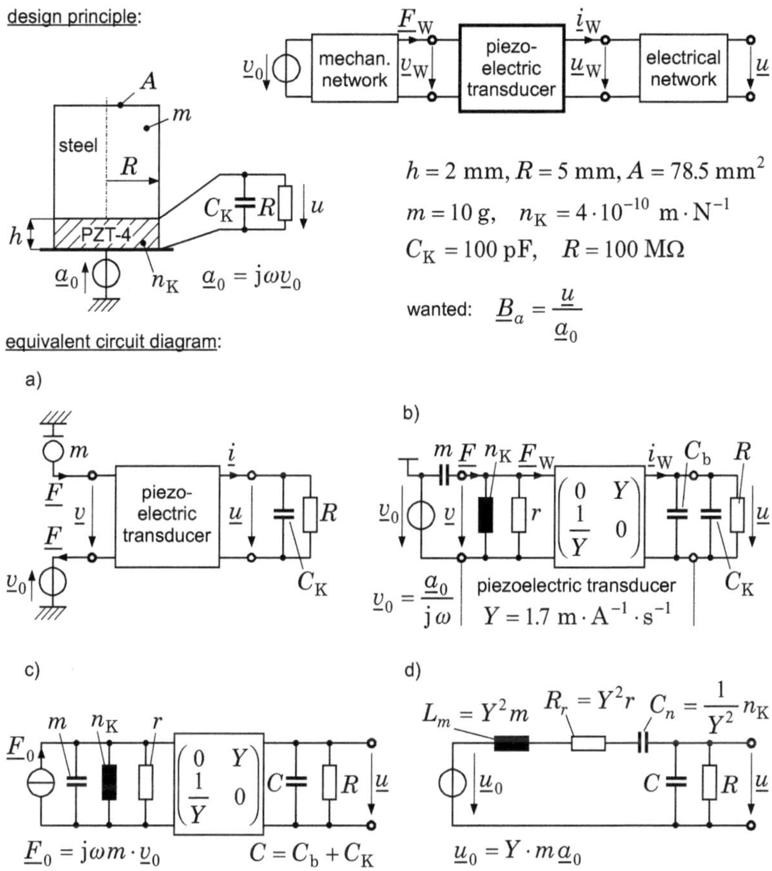

Fig. 9.44. Design and circuit of a piezoelectric accelerometer with PZT thickness element

The following considerations are done separately and approximately for very low frequencies (neglecting L_m and R_r) and for high frequencies (neglecting R). The transfer functions for both approximations $\underline{B}_{a1}$ and $\underline{B}_{a2}$ are shown in Fig. 9.45. By using the material constants for PZT-4 specified in Table 9.9 and by using the relations for the free thickness oscillator listed in Table 9.4, the component values, frequency values and transfer factors are achieved. In conclusion the total frequency response consists of the high-pass filter for low frequencies and resonant low-pass for high frequencies. The transfer factor B_0 in the operating frequency range amounts to $B_0 = 2.9\,\mathrm{mV/ms^{-2}}$. The resonant frequency amounts to $f_0 = 96.5\,\mathrm{kHz}$. If the neglected coupling compliances are taken into consideration, for real sensors this value will decrease. Piezoelectric thickness elements are used for measurement of high frequen-

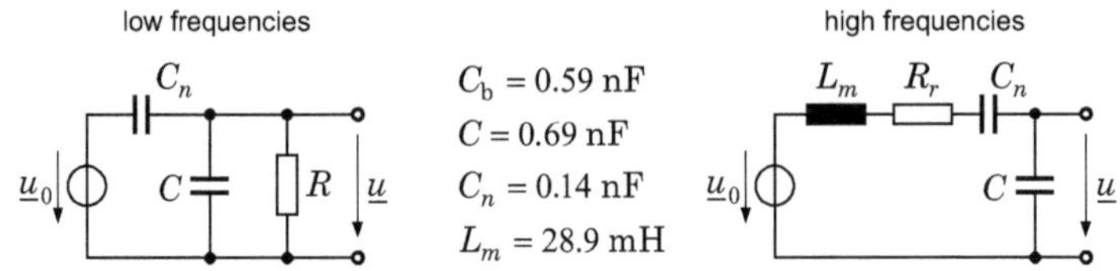

$$\underline{B}_{a1} = \frac{C_n}{C + C_n} \frac{\mathrm{j}(\omega/\omega_1)}{1 + \mathrm{j}(\omega/\omega_1)} Y \cdot m = B_0 \frac{\mathrm{j}(\omega/\omega_1)}{1 + \mathrm{j}(\omega/\omega_1)}$$

$$\omega_1 = \frac{1}{R(C + C_n)}$$

$$f_1 = 1.9 \text{ Hz}$$

$$\underline{B}_{a2} = B_0 \frac{1}{1 + \mathrm{j}(\omega/\omega_0)\frac{1}{Q} - (\omega/\omega_0)^2}$$

$$\omega_0^2 = \frac{1}{L_m \cdot C_n}(1 + C_n/C)$$

$$f_0 = 96.5 \text{ kHz}$$

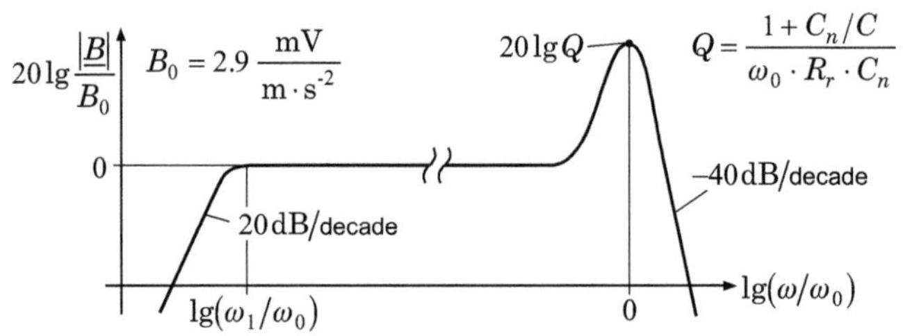

Fig. 9.45. Amplitude-frequency response of piezoelectric accelerometer with PZT thickness element

cies and high acceleration amplitudes. In order to increase the transfer factor, normally two thickness elements are used in electrical parallel connection.

Piezoelectric Microphone

As a second example, in Fig. 9.46 the basic structure and circuit diagram of a piezoelectric microphone with bimorph bending element are presented. The acoustic pressure acting on the resiliently suspended plate is transformed into a force that causes a deflection of the cantilevered bimorph. In addition to the circuit of bimorph illustrated in Fig. 9.42 including the relations for Y, C_b and n_K, mechanical and acoustic components are added. As mechanical components, the mass m and compliance n_M of the resiliently suspended plate are complemented. As acoustic components, the acoustic friction $Z_{\mathrm{a},1}$ of acoustic pressure feed and the acoustic compliances $N_{\mathrm{a},1}$ and $N^*_{\mathrm{a},2}$ of the air-filled cavities in front and behind the plate are complemented. At the electrical side, the cable capacitance C_K and input resistance R of sensor electronics are taken into account. The mechanical-acoustic transformation is described by the gyrator with $Y = 1/A$.

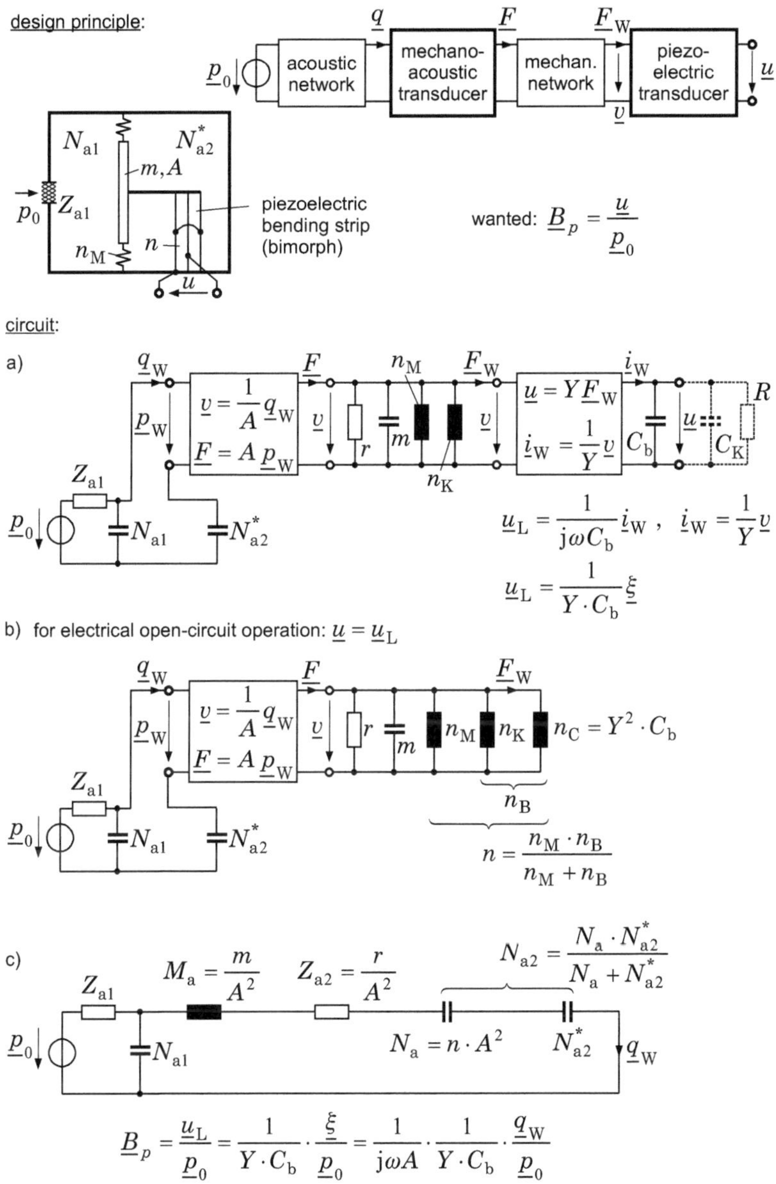

Fig. 9.46. Design and circuit diagrams of a piezoelectric microphone with bimorph sensing element

Based on the circuit diagram represented in Fig. 9.46 a) and assuming the boundary condition of electrical open-circuit operation, at first the capacitance C_b is transformed to the mechanical side (Fig. 9.46 b)). Afterwards all mechanical components are transformed to the acoustic side (Fig. 9.46 c)). In order to determine the transfer function $\underline{B}_p$, the function $\underline{q}_\mathrm{W}/\underline{p}_0$ has to be calculated from this circuit. The network analysis of Fig. 9.46 c) yields:

$$\frac{\underline{q}_\mathrm{W}}{\underline{p}_0} = \frac{\mathrm{j}\omega N_{\mathrm{a},2}}{\left(1+\mathrm{j}\frac{\omega}{\omega_0}\frac{1}{Q_1}\right)\left(1+\mathrm{j}\frac{\omega}{\omega_0}\frac{1}{Q_2}-\left(\frac{\omega}{\omega_0}\right)^2\right)+\mathrm{j}\frac{\omega}{\omega_0}\frac{N_{\mathrm{a},2}}{N_{\mathrm{a},1}}\frac{1}{Q_1}}$$

with

$$\omega_0^2 = \frac{1}{M_\mathrm{a} N_{\mathrm{a},1}}, \qquad Q_1 = \frac{1}{\omega_0 N_{\mathrm{a},1} Z_{\mathrm{a},1}}, \qquad Q_2 = \frac{1}{\omega_0 N_{\mathrm{a},2} Z_{\mathrm{a},2}}.$$

The transfer function $\underline{B}_p$ is now achieved according to:

$$\underline{B}_p = \frac{\underline{u}_\mathrm{L}}{\underline{p}_0} = \frac{1}{\mathrm{j}\omega A Y C_\mathrm{b}}\frac{\underline{q}_\mathrm{W}}{\underline{p}_0}$$

Assuming $1/Q_1 \ll 1$, this relation can be further simplified:

$$\underline{B}_p = \frac{\underline{u}_\mathrm{L}}{\underline{p}_0} \approx B_0 \frac{1}{1+\mathrm{j}\frac{\omega}{\omega_0}\frac{1}{Q_2}-\left(\frac{\omega}{\omega_0}\right)^2} \qquad \text{with} \qquad B_0 = \frac{N_{\mathrm{a},2}}{A Y C_\mathrm{b}}.$$

The amplitude-frequency response of this simplified transfer function is represented in Fig. 9.47. For low frequencies, the influence of electrical components has to be taken into account. Due to the finite internal resistance of evaluation electronics, for low frequencies an increasing discharge of the electrodes develops as with all other piezoelectric transducers, i.e. the frequency response bends for decreasing frequencies.

9.3 Piezoelectric Transducer as one-dimensional Waveguide

For higher frequencies, the mechanical properties of the piezoelectric element can not be longer described by means of the concentrated components compliance n and mass m. The *extensional waves* propagating along the rod axis now cause position-dependent inertia forces. Thus, the transition from the concentrated component mass m to a continuum of mass elements Δm occurs which are complemented by compliance elements Δn respectively. In order to simplify the model, very small lateral rod dimensions are assumed, so that inertia forces caused by transverse strain can be neglected.

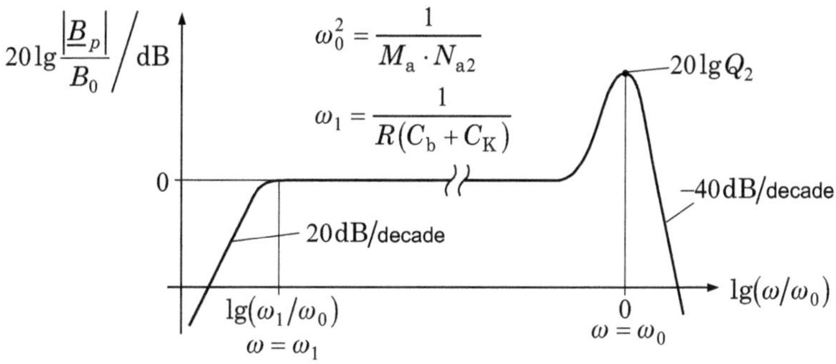

Fig. 9.47. Approximate frequency response $\underline{u}_L/\underline{p}_0$ of piezoelectric microphone with bimorph sensing element

Based on the model of rod consisting of finite rod elements, its two-port network representation as a *loss-free one-dimensional waveguide* was derived in Sect. 6.1.1. The two-port network representation and characteristic values of waveguide are specified once more in Fig. 9.48. *Wave velocity* c_D of the extensional wave propagating along the rod, *wave admittance* h_D being able to be transformed in an unvaried form through the two-port network of rod and *wave number* β are introduced as characteristic values of the waveguide. In addition, the length related compliances n' and masses m' are used which correspond to the inductance per unit length L' or capacitance per unit length C' of a homogeneous electrical conductor, respectively.

In Sect. 6.1.1, a circuit design interpretation in the form of the equivalent T or Π circuit is specified for the chain matrix represented in Fig. 9.48. By comparing the elements of circuit with the chain matrix, the relations summarized in Fig. 9.49 are achieved. For very low frequencies the admittances $\underline{h}_1$ and $\underline{h}_2$ pass into the concentrated components m and n.

9.3.1 Transition from Lumped Parameters to the Waveguide using the Example of an Accelerometer

Using the example of the piezoelectric accelerometer illustrated in Fig. 9.44, the approximate solution and exact solution of the one-dimensional waveguide can be applied in order to calculate its resonant frequency. Both solutions are compared with the result achieved with concentrated components (0th approximation). In order to calculate the resonant frequency ω_0, in Fig. 9.50 only the mechanical transducer side is considered while neglecting frictional impedance.

Based on the resonant frequency achieved by means of the *0th approximation* (concentrated components)

$$\left(\omega_0^{(0)}\right)^2 = \omega_N^2 = \frac{1}{mn_K}, \qquad m^* = m$$

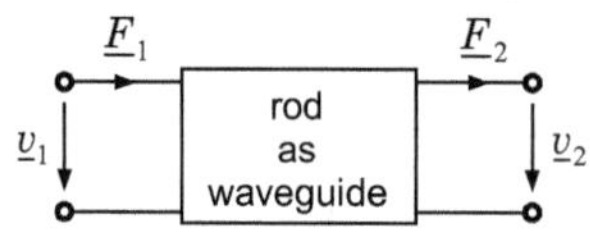

$$\begin{pmatrix}\underline{v}_1\\ \underline{F}_1\end{pmatrix}=\begin{pmatrix}\cos\beta l & \mathrm{j}h_\mathrm{D}\sin\beta l\\ \mathrm{j}\dfrac{1}{h_\mathrm{D}}\sin\beta l & \cos\beta l\end{pmatrix}\begin{pmatrix}\underline{v}_2\\ \underline{F}_2\end{pmatrix}$$

longitudinal	silicon	quartz	PZT-5A	PVDF	steel	glass	aluminum	copper
$c_\mathrm{D}/\frac{\mathrm{m}}{\mathrm{s}}$	7500 <100> 9000 <111>	5400 (X-cut)	4350	2200	5700 6000	5640 (Pyrex)	6420	5010

wave velocity

$$c_\mathrm{D}=\sqrt{\frac{E}{\rho}}$$

wavelength

$$\lambda_\mathrm{D}=\frac{2\pi}{\beta}=\frac{c_\mathrm{D}}{f}$$

wave number

$$\beta=\frac{\omega}{c_\mathrm{D}}=\omega\sqrt{\frac{\rho}{E}}$$

$$\beta=\frac{\omega}{c_\mathrm{D}}=\omega\sqrt{m'\cdot n'}$$

wave admittance

$$h_\mathrm{D}=\sqrt{\frac{n'}{m'}}=\frac{1}{A}\cdot\frac{1}{\sqrt{\rho\cdot E}}$$

wave impedance

$$z_\mathrm{D}=\frac{1}{h_\mathrm{D}}$$

Fig. 9.48. Two-port network representation of rod as one-dimensional loss-free extensional waveguide
Index D: *extensional wave*

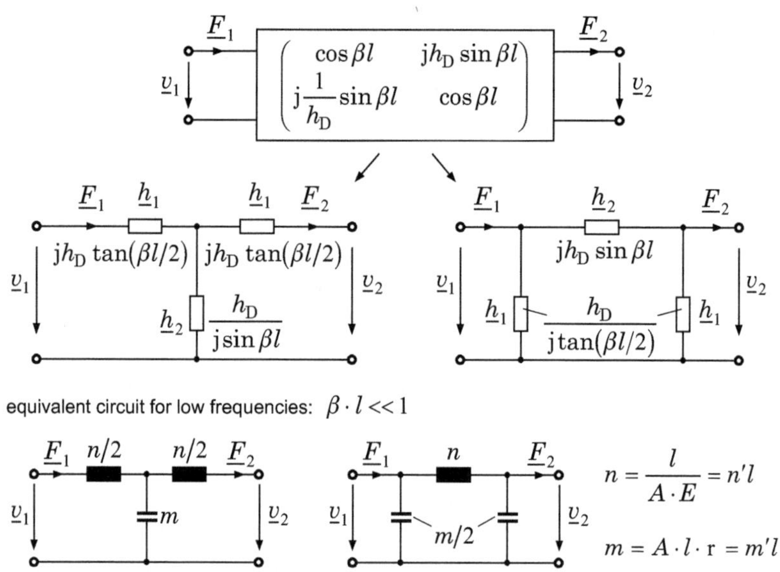

Fig. 9.49. T and Π circuit of one-dimensional waveguide
l length of the rod, A cross-sectional area of the rod, E YOUNG*'s modulus of the rod*

the *1st approximation* yields for $\beta l \ll 1$ (see T circuit in Fig. 9.49)

$$\left(\omega_0^{(1)}\right)^2 = \omega_{\mathrm{N}} \frac{1}{1+\dfrac{n}{2n_{\mathrm{K}}}}, \qquad m^* = m\left(1+\frac{n}{2n_{\mathrm{K}}}\right).$$

In order to calculate the *exact solution*, following approach

$$\omega_0^2 n_{\mathrm{K}} m^* = 1$$

is used. The effective mass m^* results from the two-port network equation of the waveguide

$$\mathrm{j}\omega m^* = \frac{\underline{F}_1}{\underline{v}_1} = \mathrm{j}\frac{1}{h_{\mathrm{D}}}\tan(\beta l), \qquad h_{\mathrm{D}} = \sqrt{\frac{n'}{m'}}.$$

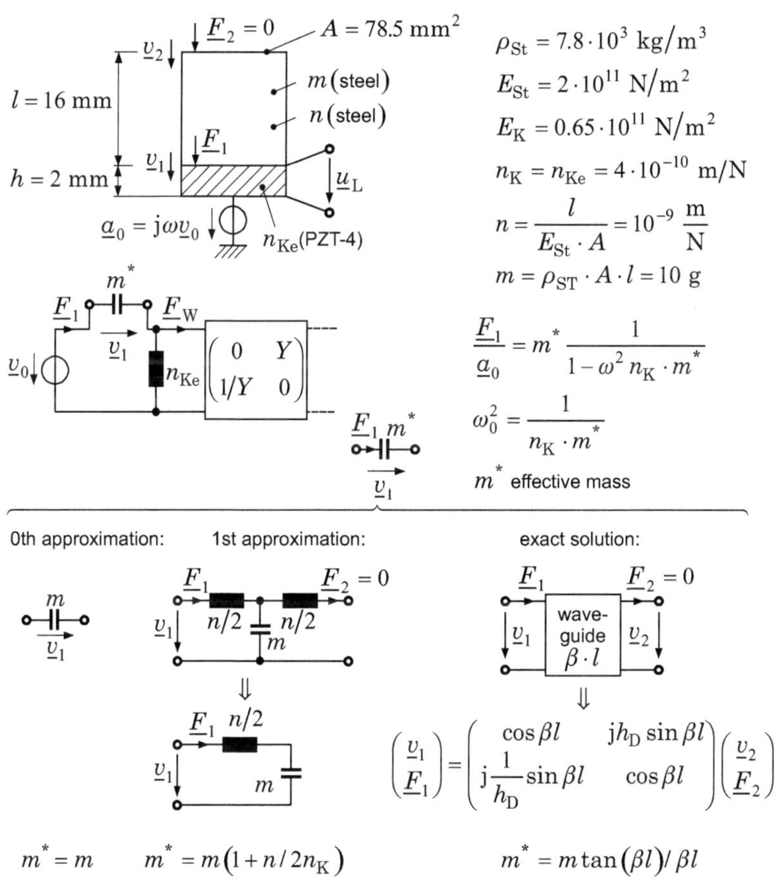

Fig. 9.50. Approximation steps for the calculation of resonant frequency of piezo-electric accelerometer with thickness element

In combination with the relations specified in Fig. 9.48

$$m^* = \frac{1}{\frac{\omega}{c_D} l} \frac{l}{c_D} \sqrt{\frac{m'}{n'}} \tan(\beta l), \qquad \frac{\omega}{c_D} = \beta, \qquad \frac{1}{c_D} = \sqrt{m'n'}$$

this equation can be simplified according to

$$m^* = m \frac{\tan(\beta l)}{\beta l}.$$

The insertion of the effective mass m^* into the basic equation yields in combination with $\beta = \omega_0 / c_D$

$$\omega_0^2 n_K m = \frac{\tan\left(\frac{\omega_0}{c_D} l\right)}{\frac{\omega_0}{c_D} l} = 1$$

and

$$\frac{\omega_0}{c_D} l \tan\left(\frac{\omega_0}{c_D} l\right) = \frac{n}{n_K}, \qquad \frac{\omega_0}{c_D} l = \frac{\omega_0}{\omega_N} \sqrt{\frac{n}{n_K}}$$

and finally results in

$$\frac{\omega_0}{\omega_N} \tan\left(\frac{\omega_0}{\omega_N} \sqrt{\frac{n}{n_K}}\right) = \sqrt{\frac{n}{n_K}}.$$

Figure 9.51 represents the resonant frequency responses normalized to the 0th approximation as a function of the compliance ratio n/n_K of seismic mass to ceramic. Compared to the resonant frequency ω_N of the 0th approximation, for which the seismic mass was assumed to be a concentrated component, the exact resonant frequency ω_0 decreases with increasing compliance n of seismic mass. It is surprising that the approximate solution of simplified T circuit and exact solution agree quite well.
For this example, following resonant frequencies are achieved in combination with $n/n_K = 2.5$:

0th approximation:	$f_0^{(0)} = f_N = 80\,\text{kHz}$
1st approximation:	$f_0^{(1)} = 53\,\text{kHz}$
exact solution:	$f_0 = 60\,\text{kHz}$

Compared to the solution with concentrated components, the approach of seismic mass as extensional waveguide has result in a significant decrease of the resonant frequency of about 25%.

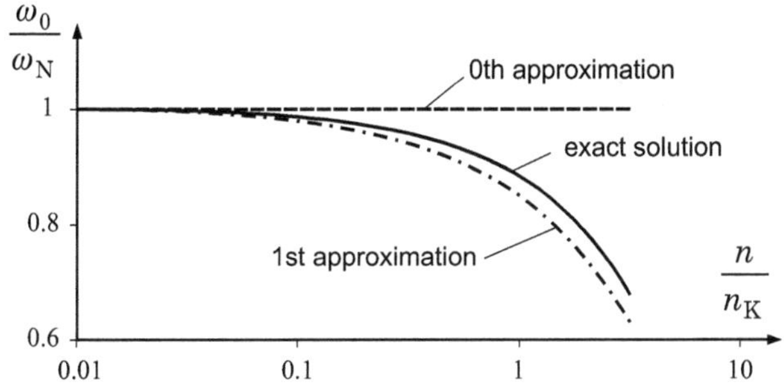

Fig. 9.51. Normalized resonant frequency response of piezoelectric accelerometer shown in Fig. 9.50

In the next step, the circuit of the *loss-free one-dimensional waveguide* is directly applied to the piezoelectric transducer as longitudinal and thickness oscillator. As a result the circuit of these transducer elements is derived.

9.3.2 Piezoelectric Longitudinal Oscillator as Waveguide

The piezoelectric longitudinal oscillator whose electrical field quantities are perpendicular to the mechanical vibration direction is represented in Fig. 9.52 as interconnection of piezoelectric transducer elements in the form of *finite network elements.*
Each finite element shows the common electrical two-port transducer network and mechanical partial two-port network represented by a Π circuit illustrated in Fig. 9.49. The same voltage $\underline{u}$ is supplied to each transducer element and due to the gyrator-like connection also all transducer forces $\underline{F}_W$ of the partial elements are equal. Therefore, the partial elements must be connected in parallel electrically and connected in series mechanically (Fig. 9.53).
Based on this arrangement of partial elements and based on the identity of transducer forces, in Fig. 9.53 the circuit diagram of the piezoelectric longitudinal oscillator is derived. In contrast to the circuit diagram of the piezoelectric transducer with concentrated components shown in Fig. 9.34, now the mechanical partial two-port network is represented by mechanical partial networks combined to a one-dimensional waveguide. In Fig. 9.54, this continuous network structure is replaced by the T circuit illustrated in Fig. 9.49. The relations specified in Fig. 9.54 apply to the admittances $\underline{h}_1$ and $\underline{h}_2$.

9.3.3 Piezoelectric Thickness Oscillator as Waveguide

With the piezoelectric thickness oscillator, the direction of electrical excitation and direction of vibration coincide. First of all, the finite network element of

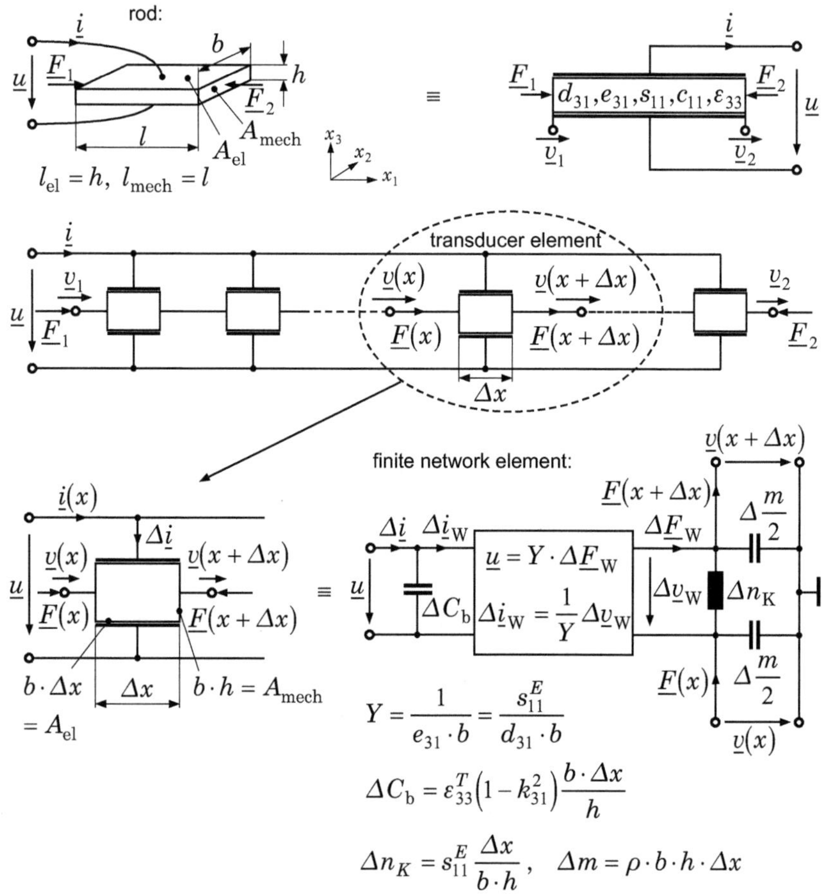

Fig. 9.52. Finite network element of piezoelectric longitudinal oscillator

the thickness oscillator is derived in Fig. 9.55. In contrast to the longitudinal oscillator, a length proportional transducer coefficient

$$\Delta Y \sim \Delta x, \qquad \Delta Y = \frac{\Delta x}{A}\frac{1}{e_{33}} = \frac{\Delta x}{A}\frac{s_{33}}{d_{33}}$$

and the partial components

$$\Delta C_{\mathrm{b}} = \frac{\varepsilon_{33} A}{\Delta x}, \qquad \Delta n_{\mathrm{K}} = \frac{1}{c_{33}}\frac{\Delta x}{A}, \qquad \Delta m = \rho A \Delta x$$

are achieved from Table 9.4. By applying the possibility of replacing the imaginary gyrator-like coupling by a real-valued transformer-like coupling or vice versa, the disadvantage of the length-related transducer coefficient can be

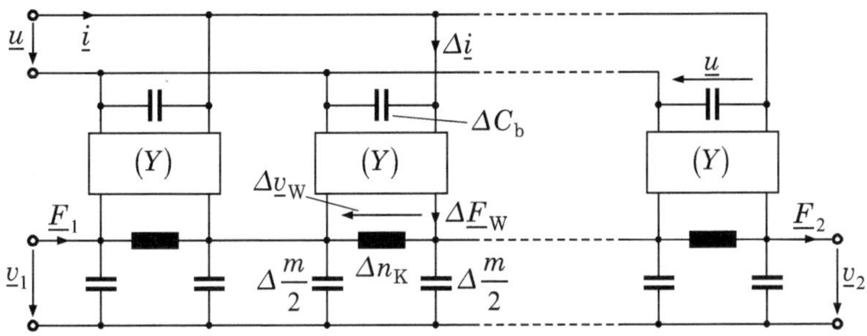

due to the identity of the transducer forces $\Delta \underline{F}_\mathrm{W} = \underline{F}_\mathrm{W} = \frac{\underline{u}}{Y}$ of all elements it applies:

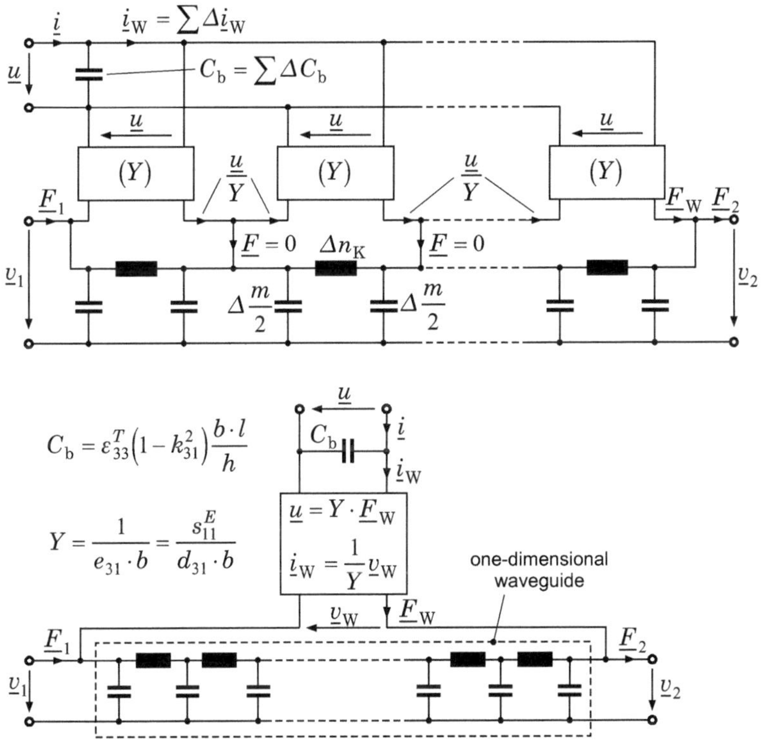

Fig. 9.53. Piezoelectric longitudinal oscillator as one-dimensional waveguide

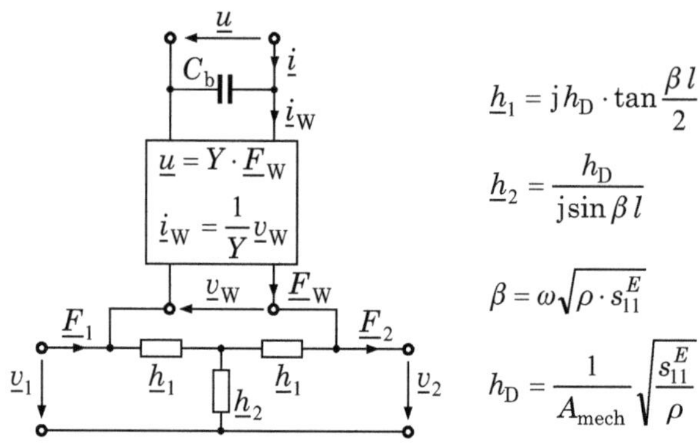

Fig. 9.54. Circuit of piezoelectric longitudinal oscillator

avoided by means of transition to an imaginary transformer-like two-port transducer network with

$$\Delta X = \frac{1}{\omega}\frac{1}{\Delta C_b \Delta Y} = \frac{1}{\omega}\frac{e_{33}}{\varepsilon_{33}}, \qquad \Delta n_C = \Delta Y^2 \Delta C_b.$$

This possibility has already been demonstrated with the derivation of the circuit diagram of the electrostatic transducer presented in Sect. 9.1.1. Due to the identical current for each partial element, an electrical series connection of the elements is achieved in contrast to the longitudinal oscillator shown in Fig. 9.53. This series connection is represented in Fig. 9.56 for the finite network elements with imaginary transducer coefficients jX. The partial capacitances ΔC_i add up to the series capacitance C_b. Since also here the transducer forces $\underline{F}_W$ match due to mechanical series connection, the mechanical partial two-port networks can be combined to a cascade connection of a one-dimensional waveguide.
At last, in Fig. 9.57 the combined imaginary transformer-like two-port transducer network is replaced by the equivalent real-valued gyrator-like two-port network. The resultant negative field compliance $n_C = -Y^2 C_b$ can be transformed as a negative series capacitance $-C_b$ to the electrical side. Thus, the circuit of the thickness oscillator differs topologically only in this additional capacitance compared to the longitudinal oscillator.
In summary, both the piezoelectric longitudinal and piezoelectric thickness oscillator can be described by a gyrator-like two-port transducer network. Its quantities C_b and Y can be gathered from Fig. 9.54 and Fig. 9.57. The homogeneous loss-free mechanical waveguide can either be represented by a three-port network with T or Π structure. The appropriate equivalent parameters are summarized in Fig. 9.49. The particular relations for the wave admittance h_D and wave number β are listed in Fig. 9.54 and Fig. 9.57. For quartz and piezoelectric ceramics, the elastic and piezoelectric coefficients can be gathered from Table 9.9.

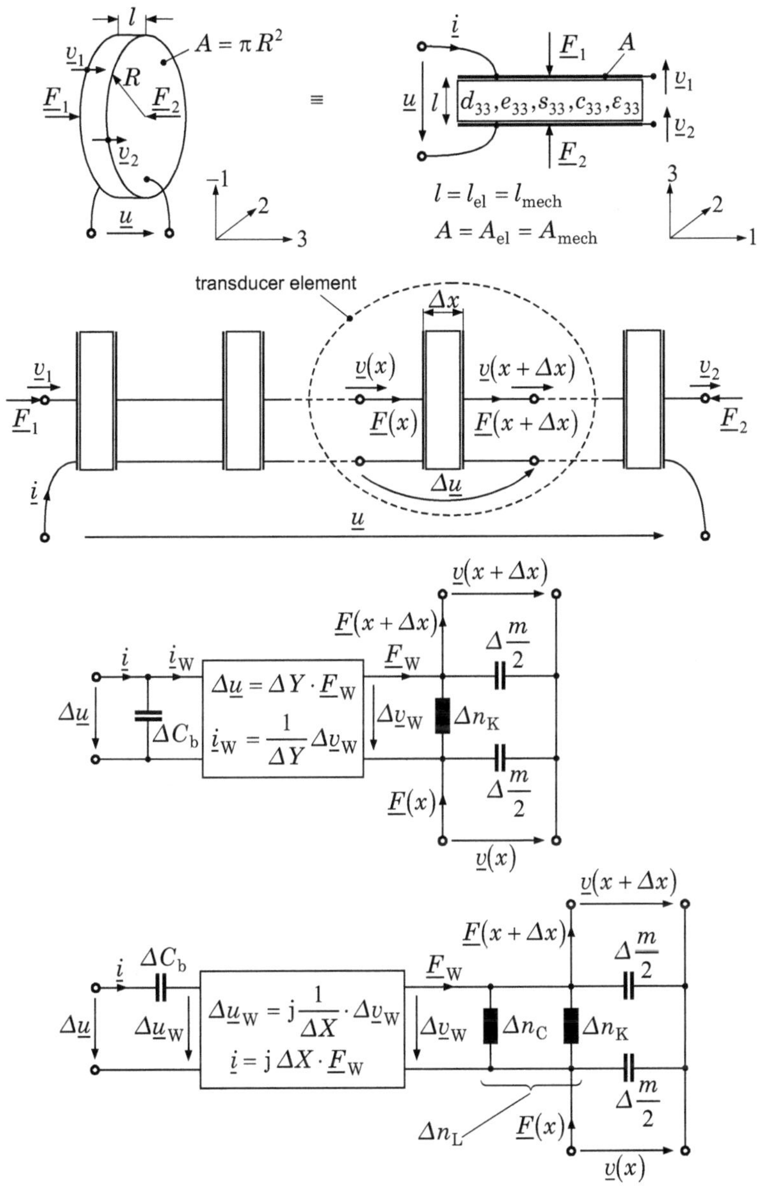

Fig. 9.55. Finite network element of piezoelectric thickness oscillator

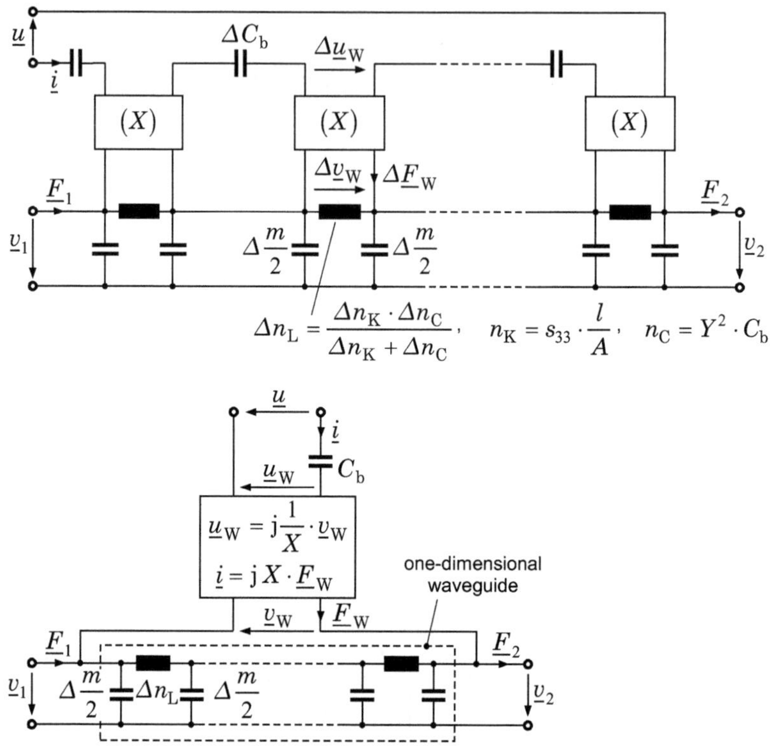

Fig. 9.56. Piezoelectric thickness oscillator as one-dimensional waveguide

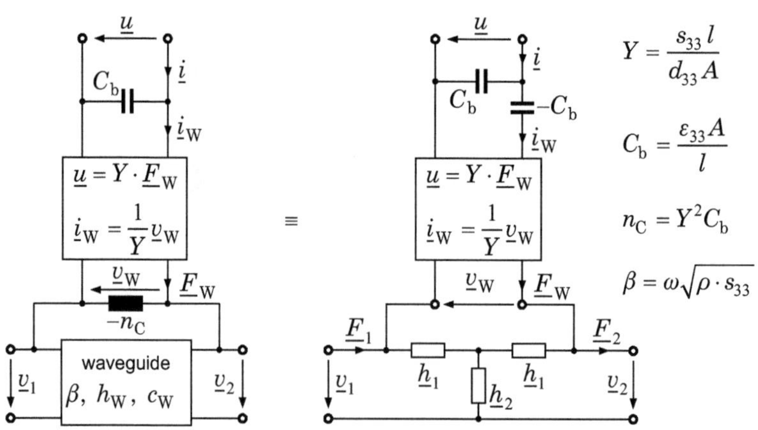

Fig. 9.57. Circuit of piezoelectric thickness oscillator

9.3.4 Sample Applications of Piezoelectric Longitudinal and Thickness Oscillators

With the circuits shown in Fig. 9.54 and Fig. 9.57 we now have the basic modules for the calculation of arbitrary *loss-free* piezoelectric longitudinal and thickness oscillators. In case of a mechanical load at both sides of the oscillators, these circuits are directly used as initial base. However, the circuits can be further simplified if the oscillator is running in a mechanical open-circuit operation at one side, e.g. as ultrasonic oscillator, or in an open circuit operation at both sides, e.g. as filter element. Both for the longitudinal and thickness oscillator the simplified circuit diagrams of oscillators running in an open-circuit operation at one and both sides including the bases of calculation for mechanical and electrical components are summarized in the Tables 9.10 and 9.11.

For the clamped-free oscillator $\underline{F}_1 = 0$ is defined and for the free-free oscillator $\underline{F}_1 = \underline{F}_2 = 0$ is defined. This results in simplified circuits at the mechanical side. In the neighborhood of the first mechanical resonance, the admittances $\underline{h}_1$ and $\underline{h}_2$ can be approximately replaced by the concentrated components n_{K}, n_{C} and m. For that purpose, the approximate two-port network representations discussed in detail in Sect. 6.1 form the basis.

For the calculation of following sample applications, it is made use of simplified circuits specified in the Tables 9.10 and 9.11. Based on the technologically important piezoelectric oscillators specified in Fig. 9.38 as a first sample calculation the free-free longitudinal quartz oscillator is analyzed.

Longitudinal Quartz Oscillator as Filter Element

The free-free longitudinal quartz oscillator with $\underline{F}_1 = \underline{F}_2 = 0$ illustrated in Fig. 9.58 is represented as simplified circuit using the approximate relations listed in the Tables 9.10 and 9.11. By the frictional impedance r connected in parallel the mechanical losses are taken into account.For the impedance $\underline{Z} = \underline{u}/\underline{i}$ of the quartz oscillator, following relation is achieved from Fig. 9.58:

$$\underline{Z} = \frac{\mathrm{j}}{\omega} \frac{\left(\dfrac{\omega}{\omega_0}\right)^2 - \left(\dfrac{\omega}{\omega_0}\right)\dfrac{1}{Q} - 1}{C_n + C_{\mathrm{b}} - \left(\dfrac{\omega}{\omega_0}\right)\dfrac{1}{Q} - \omega^2 C_n C_{\mathrm{b}} L_m}, \qquad \omega_0^2 = \frac{1}{L_m C_n}, \qquad Q = \frac{1}{R}\sqrt{\frac{L_m}{C_n}}$$

Thus, $\underline{Z}$ has two resonance points, namely

- the series resonant frequency f_{S} for electrical short-circuit operation:

$$f_S = \frac{1}{2\pi} \frac{1}{\sqrt{L_m C_n}} \qquad \text{for} \qquad \left(\frac{\omega}{\omega_0}\right)^2 - 1 = 0, \qquad Q \gg 1$$

Table 9.10. Derivation of the circuits for clamped-free and free-free longitudinal and thickness oscillators

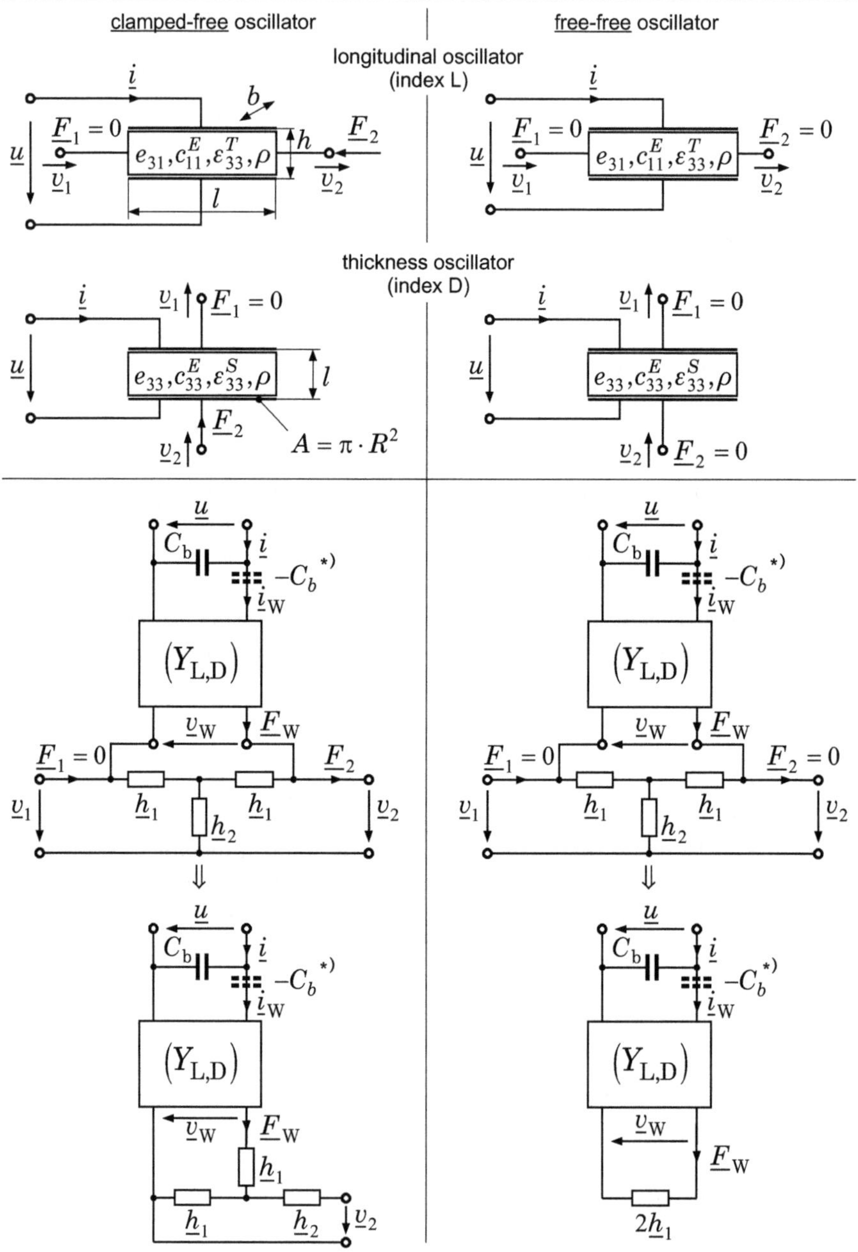

*) $-C_b$ has to be added for thickness oscillator

Table 9.11. Two-port network representation for clamped-free and free-free longitudinal and thickness oscillators

clamped-free oscillator

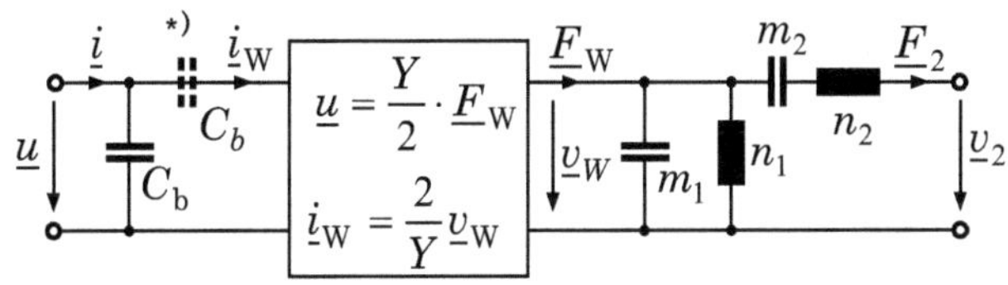

longitudinal oscillator: $n_1 = \frac{2}{\pi^2} n_{\mathrm{KL}}, \quad n_2 = \frac{1}{8} n_{\mathrm{KL}}$

thickness oscillator: $n_1 = \frac{2}{\pi^2} n_{\mathrm{KD}}, \quad n_2 = \frac{1}{8} n_{\mathrm{KD}}$

$m_1 = \frac{m}{2}, \quad m_2 = \frac{8}{\pi^2} m$

free-free oscillator

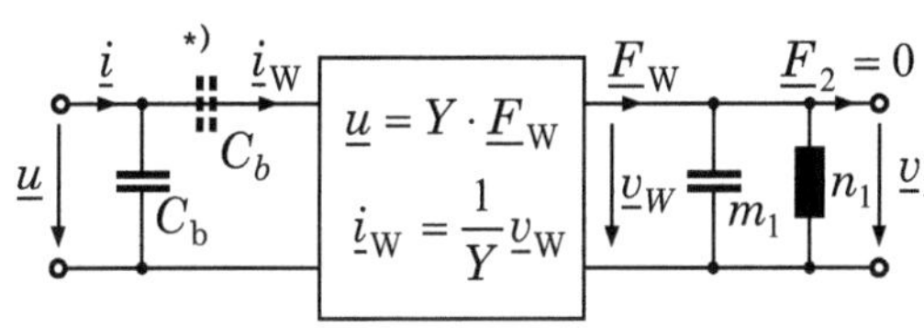

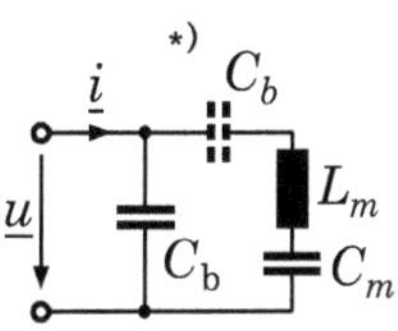

longitudinal oscillator: $n_1 = \frac{8}{\pi^2} n_{\mathrm{KL}}$

thickness oscillator: $n_1 = \frac{8}{\pi^2} n_{\mathrm{KD}}$

$m_1 = \frac{1}{8} m$

$L_m = Y^2 \cdot m_1$

$C_m = 1/Y^2 \cdot n_1$

*) C_b has to be added for thickness oscillator

$$Y_{\mathrm{L}} = \frac{1}{e_{31} \cdot b}, \quad Y_{\mathrm{D}} = \frac{l}{e_{33} \cdot A}, \quad C_{\mathrm{bL}} = \varepsilon_{33} \frac{b \cdot l}{h}, \quad C_{\mathrm{bD}} = \varepsilon_{33} \frac{A}{l}, \quad m_{\mathrm{L}} = \rho \cdot b \cdot h \cdot l,$$

$$m_{\mathrm{D}} = \rho \cdot A \cdot l, \; n_{\mathrm{KL}} = \frac{1}{c_{11}} \frac{l}{b \cdot h}, \; n_{\mathrm{KD}} = \frac{1}{c_{33}} \frac{l}{A}, \; n_{\mathrm{L}} = \frac{n_{\mathrm{KD}} \cdot n_{\mathrm{C}}}{n_{\mathrm{KD}} \quad n_{\mathrm{C}}}, \; n_{\mathrm{C}} = Y_{\mathrm{D}}^2 \cdot C_{\mathrm{bD}}$$

and

- the parallel resonant frequency f_{P} for electrical open-circuit operation:

$$f_{\mathrm{P}} = f_{\mathrm{S}} \sqrt{1 + \frac{C_n}{C_{\mathrm{b}}}} \quad \text{for} \quad C_n + C_{\mathrm{b}} - \omega^2 C_n C_{\mathrm{b}} = 0, \quad Q \gg 1$$

The operation of the quartz oscillator in series resonance is advantageous, since the resonant frequency is only dependent on the very stable material

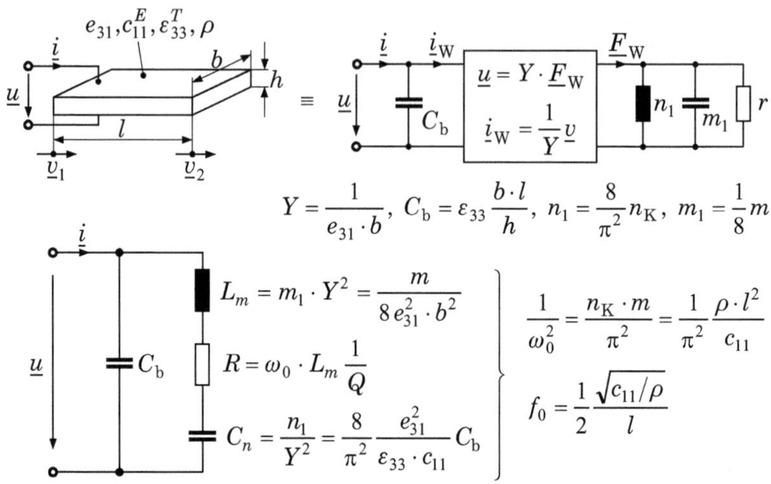

Fig. 9.58. Circuit diagram and mechanical resonant frequency of free-free longitudinal quartz oscillator

constants c_{11}^E and ρ of quartz. The difference between the resonant frequencies can be estimated according to

$$\frac{f_P - f_S}{f_S} \approx \frac{1}{2}\frac{C_n}{C_b}.$$

The component values and resonant frequencies specified in Table 9.12 are achieved for an X-cut quartz with the dimensions $l = 4\,\text{mm}$, $b = 0.5\,\text{mm}$ and $h = 0.1\,\text{mm}$. These values are compared with representative oscillator values of the X-, 5°-X- and AT-cut. The very high time stability of quartz oscillators is remarkable [89]:

Short-term stability ($T_M = 1\,\text{s}$, $\Delta\vartheta = \pm 1\,\text{K}$, temperature compensated):

$$\frac{\Delta f_S}{f_S} < 10^{-11}$$

Long-term stability ($T_M = 24\,\text{h}$, $\Delta\vartheta = \pm 1\,\text{K}$):

$$\frac{\Delta f_S}{f_S} < 10^{-9}$$

Long-term stability ($T_M = 1$ year, $\Delta\vartheta = \pm 1\,\text{K}$):

$$\frac{\Delta f_S}{f_S} < 10^{-6}$$

Table 9.12. Characteristic values of quartz oscillators as longitudinal and thickness oscillator

quantities	example	representative values		
	X-cut (longitudinal oscillator)	X-cut (longitudinal oscillator)	5°-X-cut (longitudinal oscillator)	AT-cut (thickness oscillator)
L_m / H	7.8	1.5	19.4	3.3
C_b / pF	0.8	60	7.9	5.8
C_n / pF	0.008	0.016	0.07	0.04
R / Ω	3017	60	170	5000
f_S / kHz	616	999	139	427
f_P / kHz	619	1000	140	429
Q	10^4	10^5	10^5	1800

Temperature coefficient of resonant frequency:

$$\alpha_f = 10^{-6} \text{ up to } 10^{-4}\,\mathrm{K}^{-1}$$

An ultrasonic transmitter as $\lambda/2$ thickness oscillator is considered as a second sample calculation.

Ultrasonic Transmitter in $\lambda/2$ Resonance

For the emission of ultrasonic waves in water a $\lambda/2$ thickness oscillator made of PZT-4 piezoelectric ceramics is used. The dimensions of the thickness oscillator and electrical circuit diagram are specified in Fig. 9.59.
Here, it is made use both of the circuit diagram of the clamped-free thickness oscillator represented in the Tables 9.10 and 9.11 and of the material constants of PZT-4 listed in Table 9.9. The circuit diagram is terminated by the frictional admittance h_a of the coupled water. By means of additionally integrated frictional admittances h_1 and h_2, the elastic losses of the oscillator are taken into account.
For $f \approx f_1$ only the resistance R_1 of the series resonant circuit is effective. Furthermore, for this frequency the parallel resonant circuit is negligible compared to the resistance R_a. For the calculation of the emitted power P_{ak} approximately only R_1 and R_a have to be considered in case of a voltage supply $\tilde{u}$:

$$P_{ak} = \frac{\tilde{u}_a^2}{R_a} = \left(\frac{R_a}{R_1 + R_a}\right)^2 \frac{1}{R_a}\tilde{u}^2 = \frac{\tilde{u}^2}{R_1 + R_a} \cdot \frac{R_a}{R_1 + R_a} = \eta P_{el}$$

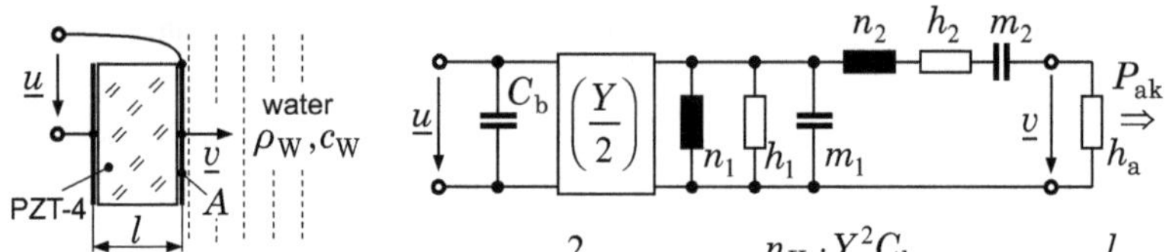

$A = \pi \cdot R^2 = 5 \cdot 10^3\ \text{mm}^2$

$l = 6.7\ \text{mm} \approx \frac{\lambda}{2}$

$\tilde{u} = 70\ \text{V}$

$\rho_W = 10^3\ \text{kg} \cdot \text{m}^{-3}$

$c_W = 1450\ \text{m} \cdot \text{s}^{-1}$

$Q_{mech} = 560$

$$n_1 = \frac{2}{\pi^2} n_L,\ n_L = \frac{n_K \cdot Y^2 C_b}{n_K + Y^2 C_b},\ n_K = s_{33} \frac{l}{A}$$

$$n_2 = \frac{1}{8} n_L,\ m_1 = \frac{m}{2},\ m_2 = \frac{8}{\pi^2} m,\ m = \rho \cdot A \cdot l$$

$$h_1 = \omega_1 \cdot n_1 \cdot Q_{mech},\ h_2 = \omega_2 \cdot n_2 \frac{1}{Q_{mech}}$$

$$h_a = \frac{1}{\rho_W \cdot c_W \cdot A} \quad \text{für} \quad R >> \lambda$$

electrical circuit diagram:

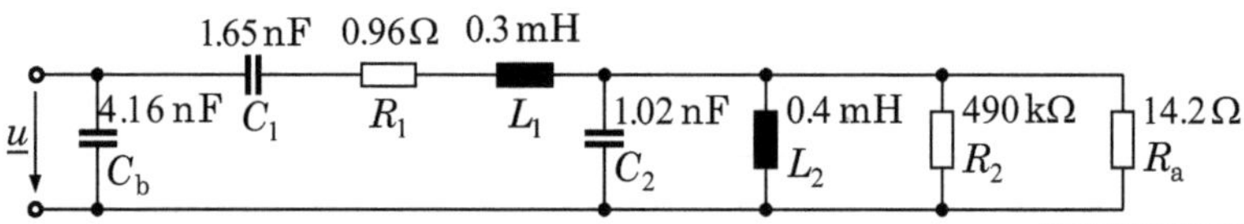

$$C_1 = \frac{4}{Y^2} \frac{2}{\pi^2} n_L = \frac{8}{\pi^2} k_{33}^2 \cdot C_b,\quad Y^2 = \frac{1 - k_{33}^2}{k_{33}^2} \frac{n_K}{C_b}$$

$$C_2 = \frac{4}{Y^2} \frac{1}{8} n_L = \frac{1}{2} k_{33}^2 \cdot C_b,\quad C_b = \varepsilon_{33} \frac{A}{l}$$

$$L_1 = \frac{Y^2}{4} m_1 = \frac{Y^2}{8} m,\quad L_2 = \frac{Y^2}{4} m_2 = \frac{2}{\pi^2} Y^2 \cdot m$$

$$f_1 = \frac{1}{2l} \sqrt{\frac{c_{33}}{\rho} \left(1 - k_{33}^2\right)} = 160\ \text{kHz}$$

$$f_2 = \frac{1}{2l} \sqrt{\frac{c_{33}}{\rho}} = 220\ \text{kHz}$$

$$R_1 = \frac{1}{Q_{mech} \cdot C_1 \cdot \omega_1},\quad R_2 = \frac{Q_{mech}}{C_2 \cdot \omega_2},\quad R_a = \frac{Y^2}{4} \rho_W \cdot c_W \cdot A$$

Fig. 9.59. Circuit diagram of a $\lambda/2$ thickness oscillator made of PZT-4 piezoelectric ceramics for ultrasonic emission in water

Thus, the efficiency results in

$$\eta = \frac{P_{ak}}{P_{el}} = \frac{R_a}{R_1 + R_a} = 0.94.$$

For a voltage supply with $\tilde{u} = 70\,\text{V}$ and thus $P_{el} \approx 320\,\text{W}$ the emitted ultrasonic power amounts to approximately 300 W. However, due to dielectric losses resulting in a heating of the transducer and due to the variation of piezoelectric constants caused by production technology of ceramics there will exist metrological differences toward lower efficiency compared with the theoretically possible value.

Piezoelectric Transformer in $\lambda/2$ Resonance

Mainly due to their small construction volume and decreasing costs, piezoelectric transformers increasingly represent an alternative to conventional magnetic transformers. However, their application is limited to the lower power range ($\leq 50\,\mathrm{W}$). The basic principle of the piezoelectric transformer is shown in Fig. 9.60.

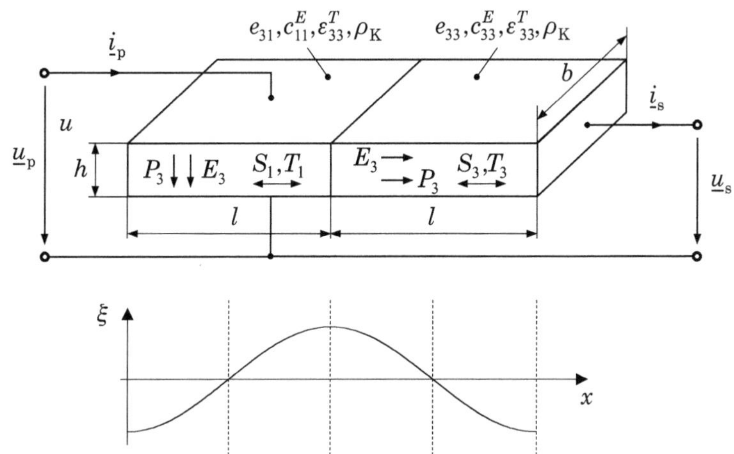

Fig. 9.60. Basic principle of a piezoelectric transformer

On the primary side of this basic configuration – ROSEN transformer [90] – a longitudinal oscillator is excited to perform mechanical vibrations in $\lambda/2$ resonance or in multiples. On the secondary side, a coupled thickness oscillator is mechanically excited on $\lambda/2$ resonance, too. The secondary voltage can be tapped at the electrodes. If both oscillators are separated by an insulating plate, the galvanic separation of primary and secondary side will be possible.

As piezoelectric material a piezoelectric ceramic is used. For the network representation shown in Fig. 9.61 both oscillators are approximately assumed as free-free longitudinal and free-free thickness oscillator according to Table 9.11. The network representation can be simplified by combining mechanical components and by transforming them to the primary side. The multiplication of both piezoelectric gyrator-like transducers results in the expected transformer-like coupling of primary and secondary side.

In combination with $\omega_0^2 = 1/\left(L_m \cdot C_n\right)$, the $\lambda/2$ resonant frequency is achieved according to

$$f_0 = \frac{1}{2l}\sqrt{\frac{c_{11}^E + c_{33}^E}{2\rho_K}}.$$

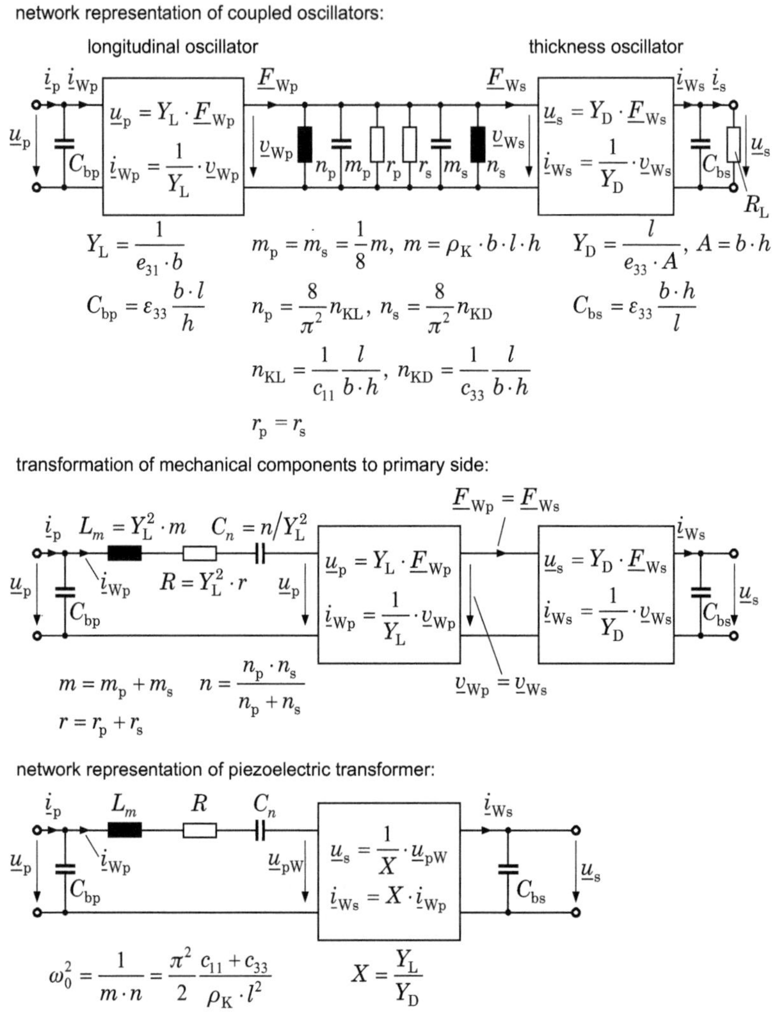

Fig. 9.61. Network representation of piezoelectric transformer

If $l = 15\,\mathrm{mm}$, $b = 15\,\mathrm{mm}$ and $h = 2\,\mathrm{mm}$ are chosen as typical dimensions, the resonant frequency of PZT-4 oscillators will result in (Table 9.9)

$$f_0 = 105\,\mathrm{kHz}.$$

The efficiency $\eta = P_{es}/P_{ep}$ amounts to 95%, thus it is higher than the efficiency of conventional transformers.
In practical applications, different geometries and directions of propagations – longitudinal oscillator, thickness oscillator, radial oscillator – are used for

piezoelectric elements. Also structures realized in multilayer technology are used in order to reduce primary voltage.

Piezoelectric Stack Actuator

Piezoelectric stack actuators are used for the operation of diesel fuel injection valves and small-power motors. The stack consists of a large number of small square PZT plates. In addition, it is mechanically prestressed (Fig. 9.62).

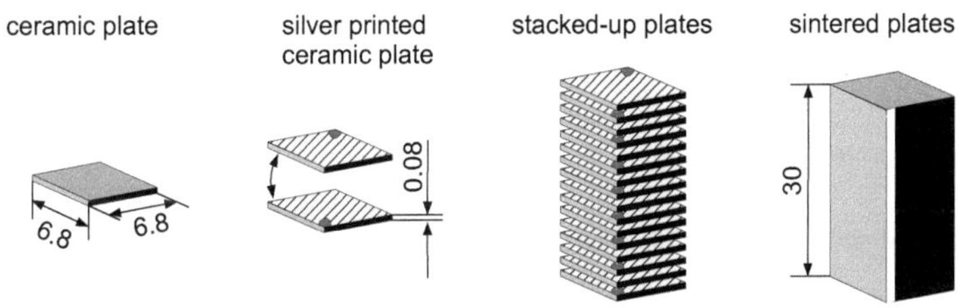

Fig. 9.62. Piezoelectric stack actuator (EPCOS)

The main characteristics of the individual small ceramic plates and the entire stack are summarized in Table 9.13. In order to achieve a network representation of the stack actuator, the circuit diagram of the free-free thickness oscillator specified in Table 9.11 is used. The internal losses of ceramics are taken into account by the frictional admittance h_i. The circuit of a single small ceramic plate is specified in Fig. 9.63. The contacting of the small plates is realized by means of silver-palladium electrodes (AgPd electrodes).

Within the stack the small piezoelectric plates are connected electrically in parallel. Due to their configuration, the polarization vectors of plates are directed in opposite directions. The elements are connected mechanically in series, i.e. the total stroke of stack equals the sum of single-element strokes. Thus, the series connection of parallel-connected spring-mass-damper elements is achieved as *canonical circuit* as shown in Fig. 9.64 [2].
The total impedance $\underline{h}_\mathrm{s}$ on the mechanical side results in:

$$\underline{h}_\mathrm{s} = \sum_{i=1}^{n} \underline{h}_i = \sum_{i=1}^{n} \left(\frac{1}{\dfrac{1}{\mathrm{j}\omega n_i} + \mathrm{j}\omega m_i + \dfrac{1}{h_i}} \right)$$

Thus, it applies for the actuator's velocity $\underline{v}_\mathrm{s}$:

$$\underline{v}_\mathrm{s} = \underline{h}_\mathrm{s} \cdot \underline{F}_\mathrm{W}, \qquad \underline{F}_\mathrm{W} = \frac{\underline{u}}{Y}$$

Table 9.13. Characteristic values of the piezoelectric stack (ANOX, Epcos)

characteristic values	typical values
dimensions of small PZT plate [mm]	6.8 x 6.8 x 0.08
stack length [mm]	30 (350 small plates)
mechanical prestress of stack [N]	800
stack compliance [$\cdot 10^{-9}\,\mathrm{m\,N^{-1}}$]	18.2
operating voltage [V]	160
field strength in PZT plate [$\mathrm{kV\,mm^{-1}}$]	2
stack capacitance [μF]	3
insulation resistance [MΩ]	$\geq$ 100
material constants of small plate	$\varepsilon_r = 1500$; $c_{11} = 346\,\mathrm{Pa}$ $d_{33} = 660 \cdot 10^{-12}\,\mathrm{m\,V^{-1}}$
operating temperature range [°C]	-55 bis 100, kurzzeitig 150

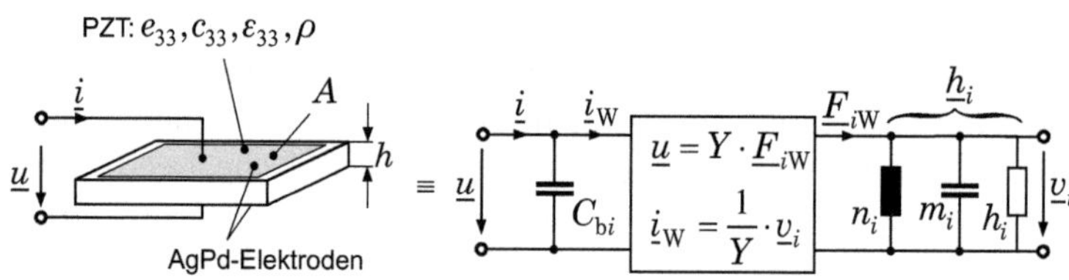

$$Y = \frac{1}{e_{33}}\frac{h}{A}, \quad C_{\mathrm{b}i} = \varepsilon_{33}\frac{A}{h}, \quad n_i = \frac{8}{\pi^2}n_{\mathrm{L}}, \quad m_i = \frac{1}{8}m$$

$$\frac{1}{\omega_0^2} = \frac{n_{\mathrm{L}} \cdot m}{\pi^2}$$

Fig. 9.63. Circuit diagram of free-free PZT thickness oscillator element

The transformation of the voltage source $\underline{u}$ and stack capacitance C_{b} to the mechanical side is performed with the specified transducer equations. In addition, the mechanical side is terminated by mechanical load consisting of the mass of valve needle m_{V} and needle friction r_{V}. Figure 9.65 shows the total mechanical circuit of the stack actuator.

The pSpice simulation of this circuit provides the transfer function $\underline{B}_{\xi} = \underline{\xi}/\underline{u}$ for open-circuit operation (Fig. 9.66). Its curve progression provides a series and parallel resonant frequency f_{s} and f_{p} of the fundamental mode. In the

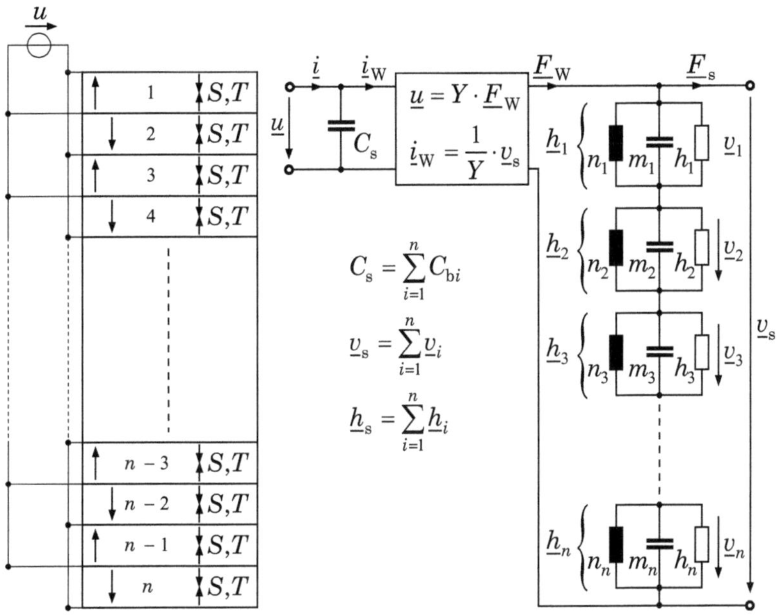

Fig. 9.64. Canonical circuit of piezoelectric stack actuator

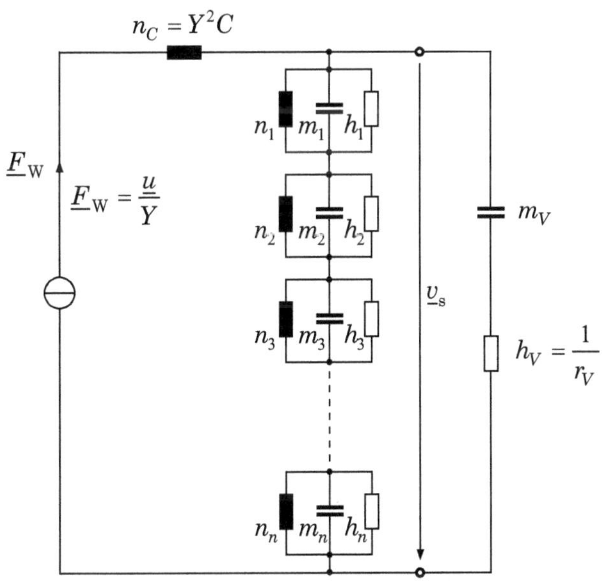

Fig. 9.65. Mechanical circuit of stack actuator

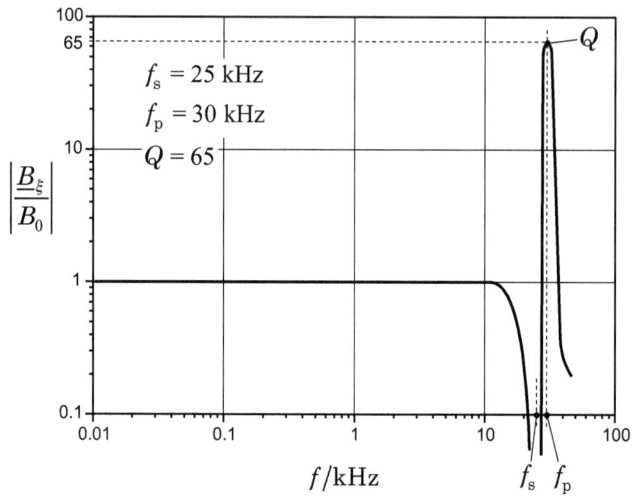

Fig. 9.66. Curve progression of transfer function of actuator deflection and its series and parallel resonant frequency

static case, a maximum stroke of $\xi_{max} = 45\,\mu$m and a maximum force (locking force) of $F_{max} = 2200$ N are achieved.

9.3.5 Piezoelectric Beam Bending Element as Waveguide

In order to generate large displacement amplitudes in the range of a few tenths of millimetres, piezoelectric bimorph bending elements as already described in Sect. 9.2.5 are used. Based on the quasi-static five-port network representation of the piezoelectric bimorph bending element shown in Fig. 9.41 and Fig. 9.67, in the following, the circuit being suitable for higher frequencies is derived. As with the longitudinal oscillator, in Fig. 9.68 the circuit is achieved by means of electrical parallel and mechanical series connection of finite bimorph bending elements shown in Fig. 9.67.

In contrast to the longitudinal oscillator, now the mechanical side is described as four-port network by means of the chain matrix of the mechanical waveguide for bending waves specified in Fig. 9.68. The cascade connection comprising finite bending elements and represented in Sect. 5.1.2 forms the basis for representation of the waveguide for bending waves.

For the technically important application of the *cantilevered beam bender*, the relations specified in Fig. 9.68 can be simplified. For selected frequency ratios the simplified circuits and appropriate components are specified in the Tables 9.14 and 9.15 according to [2]. For the calculation of following three sample applications – piezoelectric transmitter and receiver of deflections, piezoelectric accelerometer with beam bending element as well as piezoelectric multilayer

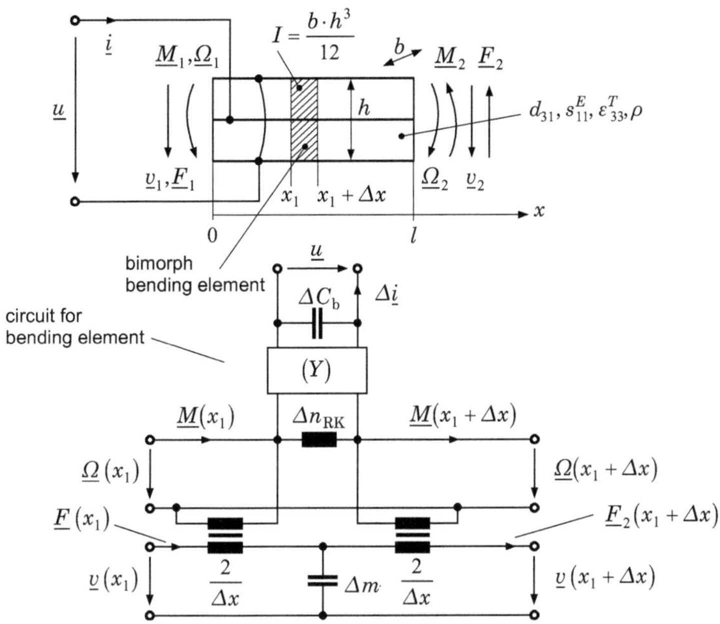

Fig. 9.67. Bending oscillator in combination with the circuit for loss-free bending element Δx

beam bending actuator – the relations specified in the Tables 9.14 and 9.15 are applied.

9.3.6 Sample Applications of Piezoelectric Beam Bending Elements

Piezoelectric Bending Actuator as Transmitter and Receiver for Deflections

The transmitter for deflections consists of an electrically actuated clamped-free parallel bimorph (Fig. 9.69). In order to calculate the transfer function $\underline{B}_u = \underline{\xi}_S / \underline{u}$, the circuit represented in Table 9.14 a) is used.
In case of a supply by means of a sufficiently low-resistance voltage source, the capacitance C_b' can be neglected for the calculation of the transfer factor. With this assumption the simplified circuit illustrated in Fig. 9.69 a) is achieved after transforming the electrical side to the mechanical.
The deflection for open-circuit operation $\underline{\xi}_L$ can be derived from the velocity for open-circuit operation $\underline{v}_L$ according to

$$\underline{\xi}_L = \frac{1}{j\omega} \underline{v}_L = n \frac{\underline{u}}{\underline{Y}}.$$

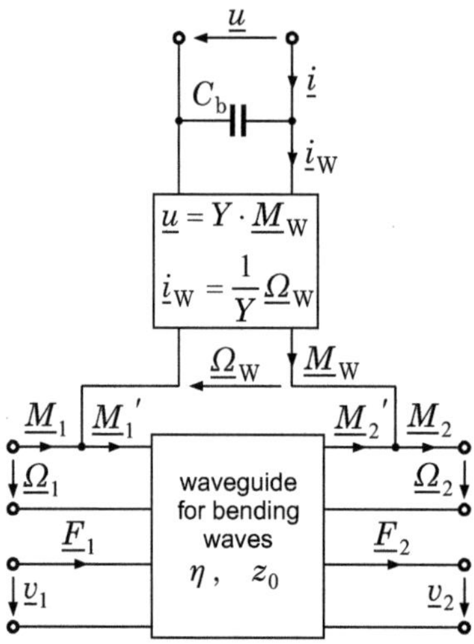

$$C_{\mathrm{b}} = 4\varepsilon_{33}\left(1-k_{31}^2\right)\frac{l\cdot b}{h}$$

$$k_{31}^2 = \frac{d_{31}}{\varepsilon_{33}\cdot s_{11}}, \quad \frac{1}{Y} = \frac{1}{2}\frac{d_{31}}{s_{11}}b\cdot h$$

$$\eta = \sqrt{\frac{\omega}{\omega_0}}, \quad \omega_0 = \frac{1}{\sqrt{m\cdot n_0}}, \quad z_0 = \sqrt{\frac{m}{n_0}}$$

$$n_0 = \frac{s_{11}\cdot l^3}{I} = n_{\mathrm{RK}}\cdot h^2, \qquad I = \frac{b\cdot h^3}{12}$$

$$n_{\mathrm{RK}} = \frac{s_{11}\cdot h}{I}\,\frac{1-k_{31}^2}{1-\frac{3}{4}k_{31}^2}$$

$$\omega_0^2 = \frac{1}{12}\,\frac{1}{\rho\cdot s_{11}}\,\frac{1-\frac{3}{4}k_{31}^2}{1-k_{31}^2}\,\frac{h^2}{l^4} = \frac{1}{m\cdot n_0}$$

chain matrix of the waveguide for bending waves:

$$\begin{pmatrix} \underline{v}_1 \\ \underline{\Omega}_1 \\ \underline{M}_1' \\ \underline{F}_1 \end{pmatrix} = \begin{pmatrix} C & -\frac{l}{\eta}S & \frac{1}{\mathrm{j}z_0\,l}c & \frac{1}{\mathrm{j}z_0}\frac{s}{\eta} \\ -\frac{\eta s}{l} & C & -\frac{\eta\cdot S}{\mathrm{j}z_0\,l^2} & -\frac{\mathrm{c}}{\mathrm{j}z_0\,l} \\ \mathrm{j}z_0\,l\,c & -\mathrm{j}z_0\,l^2\frac{s}{\eta} & C & l\frac{S}{\eta} \\ \mathrm{j}z_0\,\eta S & -\mathrm{j}z_0\,l\,c & \frac{\eta s}{l} & C \end{pmatrix} \begin{pmatrix} \underline{v}_2 \\ \underline{\Omega}_2 \\ \underline{M}_2' \\ \underline{F}_2 \end{pmatrix}$$

Rayleigh functions:

$$S(\eta) = \frac{1}{2}\left(\sinh\eta + \sin\eta\right), \qquad C(\eta) = \frac{1}{2}\left(\cosh\eta + \cos\eta\right)$$

$$s(\eta) = \frac{1}{2}\left(\sinh\eta - \sin\eta\right), \qquad c(\eta) = \frac{1}{2}\left(\cosh\eta - \cos\eta\right)$$

propagation velocity of bending waves:

$$c_{\mathrm{B}} = \sqrt{c_{\mathrm{D}}\cdot r\cdot\omega}, \quad c_{\mathrm{D}} = \sqrt{\frac{E}{\rho}}, \quad r = \sqrt{\frac{I}{A}}$$

wavelength of bending waves:

$$\lambda_{\mathrm{B}} = 2\pi\sqrt{\frac{c_{\mathrm{D}}\cdot r}{\omega}}$$

Fig. 9.68. Circuit and chain matrix of piezoelectric bending oscillator
Index B: *bending wave*

Table 9.14. Circuits and components of cantilevered bending oscillator for different frequency ranges

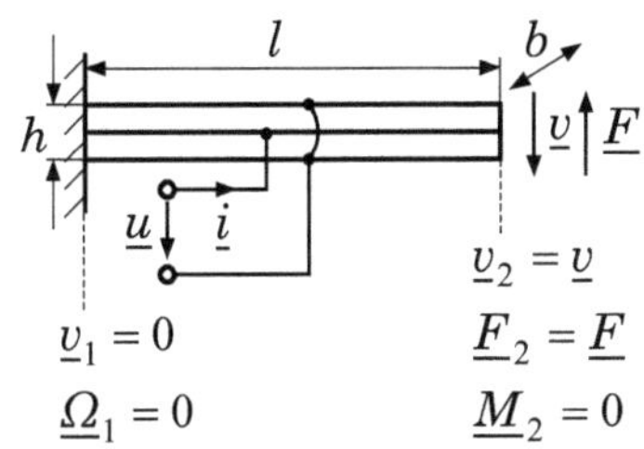

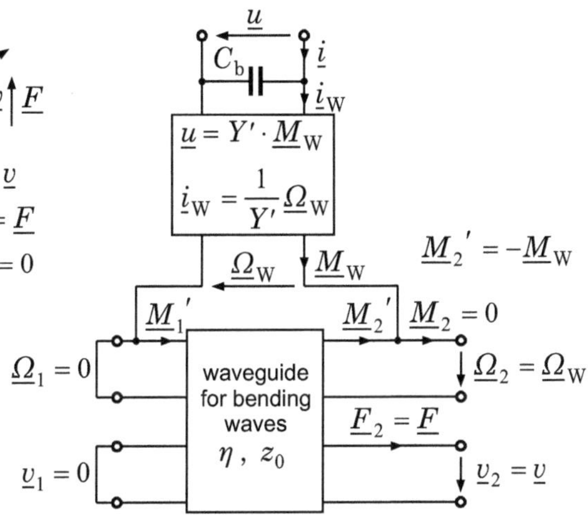

$$\eta = \sqrt{\frac{\omega}{\omega_0}}, \omega_0 = \frac{1}{\sqrt{m \cdot n_0}}$$

electrical circuits:

a)	b)	c)
η arbitrary	$\eta = \eta_1 = 1.875$ (in neighborhood of first bending resonance)	$\eta << 1$ (quasi-static)
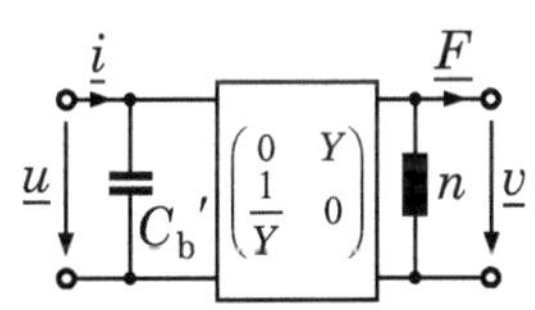	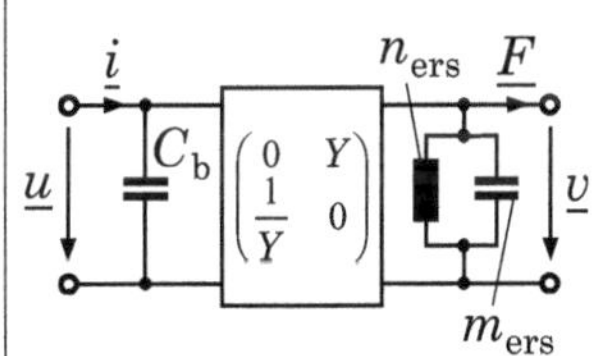	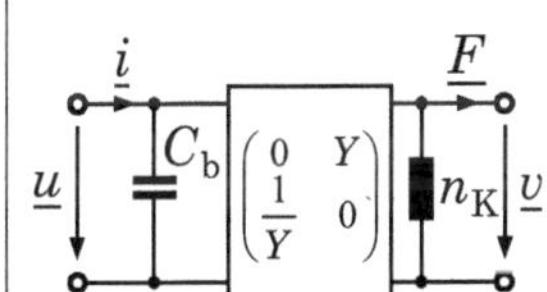
$C_b' = C_b + C_n$ $C_n = \frac{n_0}{Y'^2 \cdot l^2} \gamma$	$C_b' \approx C_b$	$C_b' \approx C_b$
$n = \frac{n_0}{\eta^3} \cdot \tau$	$n_{ers} = 0.971\, n_0/3$ $m_{ers} = m/4$	$n_K = \frac{n_0}{3}$
$Y = Y' \cdot r$ $r = \frac{l}{\eta} \cdot \varphi$	$Y = Y' \cdot r$ $r = \frac{l}{\eta} \tan \frac{\eta_1}{2} = 0.727\, l$	$Y = \frac{2}{3} Y' \cdot l$

Table 9.15. Circuits and components of clamped-free frequency oscillator for different frequency ranges

$\gamma = -\frac{1}{\eta}\left(\frac{1-\cosh\eta\cdot\cos\eta}{\sinh\eta\cdot\cos\eta-\cosh\eta\cdot\sin\eta}\right)$
$\tau = \frac{\cosh\eta\cdot\sin\eta-\sinh\eta\cdot\cos\eta}{1+\cosh\eta\cdot\cos\eta}$
$\varphi = \frac{\cosh\eta\cdot\sin\eta-\sinh\eta\cdot\cos\eta}{\sinh\eta\cdot\sin\eta}$
$\frac{1}{Y'} = \frac{1}{2}\frac{d_{31}}{s_{11}} b\cdot h = \frac{1}{2}k_{31}\sqrt{\frac{\varepsilon_{33}}{s'_{11}}}\, b\cdot h\,,\quad k_B^2 = \frac{3}{4}k_{31}^2\,,\quad n_0 = \frac{l^3\cdot s_{11}}{I} = n_{RK}\cdot h^2$
$I = \frac{b\cdot h^3}{12}\,,\qquad C_b = 4\cdot\varepsilon_{33}\frac{b\cdot l}{h}\left(1-k_{31}^2\right)$

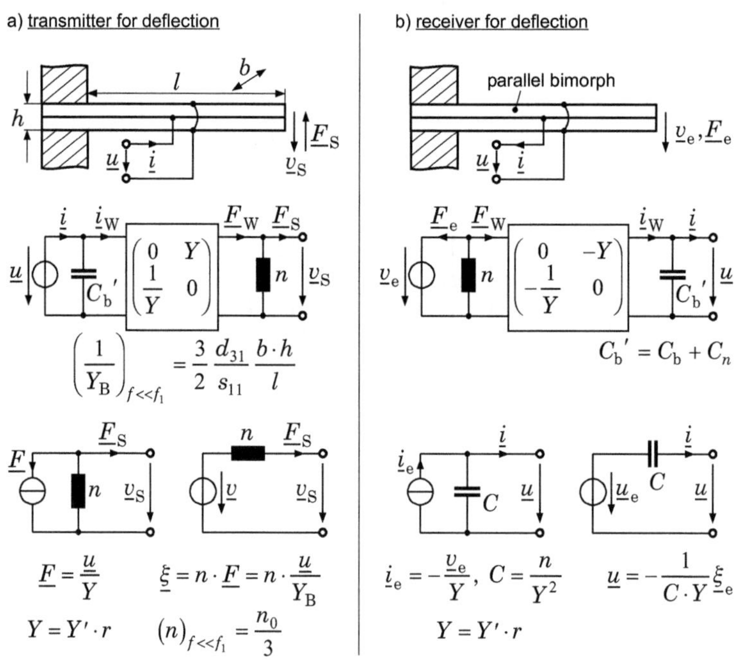

Fig. 9.69. Simplified circuit of piezoelectric transmitter and receiver for deflections

If the relations for n and Y specified in Table 9.14 are inserted, the transfer function will result in

$$\underline{B}_u = \frac{\underline{\xi}_{\mathrm{L}}}{\underline{u}} = B_{u0}\frac{1}{\eta^2}\left(\frac{\sinh(\eta)\sin(\eta)}{1+\cosh(\eta)\cos(\eta)}\right)$$

with

$$B_{u0} = \frac{1}{2}\frac{d_{31}}{s_{11}}\frac{bh}{l}n_0 = 6d_{31}\frac{l^2\left(1-k_{31}^2\right)}{\left(1+k_{\mathrm{B}}^2\right)}.$$

The normalized transfer function characteristic is shown in Fig. 9.70 a). A resonance amplification of deflection occurs already below the *first* bending resonant frequency f_1. In proximate neighborhood of the *first* bending resonant frequency the circuit illustrated in Table 9.14 b) can be used.

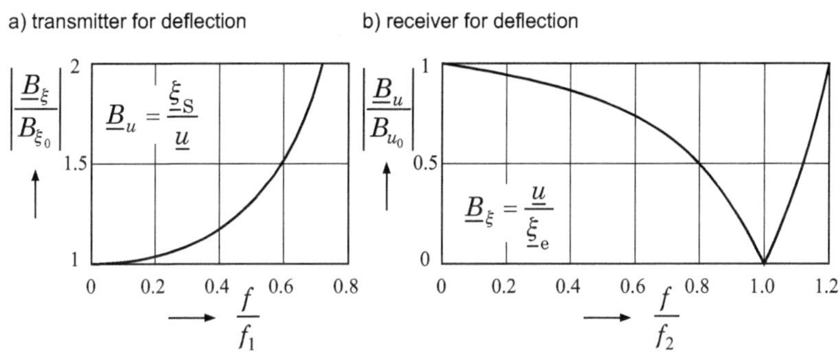

Fig. 9.70. Transfer function characteristics of piezoelectric transmitter and receiver for deflections

As an example, a transmitter for deflections made of PZT-4 ceramic is calculated in combination with the values

$$d_{31} = 12.3\cdot10^{-11}\frac{\mathrm{m}}{\mathrm{V}},\quad l = 20\,\mathrm{mm},\quad h = 0.4\,\mathrm{mm},\quad k_{31} = 0.33,\quad \hat{u} = 100\,\mathrm{V}.$$

With

$$k_{\mathrm{B}} = \frac{\sqrt{3}}{2}k_{31} = 0.29$$

the deflection for open-circuit operation results in

$$\hat{\xi}_{\mathrm{L}} = B_{u0}\hat{u} = 150\,\mu\mathrm{m}.$$

In combination with $f_1 = \eta_1^2\cdot f_0$, $\eta_1 = 1.87$ and the characteristic frequency f_0 achieved from

$$\omega_0^2 = \frac{1}{mn_0} = \frac{1}{12\rho s_{11}}\cdot\frac{1-k_{\mathrm{B}}^2}{1-k_{31}^2}\cdot\frac{h^2}{l^4},\qquad f_0 = 153\,\mathrm{Hz},$$

the bending resonant frequency yields the value $f_1 = 538\,\mathrm{Hz}$. Due to additional compliances being found in reality at the bender's clamp position, the actual resonant frequency is lower.
Transmitters for deflections with bending oscillators are used in headphones and for displacement of miniaturized mirrors but also in linear and stepper motors.

If the conversion effect is reversed, the piezoelectric bending oscillator can be operated as receiver for deflections according to Fig. 9.69 b). Now mechanical power is supplied and electrical power is consumed. Compared with the transmission operation, current and force arrows are opposed. Moreover, the coupling matrix has now negative signs. In order to calculate the transfer function $\underline{B}_\xi = \underline{u}/\underline{\xi}_\mathrm{e}$, the simplified circuit diagram being based on a velocity source with negligible internal compliance n and shown in Fig. 9.69 b) is used. The transfer function of the receiver for deflections shown in Fig. 9.69 b) yields

$$\underline{B}_\xi = -\frac{\underline{u}}{\underline{\xi}_\mathrm{e}} = -\frac{1}{CY} = -\frac{Y}{n}.$$

By inserting the relations for Y and n specified in Table 9.14, following expression can be derived:

$$\underline{B}_\xi = B_{\xi 0}\eta^2 \left(\frac{1+\cosh(\eta)\cos(\eta)}{\sinh(\eta)\sin(\eta)}\right)$$

with

$$B_{\xi 0} = -2\frac{s_{11}}{d_{31}}\frac{l}{bh}\frac{1}{n_0}$$

Using the numerical values of the considered example, the quantity $B_{\xi 0}$ amounts to

$$B_{\xi 0} = 67\ \frac{\mathrm{mV}}{\mu\mathrm{m}}.$$

The transfer function characteristic is shown in Fig. 9.70 b). For $\eta_2 = \pi$ the transfer function has a zero. The corresponding frequency amounts to

$$f_2 = \pi^2 f_0 = 1.51\,\mathrm{kHz} \qquad \text{with} \qquad f_0 = 153\,\mathrm{Hz}.$$

Piezoelectric bimorph elements are used as transducer elements in microphones. Their design principle has already been represented in Fig. 9.46. Due to their high sensitivity and good reproducibility of characteristic curve, they are also used as high-resolution position sensing elements, e.g. in force and tunneling microscopes.

Piezoelectric Accelerometer with Bending Actuator considered as having Mass

In order to calculate the transfer function $\underline{B}_a = \underline{u}_\mathrm{L}/\underline{a}$, the quasi-static circuit diagram of the clamped-free bender illustrated in Table 9.14 c) is assumed.

The additional mass m results in a decrease of resonant frequency of the bending actuator, thus the condition $\eta \ll 1$ is satisfied with sufficient approximation. The circuit of bending actuator with mass is shown in Fig. 9.71 b).

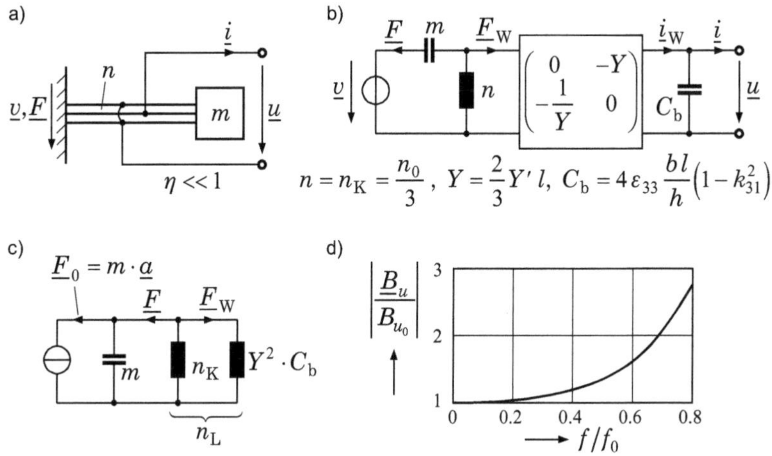

Fig. 9.71. Circuit and transfer function characteristic of an accelerometer with bending actuator considered as having mass

If in case of an open-circuit operation the capacitance C_b is transformed to the mechanical side and if the velocity source is replaced by a current source, the mechanical network represented in Fig. 9.71 c) is achieved. The open-circuit voltage is calculated from the transducer force $\underline{F}_W$ according to

$$\underline{u}_L = -Y\underline{F}_W \quad \text{with} \quad Y = \frac{4}{3}\frac{s_{11}}{d_{31}}\frac{l}{bh}.$$

By applying the current divider rule, a relation for

$$\frac{\underline{F}_W}{\underline{F}_0} = -\frac{3}{4}\frac{k_B^2}{1-\left(\frac{\omega}{\omega_0}\right)^2}, \qquad \omega_0^2 = \frac{1}{mn_L}, \qquad n_L = \frac{n_K}{3}\left(1-\frac{3}{4}k_B^2\right)$$

can be derived. This results in the transfer function for open-circuit operation

$$\underline{B}_a = \frac{\underline{u}_L}{\underline{a}} = \frac{s_{11}}{d_{31}}\frac{l}{bh}k_B^2 m\frac{1}{1-\left(\frac{\omega}{\omega_0}\right)^2} = B_{a0}\frac{1}{1-\left(\frac{\omega}{\omega_0}\right)^2}.$$

Its characteristic curve is specified in Fig. 9.71 d). In combination with d_{31} and s_{11} specified in Table 9.9 and in combination with

$$l = 20\,\text{mm}, \quad b = 10\,\text{mm}, \quad h = 0.4\,\text{mm}, \quad k_{\text{B}} = 0.29, \quad m = 4\,\text{g},$$

the transfer factor of the PZT-4 bender amounts to $B_{a0} = 168\,\text{mVm}^{-1}\text{s}^2$ and the resonant frequency amounts to $f_0 = 180\,\text{Hz}$. Thus, this value is well below the first bending resonant frequency $f_1 = 540\,\text{Hz}$.

Piezoelectric Multilayer Beam Bending Actuator

In the following, the canonical circuit of a clamped-free bending actuator consisting of n layers will be derived (Fig. 9.72). A differential homogeneous beam element as already discussed in Sect. 5.1.2 represents the starting point of following consideration. Both ends are affected both by translational quantities (velocity $\underline{v}$, force $\underline{F}$) and rotational quantities (angular velocity $\underline{\Omega}$, moment $\underline{M}$).

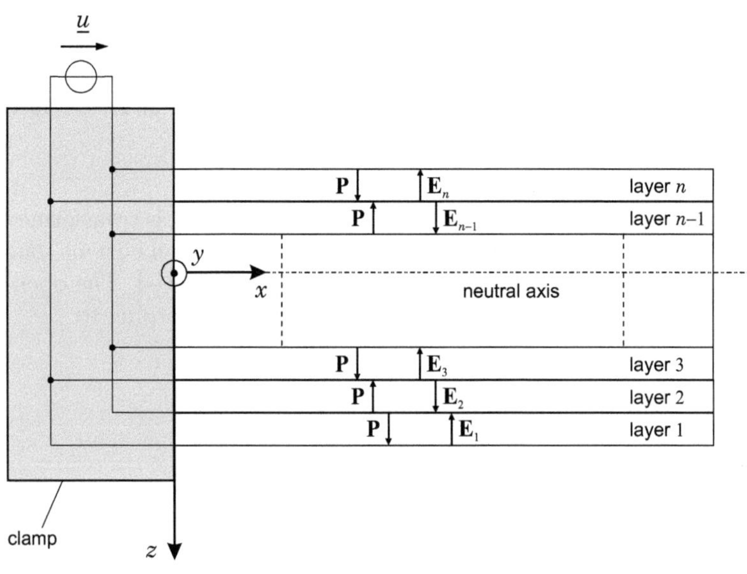

Fig. 9.72. Clamped-free piezoelectric multilayer beam bending actuator

In order to determine the dynamic behavior of the beam element, mass Δm, equivalent viscous damping Δr and an external continuous load $\underline{f}(x)$ are taken into account. Figure 9.73 illustrates this fact.

Assuming that the mass, frictional force and load can be considered to be concentrated in the middle of the beam element and taking both kinematic

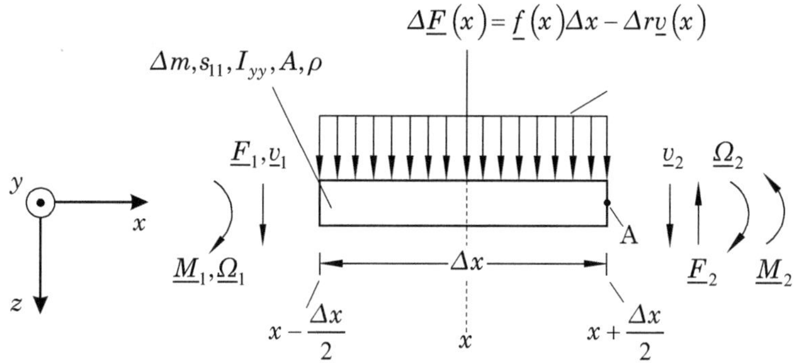

Fig. 9.73. Differential beam element taking mass Δm, friction Δr and an external continuous load $\underline{f}(x)$ into account

and dynamic conditions $\sum_i \underline{F}_i = \mathrm{j}\omega \Delta m \underline{v}(x)$ and $\sum_i \underline{M}_i = 0$ into acount, the inner structure of the differential beam element shown in Fig. 9.74 can be derived.

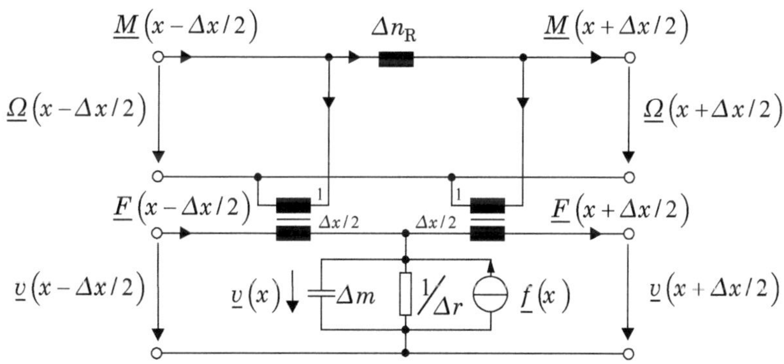

Fig. 9.74. Circuit of differential beam element taking frictional force and continuous load into consideration

The inner structure of the beam element represented in Fig. 9.74 corresponds to the circuit of a finite loss-free bending element already illustrated in Fig. 9.67, for which, however, an external continuous load $\underline{f}(x)$ and equivalent viscous damping Δr have not been taken into account. Figure 9.74 shows that translational and rotational quantities are interconnected by ideal transformers. The quantity Δn_R denotes the rotational compliance of a differential beam element. Taking both the kinematic and dynamic conditions $\sum_i \underline{F}_i = \mathrm{j}\omega \Delta m \underline{v}(x)$ and $\sum_i \underline{M}_i = 0$ already mentioned above and the individual meshes into account, the difference equations for translational and rotational velocities $\underline{v}$ and $\underline{\Omega}$ can be derived from the inner four-port network structure shown in Fig. 9.74.

The limit consideration ($\Delta x \to 0$) of both the kinematic and dynamic conditions and the derivable difference equations yields the differential equations in complex form describing the dynamic behavior of a homogeneous beam. The combination of these differential equations results in the total differential equation of a homogeneous beam in complex form

$$\frac{\mathrm{d}^4 \underline{v}(x)}{\mathrm{d}x^4} - \omega^2 \frac{\mu}{EI} \underline{v}(x) + \mathrm{j}\frac{\omega}{EI} r_{\mathrm{a}} \underline{v}(x) = \mathrm{j}\frac{\omega}{EI} \underline{f}(x)\,. \tag{9.29}$$

The explicit calculations are presented in detail in [85]. The quantities EI, μ and r_{a} in (9.29) denote the flexural rigidity of the homogeneous beam and the mass and friction per unit length.

By taking propagable modes X_m and respective corresponding wave numbers β_m of a clamped-free beam bender into account, the general solution of the differential equation (9.29) can be derived:

$$\underline{v}_m \left[\frac{EI\,(\beta_m l)^4}{\mathrm{j}\omega l^4} + \mathrm{j}\omega\mu + r_{\mathrm{a}} \right] = \frac{\int_0^l \underline{f}(x)\, X_m \mathrm{d}x}{\int_0^l X_m^2 \mathrm{d}x}. \tag{9.30}$$

A detailed derivation of (9.30) is presented in [85]. In order to achieve a representation of (9.30) within the scope of network theory, the torsional compliance per unit length $n'_{\mathrm{R}} = 1/EI$ and translational reference compliance $n_0 = l^3/EI$ are introduced as reference quantities (see also Sect. 5.1.2 and Sect. 6.1.5). Furthermore, the total mass $m = l\mu$ and coefficient of friction $r = r_{\mathrm{a}} l$ are also used as reference quantities.
Equation (9.30) can be reformulated using the defined reference quantities:

$$\underline{v}_m \left[\frac{1}{\mathrm{j}\omega \dfrac{n_0}{(k_m l)^4}} + \mathrm{j}\omega m + r \right] = \frac{l \int_0^l \underline{f}(x)\, X_m \mathrm{d}x}{\int_0^l X_m^2 \mathrm{d}x} \tag{9.31}$$

The term in brackets on the left side of (9.31) shows an analogy to the mechanical impedance $\underline{z}$ of a spring-mass-damper system with one degree of freedom when excited by a harmonic excitational force $\underline{F}$. The mechanical representation, the mechanical scheme and the corresponding network representation of a spring-mass-damper system are shown in Fig. 3.19 of Sect. 3.1.7.

In order to describe the bending actuator as an electromechanical system within the scope of network theory, the relation between difference quantities $(\underline{v}, \underline{\Omega})$ and flow quantities $(\underline{F}, \underline{M})$ is represented by means of a matrix. It is referred to that matrix as admittance matrix $\underline{\mathbf{H}}$. The general structure of interconnection is shown in matrix equation (9.32)

$$\begin{pmatrix} \underline{v}_1 \\ \underline{\Omega}_1 \\ \underline{v}_2 \\ \underline{\Omega}_2 \end{pmatrix} = \underbrace{\begin{pmatrix} \underline{h}_{11} & \underline{h}_{12} & \underline{h}_{13} & \underline{h}_{14} \\ \underline{h}_{21} & \underline{h}_{22} & \underline{h}_{23} & \underline{h}_{24} \\ \underline{h}_{31} & \underline{h}_{32} & \underline{h}_{33} & \underline{h}_{34} \\ \underline{h}_{41} & \underline{h}_{42} & \underline{h}_{43} & \underline{h}_{44} \end{pmatrix}}_{\equiv \underline{\mathbf{H}}} \begin{pmatrix} \underline{F}_1 \\ \underline{M}_1 \\ \underline{F}_2 \\ \underline{M}_2 \end{pmatrix}, \tag{9.32}$$

where following boundary conditions are defined for difference and flow coordinates:

$$\underline{v}(x)\big|_{x=0} = \underline{v}_1 \quad \wedge \quad \underline{v}(x)\big|_{x=l} = \underline{v}_2 \tag{9.33}$$

$$\underline{F}(x)\big|_{x=0} = \underline{F}_1 \quad \wedge \quad \underline{F}(x)\big|_{x=l} = \underline{F}_2 \tag{9.34}$$

$$\underline{\Omega}(x)\big|_{x=0} = \underline{\Omega}_1 \quad \wedge \quad \underline{\Omega}(x)\big|_{x=l} = \underline{\Omega}_2 \tag{9.35}$$

$$\underline{M}(x)\big|_{x=0} = \underline{M}_1 \quad \wedge \quad \underline{M}(x)\big|_{x=l} = \underline{M}_2 \tag{9.36}$$

The general calculation of the individual elements $\underline{h}_{ij}$ of the admittance matrix $\underline{\mathbf{H}}$ represented in (9.32) using (9.31) would go beyond the scope of this book, thus at this point it is referred to [85].

The transition from a homogeneous bending actuator to a piezoelectric multilayer beam bending actuator as shown in Fig. 9.72 represents the next step. Since the individual layers are connected in parallel, the actuator can be represented by a five-port network according to Fig. 9.68. At this point it should be noted that both the transducer constant Y and translationally blocked capacitance C_b of the multilayer beam bending actuator can be calculated from a combination of geometrical, elastic and electromechanical values of each individual layer. For more detailed calculations see [85].

For the waveguide for bending waves with terminal pairs $(\underline{\Omega}_1, \underline{M}'_1)$, $(\underline{\Omega}_2, \underline{M}'_2)$ and $(\underline{v}_1, \underline{F}_1)$, $(\underline{v}_2, \underline{F}_2)$, the admittance matrix $\underline{\mathbf{H}}$ determined in (9.32) is valid when replacing the moments $\underline{M}_1$, $\underline{M}_2$ by $\underline{M}'_1$, $\underline{M}'_2$.

Following relations have to be considered:

$$\underline{\Omega}_{\mathrm{W}} = \underline{\Omega}_2 - \underline{\Omega}_1 \tag{9.37}$$

$$\underline{M}_1' = \underline{M}_1 - \underline{M}_{\mathrm{W}} \tag{9.38}$$

$$\underline{M}_2' = \underline{M}_2 - \underline{M}_{\mathrm{W}} \tag{9.39}$$

With the achieved knowledge, in the next step the electromechanical circuit diagram of a *clamped-free* piezoelectric multilayer beam bending actuator will be developed and derived.
Taking the coordinate relations (9.37) up to (9.39) into account, the admittance matrix $\underline{\mathbf{H}}$ represents the starting point of further considerations.

$$\begin{pmatrix} \underline{v}_1 \\ \underline{\Omega}_1 \\ \underline{v}_2 \\ \underline{\Omega}_{\mathrm{W}} + \underline{\Omega}_1 \end{pmatrix} = \begin{pmatrix} \underline{h}_{11} & \underline{h}_{12} & \underline{h}_{13} & \underline{h}_{14} \\ \underline{h}_{21} & \underline{h}_{22} & \underline{h}_{23} & \underline{h}_{24} \\ \underline{h}_{31} & \underline{h}_{32} & \underline{h}_{33} & \underline{h}_{34} \\ \underline{h}_{41} & \underline{h}_{42} & \underline{h}_{43} & \underline{h}_{44} \end{pmatrix} \begin{pmatrix} \underline{F}_1 \\ \underline{M}_1 - \underline{M}_{\mathrm{W}} \\ \underline{F}_2 \\ \underline{M}_2 - \underline{M}_{\mathrm{W}} \end{pmatrix} \tag{9.40}$$

With the knowledge of propagable eigenmodes of a clamped-free beam bending actuator all elements $\underline{h}_{ij}$ of the admittance matrix $\underline{\mathbf{H}}$ can be determined explicitly. Since in the present representation the transducer operates as a transmitter for mechanical kinetic quantities, velocities and forces appearing at the actuator's tip are provided with the index S (Fig. 9.75).

There are no external bending moments affecting the bender's tip ($\underline{M}_2 = 0$). The actuator is driven by an excitation voltage $\underline{u}$, thus the matrix equation (9.40) can be simplified:

$$\begin{pmatrix} 0 \\ 0 \\ \underline{v}_{\mathrm{S}} \\ \underline{\Omega}_{\mathrm{W}} \end{pmatrix} = \begin{pmatrix} 0 & 0 & 0 & 0 \\ 0 & 0 & 0 & 0 \\ 0 & 0 & \underline{h}_{33} & \underline{h}_{34} \\ 0 & 0 & \underline{h}_{43} & \underline{h}_{44} \end{pmatrix} \begin{pmatrix} \underline{F}_1 \\ \underline{M}_1 - \underline{M}_{\mathrm{W}} \\ \underline{F}_{\mathrm{S}} \\ -\underline{M}_{\mathrm{W}} \end{pmatrix} \tag{9.41}$$

Figure 9.76 illustrates the five-port network reduced to a two-port network $\underline{\mathbf{A}}$ by means of the boundary conditions mentioned above.

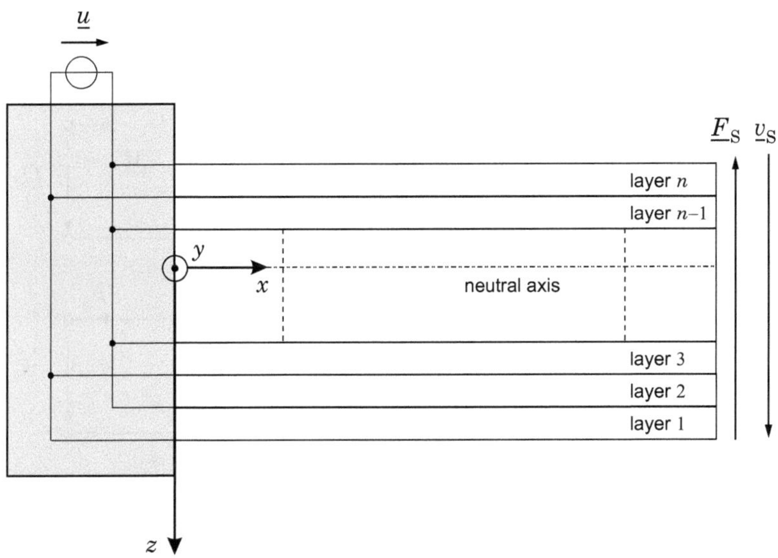

Fig. 9.75. Piezoelectric bending oscillator as transmitter for mechanical kinetic quantities (drive)

In combination with the admittance matrix (9.41), the translational velocity $\underline{v}_S$ and angular velocity $\underline{\Omega}_W$ can be formulated in dependence on the flow quantities $\underline{F}_S$ and $\underline{M}_W$:

$$\underline{v}_S = \underline{h}_{33}\underline{F}_S - \underline{h}_{34}\underline{M}_W \tag{9.42}$$

$$\underline{\Omega}_W = \underline{h}_{43}\underline{F}_S - \underline{h}_{44}\underline{M}_W \tag{9.43}$$

After some calculations, the coefficients of the associated two-port network $\underline{\mathbf{A}}$ with respect to following matrix equation

$$\begin{pmatrix} \underline{\Omega}_W \\ \underline{M}_W \end{pmatrix} = \begin{pmatrix} \underline{A}_{11} & \underline{A}_{12} \\ \underline{A}_{21} & \underline{A}_{22} \end{pmatrix} \begin{pmatrix} \underline{v}_S \\ \underline{F}_S \end{pmatrix}$$

are achieved in combination with (9.42) and (9.43). The matrix $\underline{\mathbf{A}}$ results in

$$\underline{\mathbf{A}} = \begin{pmatrix} \dfrac{\underline{h}_{44}}{\underline{h}_{34}} & \underline{h}_{43} - \dfrac{\underline{h}_{44}\underline{h}_{33}}{\underline{h}_{34}} \\ -\dfrac{1}{\underline{h}_{34}} & \dfrac{\underline{h}_{33}}{\underline{h}_{34}} \end{pmatrix} . \tag{9.44}$$

According to Fig. 9.77, the circuit representation of the two-port network $\underline{\mathbf{A}}$ results from a varied circuit representation with the admittances $\underline{h}_1$ and $\underline{h}_2$ [2].

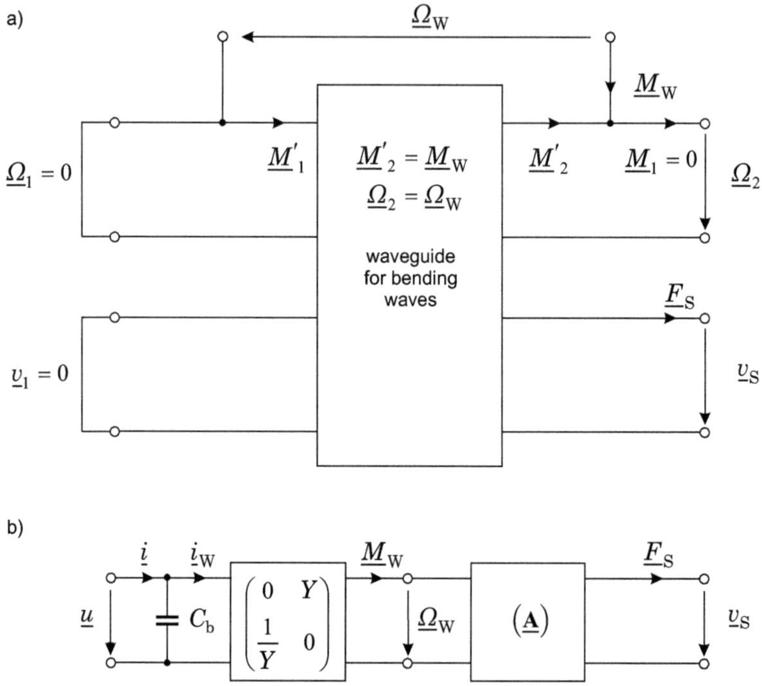

Fig. 9.76. Clamped-free piezoelectric bending oscillator
a) *appropriate five-port network representation*; b) *reduced representation consisting of a piezoelectric transducer and a two-port transducer network* ($\underline{\mathbf{A}}$)

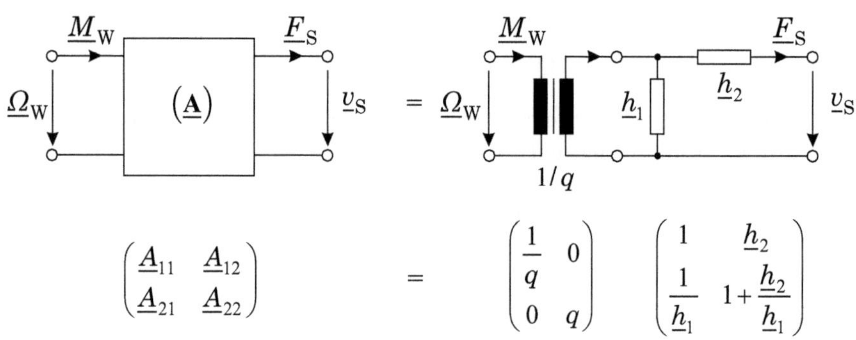

Fig. 9.77. Circuit for the mechanical partial two-port network of a piezoelectric bending oscillator and corresponding matrix notation

In combination with (9.44), the transformation ratio q and the admittances $\underline{h}_1$ and $\underline{h}_2$ can be extracted from the circuit of the mechanical partial two-port network (Fig. 9.77).

Matrix multiplication and coefficient comparison yields:

$$q = \frac{\underline{h}_{34}}{\underline{h}_{44}} \tag{9.45}$$

$$\underline{h}_1 = -\frac{\underline{h}_{34}^2}{\underline{h}_{44}} \tag{9.46}$$

$$\underline{h}_2 = \frac{\underline{h}_{34}\underline{h}_{43}}{\underline{h}_{44}} - \underline{h}_{33} \tag{9.47}$$

In the following, the piezoelectric bending actuator is assumed to be exposed to very small external loads, thus the series admittance $\underline{h}_2$ can be neglected. According to Fig. 9.76 b), the reduced two-port network representation of the piezoelectric bending actuator has the structure illustrated in Fig. 9.78.

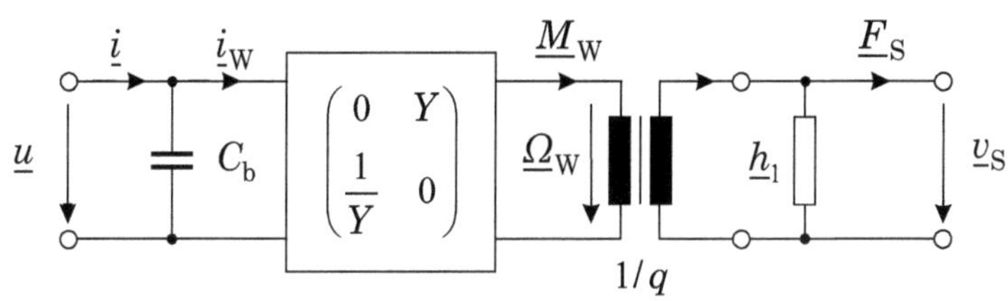

Fig. 9.78. Electromechanical circuit of a piezoelectric bending actuator taking transformation behavior of rotational and translational quantities into consideration

The relation between the electric pair of coordinates $(\underline{u}, \underline{i}_\mathrm{W})$ and mechanical pair of coordinates $(\underline{v}_\mathrm{S}, \underline{F}_\mathrm{S})$ can be derived from the series connection of gyrator and mechanical partial two-port network. Both the gyrator with the transducer constant Y and transformer with transformation ratio $1/q$ are combined to a new gyrator $\mathbf{Y}'$ with following structure:

$$\mathbf{Y}' = \begin{pmatrix} 0 & qY \\ \frac{1}{qY} & 0 \end{pmatrix} \tag{9.48}$$

By applying the gyrator-like coupling matrix (9.48), the electrical and mechanical translational coordinates can be interconnected according to

$$\begin{pmatrix} \underline{u} \\ \underline{i}_\mathrm{W} \end{pmatrix} = \begin{pmatrix} 0 & qY \\ \frac{1}{qY} & 0 \end{pmatrix} \begin{pmatrix} 1 & 0 \\ \frac{1}{\underline{h}_1} & 1 \end{pmatrix} \begin{pmatrix} \underline{v}_\mathrm{S} \\ \underline{F}_\mathrm{S} \end{pmatrix} . \tag{9.49}$$

The matrix equation (9.49) corresponds to the circuit of a clamped-free piezoelectric beam bending actuator shown in Fig. 9.79.

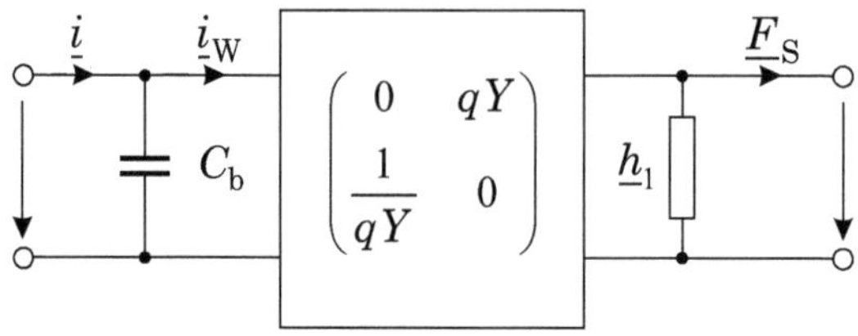

Fig. 9.79. General circuit for a clamped-free piezoelectric beam bending actuator affected by low external loads ($\underline{h}_2 = 0$)

In order to achieve a general circuit of a piezoelectric multilayer beam bending actuator taking modes of higher order into account, the definition of the mechanical admittance $\underline{h}_1$ in (9.46) is considered. By means of the transformation ratio $1/q$ according to (9.45) and in combination with the definition of the matrix element $\underline{h}_{34}$, the admittance $\underline{h}_1$ can be reformulated [85]:

$$\underline{h}_1 = -q\underline{h}_{34} = q \sum_{m=1}^{\infty} \frac{X_m(l) \dfrac{\mathrm{d}X_m}{\mathrm{d}x}(l)}{\left(\dfrac{1}{\mathrm{j}\omega \dfrac{n_0}{(\beta_m l)^4}} + \mathrm{j}\omega m + r\right) \displaystyle\int_0^l X_m^2 \,\mathrm{d}x} l \tag{9.50}$$

Within the scope of network theory, the sum notation in (9.50) corresponds to a series connection of spring-mass-damper elements connected in parallel resulting in the canonical circuit of a piezoelectric multilayer beam bending actuator, as it is illustrated in Fig. 9.80.
In a next step, the electrical side is transformed to the mechanical side. In combination with (9.51)

$$\underline{v}_\mathrm{S} = \underline{h}_1 \underline{F} \tag{9.51}$$

and by applying the calculation rule (9.50) for the mechanical admittance $\underline{h}_1$, the translational velocity $\underline{v}_\mathrm{S}$ yields

$$\underline{v}_\mathrm{S} = \frac{\underline{u}}{Y} \sum_{m=1}^{\infty} \frac{X_m(l) \dfrac{\mathrm{d}X_m}{\mathrm{d}x}(l)}{\left(\dfrac{1}{\mathrm{j}\omega \dfrac{n_0}{(\beta_m l)^4}} + \mathrm{j}\omega m + r\right) \displaystyle\int_0^l X_m^2 \,\mathrm{d}x} l. \tag{9.52}$$

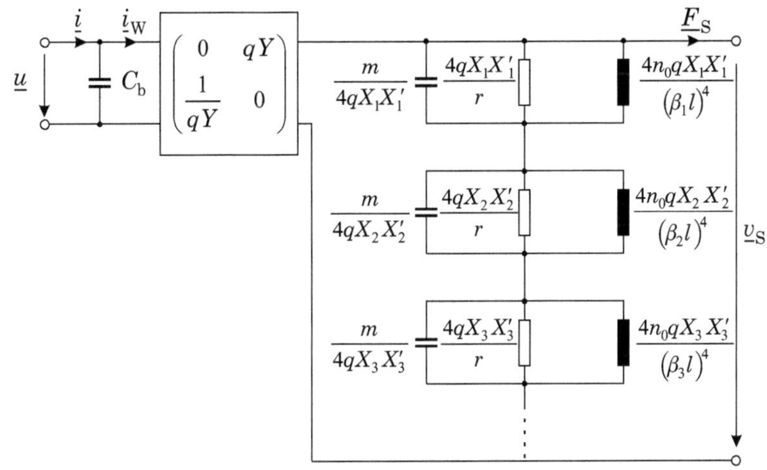

Fig. 9.80. Canonical circuit of a piezoelectric multilayer beam bending actuator

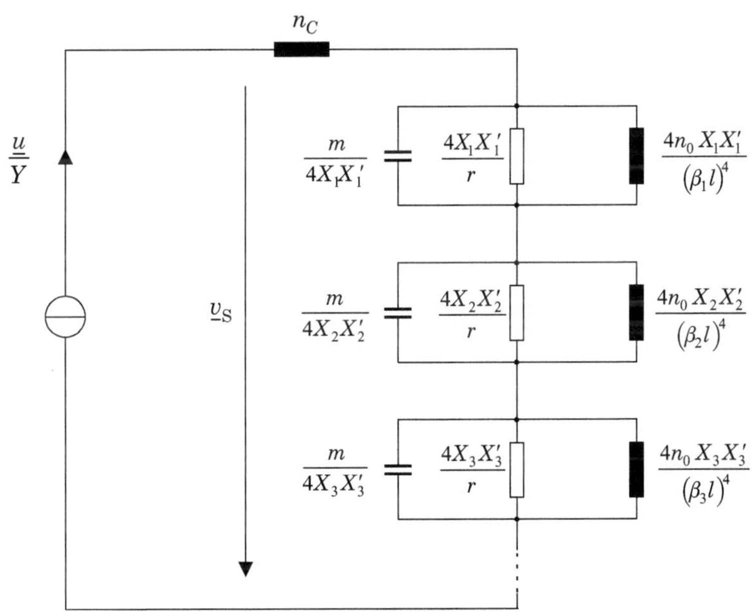

Fig. 9.81. Canonical circuit of a piezoelectric multilayer beam bending actuator after transformation to mechanical side

The transformation ratio $1/q$ is not found anymore. If the tip of bending actuator can perform unhindered motions, the circuit representation outlined in Fig. 9.81 will be achieved.

In order to verify the canonical circuit represented in Fig. 9.81, the emphasis is laid on the experimental determination of dynamic deflection characteristics $\underline{\xi}$ of the bender's tip as a function of driving voltage $\underline{u}$. A piezoelectric monomorph in multilayer technology as shown in Fig. 9.82 is used for the experimental investigations.

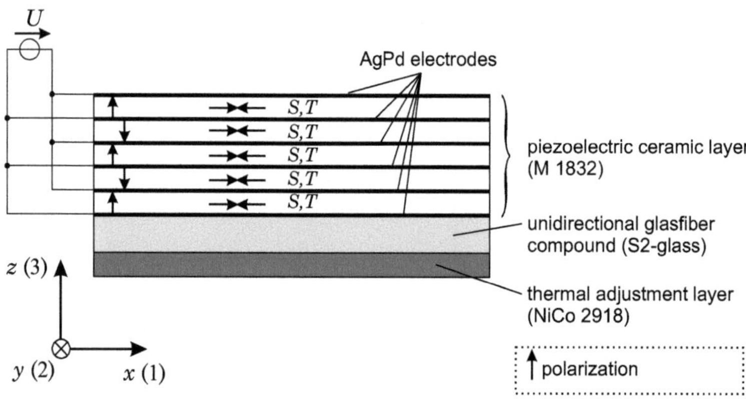

Fig. 9.82. Structure of used monomorph in multilayer technology

The geometrical and material-specific parameters of the used piezoelectric bending actuator are listed in Table 9.16. The measurement chain consists of an amplifier stage, the bending actuator and a laser triangulator. The measurement setup is shown in Fig. 9.83 [85].

Table 9.16. Geometrical and material-specific parameters of the individual layers

layer i	1	2	3-7
material	NiCo 2918	S2-glass	M 1832
l_i [mm]		20	
w_i [mm]		8	
h_i [μm]	100	200	5 x 48
$s^E_{11,i}$ [$\cdot 10^{-12}$ m^2 N^{-1}]	6.369	17.857	14.144
$d_{31,i}$ [$\cdot 10^{-12}$ m V^{-1}]	—	—	–350
$\varepsilon^T_{33,i}$	—	—	4500
ρ_i [kg m^{-3}]	8300	1998	8100

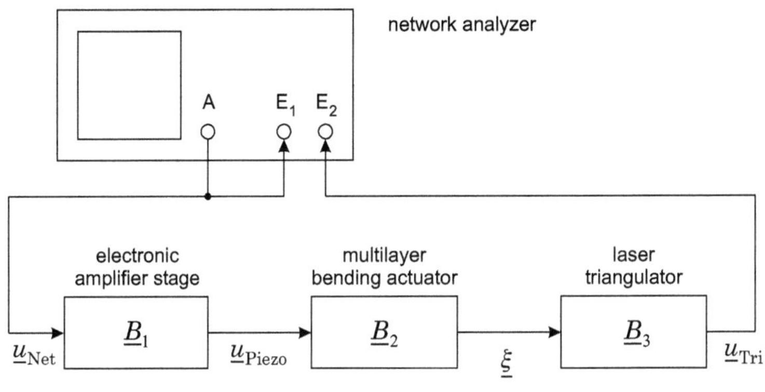

Fig. 9.83. Measurement setup for experimental verification of the canonical circuit of a clamped-free piezoelectric multilayer beam bending actuator

The measurement of deflection $\underline{\xi}$ is effected at the actuator's tip ($x = l$). The free length of the actuator amounts to $l = 19.2\,\text{mm}$. The investigated frequency range extends from 20 Hz up to 10 kHz. The voltage amplitude $\hat{u}_{\text{Net}}$ is adjusted to 500 mV.

The total transfer function $\underline{B}_{\text{G}}$ of measurement chain consists of the transfer function of the amplifier stage $\underline{B}_1$, the bending actuator $\underline{B}_2$ and the laser triangulator $\underline{B}_3$. It can be written

$$\underline{B}_{\text{G}} = \underline{B}_1 \cdot \underline{B}_2 \cdot \underline{B}_3. \tag{9.53}$$

According to the input and output quantities represented in Fig. 9.83 the transfer function (9.53) can be reformulated:

$$\underline{B}_{\text{G}} = \frac{\underline{u}_{\text{Piezo}}}{\underline{u}_{\text{Net}}} \cdot \frac{\underline{\xi}}{\underline{u}_{\text{Piezo}}} \cdot \frac{\underline{u}_{\text{Tri}}}{\underline{\xi}} = \frac{\underline{u}_{\text{Tri}}}{\underline{u}_{\text{Net}}} \tag{9.54}$$

Both the measured and analytically calculated total transfer function $\underline{B}_{\text{G}}$ are shown in Fig. 9.84 within a frequency range from 20 Hz up to 10 kHz. The coefficient of friction r of the piezoelectric bending actuator was determined on the basis of a free damped oscillation and amounts to $r = 0.078\,\text{Ns/m}$ [85].

The measurement results show a very good coincidence with the analytical calculations. Thus, the conclusion can be drawn that the developed structure of the canonical circuit diagram of a clamped-free piezoelectric bending actuator can be utilized for its description within the scope of network theory.

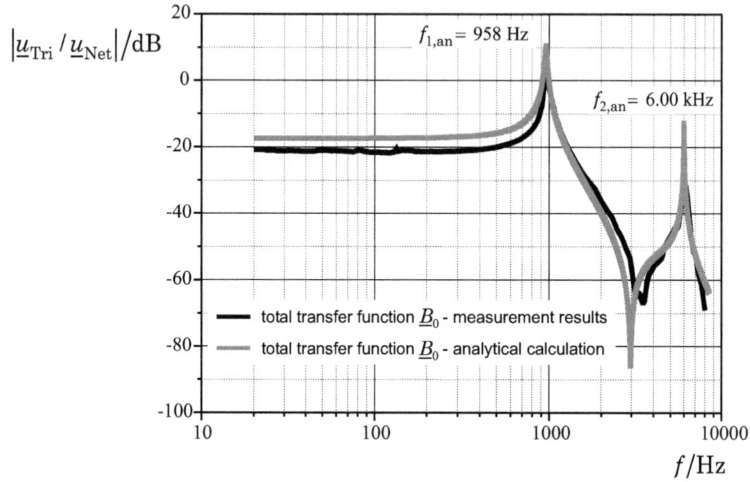

Fig. 9.84. Measured and analytically calculated total frequency response $|\underline{B}_G|$ based on the canonical circuit of a clamped-free piezoelectric multilayer beam bending actuator

10

Reciprocity in Linear Networks

In each case, *reciprocity relations* represent connections between two transfer factors and transfer impedances or admittances of a linear system, respectively. These relations provide a basis for a series of measuring methods and allow in many cases also for the calculation of relations which would be only achieved with large calculating effort otherwise. In networks of a physical structure they are identical to the symmetry of the ρ- and γ-matrices with respect to the main diagonal represented in Sect. 2.2 and Chap. 4. In this chapter, special conclusions are drawn for two-port networks, i.e. for the *interconnection of two pairs of coordinates*. By means of the well-known characteristics of tranducer two-port networks, it is particularly demonstrated that also reciprocity relations exist for systems which comprise different physical structures. These relations are of particular practical importance.

10.1 Reciprocity Relations in Networks with only One Physical Structure

In Chap. 4 it has been found out for an abstract linear n-pole that the ρ- and γ-matrices which interconnect the μ- and λ-vectors of a physical structure are symmetric to the main diagonal. It is usually referred to this characteristic of linear networks as *reciprocity* or *reversibility*. Here, it must be again pointed out that the existence of such connections within a physical structure has nothing to do with energy or power considerations, but only results from validity of the *law of superposition* in a structure characterized by (10.1) up to (10.4). For the special case of a two-port network, these relations yield in combination with the sign rules shown in Fig. 10.1:

$$\underline{\rho}_{21} = \frac{\underline{\mu}_{2\mathrm{L}}}{\underline{\lambda}_1} = \frac{\underline{\mu}'_{1\mathrm{L}}}{\underline{\lambda}'_2} = -\underline{\rho}_{12} \tag{10.1}$$

A. Lenk et al., *Electromechanical Systems in Microtechnology and Mechatronics*,
Microtechnology and MEMS, DOI 10.1007/978-3-642-10806-8_10,

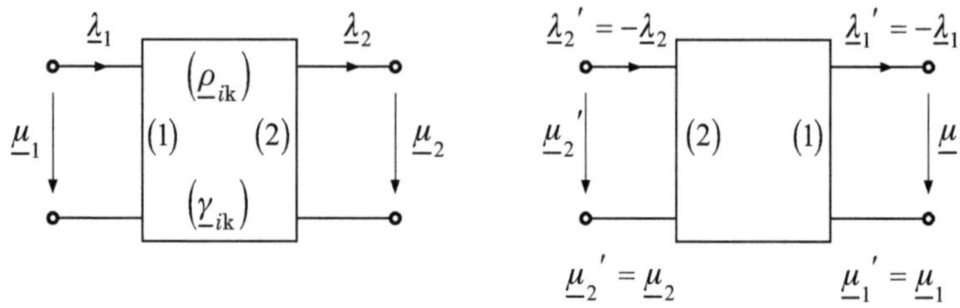

Fig. 10.1. Inversion of the general two-port network

$$\underline{\gamma}_{21} = \frac{\underline{\lambda}_{2K}}{\underline{\mu}_1} = \frac{\underline{\lambda}'_{1K}}{\underline{\mu}'_2} = -\underline{\gamma}_{12} \tag{10.2}$$

The right side of both equations corresponds in each case to the transfer of side 1 to side 2 in Fig. 10.1, the left side of both equations corresponds to the inverse transfer direction. The indices K and L denote short-circuit ($\underline{\mu} = 0$) and open-circuit operation ($\underline{\lambda} = 0$). The quantities $\underline{\rho}_{iK}$ and $\underline{\gamma}_{iK}$ characterize the connection between the coordinates $\underline{\lambda}_1$, $\underline{\lambda}_2$, $\underline{\mu}_1$ and $\underline{\mu}_2$:

$$\begin{pmatrix} \underline{\mu}_1 \\ \underline{\mu}_2 \end{pmatrix} = \begin{pmatrix} \underline{\rho}_{11} & \underline{\rho}_{12} \\ \underline{\rho}_{21} & \underline{\rho}_{22} \end{pmatrix} \begin{pmatrix} \underline{\lambda}_1 \\ \underline{\lambda}_2 \end{pmatrix} \tag{10.3}$$

$$\begin{pmatrix} \underline{\lambda}_1 \\ \underline{\lambda}_2 \end{pmatrix} = \begin{pmatrix} \underline{\gamma}_{11} & \underline{\gamma}_{12} \\ \underline{\gamma}_{21} & \underline{\gamma}_{22} \end{pmatrix} \begin{pmatrix} \underline{\mu}_1 \\ \underline{\mu}_2 \end{pmatrix} \tag{10.4}$$

By means of these equations and the conditions of (10.1) and (10.2), it can be verified easily that also following both reciprocity relations are valid being equivalent to the relations in (10.1) and (10.2):

$$\frac{\underline{\mu}_{2L}}{\underline{\mu}_1} = \frac{\underline{\lambda}'_{1K}}{\underline{\lambda}'_2} \qquad \frac{\underline{\lambda}'_{2K}}{\underline{\lambda}'_1} = \frac{\underline{\mu}_{1L}}{\underline{\mu}_2} \tag{10.5}$$

The indices K (short-circuit operation, $\underline{\mu} = 0$) and L (open-circuit operation, $\underline{\lambda} = 0$) refer in each case to the coordinate $\underline{\mu}$ and $\underline{\lambda}$ belonging to the same terminal pair (e.g. $\underline{\lambda}'_{iK} \to \underline{\mu}'_1 = 0$). Here, the right sides of both equations correspond again to the left part, the left sides correspond to the right part of Fig. 10.1.

In addition to (10.3) and (10.4), for the analysis of the transient characteristic a third form of two-port network equations – *chain matrix* – is common practice which connects the input quantities with the output quantities:

$$\begin{pmatrix} \underline{\mu}_1 \\ \underline{\lambda}_1 \end{pmatrix} = \begin{pmatrix} \underline{\alpha}_{11} & \underline{\alpha}_{12} \\ \underline{\alpha}_{21} & \underline{\alpha}_{22} \end{pmatrix} \begin{pmatrix} \underline{\mu}_2 \\ \underline{\lambda}_2 \end{pmatrix} \tag{10.6}$$

By rearranging these equations into the form of (10.3), (10.7) is achieved. In combination with (10.1) and (10.2), a further reciprocity relation in the form of an equation between the elements of the chain matrix can be extracted from (10.7):

$$\begin{pmatrix} \underline{\mu}_1 \\ \underline{\mu}_2 \end{pmatrix} = \begin{pmatrix} \dfrac{\underline{\alpha}_{11}}{\underline{\alpha}_{22}} & -\dfrac{\Delta\left(\underline{\alpha}\right)}{\underline{\alpha}_{21}} \\ \dfrac{1}{\underline{\alpha}_{21}} & -\dfrac{\underline{\alpha}_{22}}{\underline{\alpha}_{21}} \end{pmatrix} \begin{pmatrix} \underline{\lambda}_1 \\ \underline{\lambda}_2 \end{pmatrix} \tag{10.7}$$

If it applies $\underline{\rho}_{12} = -\underline{\rho}_{21}$, then it must be valid:

$$\Delta\left(\underline{\alpha}\right) = \underline{\alpha}_{11}\underline{\alpha}_{22} - \underline{\alpha}_{12}\underline{\alpha}_{21} = 1 \tag{10.8}$$

That is the last one of possible reciprocity relations with linear two-port networks. The specified reciprocity relations are equivalent, i.e. each of the relations in (10.1), (10.2), (10.5) and (10.8) enable the derivation of the other four remaining.

10.2 Reciprocity Relations in General Linear Two-Port Networks

By means of the coupling two-ports discussed in the Chaps. 3 up to 7, now n-poles can be designed which comprise pairs of coordinates of different physical structures at the ports. Fig. 10.2 shows an example.

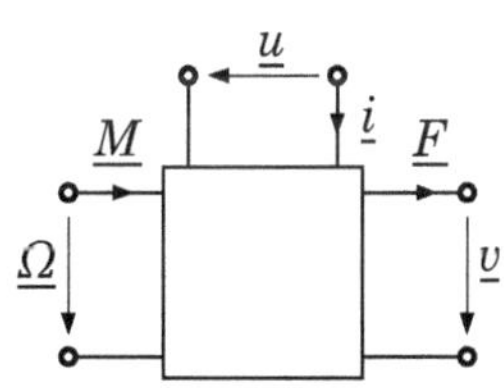

Fig. 10.2. Electromechanical three-port network

Such networks must comprise inside at least one of the transducer two-port networks represented in Fig. 2.19. The question is now whether reciprocity relations exist also for such a linear n-pole and how they appear. A formal transfer of the results of Sect. 10.1 is certainly not allowed. As substantial precondition, the law of superposition within a physical structure has been applied for derivation of these relations. However, it only allows for the superposition of uniform coordinates. At first, the case of two-port networks is considered in the following, because they are practically of prime importance and because the substantial relations are most distinguishable there. Such a

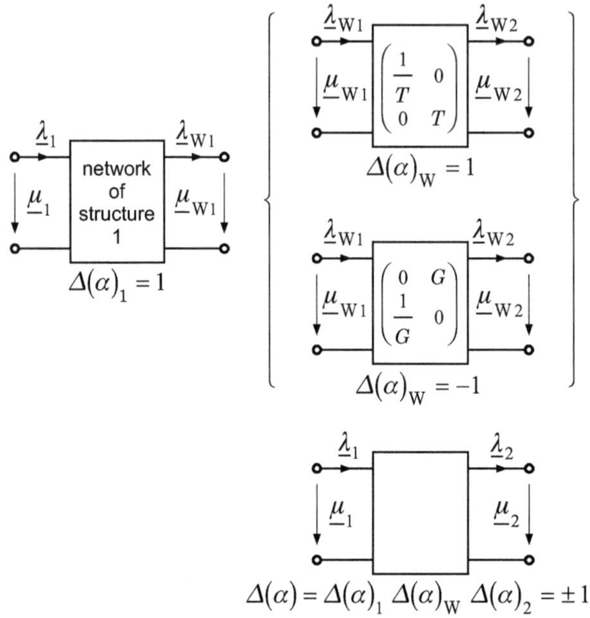

Fig. 10.3. Possible interconnections of two physical structures by a general two-port network

two-port network between different physical structures must be able to be represented by a circuit according to Fig. 10.3.

The coupling two-port network can feature only one of both forms. Thus, it can effect a transformer- or gyrator-like interconnection. The coefficients T and G can denote one of the electromechanical coupling quantities X, Y, a rod length l or an area A. This fact is used that the chain matrix of a cascade connection of several two-port networks results from the product of chain matrices of the individual two-port networks. In addition, it is valid that the determinant of the total chain matrix equals the product of the determinants of the individual chain matrices.

Thus, the chain determinant of the complete two-port network equals 1 in case of a transformer-like interconnection. For this reason, the reciprocity relations of (10.1), (10.2), (10.5) and (10.8) being equivalent with each other are valid. In the second possible case of the gyrator-like interconnection $\Delta\left(\underline{\alpha}\right) = -1$ is valid, and it has to be analyzed how the other two pairs of reciprocity relations [Equations (10.1), (10.2) and (10.5)] appear. That can be immediately determined on the basis of (10.7) and the analogous but here not specified $\underline{\gamma}$-matrix as function of the $\underline{\alpha}$-elements.

For $\Delta(\underline{\alpha}) = -1$, in (10.1), (10.2) and (10.5) only the sign changes, so that following reciprocity relations are valid in summary for the general linear two-port network:

$$\Delta(\underline{\alpha}) = \pm 1 \tag{10.9}$$

$$\frac{\underline{\mu}_{2L}}{\underline{\lambda}_1} = \pm\frac{\underline{\mu}_{1L}}{\underline{\lambda}'_2} \qquad \frac{\underline{\lambda}_{2K}}{\underline{\mu}_1} = \pm\frac{\underline{\lambda}'_{1K}}{\underline{\mu}_2} \tag{10.10}$$

$$\frac{\underline{\mu}_{2L}}{\underline{\mu}_1} = \pm\frac{\underline{\lambda}'_{1K}}{\underline{\lambda}'_2} \qquad \frac{\underline{\lambda}'_{2K}}{\underline{\lambda}_1} = \pm\frac{\underline{\mu}_{1L}}{\underline{\mu}_2} \tag{10.11}$$

For the signs of coordinates, the definitions of Fig. 10.1 are valid. In case of inversion, the directions of the flow coordinates are reversed.

Except for the normally insignificant sign, an analogy of reciprocity relation in networks with only one physical structure and in general network has been achieved. However, that must not belie the fact that in both cases completely different reasons are available for reciprocity. In case of networks of *one* physical structure, the cause was the validity of the law of superposition in combination with the system structure, whereas the analogy of the connection between one pair of transducer coordinates each originating from different physical structures is the cause of $|\Delta(\underline{\alpha})| = 1$ for transducer two-port networks.
However, this analogy was a direct consequence of the assumption that only reversible processes in terms of thermodynamics run in the transducer and that existing losses can be separated in the form of a further network. These aspects must be considered, if further physical interconnection mechanisms, as e.g. the electrothermal interconnection, are to be added in the theory which is outlined here.

10.3 Electromechanical Transducers

In this section, the results of the previous section are applied to the concrete case of electromechanical transducers and their practical importance is demonstrated by means of a technical example.
The transition from general coordinates $\underline{\mu}$ and $\underline{\lambda}$ to available coordinates $\underline{u}$, $\underline{i}$, $\underline{v}$ and $\underline{F}$ results from the considerations in Chap. 4 and Fig. 10.4:

$$\underline{u} = \underline{\mu}_1, \qquad \underline{i} = \underline{\lambda}_1, \qquad \underline{v} = \underline{\mu}_2, \qquad \underline{F} = \underline{\lambda}_2$$

In the following, it is analyzed how the interconnection is achieved by means of electric or magnetic fields.

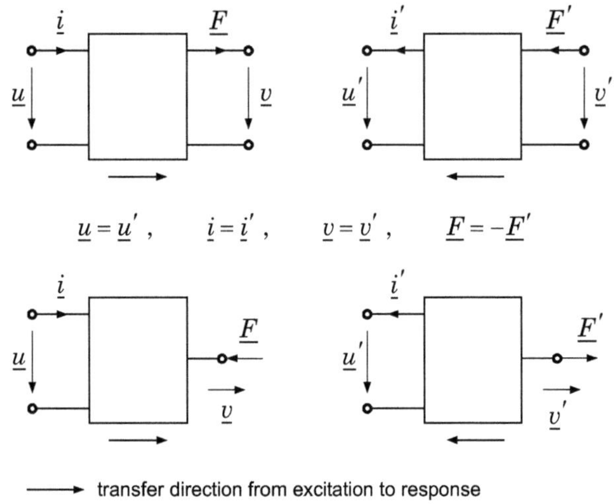

Fig. 10.4. Sign convention in case of inversion of an electromechanical transducer

In both cases, different values for $\Delta(\alpha)$ are possible:

$$\text{electrical transducers:} \qquad (\alpha) = \begin{pmatrix} 0 & Y \\ 1/Y & 0 \end{pmatrix} \rightarrow \Delta(\alpha) = -1$$

$$\text{magnetic transducers:} \qquad (\alpha) = \begin{pmatrix} 1/X & 0 \\ 0 & X \end{pmatrix} \rightarrow \Delta(\alpha) = +1$$

In combination with (10.1), (10.2), (10.5) and (10.8), the reciprocity relations result immediately in:

$$\frac{\underline{v}_{\mathrm{L}}}{\underline{i}} = \pm\frac{\underline{u}'_{\mathrm{L}}}{\underline{F}'}, \qquad \frac{\underline{F}_{\mathrm{K}}}{\underline{u}} = \pm\frac{\underline{i}'_{\mathrm{K}}}{\underline{v}'}, \qquad \frac{\underline{v}_{\mathrm{L}}}{\underline{u}} = \pm\frac{\underline{i}'_{\mathrm{K}}}{\underline{F}'_{\mathrm{K}}}, \qquad \frac{\underline{F}'_{\mathrm{K}}}{\underline{i}'} = \pm\frac{\underline{u}_{\mathrm{L}}}{\underline{v}} \tag{10.12}$$

The positive sign is assigned to two-port networks consisting of magnetic transducers and the negative sign is assigned to two-port networks consisting of electrical transducers.

The *reciprocity calibration* of electromechanical sensors represents a typical example for the application of these relations. In the following, it is assumed the task to calibrate an acceleration sensor without using a proximate electromechanical standard (calibration source for mechanical quantities). This measurement standard is indirectly represented by the resistor R existing in the measuring device, because a connection to mechanical quantities is necessary in order to define current and voltage units. It is assumed that only devices for purely electrical and purely mechanical quantities are available for absolute measurements. The measurement setup consists of an electromechanical transducer from which it is known that it comprises an interconnection by means of magnetic fields, and from which the mechanical input impedance is known in case of electrical open-circuit operation (Fig. 10.5).

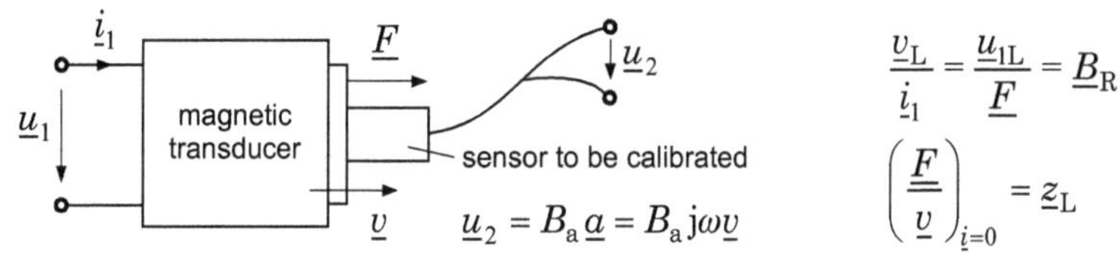

Fig. 10.5. Configuration for reciprocity calibration of an acceleration sensor

The mass of sensor is considered to be a part of the transducer system. Thus, the impedance $\underline{z}_1$ must be measured with the attached sensor. Following both experiments are now conducted successively by means of this configuration. *Experiment* I concerns a current exitation of the magnetic transducer and the determination of the transfer impedance $(\underline{u}_2/\underline{i}_1)^{\mathrm{I}}$. In combination with the relations shown in Fig. 10.5, this impedance can be expressed by $\underline{B}_{\mathrm{R}}$ and $\underline{B}_{\mathrm{a}}$:

$$\left(\frac{\underline{u}_2}{\underline{i}_1}\right)^{\mathrm{I}} = \mathrm{j}\omega \underline{B}_{\mathrm{a}} \underline{B}_{\mathrm{R}} \tag{10.13}$$

Experiment II concerns an excitation of the mechanical system with an external force $\underline{F}^{\mathrm{II}}$ during electrical open-circuit operation ($\underline{i}_1 = 0$) and the simultaneous measurement of ratio $(\underline{u}_2/\underline{u}_{1\mathrm{L}})^{\mathrm{II}}$. This ratio can also be reduced to $\underline{B}_{\mathrm{a}}$, $\underline{B}_{\mathrm{R}}$ and $\underline{z}_{\mathrm{L}}$, because in case of the known impedance $\underline{z}_{\mathrm{L}}$, with the force $\underline{F}^{\mathrm{II}}$ also the velocity $\underline{v}^{\mathrm{II}} = \underline{F}^{\mathrm{II}}/\underline{z}_{\mathrm{L}}$ is well-known:

$$\underline{u}_{1\mathrm{L}}^{\mathrm{II}} = \underline{B}_{\mathrm{R}} \underline{F}^{\mathrm{II}}, \qquad \underline{u}_2^{\mathrm{II}} = \mathrm{j}\omega \underline{B}_{\mathrm{a}} \underline{v}^{\mathrm{II}}$$

$$\left(\frac{\underline{u}_{1\mathrm{L}}}{\underline{u}_2}\right)^{\mathrm{II}} = \frac{\underline{B}_{\mathrm{R}}}{\mathrm{j}\omega \underline{B}_{\mathrm{a}}} \frac{\underline{F}^{\mathrm{II}}}{\underline{v}^{\mathrm{II}}} = \frac{\underline{B}_{\mathrm{R}}}{\mathrm{j}\omega \underline{z}_{\mathrm{L}} \underline{B}_{\mathrm{a}}} \tag{10.14}$$

By multiplying (10.13) and (10.14), $\underline{B}_{\mathrm{R}}$ can be eliminated. Thus, $\underline{B}_{\mathrm{a}}$ is achieved as a function of both measured ratios and as a function of impedance $\underline{z}_{\mathrm{L}}$:

$$|\underline{B}_{\mathrm{a}}| = \sqrt{\frac{|\underline{z}_{\mathrm{L}}|}{\omega^2} \left(\frac{\underline{u}_2}{\underline{i}_1}\right)^{\mathrm{I}} \left(\frac{\underline{u}_{1\mathrm{L}}}{\underline{u}_2}\right)^{\mathrm{II}}} \tag{10.15}$$

Generally, the system is dimensioned in such a way that $\underline{z}_{\mathrm{L}} = \mathrm{j}\omega m$ can be assumed as sufficient approximation.

The mass m can be easily determined by weighing. By means of a well-known resistor R_0, the current $\underline{i}_1$ is transformed into a voltage $\underline{u}_i$. Thus, the following relation is achieved for $|\underline{B}_{\mathrm{a}}|$:

$$|\underline{B}_{\mathrm{a}}| = \sqrt{\frac{mR_0}{\omega}\left(\frac{\underline{u}_2}{\underline{u}_i}\right)^{\mathrm{I}}\left(\frac{\underline{u}_{1\mathrm{L}}}{\underline{u}_2}\right)^{\mathrm{II}}}$$

10.4 Mechanical-Acoustic Transducers

The transition from general coordinates to the special case of mechanical-acoustic transducer results from the relation:

$$\underline{v} = \underline{\mu}_1, \qquad \underline{F} = \lambda_1, \qquad \underline{p} = \underline{\mu}_2, \qquad \underline{q} = \lambda_2$$

Thus, the internal transducer two-port network appears as gyrator according to Sect. 5.2:

$$\begin{pmatrix} \underline{v}_{\mathrm{W}} \\ \underline{F}_{\mathrm{W}} \end{pmatrix} = \begin{pmatrix} 0 & 1/(\varphi A) \\ \varphi A & 0 \end{pmatrix} \begin{pmatrix} \underline{p}_{\mathrm{W}} \\ \underline{q}_{\mathrm{W}} \end{pmatrix} \rightarrow \Delta(\alpha) = -1$$

As a result, the reciprocity relations are:

$$\frac{\underline{p}_{\mathrm{L}}}{\underline{F}} = -\frac{\underline{v}'_{\mathrm{L}}}{\underline{q}'}, \qquad \frac{\underline{q}_{\mathrm{K}}}{\underline{v}} = -\frac{\underline{F}'_{\mathrm{K}}}{\underline{p}'}, \qquad \frac{\underline{p}_{\mathrm{L}}}{\underline{v}} = -\frac{\underline{F}'_{\mathrm{K}}}{\underline{q}'}, \qquad \frac{\underline{q}'_{\mathrm{K}}}{\underline{F}'} = -\frac{\underline{v}_{\mathrm{L}}}{\underline{p}} \tag{10.16}$$

The relation of the individual quantities to the geometrical transducer configuration is represented in Fig. 10.6. Again the arrow denotes the transfer direction from excitation to response. Instead of the coordinates $\underline{q}$ and $\underline{v}$, the volume displacements $\underline{V}$ and deflections $\underline{\xi}$ can be inserted in (10.16) due to $\underline{q} = \mathrm{j}\omega\underline{V}$ and $\underline{v} = \mathrm{j}\omega\underline{\xi}$.

By means of the following example, it is demonstrated how the application of reciprocity relations can result in calculating simplification in some cases. For a configuration according to Fig. 10.7, where a force $\underline{F}$ generates a volume change $\underline{V}$, there is the task to determine the quotient $\underline{V}/\underline{F}$. The pressure $\underline{p}$ should amount to zero ($\underline{q} = \underline{q}_{\mathrm{K}}$). In order to determine this quotient, at first the deflection function of a plate would be determined for the represented lumped load caused by a pair of forces. That represents a very complex problem. Still the volume change $\underline{V}$ would have to be extracted from the achieved solution by means of integration. However, the searched quotient is contained in the last relation of (10.16). It is identical to the quotient $(-\underline{\xi}(r)/\underline{p})_{\underline{F}=0}$. The deflection $\underline{\xi}(r)$ has to be measured at that radius, the forces affect in first case. However, this last specified quotient can be easily gathered from engineering data reference books.

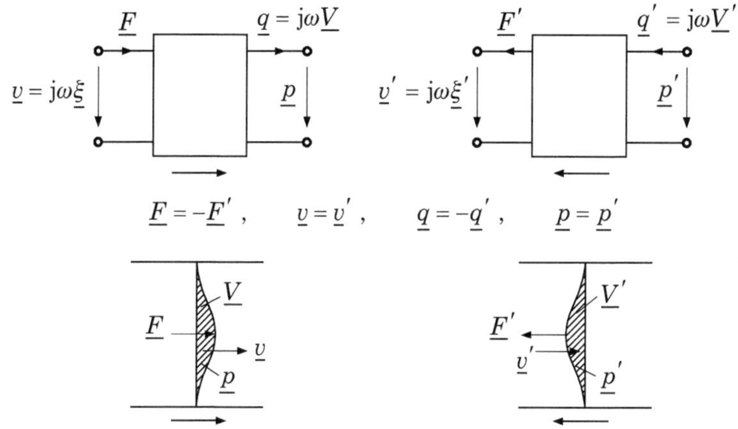

Fig. 10.6. Sign convention in case of inversion of a mechanical-acoustic transducer

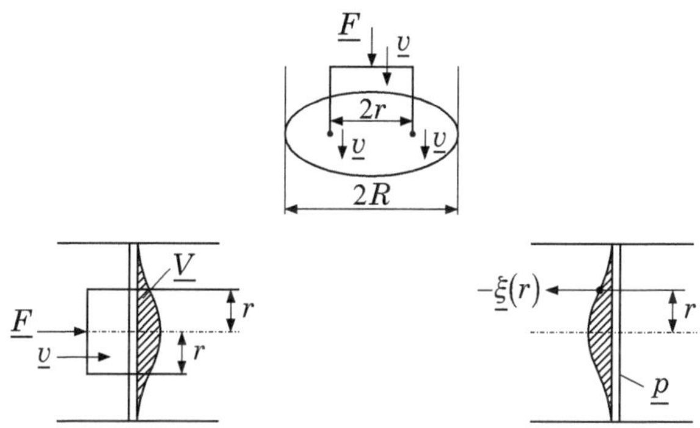

Fig. 10.7. Determination of the quotient $(\underline{V}/\underline{F})_{\underline{p}=0}$ of a mechanical-acoustic transducer by means of the reciprocity law

For a clamped plate, it can be written e.g. according to [43]:

$$-\left(\frac{\underline{\xi}(r)}{\underline{p}}\right)_{\underline{F}=0} = \frac{R^4}{64K}\left(1-\left(\frac{r}{R}\right)^2\right)^2 = \left(\frac{\underline{V}_{\mathrm{K}}}{\underline{F}}\right)_{\underline{p}=0}$$

K denotes the flexural rigidity of plate with

$$K = \frac{Eh^3}{12\,(1-\nu^2)}.$$

By means of the reciprocity relation, it succeeded to avoid an extremely troublesome and protracted calculation and to reduce it to a very simple calculation, respectively.

Part IV

Appendix

A

Characteristics of Selected Materials

A.1 Material Characteristics of Crystalline Quartz

Table A.1. Material Characteristics of Crystalline Quartz [9]

type	properties		unit	values		
α-quartz	temperature of α-β-transformation		°C	573		
	density ρ		g/cm^3	2.655		
	elastic constants	c_{11}, c_{12}, c_{13}	GPa	86.7	7.0	11.9
		c_{14}, c_{33}, c_{44}		−17.9	107.2	57.9
	MOHS hardness			7		
	coefficient of thermal expansion α		10^{-6} /K	$\alpha_{\text{a-axis}}$: 13.37; $\alpha_{\text{c-axis}}$: 7.97		
	thermal conductivity λ		W/K · m	λ_a: 5.6...7.2; λ_c: 9...13.2		
	electrical resistivity		$\Omega \cdot$ cm	$10^{14}...10^{15}$		
quartz wafer (blank)	wafer diameter (and edge length)		mm	51 (2 inches); 76 (3 inches)		
	wafer thickness		μm	500		

A. Lenk et al., *Electromechanical Systems in Microtechnology and Mechatronics*,
Microtechnology and MEMS, DOI 10.1007/978-3-642-10806-8_11,

A.2 Piezoelectric Constants of Sensor Materials

Table A.2. Piezoelectric Constants of Sensor Materials [88]

material	d_{11}	d_{14}	e_{11}	e_{14}	transformation temperature
	10^{-12} m/V		As/m^2		°C
α-quartz (left-handed)	2.3	0.67	0.171	−0.041	573
gallium ortho-phosphate ($GaPO_4$)	4.5	1.9	---	---	950
langasite ($L_3Ga_5SiO_{14}$)	6.16	−5.36	0.45	–	1470
LGN ($L_3Ga_{5,5}Nb_{0,5}O_{14}$)	6.63	−5.55	0.44	–	1510
zinc oxide (ZnO)	d_{33}: 12.3 d_{15}: −8.3	0	e_{33}: 0.96	0	---

A.3 Characteristics of Metallic Structural Materials

Table A.3. Characteristics of Metallic Structural Materials

material	DIN	state	Young's modulus $[10^9\ N/m^2]$	tensile strength $[10^6\ N/m^2]$	ultimate elongation $[10^{-2}]$	coefficient of therm. expans. $[10^{-6}/K]_{r.t.}$
tool steels						
100 Cr 6	4957	tempered		750		11.5
90 MnCrV 8	17350		210	750		12.2
C 60 W	17200			830		
X 210 CrW 12			220	470	17	
spring steels	17221	hot-rolled and annealed				
50 CrV 4			215	390	36	19
50 CrMo 4			210	640	23	19
non-corrosive steels (stainless steels)	17440					
X 5 CrNiMo 18.11			200	550	36	17.3
X 10 CrNiMoTi 18.10			200	600	35	
X 12 CrMo S 17			200	600	30	
hot-forming steels	4390					
40 NiCrMo 15	7740	forged annealed	230	560	28	
210 CrW 46		forged annealed	210	470	17	
special materials						
Hastelloy B	Ni62Mo28FeS/CrMn Si		180-220	600	< 60	10.3
Hastelloy C	Ni57/Mo17/Cr16/FeWMn		170-220	550	< 52	10.8
Monel	Ni65/Cu33/Fe2		185	550	< 40	13.9
Incoloy	Ni32,5/Cr21/Fe		196	400	< 45	14.2

A.4 Material Characteristics of Silicon and Passivation Layers

A.4.1 Comparison of Main Characteristics of Silicon, Silicon Dioxide and Silicon Nitride Layers

Table A.4. Comparison of Main Characteristics of Silicon, Silicon Dioxide and Silicone Nitride Layers [91]

characteristics	unit	silicon	silicon dioxide	silicon nitride
density	kg/m^3	2929	2200	2400
Young's modulus	GPa	[100] 130 [110] 169 [111] 188	73	290
tensile strength / yield strength *)	GPa	Si-wafer: 2...3 micro-structures: 0.5...1	*) 8.4	*) 14.0
fracture strength	GPa	4...6		
residual stress	GPa	---	–0.3	1.2
hardness	kg/mm^2	MOHS: 7.0 KNOOP: 1150	---	---
thermal conductivity	$W/K \cdot m$	r.t. 156 300°C 66	2.1	19
coefficient of thermal expansion	10^{-6} /K	r.t. 2.31 100°C 3.05 200°C 3.55 300°C 3.84 600°C 4.18	0.5	0.8
specific heat capacity	$Ws/kg \cdot K$	r.t. 695 300°C 836	1000	170
el. breakdown strength	MV/cm	---	8...12	3...13
insulating resistance	Ω/cm	---	10^{16}	$10^{10}...10^{16}$

A.4.2 Characteristics of Silicon Dioxide Layers

Table A.5. Characteristics of Silicon Dioxide Layers [92]

deposition process	unit	PECVD [1)]	SiH_4+O_2	TEOS [2)]	$SiCl_2H_2$+N_2O	thermal
composition		$SiO_{1.9}$(H)	SiO_2+O_2	SiO_2	SiO_2(Cl)	SiO_2
deposition temperature	°C	200	450	700	900	1100
density	g/cm^3	2.3	2.1	2.2	2.2	2.2
resistivity	$\Omega \cdot cm$	10^{16}				
dielectric strength	V/μm	300...600	800	1000	1000	1000
intrinsic mech. stress	MPa [3)]	−300...+300	+300	−100	−100	−300
etch rate (H_2O:HF = 100:1)	$\frac{nm}{min}$	40	6	3	3	3

[1)] PECVD: Plasma-enhanced Chemical Vapor Deposition
[2)] TEOS: Tetraethyl orthosilicate (silicon ethoxide)
[3)] (+) tensile stress, (–) compressive stress

A.4.3 Characteristics of Silicon Nitride Layers

Table A.6. Characteristics of Silicon Nitride Layers [92]

deposition process	unit	LPCVD [1)]	PECVD [2)]
composition		$SiO_3N_4(H)$	SiN_xH_y
deposition temperature	°C	700...800	250...350
Si/N ratio		0.75	0.6...1.2
density	g/cm^3	2.9...3.1	2.4...2.8
resistivity	$\Omega \cdot cm$		
dielectric strength	V/μm	1100	500
intrinsic mech. stress	MPa [3)]	+1000	−300...+500

[1)] LPCVD: Low Pressure Chemical Vapor Deposition
[2)] PECVD: Plasma-enhanced Chemical Vapor Deposition
[3)] (+) tensile stress, (−) compressive stress

A.5 Characteristics of Ceramic Structural Materials

Table A.7. Characteristics of Ceramic Structural Materials [9, 91]

characteristics $T = 300$ K	unit	Al_2O_3 (96%)	Al_2O_3 (99.5%)	AlN	BeO (99.5%)	BN	SiC	ZrO_2	Al_2TiO_5
dielectric constant		9...10	9.8	10	6.9	4.1	15...45		
dielectric loss factor at 1 MHz	10^{-4}	1	3	5	2	45	500		
el. breakdown strength	kV/mm	10...28	9...33	10			0.07		
electrical resistivity	$\Omega \cdot$ cm	$10^{12}...10^{14}$	10^{14}	$4 \cdot 10^{11}$	10^{15}	10^{11}	10^{13}		
thermal conductivity	W/K · m	21...24	35	110...170	250	25	90...270	2.5	2.0
specific heat capacity	J/kg · K	795	795	738					
cofficient of thermal expansion	10^{-6}/K	6.8...8	8	3.8	7.5	≈ 0	3.7	10	1.0
Young's modulus	GPa	340	370	300...310	$\approx$ 300	43	380	200	18
flexural strength	MPa	500	380	270...360	220	53		950	40
maximum working temperature	°C	1700	1750	>1000		>1000			
density	g/cm³	3.85						5.95	3.2

A.6 Material Characteristics of Selected Polymers

Table A.8. Material Characteristics of Selected Polymers [9]

characteristics	unit	PVC hard	PMMA	PVDF	PTFE	ABS	PFA	PUR	poly-imide
density	g/cm^3	1.38	1.18		2...2.3	1.12		1.1...1.3	1.42
Young's modulus	GPa	2...3	3.3	2.6	0.3...0.6	2.5...3.0	0.69	0.4...2	3.0
thermal conductivity	$W/K \cdot m$	0.16	0.19	0.19	0.24	0.19	0.22	0.35	0.4...0.5
specific heat capacity	$J/kg \cdot K$	1000	1500	960	1040	1420	1080	1880	---
coefficient of thermal expansion	10^{-6} /K	80	80	106	100	85	130...200	170...200	35...100
glass temperature	°C	---	106	177	---	---	305		---
electrical resistivity	$\Omega \cdot cm$	10^{15}	10^{15}	10^{18}	10^{18}	10^{13}	10^{18}	10^{18}	10^{18}
dielectric constant at 1 MHz		3	2.9	11*)	2.1	2.9...4.1	---	3.4	3.4
tan δ	10^{-2}	2	2		0.01	1...8		4.7	0.005
refractive index		1.5	1.49						

*) at 100 kHz;

ABS: polyacrylonnitrile butadiene styrene,
PFA: perfluorinated alkoxyl copolymer of PTFE (Teflon-PFA),
PMMA: polymethyl methacrylate,
PTFE: polytetraflouroethylene (Teflon),
PVC: polyvinyl chloride,
PUR: polyurethane,
PVDF: polyvinylidene fluoride.

A.7 Characteristics of Plastics as Structural Materials

Table A.9. Characteristics of Plastics as Structural Materials [9]

characteristics	unit	epoxy resin base	epoxy resin +Ag [1)]	silicone resin base	silicone resin+Ag [1)]	poly-imide	silicone rubber
processing temperature range	°C	20...170	50...150	150	50...150	20	20
tensile (shear) strength	MPa	7...75	6...15	1...7	1.3	55...70	---
electrical resistivity	$\Omega \cdot$ cm	$3 \cdot 10^{15}$	$2 \cdot 10^{-4}$	10^{14}	10^{-4}	10^{18}	$10^{14} \dots 10^{15}$
el. breakdown strength	kV/mm	$1.9 \cdot 10^{5}$	---	$20 \cdot 10^{5}$	---	---	$23 \cdot 10^{5}$
dielectric constant at 1 MHz		≈ 4	---	≈ 3.5	---	3.4	2.8
dielectric loss factor at 1 MHz		$90 \cdot 10^{-4}$	---	$25 \cdot 10^{-4}$	---	$50 \cdot 10^{-4}$	$20 \cdot 10^{-4}$
coefficient of thermal expansion	10^{-6} /K	20...100	≈ 30	20...240	95	35...70	300
thermal conductivity	W/K · m	0.65	0.8...3.7	0.4	4.2	1.3	0.2

[1)] filled with silver

A.8 Composition and Material Characteristics of Selected Glasses

Table A.10. Composition and Material Characteristics of Selected Glasses [9]

glass sort	unit	fused quartz glass	soda-lime glass	alkali zinc boro-silicate glass	boro silicate glass	aluminum boro-silicate glass	lithium aluminum silicate glass
SiO_2 content	%	100	68	65	81	50	≈ 70
Na_2O content	%		16	6	4	< 0.2	≈ 15
Al_2O_3 content	%		3	4	2	11	
B_2O_3 content	%		2	10	13	13	
other			CaO: 6 rest: K_2O, BaO, MgO	K_2O: 7 ZnO: 5 rest: TiO_2		BaO: 25 rest: As_2O_3	rest: pre-dominantly Li_2O_3
density	g/cm³	2.2	2.47	2.51	2.23	2.76	2.37
Young's modulus	GPa	78	70	74	63	68	78
flexural strength	GPa		0.05		0.025	0.08	0.06
softening temperature	°C	1500	696	720	820	835	
specific heat capacity	J/kg · K				754		879
coefficient of thermal expansion	10^{-6} /K	0.49	9.2	7.4	3.25	4.5	9
thermal conductivity	W/K · m	1.4	> 0.8	> 0.8	1.15	> 0.8	1.35
electrical resistivity	Ω · cm	$>10^{16}$	$>10^{5}$	$>10^{7}$	$>10^{7}$	$>10^{12}$	$>10^{12}$
dielectric contstant at 1 MHz		3.826	6.5	6.7	4.6	5.8	6.5

A.9 Material Characteristics of Metallic Solders and Glass Solders

Table A.11. Material Characteristics of Metallic Solders and Glass Solders [9]

type	composition	processing temperature range	coefficient of thermal expansion	electrical resistivity	tensile (shear) strength
	weight-%	°C	10^{-6}/K	$\Omega \cdot$ cm	MPa
soft solders	60 Sn; 40 Pb	180...188	25.6	$1.47 \cdot 10^{-5}$	25...40
	65 Sn; 25 Ag; 10 Sb	240...310			
	25 Pb; 50 Bi; 12,5 Sn; 12,5 Cd	60.5...70	20.6	$(4.6...6.6) \cdot 10^{-5}$	45...47
	80 Au; 20 Sn	280	16	$1.6 \cdot 10^{-5}$	275
hard solders	97 Au; 3 Si	362	12		
	55,4 Al; 44,6 Ge	424...499	18	$(1.0...1.1) \cdot 10^{-5}$	
glass solder crystallizing	75...82 PbO; 7...14 ZnO; 6...12 B_2O_3	450...500	84	$(1...5) \cdot 10^{5}$	
non-crystallizing	75...82 PbO; <7 ZnO; >12 B_2O_3	380			
active solders	70 Ag; 27 Cu; 3 Ti	850...950			120...170
	60 Ag; 24 Cu; 15 In; 1 Ti	700...750			

A.10 Sound Velocity and Characteristic Impedance

Table A.12. Sound Velocity and Characteristic Impedance [93]

material	density [kg/m³]	sound velocity [m/s]		characteristic impedance longitudinal [Ns/m³]
		longitudinal	transversal	
metals:				
aluminum (rolled)	2700	6420	3040	17.3
lead (rolled)	11400	2160	700	24.6
gold	19700	3240	1200	63.8
silver	10400	3640	1610	37.9
copper (rolled)	8930	5010	2270	44.7
copper (annealed)	8930	4760	2325	42.5
magnesium	1740	5770	3050	10.0
brass (70% Cu, 30% Zn)	8600	4700	2110	40.4
steel (stainless)	7900	5790	3100	45.7
steel (1% C)	7840	5940	3220	46.6
zinc (rolled)	7100	4210	2440	29.9
tin (rolled)	7300	3320	1670	24.2
non-metals:				
glass (flint glas)	3600	4260	2552	15.3
glass (crown glass)	2500	5660	3391	14.2
fused quartz glass	2200	5968	3764	13.1
plexiglass	1180	2680	1100	3.16
polyethylene	900	1950	540	1.76
polystyrene	1060	2350	1120	2.49

B

Signal Description and Transfer within Linear Networks

B.1 Fourier Expansion of Time Functions

B.1.1 Estimate of Approximation Error with Numerical Analyses of Fourier Series

With numerical analyses it is not possible to implement the summation over i from 0 to ∞. However, since the a_i and b_i converge to zero only for $i \to \infty$, a summation up to $i = M \gg 1$ generates only an *approximation error* that can be estimated with (B.1).

$$\varepsilon^2 = \frac{\overline{(\tilde{x}(t) - x(t))^2}}{\overline{x(t)^2}} = \frac{\frac{1}{2}\left(\sum\limits_{i=1}^{M} c_i^2 - \sum\limits_{i=1}^{\infty} c_i^2\right)}{\frac{1}{2}\sum\limits_{i=1}^{\infty} c_i^2} = \frac{\sum\limits_{i=M}^{\infty} c_i^2}{\sum\limits_{i=1}^{\infty} c_i^2} \tag{B.1}$$

The errors generated by finite upper bounds of summation can be perceived as *band limitation errors*. The inaccurate function $\tilde{x}(t)$ results from the Fourier sum of the exact function $x(t)$ by means of ideal low-pass filtering with $\omega_g = \omega_0 \cdot \mathrm{M}$.
The errors in the time domain generated in such a way appear particularly at step discontinuities or steep edges of the function $x(t)$. In order to demonstrate this special characteristic of the approximative Fourier series, the periodic iteration of a square-wave pulse of width τ represented in the left part of Fig. B.1 is subjected to different analysis operations.
The calculation of the Fourier coefficients a_i, b_i and c_i is made according to

$$a_i = \frac{2\hat{x}\tau}{T}\frac{\sin(2\pi i\tau/T)}{2\pi i\tau/T} \tag{B.2}$$

$$b_i = \frac{2\hat{x}\tau}{T}\frac{1 - \cos(2\pi i\tau/T)}{2\pi i\tau/T} \tag{B.3}$$

A. Lenk et al., *Electromechanical Systems in Microtechnology and Mechatronics*, Microtechnology and MEMS, DOI 10.1007/978-3-642-10806-8_12,

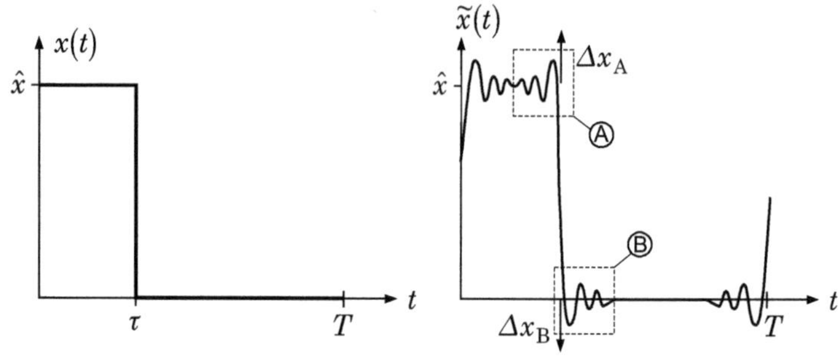

Fig. B.1. Periodic impulse function

and

$$c_i = \sqrt{a_i^2 + b_i^2} = \frac{2\hat{x}\tau}{T} \frac{\sin(\pi i\tau/T)}{\pi i\tau/T}. \tag{B.4}$$

Equation (B.1) enables the estimation of remaining error ε for a predetermined bound of summation.
If the summation in (B.6) is replaced by an integration and the denominator πi is replaced by the initial value of i in each interval between zeros of c_i, the error approximately will result in:

$$\varepsilon^2 = \frac{\sum\limits_{i=\mathrm{M}}^{\infty} c_i^2}{x(t)^2} = \frac{1}{\hat{x}^2\tau/T} \frac{4\hat{x}^2\tau^2}{T^2} \left(\frac{\sin(\pi i\tau/T)}{\pi i\tau/T}\right)^2 \tag{B.5}$$

$$\varepsilon \approx \sqrt{\frac{2}{\pi^3} \frac{1}{\tau/T} \frac{1}{\mathrm{M}}} \quad \text{for} \quad \mathrm{M} \gg 1. \tag{B.6}$$

For $\tau/T = 0.25$ and $M = 64$ the error ε amounts to 6.3%.

The approximate time function $\tilde{x}(t)$ occuring for finite M is qualitatively represented in the right part of Fig. B.2. The deviation $\Delta x(t)$ between $x(t)$ and $\tilde{x}(t)$ each represented in the boxes A and B is specified for the operations with $x(t)$ cited in the following. For reasons of symmetry, the behavior of the error in the neighborhood of $t = 0$ corresponds with that in box A, in the neighborhood of $t = T$ it corresponds with that in box B.
For the special case $\tau = 0.5$, the errors $\Delta x(t)$ are represented in the left part of Fig. B.2 for the upper bounds of summation M = 64 and 24. For reasons of symmetry, the errors in the boxes A and B coincide, too.
The behavior of $\tilde{x}(t)$ for very large M can already be estimated by means of these two cases. In the neighborhood of step discontinuities the time intervals between two zeros of oscillations of $\tilde{x}(t)$ become very small and the oscillations concentrate on an even smaller neighborhood of the step discontinuities for an approximately constant peak value at the step discontinuity.

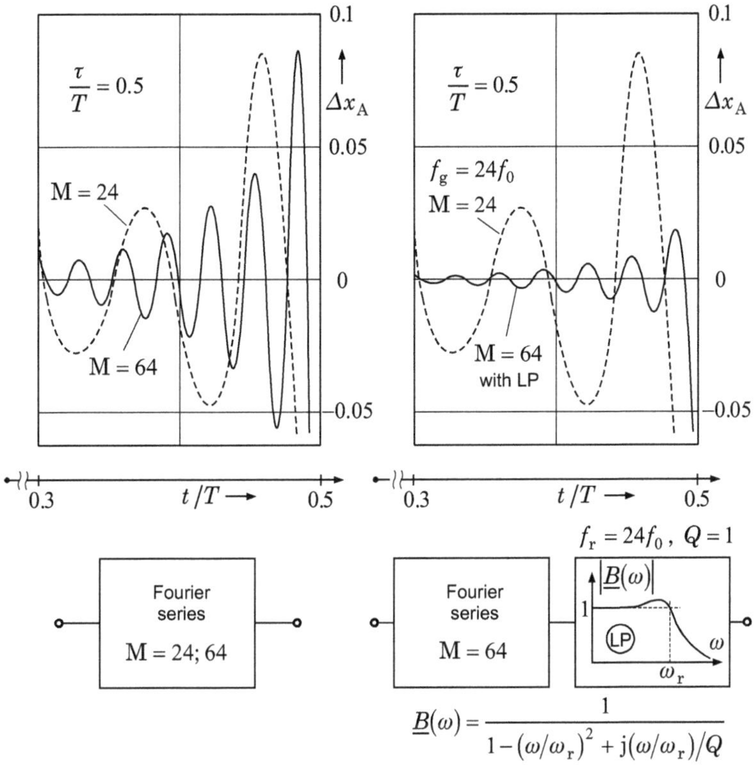

Fig. B.2. Signal operations with model function $x\,(t)$ represented in the left part of Fig. B.1 for $\tau = 0.5T$

This characteristic of the *approximative Fourier series* in the neighborhood of step discontinuities is called GIBBS phenomenon. It is a consequence of the application of the Fourier expansion to physically unreal processes. The transfer factor of real dynamic systems disappears for $\omega \rightarrow \infty$. Therefore, the same must be valid for the Fourier coefficients of real physical processes, since they can only be generated by real systems.

The *resonant low-pass* represents a typical real band limitation which often occurs also with technical systems discussed in the individual chapters of the present book. The response of such a low-pass filter with a resonant frequency of $\omega_{\text{res}} = 24\omega_0$ to the model process $x\,(t)$ shown on the right of Fig. B.1 is illustrated in the right part of Fig. B.2. In comparison to that, the approximate function $\tilde{x}\,(t)$ of the impulse function $x\,(t)$ with ideal band limitation $\mathrm{M_g} = 24$ is plotted in Fig. B.2. It can be realized that the real band limitation being approximately equivalent with respect to the cut-off frequency shows considerably smaller deviations between $\tilde{x}\,(t)$ and $x\,(t)$ than a physically unreal and ideal band limitation does.

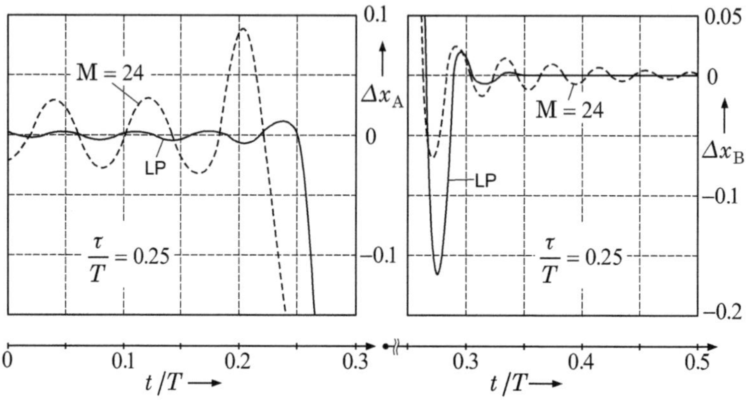

Fig. B.3. Signal operations with model function $x(t)$ represented in the left of Fig. B.1 with $\tau = 0.25T$

Finally, Fig. B.3 shows the same comparison for a model function illustrated in Fig. B.1 with $\tau = 0.25T$. Here, the errors differ in the boxes A and B.

B.1.2 Sample Application for the Periodic Iteration of Singular Processes

In the following, an *example* is given in order to provide a clear and quantitative interpretation of the general considerations. For that, the sample function $x_p(t)$ shown in Fig. B.4 is used. For simplification it is bandlimited and zero-mean. However, for a finite upper bound of summation K being inevitable with numerical analyses, band limitation errors for $\tilde{x}(t)$ must be expected due to the mentioned expansion of the spectrum of the α_k, β_k up to ∞.

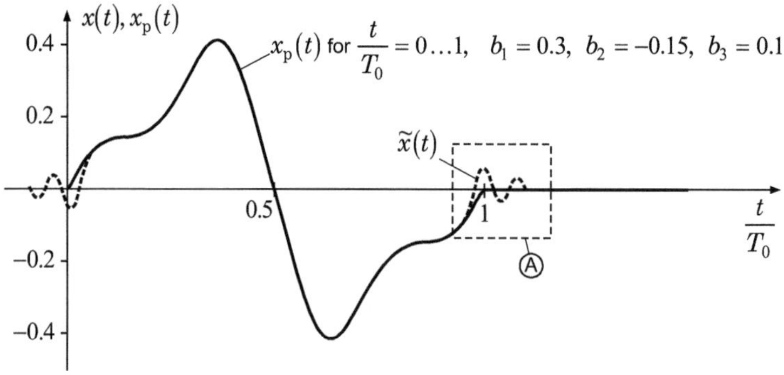

Fig. B.4. Sample function $x_p(t)$

With this sample function, the procedure represented in Fig. 2.6 (Sect. 2.1.3) was now accomplished for different L. Figure B.5 shows the α_k, β_k for increasing L. The facts already deducible from (2.51) and (2.52) show that also with infinitesimal a_i the coefficients α_k appear.

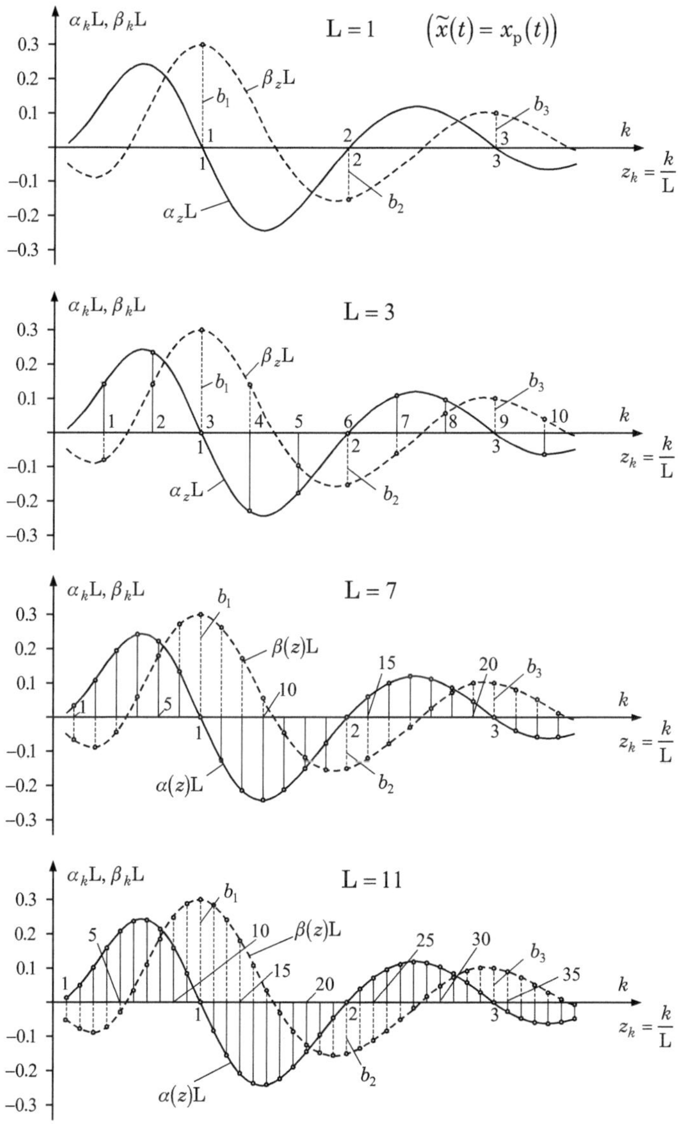

Fig. B.5. Fourier coefficients of the sample function shown in Fig. B.4 iterated with different break lengths

Figure B.6 finally shows the differences Δx between $x(t)$ and $\tilde{x}(t)$ of section A represented in Fig. B.4. Here, it must be noticed that the represented numerical values were calculated for finite upper bounds of summation K. The occurring deviations Δx are not solely determined by insertion of the break, but also by the finite K. However, it is remarkable that for L = 7 the deviation Δx exceeds the value 0.001 only during 6% of the interval time.

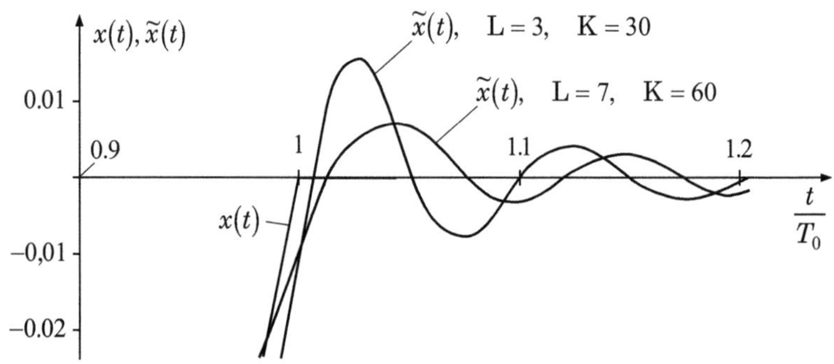

Fig. B.6. Details to the behavior of the smoothing function $x(t)$ represented in section A of Fig. B.4

Considered as a whole, it appears that a function series with two free parameters K, L has been defined in combination with (2.46) and (2.48) being valid for band- and time-limited functions $x_p(t)$ according to Fig. 2.6. It converges to a limit for an increase of both parameters independently of their order. This fact can be verified by means of numerical analyses up to the resolution limit of calculating devices. In case of cancelation of band limitation for $x_p(t)$, a further parameter M is added, but the conclusions mentioned above will not change.

B.2 Ideal Impulse and Step Functions

B.2.1 Problem Definition

The description of singular processes by Fourier and Laplace integrals allows also for another abstraction which is based on the transfer characteristics of time-limited processes of short duration $T = \tau$ – *impulses* – passing real linear systems. The experience shows that in case of very short impulses, a real linear system is not able to realize which time response the impulse has within peak time τ. The system responds only to the integral over the impulse function, the so-called impulse area I. The question is, why this fact is existent and what is the meaning of „very short" with respect to the transfer characteristics of the system. At first, this question is answered for the mentioned impulse processes, afterwards it is transfered to step processes.

B.2.2 Ideal Impulses and their System Response

As model for the ideal *impulse function* a sequence of functions is chosen which is based on the function $x\,(t/\tau)$ represented in the left part of Fig. B.7. The reference time τ with the meaning of an equivalent impulse duration is chosen in such a way that its product with $x_0 = x\,(\tau)$ results in the impulse area I.

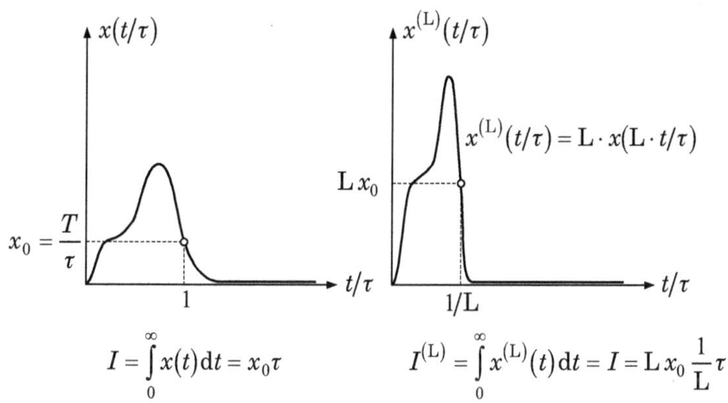

Fig. B.7. Model for a sequence of functions in order to define the ideal impulse function

According to the left part of Fig. B.7, the impulse area I can be determined according to

$$I = \int_0^\infty x\,(t)\,\mathrm{d}t = x_0\tau. \tag{B.7}$$

A reduction of the equivalent impulse duration by the factor L enables in combination with the impulse function

$$x^{(\mathrm{L})}\,(t/\tau) = \mathrm{L}\cdot x\,(\mathrm{L}\cdot t/\tau) \tag{B.8}$$

the calculation of the impulse area according to

$$I^{(\mathrm{L})} = \int_0^\infty x^{(\mathrm{L})}\,(t)\,\mathrm{d}t = I = \mathrm{L}x_0\frac{1}{\mathrm{L}}\tau. \tag{B.9}$$

It is assumed that the Fourier transform of $x\,(t)$ exists, thus the condition of (2.70) is satisfied. Furthermore, it is assumed that the impulse area amounts to $I \neq 0$. Thus, the Fourier transform of $x(t)$ results from (2.46). It can already be noticed that $\underline{c}(0) \neq 0$ is valid. Three typical examples of impulse-like processes are represented in Fig. B.8.

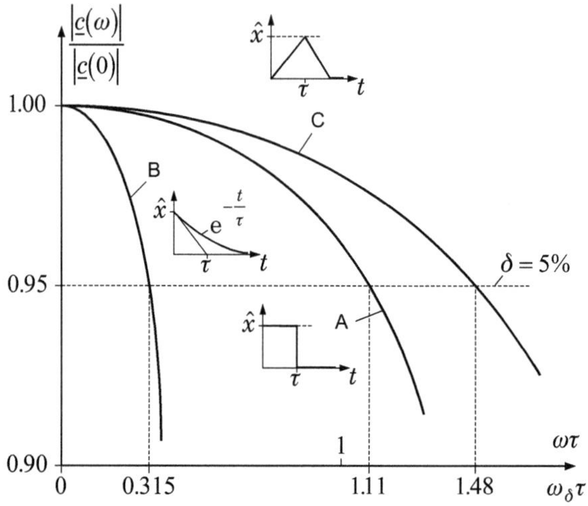

Fig. B.8. Fourier spectral densities of typical impulse time functions

The curve shapes A, B and C are associated with following spectral densities $\underline{c}(\omega)$:

$$\text{A}: \quad |\underline{c}(\omega)| = \hat{x}\tau \frac{\sin(\omega\tau/2)}{(\omega\tau/2)} \tag{B.10}$$

$$\text{B}: \quad |\underline{c}(\omega)| = \hat{x}\tau \frac{1}{\sqrt{1 + (\omega\tau)^2}} \tag{B.11}$$

$$\text{C}: \quad |\underline{c}(\omega)| = \hat{x}\tau \frac{\sin^2(\omega\tau/4)}{(\omega\tau/4)^2} \tag{B.12}$$

Figure B.8 shows that for the examples represented in the range of $\omega_\delta\tau = 0.3 \ldots 1.5$ the spectral density $\underline{c}(\omega)$ differs around less than 5% from $\underline{c}(0) = I$. If such an impulse affects as input time function $x(t)$ a linear system defined by the transfer function $\underline{B}(\omega)$, the resulting output quantity can be expressed according to (2.72). The transfer functions of real linear systems offer upper cut-off frequencies ω_g, for which $\underline{B}(\omega_\text{g})$ falls below the resolution limit of the used calculation or measuring devices. If now the effective impulse duration τ illustrated in Fig. B.7 is adjusted in such a way that the failure cut-off frequency ω_δ equals ω_g for $\underline{c}(\omega)$ shown in Fig. B.8, the situation represented in Fig. B.9 using the example of a square-wave impulse will arise.

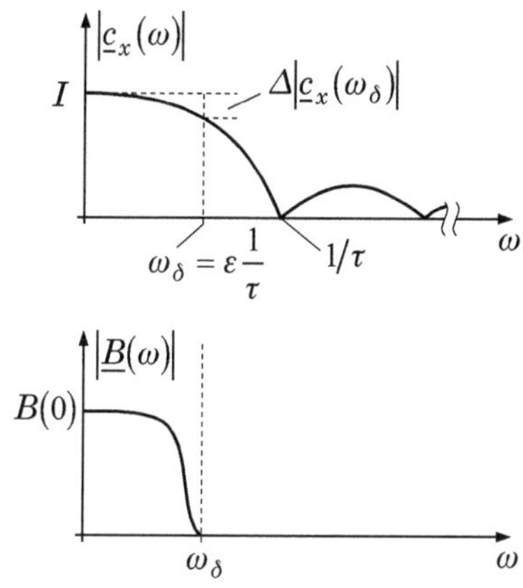

Fig. B.9. Requirements on the impulse duration for a band-limited transfer function

The error δ in Fig. B.9 results in

$$\delta = \frac{\Delta\left|\underline{c}_x(\omega_\delta)\right|}{\underline{c}_x(0)} = 1 - \frac{\sin(\varepsilon/2)}{\varepsilon/2}. \tag{B.13}$$

Within the transfer range of $\underline{B}(\omega)$, the *spectral density* $\underline{c}(\omega)$ is independent of ω within the limit of permitted error δ and equals the impulse area I. In this case, the system is not able to realize any longer which concrete input function $x(t)$ generated the output function $y(t)$. It only responds to the impulse area I.

This fact suggests the transition represented in the left part of Fig. B.7 and described by (B.8) from $x(t)$ over the sequence $x^{(\mathrm{L})}(t)$ to an ideal impulse. With increasing L the effective impulse duration becomes shorter and the amplitude becomes higher. The impulse area I remains constant with the approach described in (B.8).

The Fourier transform $\underline{c}(\omega)$ of a sequence element takes following form:

$$\underline{c}^{(\mathrm{L})}(\omega) = \int_0^\infty x^{(\mathrm{L})}(t/\tau)\,\mathrm{e}^{-\mathrm{j}\omega t}\,\mathrm{d}t = L\int_0^\infty x(\mathrm{L}t/\tau)\,\mathrm{e}^{-\mathrm{j}\omega t}\,\mathrm{d}t \tag{B.14}$$

$$\underline{c}^{(\mathrm{L})}(\omega) = \int_{t'=0}^\infty x(t'/\tau)\,\mathrm{e}^{-\mathrm{j}\omega t'/\mathrm{L}}\,\mathrm{d}t, \qquad t' = \mathrm{L}t$$

$$\underline{c}^{(\mathrm{L})}(\omega) = \underline{c}(\omega\mathrm{L}) \tag{B.15}$$

This transition is represented in Fig. B.10. With increasing L the range for which $\underline{c}(\omega)$ can be considered to be $\underline{c}(\omega) = I$ extends by factor L.

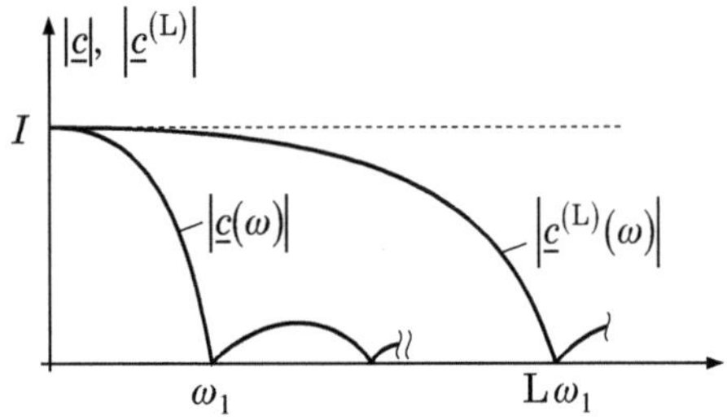

Fig. B.10. Transition from $\underline{c}(\omega)$ to $\underline{c}^{(\mathrm{L})}(\omega)$

The elements of this function sequence show following characteristics:

$$\underline{c}(0) = I \qquad \text{and} \qquad \lim_{T\to\infty} c^{(\mathrm{L})}(\omega) = 0 \tag{B.16}$$

For $\mathrm{L} \to \infty$, it is referred to the limiting value of the function sequence $x^{(\mathrm{L})}(t/\tau)/I$ as normalized ideal impulse function or DIRAC delta function

$$\delta(t) = \lim_{\mathrm{L}\to\infty} x^{(\mathrm{L})}(t)/I, \tag{B.17}$$

respectively. In connection with that, it is simplified according to

$$c\{\delta(t)\} = \lim_{\mathrm{L}\to\infty} \underline{c}^{(\mathrm{L})}(t)/I = 1 \tag{B.18}$$

and thus it follows

$$\delta(t) = \frac{1}{2\pi}\int_{-\infty}^{\infty} \mathrm{e}^{-\mathrm{j}\omega t}\,\mathrm{d}\omega. \tag{B.19}$$

However, in the strict sense the limiting process $\omega_{\mathrm{K}} \to \infty$ in (B.19) can only be made with a sequence element $c^{(\mathrm{L})}(\omega)$, because only this tends to zero for ω. Therefore, (B.19) must be written in detail in the following way:

$$\delta(t) = \lim_{\mathrm{L}\to\infty}\left[\lim_{\mathrm{K}\to\infty}\int_{-\omega_{\mathrm{K}}}^{\omega_{\mathrm{K}}} c^{(\mathrm{L})}(\omega)\int \mathrm{e}^{-\mathrm{j}\omega t}\,\mathrm{d}\omega\right] \tag{B.20}$$

The limiting value $\delta(t)$ can only be determined with the specified order of both limiting processes concerning the two parameters K and L. Taking (B.19) into account, the response of a linear system to an ideal step $x_{\mathrm{impulse}}(t) = I\delta(t)$ can be specified by means of (2.76). However, if $\underline{B}(\omega)$ has an upper cut-off frequency ω_{g} in terms of Fig. B.9, the integration in (2.76) can be restricted to $\pm\omega_{\mathrm{g}}$. Then the limiting value for $\underline{c}_x(\omega)$ from (B.18) can also be used. This results in:

$$y(t) = g(t)\,I = \frac{I}{2\pi}\int_{-\omega_{\mathrm{g}}}^{\omega_{\mathrm{g}}} B(\omega)\,\mathrm{e}^{-\mathrm{j}\omega t}\,\mathrm{d}\omega \tag{B.21}$$

with

$$g(t) = \frac{1}{2\pi} \int_{-\omega_g}^{\omega_g} B(\omega)\, e^{j\omega t}\, d\omega \qquad \text{(B.22)}$$

and

$$B(\omega) = \int_{0}^{\infty} g(t)\, e^{-j\omega t}\, dt. \qquad \text{(B.23)}$$

It is referred to the function $g(t)$ as *normalized impulse response* of a system. The inverse transform in (B.22) shows that the system characteristics are completely described both by $g(t)$ and $\underline{B}(\omega)$. Equation (B.22) allows also for the experimental determination of $\underline{B}(\omega)$ by means of a DIRAC delta function and provides a basis for numerous measuring methods.

Finally, for the determination of $g(t)$ a numerical example shall explain the relations with application of a square-wave impulse.

As transfer function $\underline{B}(\omega)$

$$\underline{B}(\omega) = \frac{1}{\left(1 - (\omega/\omega_0)^2\right)^2 + j(\omega/\omega_0)\,1/Q}, \qquad \text{(B.24)}$$

$$\text{with} \quad Q = 1, \qquad f_0 = 100\,\text{Hz},$$

a resonant low-pass is available. In consideration of $|B(\omega_g)| \leq 10^{-3}$, the value of the cut-off frequency amounts to:

$$\omega_g \approx 30\omega_0, \qquad f_g = 3\,\text{kHz}$$

Assuming an approved error $\delta = 0.01$ of the Fourier spectral density, it follows in combination with (B.10):

$$\delta = 10^{-2} \Rightarrow \varepsilon = \omega_g \tau = 0.490$$

From this, the approved impulse duration finally results in:

$$\omega_g = \omega_\delta \Rightarrow \tau = \frac{0.49}{30\omega_0} = 26\,\mu\text{s}$$

An impulse with such a duration allows for the determination of the normalized impulse response $g(t)$ with errors in the %-range.

B.2.3 The Ideal Step Function and its System Response

For the analysis of transient processes of systems, the step function represented in Fig. B.11 is useful. The normalized step function $s(t)$ is deduced from $x_\mathrm{s}(t)$. $x_\mathrm{s}(t)$ and $s(t)$ do not have a Fourier transform, because the condition of (2.70) is not fulfilled.

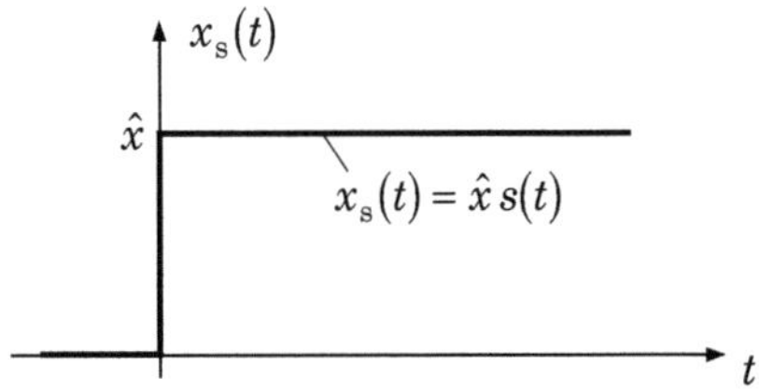

Fig. B.11. Ideal step function

Only the Laplace transform is applicable for spectral description of this function. According to (2.84) it can be written:

$$\mathcal{L}\{x_\mathrm{s}(t)\} = \int_0^\infty \hat{x}\,\mathrm{e}^{-pt}\,\mathrm{d}t = \frac{\hat{x}}{p} \tag{B.25}$$

If the step $x_\mathrm{s}(t)$ affects a linear system defined by the transfer function $\underline{B}(p)$, $y(t)$ will result according to (2.85) in

$$y(t) = \frac{\hat{x}}{2\pi} \int_{\sigma-\mathrm{j}\omega}^{\sigma+\mathrm{j}\omega} \frac{1}{p}\,\mathrm{e}^{pt}\,\mathrm{d}p = w(t)\,\hat{x}. \tag{B.26}$$

It is referred to the function $w(t)$ as *normalized step response*. Just like $\underline{B}(p)$ it denotes the transfer characteristics of the system.
If it is sought after a numerical solution for $w(t)$ without applying the process of Laplace transform, a periodic square wave function $x_\mathrm{p}(t)$ with $\tau = T/2$ according to Fig. B.12 can be used instead of $x_\mathrm{s}(t)$.
A Fourier series expansion can be specified for the zero point free part of $z(t)$. The series expansion results in an approximate function $z_\mathrm{p}(t)$. If $x(t)$ is considered to be an input function of a linear system, the output function $y_\mathrm{p}(t)$ will result in:

$$y_\mathrm{p}(t) = \frac{1}{2}\hat{x}\underline{B}(0) + \frac{1}{2}\hat{x}\sum_{i=1}^{\infty} \underline{B}(\omega_i)\,\underline{c}_x(\omega_i)\,\mathrm{e}^{\mathrm{j}\omega_i t} \tag{B.27}$$

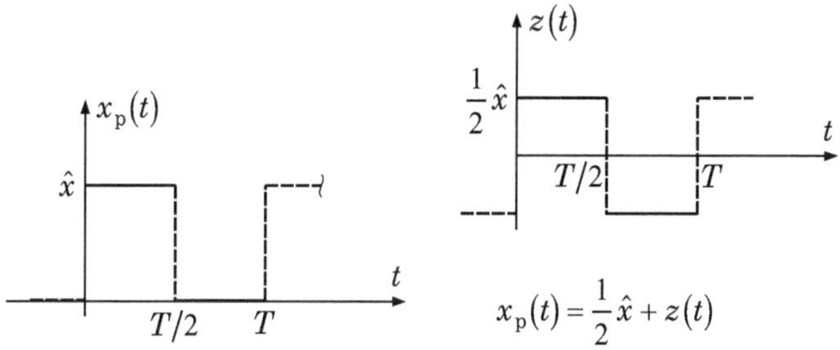

Fig. B.12. Periodic square wave function as approximate model for $x_s(t)$

With real systems, the time-dependent part of $y_p(t)$ disappears for sufficiently large t $(t > T_g)$. If $T > 2T_g$ is chosen as oscillation period, $y_p(t)$ will correspond within the interval $0 \ldots T/2$ with $y(t)$ from the exact solution of (B.26). This results in following approximate equation for $w(t)$ on adherence to these conditions which can be easily verified numerically:

$$w(t) = \frac{1}{2}\underline{B}(0) + \frac{1}{2}\sum_{i=1}^{\infty} \underline{B}(\omega_i)\, \underline{c}_{ix}(\omega_i)\, e^{j\omega_i t} \qquad \text{with} \qquad \underline{c}_{ix} = \hat{x}\frac{1 - e^{j\omega_i T}}{j\omega_i T} \tag{B.28}$$

B.3 The Convolution Integral

In combination with (B.22), it was shown in the preceding section that the system transfer function can be determined from the normalized impulse response $g(t)$. Thus, using (2.77) and (B.22) it is basically possible to reformulate (2.76) (see Sect. 2.1.3) in such a way that the system response $y(t)$ depends only on $g(t)$ and $x(t)$. Due to the problems concerning the order of limiting processes already discussed before, it may be expected that some caution is necessary with the required transformations. These complications can be avoided by means of the direct and concrete approach represented in the following.

If it is sought after the response of a system with well-known impulse response to an input function $x(t)$, $x(t)$ can be split into differential square-wave impulses according to Fig. B.13.

Each subfunction $\Delta x_n(t)$ of

$$x(t) = \sum_{n=-\infty}^{+\infty} \Delta x_n(t) \tag{B.29}$$

is transfered according to

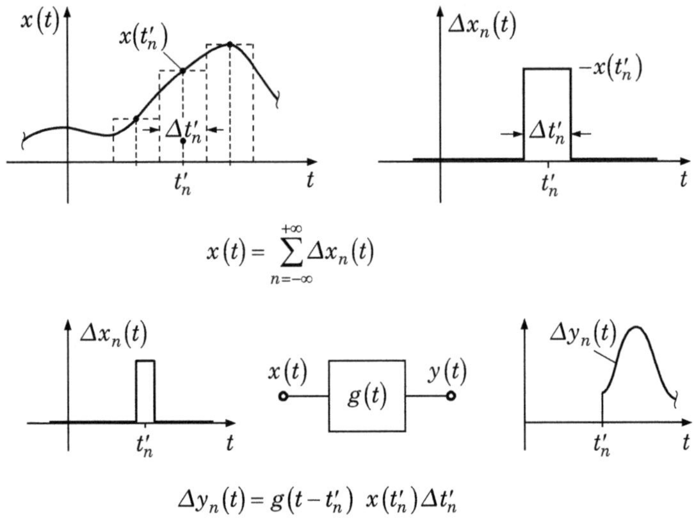

Fig. B.13. Structure of the response time function of a linear system based on the summation of square-wave impulses of $x(t)$ rated with $g(t)$

$$\Delta y_n(t) = g(t - t'_n)\, x(t'_n)\, \Delta t'_n \tag{B.30}$$

into a subfunction $\Delta y_n(t)$. The output time function $y(t)$ equals the sum of all $\Delta y_n(t)$:

$$y(t) = \sum_{n=-\infty}^{+\infty} g(t - t'_n)\, x(t'_n)\, \Delta t'_n \tag{B.31}$$

At the limit $\Delta t_n \to 0$ this sum passes into an integral of the form

$$y(t) = \int_{t=-\infty}^{\infty} g(t - t')\, x(t')\, \mathrm{d}t' \tag{B.32}$$

Due to $g(t) = 0$ for $t < 0$, this equation can be transfered into following form:

$$t - t' = \tau \Rightarrow y(t') = \int_{\tau=0}^{\infty} g(\tau)\, x(t' - \tau)\, \mathrm{d}\tau \tag{B.33}$$

The equations (B.32) and (B.33) solve the initially specified task to determine the output function $y(t)$ directly from $g(t)$ and $x(t)$. However, they can also be formulated in a second form which can be formally deduced from the relation

$$g(t) = \frac{\mathrm{d}s(t)}{\mathrm{d}t} \tag{B.34}$$

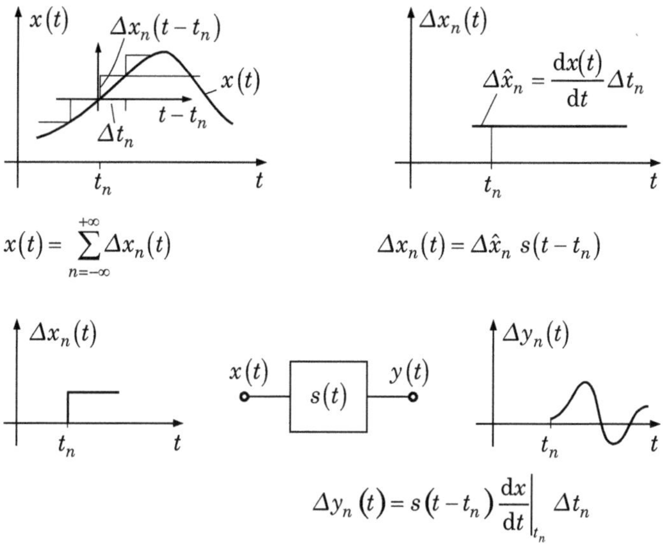

Fig. B.14. Structure of the response time function of a linear system based on summation of differential step functions

In addition, the derivation of this second variant is also possible with a simple model represented in Fig. B.14 and being similar to model illustrated in Fig. B.13.
In accordance to (B.31) and (B.33), it can finally be written:

$$y\left(t\right)=\sum_{n=-\infty}^{+\infty}s\left(t-t_{n}\right)\left.\frac{\mathrm{d}x}{\mathrm{d}t}\right|_{t_{n}}\cdot\Delta t_{n}\Rightarrow\int_{0}^{\infty}s\left(t-\tau\right)\left.\frac{\mathrm{d}x}{\mathrm{d}t}\right|_{t=\tau}\cdot\mathrm{d}\tau \tag{B.35}$$

References

1. A. Lenk. *Elektromechanische Systeme, Band 1: Systeme mit konzentrierten Parametern.* 2. Auflage. VEB Verlag Technik, Berlin, 1971.
2. A. Lenk. *Elektromechanische Systeme, Band 2: Systeme mit verteilten Parametern.* 2. Auflage. VEB Verlag Technik, Berlin, 1977.
3. A. Lenk. *Elektromechanische Systeme, Band 3: Systeme mit Hilfsenergie.* VEB Verlag Technik, Berlin, 1975.
4. G. Gerlach; W. Dötzel. *Einführung in die Mikrosystemtechnik.* Carl Hanser Verlag, Leipzig, München, Wien, 2006.
5. J. W. Gardner; Y. K. Varadan; O. O. Awadelkarim. *Microsensors, MEMS and Smart Devices.* John Wiley & Sons, New York, 2001.
6. W. Menz; J. Mohr. *Mikrosystemtechnik für Ingenieure.* 2. Auflage. VCH Wiley, Weinheim, 1997.
7. W. Ehrfeld. *Handbuch Mikrotechnik.* Carl Hanser Verlag, München, 2002.
8. VDI Richtlinie 2206. *Entwicklungsmethodik für mechatronische Systeme, Ausgabe 2004-06.* Beuth-Verlag, Berlin, 2006.
9. G. Gerlach; W. Dötzel. *Grundlagen der Mikrosystemtechnik.* Carl Hanser Verlag, München, 1997.
10. W. Krause. *Gerätekonstruktion.* 3. Auflage. Carl Hanser Verlag, München, Wien, 2000.
11. W. Cauer. *Theorie der linearen Wechselstromschaltungen.* Akademie-Verlag, Berlin, 1954.
12. K. Küpfmüller. *Die Systemtheorie der elektrischen Nachrichtentechnik.* S. Hirzel-Verlag, Stuttgart, 1952.
13. H. Barkhausen. *Einführung in die Schwingungslehre.* S. Hirzel Verlag, Leipzig, 1932.
14. K. W. Wagner. *Einführung in die Lehre von Schwingungen und Wellen.* Dietrich'sche Verlagsbuchhandlung, Wiesbaden, 1947.
15. H. Hecht. *Schaltschemata und Differentialgleichungen elektrischer und mechanischer Schwingungsgebilde.* Johann Ambrosius Barth-Verlag, Leipzig, 1950.
16. W. Reichardt. *Grundlagen der technischen Akustik.* Akademische Verlagsgesellschaft Geest und Partig KG, Leipzig, 1968.
17. L. Cremer; M. Heckl; E. E. Ungar. *Structure Borne Sound.* Springer-Verlag, Berlin, 1973.

A. Lenk et al., *Electromechanical Systems in Microtechnology and Mechatronics,*
Microtechnology and MEMS, DOI 10.1007/978-3-642-10806-8,

18. J. C. Maxwell. *Über physikalische Kraftlinien.* Ostwalds Klassiker der exakten Wissenschaften Nr. 102, herausgegeben von L. Boltzmann, Akademische Verlagsgesellschaft m.b.H., Leipzig, 1900.
19. K. Simonyi. *Theoretische Elektrotechnik.* VEB Deutscher Verlag der Wissenschaften, Berlin, 1971.
20. M.-H. Bao. *Handbook of Sensors and Actuators.* Elsevier, Amsterdam, 2000.
21. A. Lenk; G. Pfeifer; R. Werthschützky. *Elektromechanische Systeme.* Springer-Verlag, Heidelberg, 2001.
22. R. Isermann. *Mechatronische Systeme.* 2. Auflage. Springer-Verlag, Berlin, Heidelberg, New York, 2007.
23. E. Philippow. *Taschenbuch der Elektrotechnik.* VEB Verlag Technik, Berlin, 1986.
24. K. Lunze. *Einführung in die Elektrotechnik.* 13. Auflage. Verlag Technik, Berlin, 1991.
25. A. Prechtl. *Vorlesungen über die Grundlagen der Elektrotechnik, Band 2.* Springer-Verlag, Wien, New York, 1995.
26. H. Clausert; G. Wiesemann. *Grundgebiete der Elektrotechnik, Band 2.* Oldenbourg Verlag, München, Wien, 2005.
27. R. Courant; D. Hilbert. *Methoden der mathematischen Physik.* Springer-Verlag, Berlin, 1968.
28. H. v. Mangold; K. Knopp. *Eine Einführung in die höhere Mathematik, Band 3.* S. Hirzel-Verlag, Stuttgart, 1951.
29. E. Philippow. *Grundlagen der Elektrotechnik.* VEB Verlag Technik, Berlin, 1988.
30. G. Wunsch. *Systemanalyse, Band 1.* VEB Verlag Technik, Berlin, 1968.
31. E. Philippow. *Grundlagen der Elektrotechnik.* 9. Auflage. Verlag Technik, Berlin, 1993.
32. V. Oppenheim; A. S. Willsky. *Signale und Systeme.* 2. Auflage. VCH-Verlag, Weinheim, 1992.
33. R. Unbehauhen. *Systemtheorie - Eine Darstellung für Ingenieure.* Akademie-Verlag, Berlin, 1980.
34. R. Unbehauhen. *Grundlagen der Elektrotechnik, Band 2.* 4. Auflage. Springer-Verlag, Berlin, 1993.
35. G. Fritzsche. *Signale und Funktionaltransformationen.* VEB Verlag Technik, Berlin, 1985.
36. A. M. Abramowitsch; I. A. Stegum. *Handbook of Mathematical Functions.* Dover Publications Inc., New York, 1965.
37. J. M. Rhyshik; I. S. Gradstein. *Summen-, Produkt- und Integral-Tafeln.* VEB Deutscher Verlag der Wissenschaften, Berlin, 1962.
38. F. H. Lange. *Signale und Systeme, Band 1.* VEB Verlag Technik, Berlin, 1965.
39. K. W. Wagner. *Operatorenrechnung.* Johann Ambrosius Barth Verlag, Leipzig, 1940.
40. G. Wunsch. *Moderne Systemtheorie.* Akademische Verlagsgesellschaft Geest & Portig KG, Leipzig, 1962.
41. R. Paul. *Grundlagen der Elektrotechnik, Band 2.* Springer-Verlag, Berlin, 1990.
42. H. W. Bode. *Network analysis and feedback amplifier design.* van Nostrand, New York, 1945.
43. H. Göldner; F. Holzweißig. *Leitfaden der Technischen Mechanik.* VEB Fachbuchverlag, Leipzig, 1989.

44. E. Skudrzyk. *Die Grundlagen der Akustik.* Springer-Verlag, Wien, 1954.
45. E. Baumann; R. Lippold. *Breitbandige Ersatzschaltungen für akustische Masse- und Federelemente.* Unveröffentliches Manuskript. TU Dresden, Institut für Akustik und Sprachkommunikation, Dresden, 1991.
46. G. Wunsch. *Theorie und Anwendung linearer Netzwerke, Band 1.* Akademische Verlagsgesellschaft Geest und Portig KG, Leipzig, 1961.
47. S. Leschka; G. Pfeifer. *Resonant air coupled piezoelectric film transducers - a design using network methods.* 10th International Congress on Sound and Vibration, Stockholm, July 2003.
48. S. Leschka. *Entwurfsmethoden und Leistungsgrenzen elektromechanischer Schallquellen für Ultraschallanwendungen in Gasen im Frequenzbereich um 100 kHz.* Dissertation. TU Dresden, Dresden, 2004.
49. S. Sindlinger. *Einfluss der Gehäusung auf die Messunsicherheit von mikrogehäusten Drucksensoren mit piezoresistivem Messelement.* Dissertation. TU Darmstadt, Darmstadt, 2007.
50. C. Wohlgemuth. *Entwurf und galvanotechnische Fertigung metallischer Trennmembranen für mediengetrennte Drucksensoren.* Dissertation. TU Darmstadt, Darmstadt, 2008.
51. P. C. Eccardt. *Coupled Finite Element and Network Simulation for Microsystem Components.* Proc. MICRO SYSTEM Technologies '96, S. 145-150. VDI-Verlag, Berlin, 1996.
52. E. Starke; G. Pfeifer. *Die Finite-Elemente-Methode als Möglichkeit der effizienten Modellierung elektroakustischer Geräte.* Proc. 25th CADFEM User's Meeting - Internationale FEM Technologietage & German ANSYS Conference. Dresden, Oktober 2007.
53. E. Starke; G. Pfeifer. *Modellierung eines Mikrofons mit dünnem akustischen Dämpfungsgewebe in Ansys: Ein Beispiel für die Verbindung von Finite-Elemente- und Netzwerkmethoden auf Anwenderebene.* Proc. 24th CADFEM User's Meeting - Internationale FEM Technologietage & German ANSYS Conference. Stuttgart, Oktober 2006.
54. E. Starke. *Kombinierte Simulation - eine weitere Methode zur Optimierung elektromechanischer Systeme.* Dissertation. TUDpress, Dresden, 2009.
55. K.-J. Bathe. *Finite-Elemente-Methoden.* Springer-Verlag, Berlin, Heidelberg, New York, 2002.
56. R. Unbehauen. *Elektrische Netzwerke.* Springer-Verlag, Berlin, Heidelberg, New York, 1990.
57. M. Heckel; H. A. Müller. *Taschenbuch der Technischen Akustik.* Springer-Verlag, Berlin, Heidelberg, New York, 1994.
58. G. Pfeifer; R. Werthschützky. *Drucksensoren.* VEB Verlag Technik, Berlin, 1989.
59. W. Göpel; J. Hesse; J. N. Zemel. *Sensors, A comrehensive Survey.* Vol.7: Mechanical Sensors. VCH Verlagsgesellschaft, Weinheim, 1993.
60. H. R. Tränkler; E. Obermeier. *Sensortechnik.* Springer-Verlag, Berlin, Heidelberg, New York, 1998.
61. A. Lenk. *Grundlagen der Akustik.* Skript zur Vorlesung. TU Dresden, Institut für Technische Akustik, Dresden, 1995.
62. T. Obier. *Leistungsgrenzen elektrodynamischer Schallquellen bei tiefen Frequenzen dargestellt am Beispiel eines neuen Gleitlagerlautsprechers.* Dissertation. TU Dresden, Dresden, 1999.

63. U. Marschner; H. Grätz; B. Jettkant; D. Ruwisch; G. Woldt; W.-J. Fischer; B. Clasbrummel. *Integration of wireless lock-in measurement of hip prosthesis vibrations for loosening detection.* Sensors & Actuators A, Special Issue Eurosensors XXII, Dresden, September 2008.
64. G. Qi; W. Mouchon; T. Tan. *How much can a vibrational diagnostic tool reveal in total hip arthroplasty loosening.* Clinical Biomechanics, Vol. 18, pp. 444-458, 2003.
65. H. D. Stölting; E. Kallenbach. *Handbuch elektrischer Kleinantriebe.* Carl Hanser Verlag, München, Wien, 2002.
66. H. Janocha. *Actuators.* Springer-Verlag, Berlin, Heidelberg, New York, 2004.
67. H. F. Schlaak. *Mikrorelais in Mikrotechnologie - Grenzen und Potentiale.* 15. VDE-Fachtagung Kontaktverhalten und Schalten. VDE-Verlag, Karlsruhe, September 1999.
68. C. Doerrer. *Entwurf eines elektromechanischen Systems für flexibel konfigurierbare Eingabefelder mit haptischer Rückmeldung.* Dissertation. TU Darmstadt, Darmstadt, 2003.
69. H. Janocha. *Adaptronics and Smart Structures.* Springer-Verlag, Berlin, Heidelberg, New York, 1999.
70. L. Kiesewetter. *Piezoelektrizität und Magnetostriktion in Translationsmotoren.* Technische Rundschau, Nr. 22, S. 104-109. 1988.
71. K. Y. Huang. *Entwicklung von Bauformen und Untersuchungen an magnetostriktiven Motoren.* Dissertation. TU Berlin, Berlin, 1994.
72. A. Aharoni. *Demagnetizing factors for rectangular ferromagnetic prisms.* Journal of Applied Physics, Vol. 83, pp. 3432-3434, 1998.
73. W. Roshen; D. Turcotte. *Planar inductors on magnetic substrates.* IEEE Transactions on Magnetics, Vol. 24, No. 6, pp. 3213-3216, November 1988.
74. W. Roshen. *Effect of finite thickness of magnetic substrate on planar inductors.* IEEE Transactions on Magnetics, Vol. 26, No. 1, January 1990.
75. W. Fischer; S. Sauer; U. Marschner; B. Adolphi; C. Wenzel; B. Jettkant; B. Clasbrummel. *Galfenol resonant sensor for indirect wireless osteosynthesis plate bending measurements.* IEEE SENSORS 2009 Conference, Canterbury New Zealand, October 2009.
76. C. Wenzel; B. Adolphi; U. Merkel; U. Marschner; H. Neubert; W.-J. Fischer. *Resonant bending sensor based on sputtered galfenol.* Sensors & Actuators A, Special Issue Eurosensors XXII, Dresden, September 2008.
77. J. Atulasimha; A. B. Flatau; I. Chopra; R. A. Kellogg. *Effect of stoichiometry on sensing behavior of iron-gallium.* D. C. Lagoudas, Ed., Vol. 5387, No. 1, SPIE, 2004.
78. S. Rehfuß; D. Peters; R. Laur. *Integration of a permeable layer into micro coil modelling and first steps of automatic optimization.* Technical Proceedings of the Third International Conference on Modelling and Simulation of Microsystems, ISBN 0-9666135-7-0, pp. 197-200, San Diego, 2000.
79. L. Lin; R. T. Howe; A. P. Pisano. *Microelectromechanical filters for signal processing.* Journal of Microelectromechanical Systems, Vol. 7, pp. 286-293, 1998.
80. C. Thielemann. *Kapazitive Silizium-Mikrophone für Hör- und Ultraschall.* Dissertation. TU Darmstadt, Darmstadt, 1998.
81. H. F. Schlaak; P. Lotz; M. Matysek. *Muskeln unter Hochspannung - Antriebe mit elektroaktiven Polymeren.* thema Forschung, Vol. 2, S. 68-73. TU Darmstadt, Darmstadt, 2006.

82. M. Jungmann. *Entwicklung elektrostatischer Festkörperaktoren mit elastischen Dielektrika für den Einsatz in taktilen Anzeigefeldern.* Dissertation. TU Darmstadt, Darmstadt, 2004.
83. P. Lotz; V. Bischof; M. Matysek; H. F. Schlaak. *Integrated Sensor-Actuator-System based on dielectric Polymer Actuators for Peristaltic Pumps.* Actuator 2006, 10th International Conference on New Actuators, Conference Proceedings, Bremen, 2006.
84. W. P. Mason. *Piezoelectric Crystals and their Application to Ultrasonics.* D. van Nostrand Comp., New York, 1956.
85. R. G. Ballas. *Piezoelectric Multilayer Beam Bending Actuators and Aspects of Sensor Integration.* Springer-Verlag, Berlin, Heidelberg, New York, 2007.
86. W. G. Cady. *Piezoelectricity - An Introduction to the Theory and Applications of Electromechanical Phenomena in Crystals.* McGraw-Hill, London, 1946.
87. J. Tichy; G. H. Gautschi. *Piezoelektrische Messtechnik.* Springer-Verlag, Berlin, Heidelberg, New York, 1980.
88. G. H. Gautschi. *Piezoelectric Sensorics.* Springer-Verlag, Berlin, Heidelberg, New York, 2002.
89. R. Lerch. *Elektrische Messtechnik.* 2. Auflage. Springer-Verlag, Wien, 2006.
90. C. A. Rosen. *Ceramic Transformers and Filters.* Proc. Electronic Comp. Symp., pp. 205-211, 1956.
91. J. Mehner. *Entwurf in der Mikrosystemtechnik.* University Press, Band 9, Dresden, 2000.
92. G. Gerlach; W. Dötzel. *Introduction of Microsystem Technology.* John Wiley & Sons, Ltd., West Sussex, 2008.
93. H. Kuttruff. *Physik und Technik des Ultraschalls.* Hirzel Verlag, Stuttgart, 1988.

Index

α-quartz, 360–362
absorber, 217–224
abstract
  components, 130
  network, 129–136
acceleration
  amplitude, 73, 89, 368
  source, 72
  transfer function, 253
acceleration,
  prosthesis, 264
acoustic
  circuit, 111
  compliance, 110
  damping elements, 217
  friction, 112
  low-pass system, 199
  mass, 110, 112
  power, 257
  pressure level, 126, 127
  reciprocity, 224
  resonator, 107
  scheme, 113
  spring, 111
  subsystem, 4, 205
  system point, 44
  systems, 234
  transformer, 114
  wave impedance, 218
acoustic friction, 112
  component, 112
acoustic mass, 112
  component, 112
acoustic spring, 111
  component, 111
acoustic waveguide, 218
  one-dimensional, 218
active one-port, 65
actuator,
  magnetostrictive, 296
  polymer, 341
adiabatic
  change of state, 110
  exponent, 110
admittance, 76
  complex, 76
  frictional, 68, 85
  function, 178
  general, 76
  matrix, 188
  torsional friction, 105
  transformation, 179
  translational, 191
  wave, 371
AgPd electrodes, 389
algebraic set of equation, 18
AMPÈRE's law, 254, 305
amplitude,
  acceleration, 73, 89, 368
  complex, 17, 18
  differential, 35
  imbalance, 91
  pressure, 122, 123
amplitude-frequency response, 79
angle
  coordinates, 100

source, 101
angular velocity, 42, 91, 100
anisotropic dielectric, 313
approximate
function, 21, 439, 448
approximation error, 437
approximative Fourier series, 22
assigned frequency, 86
average bending stiffness, 302

band limitation error, 437
bandlimited time function, 30
BARKHAUSEN, 11
basic function, 80
bed volume, 200, 201
bending
angle, 141
bending waves wavelength, 394
BERNOULLI hypothesis, 303
bimorph, 358–360
BIOT-SAVART, 47
force, 54, 235
law, 53, 235
BODE diagram, 79

canal elements,
real acoustic, 110
virtual, 108
canal-like cavity, 107, 108
canonical circuit, 389, 408, 409
capacitance matrix, 48
capacitive diaphragm transducer, 195
capillary system, 199
CARNOT cycle, 56
Cartesian coordinate system, 40
cascade connection, 143, 144
CAUER, 11
cavity,
canal-like, 107, 108
volumetric, 107
center
frequency, 330
of mass, 65
chain matrix, 145, 188, 206, 218, 371, 414, 415
change of state,
adiabatic, 110
virtual, 52
characteristic frequency, 79, 397
characteristic,
directional, 222
force, 269, 280, 316
quadratic, 269
charge density, 335
distribution, 314, 315
circuit,
acoustic, 111
canonical, 389, 408, 409
plate transducer, 154
circular function, 16
harmonic, 16
coefficients,
Fourier, 32, 437, 439
transducer, 58
combined simulation, 202, 203
compensation method, 230, 233
complete solution system, 186
complex
admittance, 76
amplitude, 17, 18
frequency, 20
impedance, 76
load, 197
network coordinate, 76
partial oscillation, 35
plane, 129
propagation constant, 217
radiation impedance, 210–213
time function, 17
transfer function, 23
velocity, 74, 75
wave impedance, 217
YOUNG's modulus, 88
compliance, 67
acoustic, 110
equivalent, 175
field, 272, 319, 378
rotational, 100
compliance per unit length, 168
compliance,
interface, 265
component,
acoustic friction, 112
acoustic mass, 112
acoustic spring, 111
friction, 67, 68
lever, 73–75
mass, 68

moment of inertia, 101–104
spring, 66, 67
torsion spring, 101–104
torsional friction, 101–104
components,
abstract, 130
concentrated, 61, 62
one-port, 15, 38
time-invariant, 36
concentrated, 62
charges, 242
components, 61, 62
frictional elements, 62
mass, 62
springs, 62
condenser microphone, 334–337
conductance matrix, 38, 39
cone, 194
constant,
elastic, 352
magnetic, 286
piezomagnetic, 286
spring, 67
continuous function, 22
coordinate
of motion, 65, 112
representation, 63
coordinate system,
Cartesian, 40
cylindrical, 100
coordinates,
angle, 100
difference, 16, 56
flow, 16
force, 40
general, 130, 417
generalized, 41, 42
mechanical, 48
position, 40
source, 72
core
admittance, 135
impedance, 135
COULOMB, 47
equations, 47, 49, 235
force, 47, 234, 235, 314, 331
COULOMB's law, 48
coupled simulation, 203
coupling
factor, 246
function, 40
two-port network, 138
coupling element,
gyrator-like, 181
transformer-like, 180
coupling four-port network, 139
coupling systems, 40
generalized, 50
CREMER, 11
CURIE temperature, 296, 361
current density, 273
current,
leg, 39
short-circuit, 40
cut-off frequency, 123
cylindric coil, 304
cylindrical coordinate system, 100

damping, 81
DC magnetron sputtering, 309
deflection
function, 150, 155, 420
method, 230, 232
receiver, 393, 396–398
transmitter, 393, 396–398
demagnetization
factor, 304
field, 304
density,
charge, 314, 315, 335
current, 273
field energy, 267
derivative element, 81
dermis, 265
dielectric polymers, 341
dielectric,
anisotropic, 313
isotropic, 313, 339, 341
difference coordinates, 16, 56
difference,
pressure, 108
velocity, 65, 67
differential
amplitude, 35
bending element, 141–143
equations of state, 14
pressure sensor, 198
variation, 239

dipole bass loudspeaker, 209
DIRAC delta function, 446
directional characteristic, 222
displacement, 40, 41
  virtual, 238
dissipation factor, 88
distal
  end, 264
  excitation force, 265
  excitation velocity, 265
distributed parameters, 13, 14
distribution,
  charge density, 314, 315
  field, 239
domain,
  frequency, 18
  time, 18
dynamic
  system modeling, 5

eccentricity, 91
effect,
  frictional, 68
  magnetoelastic, 293, 309
  magnetostrictive, 293, 297
  piezoelectric, 345–347
  piezomagnetic, 286
  pyroelectric, 355, 361
EINSTEIN
  summation convention, 352
elastic constant, 352
elastomechanical systems, 153
electret, 216, 224, 335
electric
  field energy, 12, 36
  field force, 239
  field strength, 47
electrical
  network, 36–40
  power, 230, 253, 255, 279
  reference point, 50, 235
  resistance, 68
  transducer, 4, 57
electroacoustic systems, 61, 146
electroactive polymers, 341
electrodynamic
  auto focus system, 255
  drive system, 251–255
  loudspeaker, 255–258
  transducer, 247–251
  vibration calibrator, 258–264
electromagnetic
  actuating drive, 282–285
  transduction coefficient, 304
  vibrating conveyor, 278–282
electromechanical
  coupling factor, 246
  relay, 275–277
  system, 4, 51
electroplating, 309
electrostatic
  diaphragm transducer, 331–334
  polymer actuator, 341–345
  sensing probe, 325–327
  solid body transducer, 339, 340
  transducer, 313
electrostatic field, 48–50
  quantity, 47
electrostriction, 340
element,
  derivative, 81
  far field, 211, 213
  integrating, 81
  rod, 166, 184
energy,
  inner, 39
  kinetic, 173
  potential, 173
enthalpy, 54
envelope function, 30
equations of state, 348, 350–352
  piezoelectric, 348, 350–352
equations,
  COULOMB, 47, 49, 235
  LAGRANGE, 40
  MAXWELL, 11
  PARSEVAL, 22
equivalent
  compliance, 175
  mass, 173
  parameter, 172, 174
  system, 175
error,
  approximation, 437
  band limitation, 437
  linearity, 270
  numerical, 213
  reduced, 261

remaining, 438
excitational time function, 16
exhaust gas muffler, 126
expansion,
Fourier, 20, 437
series, 21, 48, 54, 177
extensional
waveguide, 301, 374
waves, 166, 189, 370

factor,
demagnetization, 304
dissipation, 88
structure, 219
failure cut-off frequency, 444
far field, 127, 197
elements, 211, 213
FARADAY, 11
FE
analysis, 205
method, 204
model, 204, 205, 207
femur
-prosthesis, 264, 265
condyle, 265
excitation, 264
excitation model, 264
ferroelectric polymers, 341
ferroelectrics, 360, 361
field
compliance, 272, 319, 378
distribution, 239
field energy,
density, 267
electric, 12, 36
magnetic, 12, 36
field force,
electric, 239
magnetic, 54, 235, 287
nonlinear, 238
field,
demagnetization, 304
electric, 47
electrostatic, 48–50
filter
frequency response, 183
systems, 177
filter,
mechanical, 181–183
micromechanical, 329–331
finite
network element, 195, 196
spring element, 166
five-port network, 358, 359
flexural
cantilever waveguide, 327
vibrations, 183
flow coordinates, 16
flow,
laminar, 117
signal, 229
fluid-mechanical
network, 43, 44
system, 42
force
-loaded reference point, 145
characteristic, 269, 280, 316
source, 72
vector, 54
force coordinates, 40
generalized, 41
force,
BIOT-SAVART, 54, 235
COULOMB, 47, 234, 235, 314, 331
LORENTZ, 247
short-circuit, 91, 253, 255, 279, 319
force-loaded
reference point, 145
Fourier
coefficients, 32, 437, 439
expansion, 20, 437
spectral density, 447
sum, 437
transform, 25
Fourier series, 16, 20
approximative, 22
interpolative, 21
frequency
-domain, 18
function, 87, 129, 177
limit, 108, 124
locus, 252
frequency,
center, 330
characteristic, 79, 397
complex, 20
cut-off, 123

operating, 10, 89, 94, 317, 334, 339, 355, 367
reference, 81
friction, 67, 68
component, 67, 68
torsional, 101–104
frictional
admittance, 68, 85
impedance, 67
frictional effect, 68
viscous, 68
function sequence, 26
function,
admittance, 178
approximate, 21, 439, 448
basic, 80
continuous, 22
deflection, 150, 155, 420
DIRAC delta, 446
envelope, 30
frequency, 87, 129, 177
impedance, 172
impulse, 439, 443, 446
input, 20, 32, 445, 448, 449
LAGRANGE, 11
model, 25, 440
output, 20
periodic, 20
position, 151, 193
sample, 440, 441
smoothing, 442
square wave, 448, 449
state, 51, 52
step, 448
time-limited, 20, 26, 442
trigonometric, 141
zero-mean, 27
functional approach, 204
functional transformation, 18, 31
modified, 35
fundamental network, 56, 58

Galfenol, 309
gallium orthophosphates, 360
gas constant, 109
general
admittance, 76
coordinates, 130, 417
impedance, 76
generalized
coordinates, 41, 42
coupling systems, 50
force coordinates, 41
network, 49
position coordinates, 50
vector, 49
gyrator, 147, 148
gyrator-like coupling element, 181

HAMILTON, 10
harmonic
analysis, 205
circular function, 16
HECHT, 11
HELMHOLTZ, 11
resonator, 107
high-pass, 81
hip
prosthesis, 267
replacement, 267
hybrid matrix, 206
hydraulic systems, 108, 146

ideal
gas, 109
lever, 64
piston transducer, 147
rod, 138, 139
transformer, 36
imaginary transducer constant, 272
imbalance amplitude, 91
impedance, 76
complex, 76
frictional, 67
general, 76
radiation, 210–213
translational, 189
impedance function, 172
impedance matrix, 188
impulse area, 442
impulse function, 439, 443, 446
periodic, 438
inchworm motor, 297–299
independent position coordinates, 40
individual transfer function, 82
induced
polarization, 340
voltage, 248

inductance, 78
inertial system, 69
inner energy, 39
input
  function, 20, 32, 445, 448, 449
  time function, 444
integrating element, 81
interface
  compliance, 265
  length, 266
internal polarization, 340
interpolating envelope curve, 30
interpolative
  form, 21
  Fourier series, 21
isomorphism, 77
isotropic dielectric, 313, 339, 341

KÜPFMÜLLER, 11
kinetic
  energy, 173
  reactive power, 174
KIRCHHOFF, 11

LAGRANGE, 10
  equations, 40
  formalism, 41
  function, 11
$\lambda/4$-waveguide, 181
$\lambda/2$-waveguide, 180
laminar flow, 117
large-signal behavior, 286
law,
  BIOT-SAVART, 53, 235
  superposition, 413
lead zirconate titanate, 361
leg
  conductance, 38, 39
  current, 39
  resistance, 39
  voltages, 39
lever, 73–75
  component, 73–75
  ideal, 64
limit,
  frequency, 108, 124
  stability, 280
linear
  combination, 16, 41
  network, 16, 58
  system, 19, 25, 444
linearization, 267, 268, 278
lithium
  niobate, 360
  tantalate, 360
longitudinal oscillator, 375
  piezoelectric, 375
loop rule, 74
loose region, 266
LORENTZ force, 247
loss-free
  spring, 88
  two-port networks, 229
low-pass, 81
  resonant, 81
lumped parameters, 13, 14, 61

magnetic
  constant, 286
  field energy, 12, 36
  field force, 54, 235
  field leakage, 273
  field strength, 287
  flux, 268
  flux rate, 303
  induction, 248
  interaction, 235
  resistance, 303
  scanning systems, 248
  susceptibility, 304
  transducer, 241
  unimorph, 302
  voltage, 302
magnetoelastic effect, 293, 309
magnetostatic field equations, 15
magnetostriction, 285
magnetostrictive
  actuator, 296
  effect, 293, 297
  traveling wave motor, 297
  unimorph, 302
mass per unit length, 168, 371, 402
mass,
  acoustic, 110, 112
  component, 68
  concentrated, 62
  equivalent, 173
  rotating, 99–103, 106

matrix,
  admittance, 188
  capacitance, 48
  chain, 145, 188, 206, 218, 371, 414, 415
  conductance, 38, 39
  hybrid, 206
  impedance, 188
  reciprocal, 39
  system, 41
  transducer, 242
  transformation, 41
MAXWELL, 11
  equations, 11
  stress, 339, 340, 342
mean square deviation, 22
mechanical
  coordinates, 48
  filter, 181–183
  one-port, 92
  point systems, 153
  prestress, 296, 331
  quality factor, 87, 90
  scheme, 86, 93
  strain, 88
  stress, 88
  subsystem, 229, 365
  system force, 234
  transducer, 137
medical-lateral direction, 266
micromechanical filter, 329–331
microphone capsule, 216
microphone,
  piezoelectric, 368–371
  silicon, 337–339
model
  function, 25, 440
  time function, 16
modified functional transformation, 35
moment of inertia, 102
  component, 101–104
moment source, 104
monomorph, 358
motion coordinate, 65, 112
moving coil, 248

N-port, 51
N-port, 51
network,
  abstract, 129–136
  electrical, 36–40
  five-port, 358, 359
  fluid-mechanical, 43, 44
  fundamental, 56, 58
  generalized, 49
  linear, 16, 58
  rotational, 43
  time-invariant, 36
  translational, 42
neutral axis, 141
nodal rule, 74
node
  currents, 38, 39
  voltages, 38, 39
nonlinear field force, 238
normalized quantities, 97
numerical error, 213

one-dimensional
  acoustic waveguide, 218
  waveguide, 165
one-port components, 15, 38
one-port,
  active, 65
  mechanical, 92
  rotational, 101
open-circuit
  transfer function, 399
  voltage, 40
operating
  frequency, 10, 89, 94, 317, 334, 339, 355, 367
  point, 14, 231
orthogonality, 22
oscillator,
  thickness, 353, 354
output
  function, 20
  signal, 181, 316
  time function, 450

parametric transducer, 309
PARSEVAL
  equations, 22
partial oscillation, 32
  complex, 35
passive
  transducer, 229, 230

vibration absorber, 94–99
periodic
function, 20
impulse function, 438
time, 22
time function, 21
permeability change, 310
permittivity, 339
phase
angle, 16
velocity, 189
phase-frequency response, 79
Π-circuit, 165, 170, 171
piezoelectric
accelerometer, 365, 367, 368
ceramics, 346, 360, 361, 363
charge constant, 352
effect, 345–347
equations of state, 348, 350–352
flexural vibrator, 358
force constant, 351
longitudinal effect, 350
longitudinal oscillator, 375
microphone, 368–371
multilayer beam bending actuator, 400
signal generator, 206
stack actuator, 389–392
thickness oscillator, 353, 354
transducer, 313, 345, 349
transformer, 387, 388
transverse effect, 354
piezomagnetic
bending element, 303
constant, 286
effect, 286
transducer, 247, 285, 286
ultrasonic transmitter, 299–301
piston transducer,
ideal, 147
real, 149
pistonphone, 121–123
planar coil, 305, 306
plane wave, 217, 218
plate transducer, 149, 150, 313, 314
plate transducer circuit, 154
point charge, 47
polar diagram, 222, 224
polarization,
induced, 340
internal, 340
polymer actuator, 341
polymers,
dielectric, 341
electroactive, 341
ferroelectric, 341
porosity, 219
position coordinates, 40
generalized, 50
independent, 40
position function, 151, 193
potential
reactive power, 174
potential energy, 173
power,
acoustic, 257
electrical, 230, 253, 255, 279
pressure
amplitude, 122, 123
difference, 108
transfer function, 200
prosthesis
acceleration, 264
loosening, 264
shaft, 264
shaft acceleration, 266
velocity, 265
vibrations, 266
proximal end, 264
PVDF, 341, 360, 364, 365
pyroelectric effect, 355, 361
pyroelectricity, 361
PZT, 361, 363

quadratic
characteristic, 269
transduction coefficient, 304
quality factor, 81
mechanical, 87, 90
quantities,
normalized, 97
reference, 186, 402
quartz, 360–362
quasi-static deformation, 176

radiation impedance, 210–213
complex, 210–213
Rayleigh functions, 186

reactive power,
kinetic, 174
potential, 174
real
acoustic canal elements, 110
acoustic volume elements, 114, 115
piston transducer, 149
spring, 88
time function, 17
reciprocal
capacitance coefficients, 48
inductance coefficients, 55
matrix, 39
reciprocity
calibration, 418
characteristics, 39
relations, 413–417
reduced error, 261
reference
frequency, 81
quantities, 186, 402
reference point,
electrical, 50, 235
rotational, 138
REICHARDT, 11
relative permeability, 293
relay, 275–277
electromechanical, 275–277
solid state, 276
reluctance principle, 275
remaining error, 438
resistance, 68
electrical, 68
magnetic, 303
resonant
bending sensor, 309
frequency, 81
low-pass, 81
resonator,
acoustic, 107
HELMHOLTZ, 107
response time function, 450
response,
amplitude-frequency, 79
phase-frequency, 79
system, 21
reversibility, 236
rigid rod, 137–140
rod element, 166, 184
rod,
ideal, 138, 139
rigid, 137–140
ROSEN transformer, 387
rotating mass, 99–103, 106
rotational
compliance, 100
network, 43
one-port, 101
reference point, 138
subsystem, 4
systems, 42
transduction coefficient, 302

sample
function, 440, 441
sample value, 21
saturation magnetostriction, 285
scheme,
acoustic, 113
mechanical, 86, 93
self capacitance, 310
sensing probe, 325–327
sensor velocity, 266
series expansion, 21, 48, 54, 177
series,
Fourier, 16, 20
Taylor, 177
shaker, 264, 265
shear modulus, 67
shock absorber, 17
short-circuit
current, 40
force, 91, 253, 255, 279, 319
signal
flow, 229
operations, 440
processing direction, 4
signal generator, 206
piezoelectric, 206
silicon
measuring element, 198, 199, 201
microphone, 337–339
microtechnology, 276
surface micromechanics, 329
SIMONYI, 12
simulation
methods, 10–14
systems, 13

singularity, 189, 191
small-signal
  behavior, 242
  transducer circuit, 270
smoothing function, 442
solenoid, 304
solid state relay, 276
sound generating systems, 120
source
  amplitude, 72
  coordinates, 72
  quantity, 62, 116, 126
source,
  acceleration, 72
  angle, 101
  force, 72
  moment, 104
  velocity, 72
specific heat capacity, 110
spectral
  lines, 30
  representation, 25, 26
spring, 66, 67
  component, 66, 67
  loss-free, 88
  real, 88
  torsion, 101–104
spring constant, 67
square
  effect, 346
  wave function, 448, 449
stability limit, 280
state
  function, 51, 52
  space, 51
  vector, 51
stationary time function, 15
step function, 448
stress,
  MAXWELL, 339, 340, 342
  mechanical, 88
strip
  diaphragm, 195, 196
  plate, 197
structure factor, 219
subcutis, 265
subsystem,
  acoustic, 4, 205
  mechanical, 229, 365
  rotational, 4
  translational, 4
support beam, 302
system
  matrix, 41
  response, 21
system force, 234
  mechanical, 234
system,
  capillary, 199
  electromechanical, 4, 51
  equivalent, 175
  linear, 19, 25, 444
  trigonal, 361
systems,
  acoustic, 234
  elastomechanical, 153
  electroacoustic, 61, 146
  filter, 177
  hydraulic, 108, 146
  rotational, 42
  simulation, 13
  translational, 42

Taylor series, 177
thermal conductivity, 114
thickness oscillator, 353, 354
  piezoelectric, 353, 354
thin film technology, 309
three-port configuartion, 148
time function,
  approximate, 438
  bandlimited, 30
  complex, 17
  excitational, 16
  input, 444
  model, 16
  output, 450
  periodic, 21
  real, 17
  response, 450
  stationary, 15
time-domain, 18
time-invariant
  components, 36
  network, 36
time-limited
  function, 20, 26, 442
  processes, 25

torsion spring, 101–104
  component, 101–104
torsional
  compliance per unit length, 402
  moment, 42, 100
torsional friction, 101–104
  admittance, 105
  component, 101–104
total
  differential, 236
  transfer function, 83, 411
transducer
  coefficients, 58
  matrix, 242
  parametric, 309
  two-port networks, 56
transducer constant, 272
  imaginary, 272
transducer,
  electrical, 4, 57
  electrodynamic, 247–251
  electrostatic, 313
  magnetic, 241
  mechanical, 137
  passive, 229, 230
  piezoelectric, 313, 345, 349
  piezomagnetic, 247, 285, 286
  plate, 149, 150, 313, 314
transduction coefficient,
  electromagnetic, 304
  quadratic, 304
  rotational, 302
  translational, 302
transfer function,
  acceleration, 253
  complex, 23
  individual, 82
  open-circuit, 399
  pressure, 200
  total, 83, 411
transformation
  admittance, 179
  matrix, 41
transformer,
  acoustic, 114
  ideal, 36
  piezoelectric, 387, 388
transformer-like coupling element, 180
translational
  admittance, 191
  axis, 65, 69
  impedance, 189
  network, 42
  reference compliance, 402
  subsystem, 4
  systems, 42
  transduction coefficient, 302
traveling wave motor, 297
trigonal system, 361
trigonometric function, 141
trimorph, 358
trochanter, 266
two-port networks,
  loss-free, 229
  transducer, 56

ultrasonic microactuator, 195–198
ultrasonic transmitter, 299–301
  piezoelectric, 299–301

vector,
  force, 54
  generalized, 49
  state, 51
velocity
  difference, 65, 67
velocity source, 72
velocity,
  angular, 42, 91, 100
  complex, 74, 75
  prosthesis, 265
  sensor, 266
  wave, 371, 372
vibration
  analysis, 264
  calibrator, 258–264
  isolation, 91
virtual
  canal elements, 108
  change of state, 52
  displacement, 238
viscosity, 117
viscous frictional effect, 68
voltage integral, 235
voltage,
  induced, 248
  magnetic, 302
  open-circuit, 40

volume elements, 114, 115
  real acoustic, 114, 115
volumetric
  cavity, 107
  flow rate, 108

WAGNER, 11
wave
  admittance, 371
  number, 168, 189
  velocity, 371, 372
wave impedance, 169
  acoustic, 218
  complex, 217
waveguide,
  acoustic, 218
  extensional, 301, 374
  flexural cantilever, 327
  $\lambda/4$, 181
  $\lambda/2$, 180
  one-dimensional, 165
wavelength, 169
  bending waves, 394

YOUNG's modulus, 67
  complex, 88

zero-mean function, 27

MIX
Papier aus verantwortungsvollen Quellen
Paper from responsible sources
FSC® C105338

If you have any concerns about our products, you can contact us on
ProductSafety@springernature.com

In case Publisher is established outside the EU, the EU authorized representative is:
Springer Nature Customer Service Center GmbH
Europaplatz 3, 69115 Heidelberg, Germany

Printed by Libri Plureos GmbH
in Hamburg, Germany